Aquatic Toxicology and Hazard Assessment: 10th Volume

William J. Adams, Gary A. Chapman, and Wayne G. Landis, editors

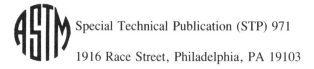 Special Technical Publication (STP) 971

1916 Race Street, Philadelphia, PA 19103

Copyright © by AMERICAN SOCIETY FOR TESTING AND MATERIALS 1988

NOTE

The Society is not responsible, as a body,
for the statements and opinions
advanced in this publication.

Peer Review Policy

Each paper published in this volume was evaluated by three peer reviewers. The authors addressed all of the reviewers' comments to the satisfaction of both the technical editor(s) and the ASTM Committee on Publications.

The quality of the papers in this publication reflects not only the obvious efforts of the authors and the technical editor(s), but also the work of these peer reviewers. The ASTM Committee on Publications acknowledges with appreciation their dedication and contribution of time and effort on behalf of ASTM.

Printed in Ann Arbor, MI
May 1988

Foreword

This publication, *Aquatic Toxicology and Hazard Assessment: 10th Volume,* contains papers presented at the Tenth Symposium on Aquatic Toxicology and Hazard Assessment, which was held in New Orleans, 4–6 May 1986. The symposium was sponsored by ASTM Committee E-47 on Biological Effects and Environmental Fate. William J. Adams, Monsanto Co., presided as symposium chairman and is a coeditor of this publication. Gary A. Chapman, Environmental Protection Agency, and Wayne G. Landis, USA Aberdeen Proving Grounds, also served as symposium cochairmen and were coeditors of this publication.

Contents

Overview 1

INTRODUCTORY PAPER

Aquatic Toxicology: Ten Years in Review and a Look at the Future—P. R. PARRISH,
K. L. DICKSON, J. L. HAMELINK, R. A. KIMERLE, K. L. MACEK, F. J. MAYER, JR., AND
D. I. MOUNT 7

BIOENGINEERING AND AQUATIC ENVIRONMENTS

Gene Probes as a Tool for the Detection of Specific Genomes in the Environment—
T. BARKAY AND G. S. SAYLER 29

Plasmid Mobility in the Ocean Environment—D. J. GRIMES, C. C. SOMERVILLE,
W. STRAUBE, D. B. ROSZAK, B. A. ORTIZ-CONDE, M. T. MACDONELL, AND R. R.
COLWELL 37

Genetically Engineered Microorganisms in the Aquatic Environment:
Environmental Safety Assessment—V. MOLAK AND J. F. STARA 43

APPLICATION OF PALEOECOLOGICAL METHODOLOGY TO THE ANALYSIS OF ENVIRONMENTAL PROBLEMS

Paleoecology and Environmental Analysis—U. M. COWGILL 53

Long-Term History of Eutrophication in Washington Lakes—R. B. BRUGAM ... 63

The Use of Subfossil Benthic Invertebrates in Aquatic Resource Management—
T. L. CRISMAN 71

Two-Hundred-Year pH History of Woods, Sagamore, and Panther Lakes in the
Adirondack Mountains, New York State—R. B. DAVIS, D. S. ANDERSON, D. F.
CHARLES, AND J. N. GALLOWAY 89

SEDIMENT ASSESSMENT AND TOXICITY TESTING

An Evaluation of the Screening Level Concentration Approach for Validation of
Sediment Quality Criteria for Freshwater and Saltwater Ecosystems—
J. M. NEFF, B. W. CORNABY, R. M. VAGA, T. C. GULBRANSEN, J. A. SCANLON, AND
D. J. BEAN 115

A Technique for Better Sediment Characterization—K. A. KNAPP AND B. C. DYSART ... 128

Interstitial Water Sampling in Ecotoxicological Testing: Partitioning of a Cationic
Surfactant—C. A. PITTINGER, V. C. HAND, J. A. MASTERS, AND L. F. DAVIDSON ... 139

Factors Affecting the Bioavailability of Hexachlorobiphenyls to Benthic
Organisms—A. E. McELROY AND J. C. MEANS   149

BIOMONITORING OF COMPLEX EFFLUENTS

Biomonitoring as an Integral Part of the NPDES Permitting Process: A Case
Study—J. M. MARCUS, G. R. SWEARINGEN, AND G. I. SCOTT   161

Environmental Persistence/Degradation of Toxicity in Complex Effluents:
Laboratory Simulations of Field Conditions—G. M. DEGRAEVE, W. H. CLEMENT,
M. F. ARTHUR, R. B. GILLESPIE, AND G. K. O'BRIEN   177

A Review of Inter- and Intralaboratory Effluent Toxicity Test Method Variability—
W. J. RUE, J. A. FAVA, AND D. R. GROTHE   190

Protocol for the Identification of Toxic Fractions in Industrial Wastewater
Effluents—A. GASITH, K. M. JOP, K. L. DICKSON, T. F. PARKERTON, AND S. A.
KACZMAREK   204

ENVIRONMENTAL MODELING AND EXPOSURE ASSESSMENT

Implications of Molecular Speciation and Topology of Environmental Metals:
Uptake Mechanisms and Toxicity of Organotins—F. E. BRINCKMAN, G. J. OLSON,
W. R. BLAIR, AND E. J. PARKS   219

Partitioning Between Dissolved Organic Macromolecules and Suspended
Particulates: Effects on Bioavailability and Transport of Hydrophobic
Organic Chemicals in Aquatic Systems—J. F. McCARTHY AND M. C. BLACK   233

Techniques for Environmental Modeling of the Fate and Effects of Complex
Chemical Mixtures: A Case Study—B. W. VIGON, G. B. WICKRAMANAYAKE, J. D.
COONEY, G. S. DURELL, A. J. POLLACK, T. SHOOK, AND M. FRAUENTHAL   247

An Integrated Fates and Effects Model for Estimation of Risk in Aquatic
Systems—S. M. BARTELL, R. H. GARDNER, AND R. V. O'NEILL   261

Evaluating Ecosystem Response to Toxicant Stress: A State Space Approach—A. R.
JOHNSON   275

SHORT-TERM INDICATORS OF CHRONIC TOXICITY

Field Utilization of Clinical Measures for the Assessment of Xenobiotic Stress in
Aquatic Organisms—D. J. VERSTEEG, R. L. GRANEY, AND J. P. GIESY   289

Ventilatory and Movement Responses of Bluegills Exposed to 1,3,5-
Trinitrobenzene—W. H. VAN DER SCHALIE, T. R. SHEDD, AND M. G. ZEEMAN   307

Uninduced Rat Liver Microsomes as a Metabolic Activation System for the Frog
Embryo Teratogenesis Assay—*Xenopus* (FETAX)—J. A. BANTLE AND
D. A. DAWSON   316

Behavioral Responses to Low Levels of Toxic Substances in Rainbow Trout (*Salmo Gairdneri*, Rich)—J. HADJINICOLAOU AND G. LAROCHE   327

LABORATORY AND FIELD COMPARISONS

An Approach for Integration of Toxicological Data—M. A. SHIRAZI AND L. N. LOWRIE   343

What Constitutes Field Validation of Predictions Based on Laboratory Evidence?—J. CAIRNS, JR.   361

Field Verification of Multispecies Microcosms of Marine Macroinvertebrates—R. J. LIVINGSTON   369

Interlaboratory Testing of a Standardized Aquatic Microcosm—F. B. TAUB, A. C. KINDIG, AND L. L. CONQUEST   384

Effects of Pulp Mill and Ore Smelter Effluents on Vertebrae of Fourhorn Sculpin: Laboratory and Field Comparisons—F. L. MAYER, JR., B.-E. BENGTSSON, S. J. HAMILTON, AND A. BENGTSSON   406

AQUATIC TOXICOLOGY

Evaluation of the Indicator Species Procedure for Deriving Site-Specific Water Quality Criteria for Zinc—T. F. PARKERTON, S. M. STEWART, K. L. DICKSON, J. H. RODGERS, JR., AND F. Y. SALEH   423

A Toxicity Assessment of Tar Sands Tailings—R. N. YONG AND R. D. LUDWIG   436

Drilling Fluid Bioassays Using Pacific Ocean Mysid Shrimp, *Acanthomysis sculpta*, a Preliminary Introduction—M. J. MACHUZAK AND T. K. MIKEL, JR.   447

In Vitro Cytotoxicity of Polychlorinated Biphenyls (PCBs) and Toluenes to Cultured Bluegill Sunfish BF-2 Cells—H. BABICH AND E. BORENFREUND   454

A Method for Evaluating Effects of Toxic Chemicals on Fish Growth Rates—N. O. CROSSLAND   463

The Aquatic Toxicity and Fate of Brass Dust—M. V. HALEY, D. W. JOHNSON, W. T. MUSE, JR., AND W. G. LANDIS   468

WASTE SITE HAZARD ASSESSMENT AND BIODEGRADATION

How Clean Is Clean? A Use of Hazard Assessment in Groundwater for Evaluation of an Appropriate Formaldehyde Spill Remedial Action Endpoint—C. A. STAPLES   483

Effects of Metals from Mine Tailings on the Microflora of a Marsh Treatment System—R. M. DESJARDINS, W. C. BRADBURY, AND P. L. SEYFRIED   491

Determination of Optimal Toxicant Loading for Biological Closure of a Hazardous Waste Site—M. A. BIANCHINI, R. J. PORTIER, K. FUJISAKI, C. B. HENRY, P. H. TEMPLET, AND J. E. MATTHEWS   503

Enhanced Biotransformation and Biodegradation of Polychlorinated Biphenyls in the Presence of Aminopolysaccharides—R. J. PORTIER AND K. FUJISAKI   517

RESEARCH BENEFICIAL TO THE STANDARDS SETTING PROCESS

An Evaluation of Appropriate Expressions of Toxicity in Aquatic Plant Bioassays as Demonstrated by the Effects of Atrazine on Algae and Duckweed—J. S. HUGHES, M. M. ALEXANDER, AND K. BALU   531

A Short-Term Chronic Toxicity Test Using *Daphnia Magna*—P. A. LEWIS AND W. B. HORNING   548

A LOOK TO THE FUTURE

Integrated Resource Management: The Challenge of the Next Ten Years—J. CAIRNS, JR.   559

Indexes   567

Overview

The last ten years have been exciting for the science of aquatic toxicology. The intent of the tenth symposium was to reflect on past accomplishments of a decade and to consider what new directions are needed for the future.

During the past ten years we have watched the field of aquatic toxicology grow from its infancy and a need to deal with acute problems to a mature field of science capable of dealing with long-term and sophisticated issues of national importance. Ten years ago there was a feeling of crisis; rivers burned, fish kills made the headlines, and legislative tools had just been put in place for effective enforcement. The science of aquatic toxicology was just emerging and was learning how to assist the nation in solving its water quality problems.

Many of the papers in the early symposia dealt with methods and case studies. Over the years many of the acute problems have been effectively eliminated. Consequently, the emphasis has changed to the long-term chronic perturbations that can have just as damaging an environmental and economic effect. An example of such an effect has been the destruction of the Rock Bass fishery in the Chesapeake Bay. Apparently a combination of eutrophication, point and nonpoint pollution, habitat destruction caused by development, and overexploitation has destroyed an important economic resource. Sediment contamination, acid precipitation, groundwater pollution, and hazardous waste sites have all gained increasing emphasis. Consequently, the prediction of long-term effects has become more and more important. More recently biotechnology has opened the door to many new scientific frontiers and offers the prospect of new chemicals including medicines, pesticides, growth hormones, and industrial chemicals. Biotechnology also holds promise as a means for hazardous waste site and effluent cleanup through the modification of microbial communities to enhance biodegradation. At the same time it offers the prospect of new effluents and the release, either accidentally or intentionally, of genetically altered microorganisms. This presents a challenge to the scientific community to find appropriate ways to assess the risk associated with biotechnology without stifling creativity.

The symposium certainly reflected the state of aquatic toxicology in 1986. The leadoff session, "Aquatic Toxicology: Ten Years in Review and a Look to the Future," may eventually be looked upon as a marker for the end of the adolescence of aquatic toxicology. *Dickson* described the change in emphasis of the papers of each volume, showing the shift from new methods to refinements and papers coupling laboratory and field experiments. Many of the speakers emphasized the rapid progress of the science and the cooperation of its participants from government, industry, and academia. The last speaker, *Mount*, looked to the future and emphasized that aquatic toxicology needed to evolve into a science able to offer alternatives, risks, and technologies to preserve what has already been accomplished coupled to a responsiveness to the future.

Sessions on biotechnology and paleolimnology set off two aspects of aquatic toxicology new to most participants. Biotechnology is in many ways a game with a different set of ground rules and detection technologies. Risk assessment must take into account the reproductive potential of the organisms and the promiscuous nature of genetic exchange among procaryotes. Paleolimnology is a way of looking at truly long-term changes in ecosystem dynamics, on a range from tens to thousands of years.

Sediments constitute an enormous problem in the assessment and evaluation of hazard. The session on sediments was dominated by papers characterizing sediments, evaluating risks, and managing dredge material. Several papers concerning sediments and dredge materials were also included in the poster session.

The poster session received an increased emphasis in the tenth symposium. The session took advantage of the opportunity for hands-on demonstrations of techniques and the chance to interact with the experimenter. Among the demonstrations was the video by *Sabourin* and *Dawson* on the use of Xenopus embryos for the screening of materials for teratogenicity. Posters on biomonitoring, sediment toxicity, microcosm research, and the evaluation of chronic toxicity were also presented.

Sessions on the biomonitoring of complex effluents, environmental monitoring and exposure assessment, short-term indicators of chronic toxicity, laboratory and field comparisons, aquatic toxicology, waste site hazard assessment and biodegradation, and research beneficial to the standards-setting process all indicate the diversity of the field. The examples below serve to illustrate the diversity of the symposium. *Van der Schalie et al.* demonstrated the benefits of using parameters other than ventilatory frequency in monitoring the effects of trinitrobenzene on bluegill sunfish. Progress on interlaboratory testing on the Standardized Aquatic Microcosm was presented by *Taub*. In the hazardous waste site arena, *Portier* presented data on the enhancement of PCB degradation by the enrichment of the bacterial inoculations using aminopolysaccharides.

This short overview reflects the state of the science of aquatic toxicology in 1986. This diverse field incorporates parts of many disciplines including ecology, chemistry, physiology, algology, and limnology, to mention only a few. Compared to the many disciplines, the number of aquatic toxicologists is relatively few.

Attendance at the symposia has remained about 200 for the last several years. However, aquatic toxicology has played a crucial and influential role in the progress made in the improvement in the environment during the last decade. The research emphasis is now moving on to new problems and to finding long-term solutions for environmental contamination.

We see several areas of basic research that need to be addressed. Aquatic toxicology is still to a large extent an empirical science. Only a small body of work exists on developing a theory of how toxicants affect ecosystems. A suitable theory would be of dramatic practical impact. Such a workable theory would help to extrapolate a series of data from a microcosm to a larger ecosystem. Short-term methods could be modified, if necessary, to more accurately present information relevant to evaluating risk to an ecosystem. Toxicity of complex effluents is a crucial research area. Most chemicals enter an ecosystem as a mixture or soon become part of the complex mix of synthetic and natural chemicals that exist in the environment. We need to learn how to assess the hazard of chemical mixtures. Clean up of waste sites is an issue of national importance. Biological methods have the potential to reduce costs and in some cases make cleanup possible in environments inaccessible to current methods. In the near future organisms with altered genomes will be entering the environment. An understanding of the potential, if any, of these organisms for disruption of ecosystem processes or the degradation of toxic materials needs to be understood. Aquatic organisms are already playing an important role in the search for alternatives to using mammals as test organisms. Research in this area will also emphasize the need for aquatic toxicologists to look closely at mechanisms and physiological parameters related to toxicity and environmental health.

Aquatic toxicology has an interesting future ahead. But we are also concerned that a new generation of scientists may not be coming forward to participate. Currently, the mechanisms of support for graduate students do not meet the need, and recruitment into the science appears to have slowed. Members of the community of aquatic toxicology need to encourage and find support for the new generation. There are a lot of existing problems yet to solve, and we need to put in place the basic research structure that will enable us to deal effectively with the key issues that will face our nation ten years from now. In short, we have an exciting future ahead and need to insure that the next generation of researchers is being developed to continue the work that has begun.

William J. Adams,

Montsanto Co., St. Louis, MO; symposium chairman and coeditor.

Gary A. Chapman,

Environmental Protection Agency, ERL-Narragansett, Pacific Division, Newport, OR; symposium cochairman and coeditor.

Wayne G. Landis,

USA Aberdeen Proving Grounds, Aberdeen Proving, MD; symposium cochairman and coeditor.

Introductory Paper

Patrick R. Parrish,[1] *Kenneth L. Dickson,*[2] *Jerry L. Hamelink,*[3] *Richard A. Kimerle,*[4] *Kenneth J. Macek,*[5] *Foster L. Mayer, Jr.,*[6] *and Donald I. Mount*[7]

Aquatic Toxicology: Ten Years in Review and a Look at the Future

REFERENCE: Parrish, P. R., Dickson, K. L., Hamelink, J. L., Kimerle, R. A., Macek, K. J., Mayer, F. L., Jr., and Mount, D. I., "**Aquatic Toxicology: Ten Years in Review and a Look at the Future**," *Aquatic Toxicology and Hazard Assessment: 10th Volume, ASTM STP 971,* W. J. Adams, G. A. Chapman, and W. G. Landis, Eds., American Society for Testing and Materials, Philadelphia, 1988, pp. 7–25.

ABSTRACT: This symposium marks the tenth time that we have gathered as a group of professional scientists who share common goals and ideas concerning the protection of our nation's aquatic resources. This tenth symposium seems like a fitting time to reflect on our origins, our successes, and our plans for the future. To that end, several people who have been instrumental in shaping the science of aquatic toxicology and hazard (risk) assessment were invited to present their views on the growth of this science and their ideas about its future. This paper is, then, a collection of those viewpoints, which are set down in writing so that others may benefit from the experience of the authors and so that newcomers to this field may benefit by knowing about the roots of aquatic toxicology and hazard assessment. The fact that the science has persisted and grown over the past ten years is a tribute to all those who have contributed their time, energy, and intellect.

KEYWORDS: aquatic toxicology, review, ASTM symposia, hazard (risk) assessment

Introduction—P. R. Parrish

The special session of this symposium was planned to allow several of the "movers and shakers" in aquatic toxicology and hazard (risk) assessment to share their thoughts about the progress of our science during the past ten years and to project their views of the future. It was difficult to limit the number of speakers—there were others who were instrumental in the formation of this symposium and in the science of aquatic toxicology and hazard assessment. Chuck Stephan, Rick Cardwell, John Eaton, Rich Purdy, Gene Kenaga, Dean Branson, Rita Comotto Bahner, Leif Marking, Al Hendricks, Gareth Pearson, Bob Foster, Bill Bishop, Barb Heidolph, Howard Alexander, Bill Peltier, and others have contributed. Symposium chairman Bill Adams and I

---

[1] U.S. Environmental Protection Agency, Environmental Research Laboratory, Sabine Island, Gulf Breeze, FL 32561.
[2] North Texas State University, Institute of Applied Sciences, P.O. Box 13078, Denton, TX 76203.
[3] Eli Lilly and Co., Greenfield Laboratories, P.O. Box 708, G993/B418, Greenfield, IN 46140.
[4] Monsanto Co., 800 North Lindbergh Boulevard, St. Louis, MO 63167.
[5] EG&G, 40 Williams St., Wellesley, MA 02181.
[6] U.S. Environmental Protection Agency, Environmental Research Laboratory, Sabine Island, Gulf Breeze, FL 32561.
[7] U.S. Environmental Protection Agency, Environmental Research Laboratory, 6201 Congdon Boulevard, Duluth, MN 55804.

agreed, however, that the special session speakers represented a little of every perspective from which our science can be viewed.

It has been ten years since the first ASTM symposium. That's three thousand, six hundred, and fifty-two days. (There were two leap years.) This two-hour session will be about 23 parts per million of that time. Not much, is it? But most of us work with chemicals whose concentrations cause effects in the parts-per-million range, so we can relate to such a small amount. Will we see an effect after tonight's session? Should we consider this a pulse exposure? Have we been prestressed and will that confound our responses (all of which will be sublethal, I trust)? Speaking personally, there has been some stress since that day in Memphis, Tennessee, in the Holiday Inn from which you could watch the tugboats pushing barges laden with grain on the Mighty Mississippi, that day when Don Mount informed and amused us with his clever cartoons in that tunnel of a room. Some others who share the platform tonight have been stressed, too. There have been job changes aplenty and one career change. Administrators have become researchers and researchers have become administrators. (Will we *ever* learn?) There have been divorces, near divorces, a couple of marriages, and more divorces. People have moved from Missouri to Florida, from Florida to Rhode Island, and from Virginia to Texas. There has been the graying of hair (I would never stoop so low as to mention the loss of same), the weakening of eyes, and the spreading of middle-aged midriffs.

But there has been joy and reward and growth and learning. Toddlers have grown to teenagers and 10-year-old boys to 20-year-old men. (I have examples sitting in the audience. And they are with their mother, my lovely wife of 23 years, who was in Memphis ten years ago and had a whale of a good time!) There have been professional accomplishments and personal achievements. We have partied together, eaten great food together, and have come to know and appreciate and, yes, to love each other. We get along well together, I think, because professionally we are alike (that's not to say, of course, that we don't occasionally disagree), but personally we are different. The diversity of this group is amazing. There's a Texan from Texas, a stubborn Dutchman from Michigan, a St. Louis native of German descent, a Polish kid who grew up in South Boston, an Oklahoman whose ancestors were the original inhabitants of this country, an Ohio farm boy, and a fifth-generation Florida cracker. From such diversity has come much of the strength of our science and of this symposium.

I choose to begin this technical meeting with personal comments and remembrances because I think that anniversaries are times to reminisce and to be retrospective. When we consider the magnitude of changes in our lives and in our science during the past ten years, it is appropriate to stop for a moment and to consider where we have been and where we are going.

Review of ASTM Symposia Proceedings–K. L. Dickson

My purpose is to analyze the types of papers that have been published in the first eight ASTM symposia proceedings. I have attempted to categorize each of the 242 papers that have been published, and the results are as indicated in Table 1.

Perspective

These papers challenge us to think about aquatic toxicology as a science. They attempt to identify our strengths and weaknesses. Examples of perspective papers are:

1. Don Mount's humorous presentation entitled "Present Approaches To Toxicity Testing—A Perspective" [1] published in the first proceedings.
2. John Zapp's paper entitled "Historical Consideration of Interspecies Relationships in Toxicity Assessment" [2] published in the third proceedings.

TABLE 1—*Types of papers published in first eight ASTM symposia proceedings.*

Type of Paper	Number	Percentage
Perspective	17	7
Methods	58	24
Results	97	40
Interpretation	13	5
Hazard assessment	11	5
Fate modeling	14	6
Regulatory	7	3
Laboratory-field validation	8	3
Bioavailability	6	2
Water quality criteria	11	5
	242	100

3. Wes Birge and J. A. Black's paper on "Research Needs for Rapid Assessment of Chronic Toxicity" [3] published in the eighth proceedings.

There is a definite increasing trend in perspective papers (Fig. 1). This is excellent and should continue; it will lead to qualitative growth in our discipline rather than to simple quantitative growth. As Ken Macek so vividly reminded us five years ago in the fifth proceedings [4], if we are to remain a viable discipline, we must grow qualitatively. I see signs that we are, in fact, growing qualitatively, and I take the increasing trend in perspective papers as a very positive sign of maturity.

Methods

It is not surprising that many (24%) of the papers published in the proceedings have dealt with toxicity testing methods. An in-depth examination of these papers reveals the trend shown in Fig. 2. At the first few symposia, there were a large number of "new methods" papers. However, the number has decreased dramatically. Interestingly, the major area of new method development in recent symposia has dealt with rapid assessment methods. Ten different rapid methods to predict chronic toxicity have been published; this is in recognition of the extreme importance of no-observed-effect concentrations in making hazard (risk) assessment decisions. The search needs to continue for rapid and inexpensive means of predicting chronic effects of toxicants. It is in this area and in the sediment toxicity area where we need to place more of our efforts. I am particularly concerned that we do not have adequate methods to assess the effects of sediment-associated chemicals, particularly in light of the movement to develop numerical sediment criteria.

Method Refinement

While the trend line for the development of new methods is definitely decreasing, there is an increasing trend in papers that refine and evaluate existing methods (Fig. 3). These papers range from studies about the effects of diet on brood size and weight of *Daphnia* to studies that evaluate the sensitivity of various chronic testing endpoints and test duration for *Ceriodaphnia* toxicity tests.

Methods refinement is an extremely important area in aquatic toxicology. If one considers the magnitude of the decisions that often rests on the use of our aquatic toxicity data, then it is

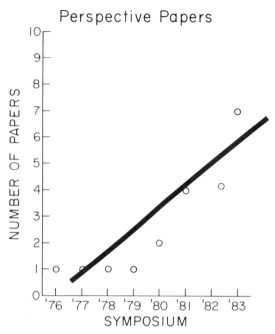

FIG. 1—*Papers on perspective published in the first eight ASTM symposia proceedings, aquatic toxicology and hazard assessment.*

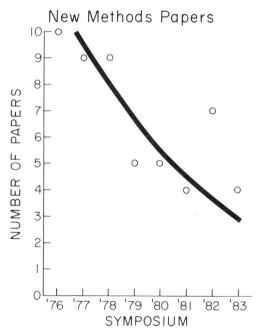

FIG. 2—*Papers on new methods published in the first eight ASTM symposia proceedings, aquatic toxicology and hazard assessment.*

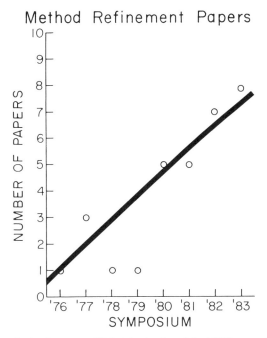

FIG. 3—*Papers on method refinement published in the first eight ASTM symposia proceedings, aquatic toxicology and hazard assessment.*

essential that we have accurate and reproducible methods for measuring effects. One area that warrants our continued efforts is the nutritional requirements of our test organisms. Several papers in the symposia proceedings clearly indicate that the nutritional/physiological health of test organisms significantly affects test results. I hope that we continue to see more and more papers that address this issue.

Results

Ninety-seven papers (40%) have reported the results of studies that assessed the effects of chemicals and effluents on aquatic life. As might be expected, there was great diversity in test materials and test organisms. There appears to be a relatively constant number of results papers published each year in the proceedings (Fig. 4). A closer look at the kinds of results papers published shows that there are more papers (14) on bioconcentration/bioaccumulation/biomagnification than on any other topic. There have been 13 papers that reported the results of studies on the effects of chemicals/effluents on the physiology of organism, and 12 papers that dealt with fate and effects of chemicals in microcosms. Microcosm studies and multispecies responses are controversial areas that need considerably more research; I hope that this topic will be a high priority of aquatic toxicologists, along with work that will allow us to better understand the modes of action of toxicants.

One area that I feel is conspicuously lacking in the symposium proceedings is results papers that assess the effect of toxicants on the structure and function of aquatic ecosystems. There are only two papers on this topic and, in the future, we should see more research conducted in this area.

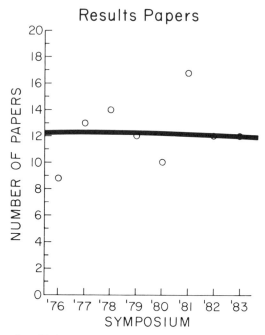

FIG. 4—*Papers on results published in the first eight ASTM symposia proceedings, aquatic toxicology and hazard assessment.*

Interpretation

Papers that I classify as interpretive have appeared in recent proceedings (Fig. 5). These are papers that provide general guidance on how to interpret and use the results of our research. Examples are:

1. Feder and Collin's paper published in the fifth proceedings entitled "Considerations in the Design and Analysis of Chronic Aquatic Tests of Toxicity" [5].
2. The paper by Javitz published at the same time on "Relationship Between Response Parameter Hierarchies, Statistical Procedures and Biological Judgement in the NOEL Determination" [6].
3. G. Fred Lee and R. A. Jones' paper entitled "Integration of Chemical Water Quality Data" [7] published in the second proceedings.

I feel that this is an extremely important area and one that should continue to get our attention. Unfortunately, I sometimes get the feeling that we are better at generating data than we are at using them!

Hazard Assessment

There has been a decreasing number of hazard (risk) assessment papers published in the proceedings (Fig. 6). Although this may reflect a satisfaction with the current approaches, it seems to me that we need to always be seeking new and better approaches, especially in consideration of the wide use of hazard assessment approaches in making decisions about the development of new chemicals, about registering pesticides, and about assessing hazardous waste sites. Therefore, I challenge all of us to rethink this area and to continue to develop and publish assessment rationales, procedures, and case studies.

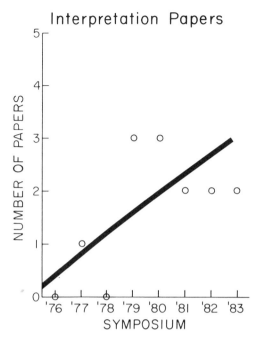

FIG. 5—*Papers on interpretation published in the first eight ASTM symposia proceedings, aquatic toxicology and hazard assessment.*

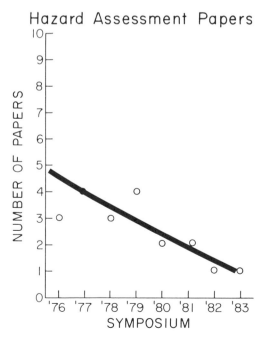

FIG. 6—*Papers on hazard (risk) assessment published in the first eight ASTM symposia proceedings, aquatic toxicology and hazard assessment.*

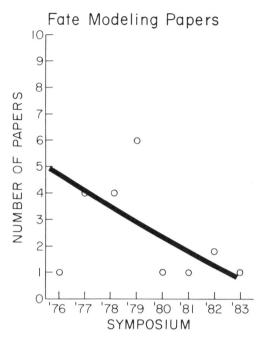

FIG. 7—*Papers on fate modeling published in the first eight ASTM symposia proceedings, aquatic toxicology and hazard assessment.*

Fate Modeling

Another area where the number of papers published has decreased is the fate modeling area (Fig. 7). I cannot explain this, and it concerns me. I think that it would be a tragedy if people interested in hazard assessment and fate modeling no longer participated in the symposia. We all recognize the importance of the interaction of aquatic toxicologists, chemists, and engineers. Thus, I hope that future symposia organizers will actively seek participation in the fate modeling area.

Laboratory/Field

There has been an increasing number of papers (Fig. 8) that report results of studies designed to answer the question, "How predictive are laboratory results of impacts in the field?" This is an extremely important area for aquatic toxicology, and it is particularly important now that the pollution control philosophy in this country is shifting from a technology-based approach to a water quality–based approach. Water quality criteria and effluent toxicity evaluations are taking on considerably more importance in today's regulatory processes. Thus, it is essential that we understand how predictive of instream impacts are our laboratory test results. A number of the studies reported in the proceedings indicate that water quality criteria are, in fact, protective of in situ aquatic life. Other studies, although not reported in the proceedings, indicate that chronicity estimates of complex effluents are also predictive of instream impacts. I hope that we continue research that addresses the efficacy of laboratory data to predict instream effects.

Finally, I challenge each of you. The number of integration papers published in the eight symposia proceedings is *zero!* Integration papers are those papers that take information we currently

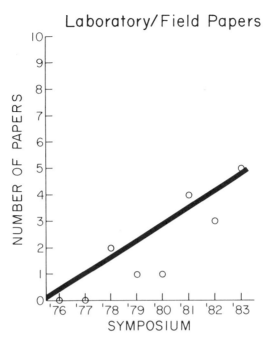

FIG. 8—*Papers on laboratory/field published in the first eight ASTM symposia proceedings, aquatic toxicology and hazard assessment.*

have and develop basic relationships. When I think of integration papers, I think of some of Gene Kenaga's papers published in the open literature where he explored the responses of test organisms to hundreds of different chemicals and made predictive correlations between species. I think of some of the U.S. EPA Environmental Research Laboratory-Duluth papers that analyzed the priority pollutant data base to address the question of range of response of different test species to chemicals. From papers such as these, guidance on how many different species that have to be tested to adequately measure the range of response of organisms to a chemical can be derived. These kinds of integration papers have not been published in the proceedings, and I hope that we will see an exponential increase in the future.

1976 to 1986: A Decade of Growth for the Aquatic Sciences Within Industry—
J. L. Hamelink

The worldwide implementation of various regulatory programs during the 1970s created a large demand for aquatic toxicologists in industry. Fortunately, a lot of other skills are available within industry that helped support the development of aquatic toxicology. I am going to describe some of the advances in aquatic toxicology that have been made with this support, as illustrated by our facilities in Greenfield, Indiana.

I joined Eli Lilly and Co. in December of 1974 to start forming a Department of Environmental Toxicology. We took over an old building that was across the road from the main toxicology building. This building was a great place to try out almost anything until we got our ideas to work. We moved into a nice 1.2-acre building in January of 1981, complete with solar panels which we use to help heat some of the fish water. Since we may use up to 30 000 gallons of tempered water a day, we can use a bunch of calories any day of the year. We do not run water through the panels; we use a less efficient but infinitely less risky system of hot-air-to-water heat exchangers.

Our fish water is not just ordinary well water. We remove all of the iron with special filters and about half of the calcium from the water before the fish get to enjoy it. Thus, we have no problems with scale or precipitates in our test waters. We originally acidified the water with CO_2 and used a big reverse osmosis unit to remove some of the calcium. It worked okay, but it took a lot of operator time. Now, we use a couple of electrodialysis units that work great and only have to be cleaned once a week. The only real problem with this kind of system is that it uses a lot of water. If you live and work over liquid limestone (like we do in the Heartlands), then you have to throw away about 30% of the water you take into the lab and strip it of iron because the electrodialysis membranes are constantly flushed with water to carry off the calcium that has been removed from the product water.

Given this good water, we do not have any problems with scale buildup in our diluters. On the other hand, a little scale wouldn't really bother any of our technologically advanced diluters which a coworker, Roger Mohr, designed for use in our lab. The Mohr manifold diluter consists of parallel pairs of toxicant solution and dilution water cells that receive the appropriate solutions through common supply lines (that is, the manifold). The measuring cells drain through a bank of paired, air-actuated, two-way, Teflon-coated ball valves. This pressurized proportional diluter also consists of two banks of dilution water and toxicant water measuring cells. Air and water flow in and out of the stainless steel tubes fitted to the top of each cell are controlled by paired, air-actuated, three-way, Teflon-coated ball valves.

Pressurized proportional diluters are great diluters. They can easily handle those difficult test solutions, like 100 ppm of suspended solids, and they are fast. They can complete a cycle in about 3 min versus about 6 min for a manifold diluter. The biggest advantage of any Mohr diluter is the control box. It is an automonitoring terminal that is connected to a small in-house computer. Thus, the activities of a diluter are automatically recorded, and if events don't occur within a programmed interval of time, the diluter automatically shuts off and goes into alarm. Besides calling up the guardhouse and causing the on/off light to flash, a single digit number gets displayed. This number facilitates repairs by telling the operator what went wrong. This feature is really nice after you've been called in during the middle of a cold winter night. Yes, we've been called in several times, but in six years of operations with the Mohr diluters, we've never lost a study. We've experienced lots of shutdowns, but we've always been able to get the units running again within an acceptable period of time, as proven by our analytical data. Consequently, I strongly advocate installation of an automatic, record-keeping, shutdown-alarm system for every diluter.

Computers and advanced electronics began to play a major role in the laboratory during the last ten years. Now we use a digitizer instead of a ruler to measure fish [8]. Another machine that we have learned to love is our autocounter. As reported by Le Blanc [9], this is the kind of device to use for counting *Daphnia magna* neonates during a life-cycle test. The neonates should be filtered out of the water and counted on filter paper because you pick up ghost images (and get wrong counts) wherever water forms an edge with glass. The machine cannot see and count *Ceriodaphnia* neonates, however, because they are too small. In fact, it can only see adult *Ceriodaphnia* on flat-black filter paper. Thus, it looks like there is no escaping the dissecting scope curse when it comes to working with *Ceriodaphnia*.

Another advance in our laboratory involved fathead minnows (*Pimephales promelas*). We hold our brood stocks in large raceways; each brood has 20 to 30 males and 40 to 60 females. Until we want eggs, they just swim around in the deep covered end of the tank and eat as much as they want. To get eggs, we put in 20 to 30 spawning tiles. After a while, the males swim out to stake a claim on one of the tiles and spawning begins.

Computers have changed a lot over the past decade. Unfortunately, most of us never see the computer and do not appreciate the advances in hardware. We spend our time with something like the digitizer and a stack of pictures or we sit in front of a terminal that is connected to a mainframe computer in some faraway place. Computers are neat! You can get all kinds of good information with them and you can get it very fast. For example, you can conduct a literature search based on just a few key words. The run will only take a few minutes, but it will be far superior to anything you could have put together by spending a few years in the stacks. You can also write or buy all kinds of neat programs, if you know what you want and how to use them. We have a computer support group within the company and an environmental fate modeler within our department who we can turn to for help. Fate modeling is used to estimate the potential concentration of contamination that would be expected in surface waters and groundwaters after widespread use of one of our new agricultural-chemical products. The modeling is based on the physicochemical properties of a new chemical and its projected use patterns. CREAMS is generally used for estimating runoff into surface waters and PRZM is used for transfers into groundwaters. These modeling efforts often serve to identify those critical parameters or dominant pathways that need some additional study, and computers allow us to complete the work with ease.

The program CLOGP3 [10] does a good job of calculating log P, the octanol/water partition coefficient, for most chemical structures. I use the program to get a structure-activity relationship (SAR) estimate of the toxicity of a new chemical to fish the first time I test it. I use the SAR estimate to design my first test with fish because then I am guaranteed feedback. If the observed LC_{50} lies within a two-log mole wide band, I do not get too excited; it is like most of the chemicals we have tested before. However, if it falls outside the band, and especially if it is toxic enough to fall below the band, I get very interested.

This calculated Log P-LC_{50} regression is an interesting relationship. It has a much shallower slope than that derived for narcotics by Veith et al. [11] or the chemicals for which we have measured log Ps, and it covers a wider range of log P (Fig. 9). The program will derive a log $P = 8$ with confidence. It will tell you that it is unrealistic, but it will do it. The shallow slope does not come from the few points that are on the far left and right hand ends of the regression. I ran the regression without them and got essentially the same values for a and b. However, I have to admit that neither regression is very strong. The correlation coefficient was only 37.4% when all the data were used. Thus, this SAR is not a strong tool but it is a useful tool and that is what computers are really good at—quickly providing some useful information.

In summary, I have seen vast improvements take place in water treatment systems, in diluters,

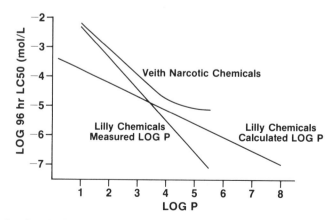

FIG. 9—*The Log* P *and 96-h* LC_{50} *for narcotic chemicals* [11] *and for Eli Lilly chemicals, calculated and measured.*

in record keeping, in biology, and in our utilization of electronic devices during the last decade. All of these developments have helped to reduce the labor and improve the accuracy of our daily activities in aquatic toxicology. In the decade to come, I expect the use of SARs and *Ceriodaphnia* as a test organism to expand.

Ten Good Years—R. A. Kimerle

On this, the tenth anniversary of ASTM's first Aquatic Toxicology and Hazard Assessment Symposium, I would like to share some of the good news and a little of the bad news of the past ten years.

The subject of environmental science and aquatic hazard assessment has grown significantly in the past decade. Concurrently, we have improved our ability to obtain useful data and to provide a basis for making decisions on the safety of chemicals in the environment. As an environmental scientist whose professional career has been entwined in many of the significant events of this period of growth, I see the benefits and accomplishments of methods development groups and how they have enabled progress to be made. Lately, however, I have also come to sense that in some areas there is less reliance on past accomplishments in aquatic toxicology and too much "reinventing the wheel." Opportunities also exist to further advance the science into new areas.

One of the important accomplishments on methods development prior to ASTM's entry into environmental toxicological methods with Committee E35 on Pesticides (and more recently Committee E47 on Biological Effects and Environmental Fate) was that of the ad hoc committee that wrote the heavily referenced "Methods For Acute Toxicity Tests With Fish, Macroinvertebrates, and Amphibians" [*12*]. This committee was composed of environmental scientists from government, industry, and universities. A unique feature of this group was the commitment they brought from their employers to make the consensus method writing activity successful. Representatives from ASTM and *Standard Methods* approached this ad hoc group and asked them to join their own long-established methods development format and build new strengths in the environmental toxicology area. Eventually, the group joined Committee E35 (and subsequently Committee E47), under which an extensive number of methods have been developed and are used in the scientific and regulatory communities.

One of the problems sometimes brought up concerning the development of consensus methods has been the slow pace of draft after draft prior to final approval. Certainly, new effort needs to be made to accelerate the rate of issuing final standard practices. The act of environmental professionals coming together and working on methods has enhanced the science. The real danger of not improving our overall productivity is dwindling support by agencies and individuals who really need to be involved in the emerging science. When people are not involved, they tend to ignore past progress and try to solve their environmental problems in a manner contrary to the proven direction of the science.

All things considered, the ASTM activities and related spinoffs have well served the needs of the environment and our society. From a personal viewpoint, I look back and can enthusiastically say that it has been rewarding and fun.

Aquatic Toxicology: Observations From a True Observer—K. J. Macek

I am now on the outside of the science of aquatic toxicology, looking in. I am involved in cancer treatment. Nevertheless, I am still burdened with an aquatic toxicology albatross called regulatory judgement, the process that results in decisions and actions. The term regulatory judgement, in my opinion, is what is called in English grammar class an oxymoron. It means a rhetorical figure in which an epigrammatic effect is created by the conjunction of incongruous or contradictory terms. But more on regulatory judgement in a minute.

It just does not seem that it was 10 years ago when Don Mount [1], in an hour, destroyed at least half-a-dozen myths about aquatic toxicology and delivered the science from a kind of bondage, making it free to grow. This was a milestone in the science.

Less than a year later, John Sprague [13] uttered a pearl of wisdom near and dear to my own heart, "A fish is a fish is a fish." This, too, was a milestone. Shortly thereafter, Nick D'Oude, at the first Pellston Workshop, Douglas Lake, Michigan [14], drew a figure on a blackboard that provided a sense of organization to the activities of aquatic toxicology and crystallized the relationships between the roles of the aquatic toxicologist, the environmental chemist, and the regulator in the hazard (risk) assessment process. The graph was a little misleading, however. It suggested that we would be equally precise at determining both effect concentrations and environmental concentrations at any time. *This is not the case*. Aquatic toxicology has proceeded much faster. We can measure toxicity. We know what it is because of the responses of test organisms. To paraphrase Sprague: "Toxicity is toxicity is toxicity," or toxicity equals a constant. (Remember that in Ken Dickson's analysis of symposia publications that methods papers decreased dramatically.) The shift from aquaria to microcosms to field studies is not concerned with toxicity; it is concerned with the real variable in hazard assessment, the exposure assessment.

The challenge of the next decade is for environmental chemists to define exposure as precisely as you people have defined aquatic toxicity—this is a scientific challenge. The practical challenge is to educate the regulators on how to reasonably interpret the inputs from aquatic toxicology and exposure assessment to their hazard assessment and to regulatory decisions. Based on past examples of hazard assessment disasters, regulators will continue to be an albatross to reasonable environmental decisions until they are better educated as to the relevance and significance of the relationship of aquatic toxicology and environmental concentrations of toxicants.

My view of the future is that the education process will occur, if slowly, as we move to gain more experience and begin to integrate the results of environmental toxicological and chemical studies.

Progress in Aquatic Toxicology: A Mile or a Millimetre?–F. L. Mayer, Jr.

In the beginning, our goals and objectives were very obvious. We developed and improved much needed toxicological tests. Today, we have several types of tests—acute and chronic, with three types of chronic tests (full life-cycle, partial life-cycle, and abbreviated), several clinical tests, including biochemistry, physiology, immunology, histopathology, and behavior, and three major areas in residue dynamics (bioconcentration, biomagnification, and bioaccumulation). We also have tests for all levels of biological organization (subcellular, cell, tissue, organ, organism, population, community, and ecosystem). In addition, we have integrated these tests into hazard assessments. We have even rediscovered the concept of dose, which was confirmed by Paracelsus in 1564 when he stated that "No substance is a poison by itself, it is the dose that makes a substance a poison" [14].

I think that we are floundering around a bit (Fig. 10). Some of us are continuing to develop and improve methods that are very important, some of us are going around in circles, and some of us are ricochetting off into God knows where (Fig. 11). I think the reason for this situation is that in the beginning, the problems with contaminants were more obvious—there were flagrant chemical spills, fish kills, and declining numbers of birds of prey. Raw sewage was dumped into the environment and rivers caught on fire. The tools and data we needed to address those problems were evident and simple, and we did a very good job in developing solutions. The problems we are addressing today and those that we'll address in the future are less obvious and the approaches may be more complex. We are confronted with such things as effects of complex chemicals from the population to the ecosystem level.

T. H. Huxley, around the turn of the century, said, "Science is, I believe, nothing but trained

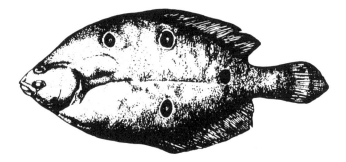

FIG. 10—Flounder—*to struggle with stumbling or plunging movements; to struggle clumsily or helplessly in embarrassment or confusion.*

and organized common sense, differing from the latter only as a veteran may differ from a raw recruit: and its methods differ from those of common sense only so far as the guardsman's cut and thrust [with his sword] differ from the manner in which a savage yields his club'' [*15*]. I think that we have tended more towards the savage and his club in the past, but I feel that it was necessary. To address the complex problems of today and the future, we are going to have to lean more towards the sword to develop approaches to those problems.

For the future, I would like to propose additional emphasis on two major areas, neither of which is new. The first area deals with integrative approaches and concentrates less on specific, single tests. We have progressed to some degree by developing hazard (risk) assessment approaches that relate effects to exposures (Fig. 12). However, we need to reevaluate what constitutes effects. Toxicology is a multidisciplinary science and consists of more than looking at survival, growth, and reproduction. To understand and solve complex or multivariate problems, we are going to require more complex or multivariate approaches that are available through a more complete use of toxicology. It is inherent in this approach that more advanced statistics be used—multiple regression, step-wise multiple regression, discriminate function analysis, and so on. We are going to have to quit dwelling on single, specific tests and endpoints for all situations. We need to learn to match the tool to the problem. We do not necessarily have to get away from our solve-all-problems approach through the "test-of-the-month club" and "magic bullets." At least, however,

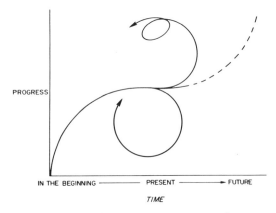

FIG. 11—*Graph of progress among researchers.*

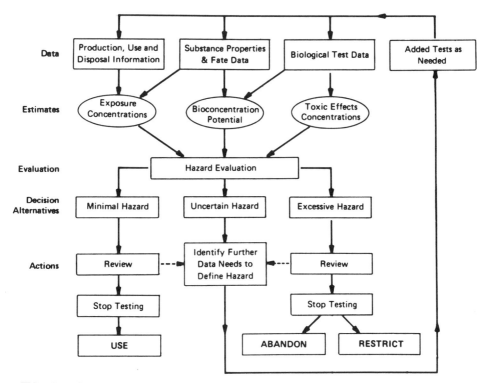

FIG. 12—*Schematic diagram of the environmental hazard evaluation process (modified from American Society of Testing and Materials Hazard Evaluation Task Group, J. R. Duthie, Chairman)* [16].

we should switch the wording to "approach-of-the-month club" and "fragmentation bombs"—this would reflect our potentially expanded thinking.

The second major area of renewed emphasis deals with using all of our data and using it better. We must use statistics as a tool—we should control the tools and not let them control us. I am going to use acute toxicity test data as an example. In acute toxicity tests, we seldom use any data other than the estimated concentration for 50% kill at the end of the test. We then use that number in all sorts of manipulations. We place all sorts of confidence and probability limits around that point, and then we predict something, such as chronic toxicity, from that point. This is ridiculous. We cannot predict from a single point. Some crude estimates can be made, but a single point has no direction. We should use all of our data to add the directional component and relax our attitude some on extrapolating within and outside of our data if we are truly going to predict.

However, there are reasons why we seldom extrapolate outside of our data. We get hung up on confidence limits, because the limits are statistically smallest around the LC_{50}. We have spent so much time in perfecting analyses for determining the LC_{50} and its confidence limits, and the analyses have become so precise, that it is probably beyond the ability of the organisms to respond to that degree of precision. We then constrain ourselves with that same precision for extrapolation and predicting. We do not predict outside of our data because the confidence limits are too broad and, therefore, the predicted values are no good. This is only true in part. As we move from the average of our data, we become less precise in our predictions, but not necessarily less accurate. It is inherent in the mathematics of statistical regression that the confidence intervals are going to diverge from the line at either end. There is nothing you can do about that. However, do not let the precision of observed data control your ability to derive accurate predicted data. The repeatability of accuracy of prediction is the confidence limits you need to establish.

So, can we predict? It appears that if we use all of the data in acute tests—all concentrations, all responses, all times—we *can* accurately predict chronic no-effect concentrations [*17*]. This ability to predict should come as no surprise. We have been doing it in an indirect manner for some time. We have moved back slowly from the full life-cycle test to the abbreviated or rapid estimator test. We have done this because the partial life-cycle estimates the full life-cycle, and the abbreviated test estimates all of the above. The toxicological element here is the time course of events, and the four-day acute test is not that different in time from the seven-day abbreviated test.

In summary, I think that we will find simple answers to a lot of things. Maybe our thought processes will have to become more complex to get there. The road will not be an easy one to follow, but it will be worth it for our future generations. And so, returning to the original question, how far have we progressed, a mile or a millimeter? It obviously depends on your standard of measure. Overall, we have come a great distance in some areas but have a long way to go in others.

The Life Cycle of Aquatic Toxicology—D. I. Mount

Professions go through life cycles, like organisms, but professions don't have to become senescent and die. The members of the profession can control its destiny. Professions do, like organisms, live in a larger context—an environment—that affects them, too. Let's examine our "organism" and its "environment."

In the early 1950s, aquatic toxicology was in its infancy; it was the era of pickle jar biology. By the late 1950s, there was some maturity; facial hair had been developed. Our science went to college in the 1960s, and in the 1970s we had our oral defense. Now, in the 1980s, we are earning a living.

We have dramatically improved water quality in the United States in recent years. One has only to remember areas like the Kanawha River in the 1960s and the Cuyahoga River that literally burned. Now fish are back in many of these rivers and so are local fisheries. Despite the fact that we have used single-species tests, we have done a lot to improve water quality. But we will not admit it! We seem to want to hide the progress that we have made for fear further progress will stop! What we should do is to continue to mature as a profession and to apply what we know. We now have or are developing multispecies tests, shorter tests to estimate chronic effects, surrogate species, and endpoints other than lethality, growth, and reproduction. We recognize the importance of diet in influencing response to toxicants. These were all tools of our trade needed for the problems we face as we proceeded through the period of environmental awakening.

Aquatic toxicology had its beginnings in a crisis atmosphere—fish kills, pesticides, declining fish-eating birds, and eutrophied lakes. Our job was to ascertain how bad it all was and to point the finger at whomever we could to correct the problem.

I believe that we have cleaned up most of the obvious environmental "messes." Any good corporate citizen—and there are many of them—knows today that environmental protection is part of doing business. And that attitude and awareness puts us in a new arena. The issue that we must face today is "Do we need more environmental protection, where do we need it, how much do we need, and who will benefit?"

If the crisis is passing, then we need to reexamine what tools we need for the job at hand. We need, in this now emerging period, a positive rather than a negative stance. Rather than, as before, trying to force as much treatment of waste by our testing as we could, we need to determine by our testing how much of what type of treatment we really need—yes, how much is environmentally sound. It is not a risk-free world in which we live, and removing the last traces of toxicants from air and water only shifts the environmental damage from one sector of the environment to another. The attitude that I call "out-of-water-at-any-price" is no longer useful. We must consider *all* options in deciding where and how to provide environmental protection, and disposal of certain wastes in certain waters may be the option of choice. We must recognize and factor into our environmental protection formula the cost of environmental toxicity and chemical testing—dischargers do not have infinite resources. Decisions on required tests cannot be made in an economic vacuum; the national deficit affects us all.

We must change our thinking and the terms in which we express our thoughts, too. The "most sensitive species" must become the "most representative species," a "worst-case scenario" must be changed to "unacceptable harm" or "probable acceptable risk," and the "most sensitive life stage" must become the "life stage at greatest risk." Finally, we must start emphasizing allowable concentrations of toxicants at the needed level rather than to continue driving allowable concentrations lower and lower.

We are past most crises, those flagrant and willful violations of common sense that produced gross environmental insults we suddenly recognized in the 1960s. We cannot continue to remain healthy as a science if we base our actions on past crises. We must change. Aquatic toxicology like any species must migrate, mutate, or die when the "environment" changes—and the environment is changing. We as a profession have been very successful at developing the tools we need. How successfully we decide what the needs are will become obvious in the coming decade.

Conclusions—P. R. Parrish

1. The ASTM symposia on aquatic toxicology and hazard assessment have provided a forum for professional growth and personal enrichment. The contributions of a group of diverse individuals, six of which were chosen to present their views of our science at this special session of the tenth symposium, have been invaluable. Their role in the success of the symposia is gratefully acknowledged.

2. An analysis of the publications from the first eight ASTM symposia reflects the maturation of the participants and the science. Publications on methods have decreased while more encompassing papers on perspective (those that challenge us to think about aquatic toxicology as a science) have increased. Methods refinement papers have increased, but results papers have remained constant. Interpretation and laboratory/field papers are appearing more often, while hazard assessment and fate modeling papers are decreasing. Most interesting, the number of integration papers is zero, a deficiency that must be overcome in the future.

3. Vast improvements have occurred in aquatic toxicity testing laboratories during the past ten years. Water treatment systems, diluter efficiency and dependability, data recording and retrieval, test animal culture and maintenance techniques, and utilization of electronic devices—especially computers—have all enhanced the quality of work performed by aquatic toxicologists.

4. The growth of environmental science and aquatic hazard assessment, although slow, has progressed through the commitment of environmental scientists from government, industry, and universities. Consensus test methods have developed primarily through the efforts of The Committee on Methods for Acute Toxicity Tests with Fish, Macroinvertebrates, and Amphibians, ASTM Committee E35 on Pesticides, and subsequently ASTM Committee E47 on Biological Effects and Environmental Fate. A danger of decreasing productivity exists because of our failure to effectively recognize and apply methods already developed and because of the dwindling support of agencies and individuals.

5. The concept that our ability to predict effect concentrations and environmental concentrations of a chemical with equal precision is not correct. Toxicity of chemicals can be accurately defined by performing tests of increasing complexity. Fate of chemicals, however, is much more difficult

to determine. The challenge for the next decade is for environmental chemists to define exposure as precisely as toxicity has been measured in the past decade. Regulators must be educated as to the relevance and significance of the relationship between aquatic toxicology and environmental concentrations of toxicants in order to perform hazard (risk) assessments.

6. Aquatic toxicology is a multidisciplinary science that has advanced to the point where advanced, multivariate statistics must be used to evaluate effects beyond survival, growth, and reproduction. By using all data and by using them in a statistically sound manner, we may be able to predict chronic no-effect concentrations from acute data. Predictions outside known data may decrease precision of estimates but not necessarily the accuracy of the estimates.

7. Since the time when aquatic toxicology began in a crisis atmosphere of fish kills, lake eutrophication, declining populations of fish-eating birds, and burning rivers, we have cleaned up most of the obvious environmental "messes." And although simplistic, single-species laboratory toxicity tests have been used to generate data on which environmental regulations were set, we have dramatically improved the quality of U.S. surface waters. We do not live in a risk-free world, and removing the last traces of toxicants from air and water may only shift damage from one environmental sector to another. We must consider all options in deciding how and where to dispose of wastes created by our society, and disposal of certain wastes in certain waters must be considered. Further, the cost of required environmental toxicity and fate studies must be included in any environmental protection formula.

References

[1] Mount, D. I. in *Aquatic Toxicology and Hazard Evaluation (First Conference), ASTM STP 634*, F. L. Mayer and J. Hamelink, Eds., American Society for Testing and Materials, Philadelphia, 1977, pp. 5–14.
[2] Zapp, J. A., Jr. in *Aquatic Toxicology, ASTM STP 707 (Third Conference)*, J. G. Eaton, P. R. Parrish, and A. C. Hendricks, Eds., American Society for Testing and Materials, Philadelphia, 1985, pp. 2–10.
[3] Birge, W. J. and Black, J. A. in *Aquatic Toxicology and Hazard Assessment (Eighth Symposium), ASTM STP 891*, R. C. Bahner and D. J. Hansen, Eds., American Society for Testing and Materials, Philadelphia, 1985, pp. 51–60.
[4] Macek, K. J. in *Aquatic Toxicology and Hazard Assessment (Fifth Symposium), ASTM STP 766*, J. G. Pearson, R. B. Foster, and W. E. Bishop, Eds., American Society for Testing and Materials, Philadelphia, 1982, pp. 3–8.
[5] Feder, P. I. and Collins, W. J. in *Aquatic Toxicology and Hazard Assessment (Fifth Symposium), ASTM STP 766*, J. G. Pearson, R. B. Foster, and W. E. Bishop, Eds., American Society for Testing and Materials, Philadelphia, 1982, pp. 32–68.
[6] Javitz, H. S. in *Aquatic Toxicology and Hazard Assessment (Fifth Symposium), ASTM STP 766*, J. G. Pearson, R. B. Foster, and W. E. Bishop, Eds., American Society for Testing and Materials, Philadelphia, 1982, pp. 17–31.
[7] Lee, G. F. and Jones, R. A. in *Aquatic Toxicology, ASTM STP 667*, L. L. Marking and R. A. Kimerle, Eds., American Society for Testing and Materials, Philadelphia, 1979, pp. 302–321.
[8] Sauter, S. and Harrison, J. L. in *Aquatic Toxicology and Hazard Assessment (Eighth Symposium), ASTM STP 891*, R. C. Bahner and D. J. Hensen, Eds., American Society for Testing and Materials, Philadelphia, 1985, pp. 321–327.
[9] Le Blanc, G. A., *Bulletin of Environmental Contamination and Toxicology*, Vol. 23, 1979, pp. 837–839.
[10] Weininger, D. and Leo, A., CLOP Version 3.2, Medicinal Chemistry Project, Pomona College, Claremont, CA, 1984.
[11] Veith, G. D., Call, D. J., and Brooke, L. T., *Canadian Journal of Fisheries and Aquatic Sciences*, Vol. 40, 1983, pp. 743–748.
[12] The Committee on Methods for Toxicity Tests with Aquatic Organisms, "Methods for Acute Toxicity Tests with Fish, Macroinvertebrates, and Amphibians," EPA-660/3-75-009, U.S. Environmental Protection Agency, Corvallis, OR, 1975.
[13] Sprague, J. B. and Fogels, A. in *Proceedings of the 3rd Aquatic Toxicity Workshop*, Technical Report

Number EPS-5-AR-77-1, Canadian Environmental Protection Service, Halifax, Nova Scotia, 1977, pp. 107–118.
- [14] Hayes, W. J. Jr., *Toxicology of Pesticides*, Williams & Wilkins Co., Baltimore, MD, 1975.
- [15] Hardin, G., *Biology: Its Human Implications*, W. H. Freeman and Co., San Francisco, CA, 1953, p. 4.
- [16] Cairns, J., Jr., Dickson, K. L., and Maki, A. W. in *Estimating the Hazard of Chemical Substances to Aquatic Life, ASTM STP 657*, J. Cairns, Jr., K. L. Dickson, and A. W. Maki, Eds., American Society for Testing and Materials, Philadelphia, 1978.
- [17] Mayer, F., Buckler, D., Ellersieck, M., and Krause, G., *Abstracts*, Fifth Annual Meeting, Society of Environmental Toxicology and Chemistry, Arlington, VA, 1984, p. 169.

Bioengineering and Aquatic Environments

Tamar Barkay[1] and Gary S. Sayler[2]

Gene Probes as a Tool for the Detection of Specific Genomes in the Environment[3]

REFERENCE: Barkay, T. and Sayler, G. S., "**Gene Probes as a Tool for the Detection of Specific Genomes in the Environment,**" *Aquatic Toxicology and Hazard Assessment: 10th Volume, ASTM STP 971,* W. J. Adams, G. A. Chapman, and W. G. Landis, Eds., American Society for Testing and Materials, Philadelphia, 1988, pp. 29–36.

ABSTRACT: Gene probes hold great promise as a tool in environmental sciences. They may be used to detect specific genotypes, to follow gene flow process, to delineate complex taxonomic aggregates, and to monitor genetically engineered organisms in the environment. The sensitivity of the method is currently limited by experimental procedures, and its specificity depends on the nature of the DNA sequences used as probes and the efficacy of lysing methods. Variable genetic determinants which code for the same trait determine the universality of gene probes. Finally, the method is highly feasible in terms of cost, speed, and expertise. Current and future developments in molecular microbial ecology are likely to contribute toward the improvement of the probing methodology for the full realization of its potential in environmental sciences.

KEYWORDS: gene probes, detection, sensitivity, specificity, universality, feasibility

Gene Probes in Environmental Studies—What For?

Recent advances in molecular biology have resulted in a powerful methodology for the identification of specific DNA sequences. The method is based on the propensity of two complimentary denatured DNA strands to reanneal by forming base pairs. If one such strand (that is, the probe) is radioactively or otherwise labeled, the identity of a strand of an unknown origin (that is, target) is revealed by virtue of its ability to bind (that is, hybridize) to the labeled probe, [1]. This principle has served extensively in the cloning of specific eucaryotic and procaryotic DNA sequences, and more recently was applied to the detection of microbial contaminants of food products [2,3] and of drinking water [4]. This paper discusses the potential of gene probes to become a tool in basic, as well as applied, microbial ecology.

There are several areas of research in microbial ecology that could greatly benefit by detecting specific organisms through the homology of their DNA and RNA with known gene probes.

1. *Detecting functional groups of microorganisms.* The phenotypic characterization of organisms freshly isolated from the environment is important to define the microbial metabolic potential of that environment. Traditionally, this has been carried out by using selective growth media, whereby organisms were detected by their growth of specific substrates or in the presence of inhibitors or

---

[1] Research microbiologist, Environmental Research Laboratory, Environmental Protection Agency, Gulf Breeze, FL 32561.
[2] Professor, Department of Microbiology and the Graduate program in Ecology, University of Tennessee, Knoxville, TN 37996.
[3] Contribution No. 578 of the U.S. Environmental Research Laboratory, Gulf Breeze, Florida.

by their ability to carry out specific reactions resulting in a color change. These methods suffer from a high degree of false positive and negative results due to cross-feeding between organisms and to the limited opportunity for growth provided by any laboratory growth medium, respectively. This has been a major hindrance in studying organisms which could carry out the biodegradation of xenobiotic compounds [5]. Gene probes constructed from the DNA sequences coding for the function of interest clearly distinguish potentially active organisms by the presence of the specific gene systems in their genomes. Such DNA probes have been constructed for the detection of toluene and naphthalene [5], 2,4-dichlorophenoxyacetic acid [6], and 4-dichlorobiphenyl [7] biodegradative microorganisms. In addition, organisms whose cultivation is laborious and whose biochemical characteristics are not well-defined may be distinguished by species-specific DNA probes. An example of this case was recently presented by Kuritza et al. [8], who detected species of *Bacteroides* in order to study polysaccharide degradation in the human colon.

2. *Gene flow in natural microbial communities.* The processes and mechanisms of evolution of specific microbial activities and how they are influenced by the changing environment is a basic field of study in microbial ecology. So far, these studies have been based on the evolving microbial phenotypes. Gene probes provide a tool to look at the genetic level of the underlying molecular events of evolution in the natural environment. It was with this idea in mind that a DNA gene probe was developed for the bacterial mercury-resistance gene system [9]. However, the biodegradative gene probes just mentioned could be utilized to follow gene flow in impacted environments. That approach has been taken by Sayler et al. [5], who showed a positive correlation between synthetic oil treatment of microcosm sediments and the distribution of the toluene degradative (TOL) plasmid in the indigenous microbial community.

3. *The detection of genetically engineered organisms (GEMs).* The prospect of utilizing biotechnological approaches to improve the management of ecosystems presents a serious question regarding the safety of releasing GEMs into the environment. In order to evaluate their effects, the organisms must be detected in complex environmental samples and be differentiated from abundant indigenous flora. This could be achieved by using gene probes specific to the genome of the introduced species. Although some use of this approach has already been documented [10], its applicability remains to be evaluated.

4. *The detection of nonculturable organisms.* The majority of the microorganisms in the natural environment cannot be cultured on conventional laboratory media. Some are fastidious microorganisms whose growth requirements have not been defined, and, in addition, even culturable organisms enter unculturable life stages in response to harsh environments [11]. Gene probes provide a means by which unculturable organisms can be detected by hybridization with total microbial DNA isolated from environmental samples [12]. Low recovery of microbial DNA from environmental samples is a major stumbling block in the application of this approach in microbial ecology, and several laboratories are engaged currently in research to improve available procedures of DNA isolation and purification [13,14].

Any monitoring method has to fulfill several criteria in order to be successful. A method has to be specific, reproducible, sensitive, universal, and feasible in laboratories with diverse expertise and resources. The following is a discussion of these criteria in regard to gene probe hybridization as a method for environmental monitoring.

Specificity

The reliable use of a gene probe requires a right degree of specificity to avoid false negative results due to slightly altered target DNA and false positive results due to unrelated sequences. Genes coding for the same function in different organisms may be heterologous if they have evolved independently, for example, the *mer* gene probe prepared by Barkay et al. [9] from

Shiegella flexneri, hybridized with mercury-resistant, Gram-negative organisms but not with Gram-positive ones, although in both groups resistance was mediated by the same mechanism. On the other hand, as new gene functions evolve by gene duplication and mutation, related sequences may code for completely different functions [15]. In addition, repeated DNA sequences (that is, insertion elements, transposons) may result in false positive hybridization signals [5]. This problem was evaluated by hybridization of a common cloning vector, pACYC184 coding for chloramphenicol and tetracycline resistances, with a collection of Gram-negative enteric organisms. Approximately 35% of the 108 strains tested were hybridization positive, although none was resistant to chloramphenicol and tetracycline (unpublished results). Low specificity of gene probes may lead to high background hybridization in environmental studies. This was demonstrated by a study of groundwater aquifer microcosms in which five gene probes were utilized to follow four introduced bacterial strains [16]. The specificity of the probes ranged (in increased order) from that of a whole genomic probe to detect strain ABS10 through a cloned chromosomal fragment to detect strain AHS24, to a cryptic plasmid probe for strain AOS23, and to a cloned restriction fragment of the TOL plasmid for the detection of *Pseudomonas putida* (TOL$^+$). The latter strain also contained the resistance plasmid RK2, which was detected by hybridization with a whole RK2 probe. The results (Table 1), which summarize eight microcosm samplings over an eight-week period, indicate that whereas an increase in the abundance of hybridization-positive target colonies occurred with all probes (inoculated versus natural microcosms), the background hybridization related to the degree of the probe specificity (natural microcosms). Thus, high background hybridizations were observed with the less specific probes for RK2, ABS10, AHS24, AOS23, while TOL$^+$ colonies were found only in inoculated microcosms (Table 1).

There are several means by which the specificity of gene probes can be improved for environmental monitoring:

1. Elimination of unrelated sequences, if the molecular details of the target genes are available, can produce highly specific DNA probes [9].
2. The stringency under which the hybrids are formed can be increased (that is, more homology between the probe and the target sequence is needed for hybridization to occur) to further specificity [1].
3. Whole genomic libraries can be screened to select those cloned fragments which hybridize only with the desired target organisms [3,8].

Sensitivity

The application of gene probes to environmental studies requires high sensitivity since specific organisms are to be detected in physically as well as biologically complex and heterogeneous

TABLE 1—*Detection by DNA:DNA colony hybridization of introduced organisms in groundwater aquifer microcosms.*

Treatment	Positive Target Colonies (Fraction of Total CFU)				
	TOL	RK2	ABS10	AHS24	AOS23
Natural[a] sample	0.001[b]	0.023	0.130	0.010	0.025
Inoculated	0.066	0.070	0.050	0.028	0.126

NOTE: CFU = colony-forming units.
[a] Natural aquifer material supplemented with about 1×10^6 cell/g killed ABS10 strain.
[b] 0.001 arbitrary limit of detection.

matrices. False negatives may result if target organisms are masked on isolation media by the growth of heterologous aerobic heterotrophs or if they fail to lyse (and thus, hybridize) by the conventional lysis procedures which were designed for the manipulation of *Escherichia coli*. The latter is especially important if one tries to follow gene flow in natural microbial communities where the objective is to detect specific genes in heterologous host background. Detection of specific organisms in a community genome isolated directly from the environment may be rendered insensitive due to low recoveries or the physical interference of natural substances (for example, clay particles, humic material) [13,14]. Thus, the limit of detection of each gene probe should be defined before experimental results can be interpreted.

Table 2 summarizes the sensitivity levels of several DNA gene probes. The variability in the methods which were employed by the different researchers exclude valid comparisons between observed limits of detection. Sayler et al. [5] could detect colonies of *P. putida* carrying the TOL plasmid among 5×10^6 heterologous colonies on nitrocellulose filters. However, additional data indicated that at the high density of colonies required to achieve such high sensitivity, the number of target organisms would be underestimated due to masking. The effect of masking on sensitivity was also observed by Hill et al. [2], whose efficiency of detection of virulent *Yersinia enterocolitica* declined sharply when their ratio to nonvirulent strains dropped below 1:200. A species-specific DNA probe detected *B. vulgatus* only if target DNA comprised at least 2% of the DNA isolated directly from fecal material. With this sensitivity, only the ten to twelve major organisms in feces, where numbers reach 10^{10} organisms/g, could be detected [17]. This is a rather high limit of detection and may render this methodology impractical for samples with lower population densities. Sensitivity may be improved by concentrating large volumes of samples and by growing colonies under selective conditions which eliminate many of the heterologous organisms. Echeverria et al. [4] and Fitts et al. [3] achieved sensitivities of one target organism/mL of drinking waters and 4.3 organisms/mL broth, respectively, by employing these approaches (Table 2). In the latter study, the filtered organisms were lysed and hybridized without growth into colonies [3], a strategy that could be used to overcome masking by overgrown heterologous colonies.

The susceptibility of the test organisms to lysis conditions could be verified by employing a universal probe, a DNA sequence which is commonly found in all bacteria. Hybridizing the same sample with both the universal and the specific probe would eliminate false negative results due to inappropriate lysis conditions. Lysis conditions can then be manipulated to ensure optimal results with the tested sample. That inappropriate lysis was responsible for false negative results became apparent during a study which employed the *mer* gene probe [9] to follow the distribution of the mercury-resistance genes in natural microbial communities. Consequently, alternative lysis

TABLE 2—*Sensitivity of DNA gene probes for the detection of specific organisms in environmental samples.*

Probe	Test Method	Limit of Detection	Reference
TOL plasmid	Colonies grown on nitrocellulose filters	1 target/10^6 colonies	Sayler et al., 1985 [5]
Bacteroides vulgatus	Whole DNA fraction from feces	Target 2% of total population	Kuritza and Salyers, 1985 [17]
Yersinia enterocolitica	Colonies grown on nitrocellulose filters	1 target/200 colonies	Hill et al., 1983 [2]
Enterotoxigenic *Escherichia coli*	Colonies grown on McConkey agar	1/mL (1/10^5 heterologous organisms)	Echeverria et al., 1982 [4]
Salmonella sp.	Bacterial cells on nitrocellulose filters	4.3/mL growth medium	Fitts et al., 1983 [3]

strategies were designed and the results were monitored by a ribosomal DNA gene probe. Eight bacterial strains representing a wide variety of Gram-negative organisms were grown on nitrocellulose filters and lysed by the conventional method of Grunstein and Hogness [18]. Another set of colonies was treated by a modification of this method which included the sequential treatments of colonies on the filter with sucrose lysozyme solution (1.2 μg/mL lysozyme; 25% sucrose, 0.05 M Tris; 0.001 M EDTA; pH 8.0), 0.25 M EDTA and an SDS high pH solution [3% SDS, 0.01 M Tris, 0.1 M sodium chloride (NaCl); 0.01 M EDTA adjusted to pH 12.6 with sodium hydroxide (NaOH)] prior to the NaOH treatment accounted for by the Grunstein-Hogness procedure. Colonies treated by both lysis procedures were hybridized with the ribosomal DNA probe under stringency which allowed hybridization of all the tested strains (as had been previously determined by hybridizing dot blots of clear DNA lysates with the same probe). The results (Fig. 1) indicated that whereas colonies of most of the strains were lysed by both methods, the lysozyme-SDS procedure resulted in clearer and more defined signals. Reduced smearing was due probably to the SDS treatment [19] included in the modification, but poor lysis could also account for the weak signals observed with the Grunstein-Hogness procedure. This is especially true with *Acinetobacter calcoaceticus, Moraxella phenylpyruvica*, and *Pseudomonas maltophilia*. This observation was reproducible, and thus it appears that this method is not appropriate for the preparation of colonies of these strains for hybridization. However, the lysozyme-SDS procedure failed with a colony of *Flavobacterium odoratum*, as is indicated by its weak hybridization signal (Fig. 1). Obviously, in this case, the Grunstein-Hogness method provided better results. Because no one lysis procedure was found to be completely reliable, the ribosomal DNA gene probe is routinely used to control for false negative results in studies employing the *mer* probe.

Universality

A reliable gene probe detects all the organisms it is designed to detect, regardless of their origin or physiological condition. Whereas this is of no consequence for gene probes designed to detect a specific gene in a specific organism (for example, detection of GEMs) [8,3], the universality of probes for specific genotypes [5] or gene systems [9] in mixed microbial communities needs to be assessed because several genes may code for a similar phenotype. For example, at least five different determinants code for tetracycline resistance [20], and at least two determinants code for arsenite/arsenate resistance [21] in enterobacteria. The Universality of the *mer* probe was tested by hybridization to Gram-negative, mercury-resistant bacteria isolated from fecal material, soils, and sediments (Table 3). Between 32 and 62% of the organisms did not hybridize to the *mer* probe, suggesting that there is at least one alternative mercury-resistance determinant in nature.

Reproducibility

The reproducibility of hybridization procedures is excellent as long as conditions are maintained stably. However, using them to monitor organisms in the environment is subjected to all the considerations pertinent to environmental monitoring protocols in general.

Feasibility

Gene probes are attractive as tools for routine monitoring because of the relative simplicity of their use and the speed with which results can be obtained. Very small amounts of the probe sequences (<1 μg DNA/reaction) are needed, and results are obtained within 24 hs. The required equipment and supplies are simple and inexpensive, and an experienced lab technician can be trained and qualified to carry out the procedure in a matter of days. The current major drawback of the method is the use of radioactive label, which is short-lived, hazardous, and requires special

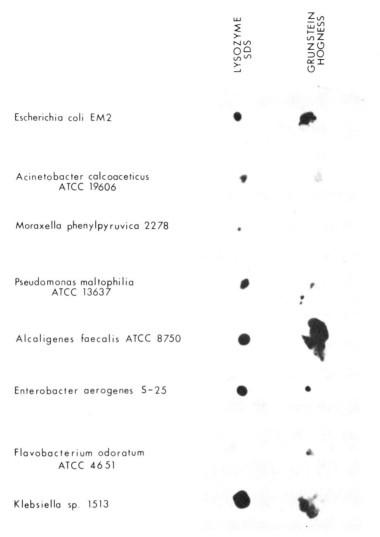

FIG. 1—*Colony hybridization of Gram-negative bacterial strains obtained with the lysozyme-SDS and the Grunstein Hogness procedures with a ribosomal DNA probe. Colonies were lysed as described in the text (lysozyme-SDS procedure) and by Maniatis et al. [19] (Grunstein Hogness procedure). DNA was labeled with ^{32}P-dCTP to a specific activity of 4830 Ci/µmol using a nick translation kit (BRL, Bethesda, MD). The autoradiogram was obtained with XRP-5 X-ray film (Eastman Kodak Co., Rochester, NY) after overnight exposure at $-70°$ C.*

disposal arrangements. Alternative labelling approaches are now available. Biotinylated, fluorescent, and enzyme-linked probes have all been suggested as a replacement for ^{32}P. The applicability and especially sensitivity of these methods require improvements before they can be used widely and reliably [1].

TABLE 3—*Hybridization of the* mer *probe with DNA of mercury-resistant bacteria isolated from variable origins.*

Origin	No. of Strains Tested	mer[a]	%
Fecal material			
Urban dwellers	26	23	88
Hospital patients	42	26	62
Rural dwellers	20	9	45
Miscellaneous[b]	7	7	100
Total	95	65	68
Soil	34	23	68
Lake sediments	21	10	48

[a] Positive hybridization.
[b] Included strains whose source was not known.

Conclusions and Future Developments

Gene probes have the potential of becoming a powerful tool in monitoring specific organisms and genes in environmental samples. This method could broaden the scope of microbial ecology by answering fundamental questions whose investigation has been limited by the availability of appropriate methodologies. A cost-effective, simple tool would be provided for monitoring the distribution and abundance of specific organisms. The latter is especially significant in light of the pressing concern with the fate and effects of GEMs in the environment. The full utilization of gene probes is currently limited by technical difficulties pertaining to their sensitivity, specificity, and universality. However, new developments in molecular microbial ecology promise an improvement. Probing specific DNA sequences directly isolated from the environment, the inclusion of concentration steps, and the enrichment of specific sequences by affinity chromatography (R. Q. Colwell, University of Maryland, personal communication) would increase sensitivity. The careful preparation of specific probes which cover all possible genetic determinants is possible through molecular genetics. This would improve the specificity and universality of the method. For example, subgenomic DNA probes for the careful detection of *mer* sequences in natural communities are now available (A. O. Summers, University of Georgia, personal communication). The direct application of probes to environmental samples by using autoradiography may become possible through a better understanding of the fate of naked DNA in the environment. Such a methodology would reduce the excessive sample handling which currently complicates the statistical analysis of the observations.

Acknowledgement

Portions of the work described in this paper were carried out at the laboratory of B. H. Olson, The University of California, Irvine.

References

[1] Meinkoth, J. and Wahl, G., *Analytical Biochemistry*, Vol. 138, 1984, pp. 267–284.
[2] Hill, W. E., Payne, W. L., and Aulisio, C. C. G., *Applied and Environmental Microbiology*, Vol. 46, 1983, pp. 636–641.
[3] Fitts, R., Diamond, M., Hamilton, C., and Neri, M., *Applied and Environmental Microbiology*, Vol. 46, 1983, pp. 1146–1151.

[4] Echeverria, P., Seriwatana, J., Chityothin, O., Chaicumpa, W., and Tirapat, C., *Journal of Clinical Microbiology*, Vol. 6, 1982, pp. 1086–1090.

[5] Sayler, G. S., Shields, M. S., Tedford, E. T., Breen, A., Hooper, S. W., Sirotkin, K. M., and Davis, J. W., *Applied and Environmental Microbiology*, Vol. 49, 1985, pp. 1295–1303.

[6] Amy, P. S., Schulke, J. W., Frazier, L. M., and Seidler, R. J., *Applied and Environmental Microbiology*, Vol. 49, 1985, pp. 1237–1245.

[7] Pettigrew, C. A. Jr. and Salyer, G. S., *Journal of Microbiological Methods*, Vol. 5, 1986, pp. 205–213.

[8] Kuritza, A. P., Shaughnessy, P., and Salyers, A. A., *Applied and Environmental Microbiology*, Vol. 51, 1986, pp. 385–390.

[9] Barkay, T., Fouts, D. L., and Olson, B. H., *Applied and Environmental Microbiology*, Vol. 49, 1985, pp. 686–692.

[10] Sayler, G. S., Harris, C., Pettigrew, C., Pacia, D., Breen, A., and Sirotkin, K. M., *Developments in Industrial Microbiology*, Vol. 27, 1987, pp. 149–165.

[11] Xu, H. S., Roberts, N., Singleton, F. L., Attwell, R. W., Grimes, D. J., and Colwell, R. R., *Microbial Ecology*, Vol. 8, 1982, pp. 313–323.

[12] Sayler, G. S. and Stacey, G., in "Biotechnology Risk Assessment," J. R. Fiksel and V. T. Covello, Eds., Pergamon Press, NY, 1986, pp. 35–55.

[13] Ogram, A., Gustin, D., and Sayler, G. S., *Abstracts of the Annual Meeting of the American Society for Microbiology*, Q143, Washington, DC, 1986, p. 307.

[14] Steffan, R. J., Goksoyr, J., and Atlas, R. M., *Abstracts of the Annual Meeting of the American Society for Microbiology*, Q144, Washington, DC, 1986, p. 308.

[15] Ohno, S., *Proceedings of the National Academy of Sciences USA*, Vol. 81, 1984, pp. 2421–2425.

[16] Sayler, G. S., Wilson, J. T., Houston, L., Pacia, D., and Jain, R. K., *Applied and Environmental Microbiology*, in press.

[17] Kuritza, A. P. and Salyers, A. A., *Applied and Environmental Microbiology*, Vol. 50, 1985, pp. 958–964.

[18] Grunstein, M. and Hogness, D., *Proceedings of the National Academy of Sciences USA*, Vol. 72, 1975, pp. 3961–3965.

[19] Maniatis, T., Fritsch, E. F., and Sambrook, J., "Molecular Cloning. A Laboratory Manual," Cold Spring Harbor, NY, 1982.

[20] Marshall, B., Tachibana, C., and Levy, S. B., *Antimicrobial Agents and Chemotherapy*, Vol. 24, 1983, pp. 835–840.

[21] Mobley, H. L. T., Silver, S., Porter, F. D., and Rosen, B. P., *Antimicrobial Agents and Chemotherapy*, Vol. 25, 1984, pp. 151–161.

D. J. Grimes,[1] C. C. Somerville,[1] W. Straube,[1] D. B. Roszak,[2] B. A. Ortiz-Conde,[1] M. T. MacDonell,[2] and R. R. Colwell[1]

Plasmid Mobility in the Ocean Environment

REFERENCE: Grimes, D. J., Somerville, C. C., Straube, W., Roszak, D. B., Ortiz-Conde, B. A., MacDonell, M. T., and Colwell, R. R., **"Plasmid Mobility in the Ocean Environment,"** *Aquatic Toxicology and Hazard Assessment: 10th Volume, ASTM STP 971*, W. J. Adams, G. A. Chapman, and W. G. Landis, Eds., American Society for Testing and Materials, Philadelphia, 1988, pp. 37–42.

ABSTRACT: Evidence of plasmid selection and genetic exchange in natural aquatic environments, including the ocean, includes: (*a*) high incidence of plasmid-containing strains in polluted areas; (*b*) presence of free DNA in natural environments; (*c*) co-existence of identical plasmids in different cohabiting strains; and (*d*) data from *in situ* plasmid transfer experiments. Current research in our laboratory regarding plasmid mobility in the ocean centers around viable but nonculturable bacteria, cloning of ecologically significant genes, genetic exchange between deep sea bacteria under pressure at low temperature, and development of 16S ribosomal DNA probes for tracking genetically engineered microorganisms that are released to natural environments.

KEYWORDS: plasmid, ocean, genetically engineered microorganism, genetic exchange, viviforms, chitinase, deep sea bacteria, somnicell

An important issue, with regard to release of genetically engineered microorganisms (GEMs) to the natural environment, is whether or not release contributes significantly to increase in genetic transfer above "background," that is, transfer occurring in nature [*1*]. It is not unreasonable to hypothesize that, within a given habitat, the genetic product will be similar, even though the process, that is, introduction of "alien" forms by release of GEMS, versus natural genetic exchange and selection (allogenic succession), may differ [*1*]. The assumption can be made, with sufficient documentation already available, that transfer of genetic elements occurs in natural environments, including the ocean.

Herein we review literature supporting the hypotheses that (*a*) exchange of genetic material between bacteria occurs in nature and (*b*) bacteria remain viable in the natural environment, even when not detectable by conventional means, and hence are capable of supplying and acquiring genetic information *in situ*. We conclude with a brief overview of our current work on plasmid transfer between baroduric bacteria and on tracking GEMs in the aquatic environment.

Evidence for Plasmid Selection and Exchange

There are several direct and indirect lines of evidence for plasmid selection and exchange in natural aquatic environments, including the ocean. Table 1 lists the four main lines of evidence.

Several publications available in the literature provide data documenting high incidence of plasmid-containing bacteria in polluted areas. Guerry [*2*] observed a greater incidence of plasmids

---

[1] Department of Microbiology, University of Maryland, College Park, MD 20742.
[2] Molecular Biosystems, Inc., San Diego, CA 92121.

TABLE 1—*Evidence for plasmid selection and/or exchange.*

1. High incidence of plasmid containing strains in polluted areas.
2. Presence of free DNA in natural environments.
3. Coexistence of identical plasmids in different cohabiting strains.
4. Data from *in situ* plasmid transfer experiments.

among bacteria isolated from polluted areas of Baltimore Harbor than from other, nonpolluted areas of the Chesapeake Bay. Similarly, Sizemore [3] detected a greater plasmid incidence among bacteria from harbor and coastal water and ship bilge than among bacteria isolated from the open Atlantic Ocean. In a later study, Hada and Sizemore [4] found a greater plasmid incidence among marine *Vibrio* species isolated from oil-impacted waters of the Northwest Gulf of Mexico compared to unpolluted control sites. Cavari et al. [5] demonstrated similar findings for the Mediterranean Sea at off-shore sites near Israel, and Wickham and Atlas [6] reported plasmid frequency to be positively correlated with increasing chemical stress in soil. Baya et al. [7] reported several observations regarding the coincidence of plasmids and antimicrobial resistance in marine bacteria isolated from polluted and unpolluted Atlantic Ocean samples. They also noted that chemically polluted areas yielded bacteria with greater numbers of plasmid bands than from unpolluted areas. Figure 1 shows the frequency of plasmid bands detected in bacteria from the various locations sampled. It was concluded from these data that bacterial isolates from toxic chemical waste-receiving waters more frequently contain plasmid DNA and demonstrate antimicrobial resistance than bacteria from noncontaminated areas [7].

The observation that high concentrations of naked DNA occur in certain marine habitats provides an indirect line of evidence supportive of the hypothesis that exchange of genetic material can occur freely in aquatic environments. Lorenz et al. [8] demonstrated that a considerable amount of DNA was tightly bound to the inorganic fraction of marine sediment. In a later study, this same group showed that sediment-adsorbed transforming DNA was more protected from the action of DNase I than was unbound DNA in saltwater [9]. Deflaun et al. [10] examined dissolved DNA in water over coral heads, in coral mucus, and in oligotrophic, offshore water. They detected DNA in all samples, but reported that coral mucus was enriched in DNA by a factor of 2 to 12 times. Similarly, Winn and Karl [11] found that as much as 75 to 90% of total DNA in the oligotrophic Pacific Ocean was present as nonreplicating DNA, that is, DNA associated with dead,

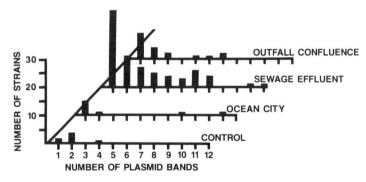

FIG. 1—*Plasmid frequency among strains of bacteria isolated from unpolluted Atlantic Ocean water samples (control), domestic sewage outfall confluence at Ocean City, Maryland (Ocean City), effluent from a wastewater treatment plant treating both toxic industrial and domestic wastes (sewage effluent), and the mixing zone of the toxic industrial-domestic waste treatment plant (outfall confluence). See Baya et al. [7] for further details.*

dormant, or debilitated cells or adsorbed to nonliving particulates. It is apparent from these studies that DNA can exist in a free state in the marine environment and may, therefore, be available for uptake and recombination by competent strains of bacteria.

Reports of identical plasmids in different cohabiting bacterial strains isolated from the same aquatic habitats are few, partially because of the difficulty of proving that such plasmids are, indeed, identical. Work done in our laboratory with isolates obtained from the gut of an amphipod collected from the Bay of Biscayne at a depth of 4300 m showed that each of two isolates, one a *Vibrio* sp. and the other a *Pseudomonas* sp., contained identical plasmids, based on agarose mobility, restriction enzyme analysis, and DNA hybridization [*12*]. The conclusion drawn from results provided in these reports is that bacteria can, and most probably do, exchange plasmids *in situ*.

More frequent are reports of *in situ* genetic exchange. Nearly every habitat examined for evidence of such transfer has yielded supporting data. Baross et al. [*13*] were the first to report data suggesting transfer of genetic markers in a marine environment. In a series of experiments employing oysters (*Crassostrea gigas*), it was observed that: (*a*) oysters typically contained high bacteriophage titers; (*b*) oysters accumulated both *Vibrio parahaemolyticus* and specific bacteriophage; (*c*) oysters facilitated contact between bacteriophage and bacteria; and (*d*) oysters supported transduction of an agarase marker into an agarase-negative strain of *V. parahaemolyticus*. It was hypothesized that the role of bacteriophage in nature may be to preserve and transfer genetic information [*13*].

Reports of *in situ* transfer of genetic markers also exist for freshwater aquatic environments. Morrison et al. [*14*] demonstrated transduction of streptomycin resistance, using *Pseudomonas aeruginosa* cultures in membrane chambers suspended in a Tennessee reservoir. Mach and Grimes [*15*] also used membrane chambers to demonstrate conjugal transfer of antibiotic resistance markers between enteric bacteria suspended in a wastewater treatment facility. Gealt et al. [*16*] extended the implications of the Mach and Grimes [*15*] study to include transfer of nonconjugative pBR plasmids from laboratory strains of *Escherichia coli* to enteric bacteria isolated from wastewater. In each case, a mobilizing strain of *E. coli* was required to mediate the transfer [*16*].

Evidence for genetic selection and exchange occurring in natural aquatic environments, therefore, ranges from circumstantial to conclusive, the latter employing carefully controlled environmental simulations. Thus, transfer of genetic material can be concluded and, in all probability, can be expected to occur in systems receiving GEMs.

Current Research on Plasmid Mobility in the Ocean

A line of research ongoing for several years in our laboratory involves the discovery of viable, but nonculturable, bacteria, referred to by Roszak [*17*] as "viviforms" and further defined by Roszak and Colwell [*18*] as "somnicells." Briefly, when selected strains of gram-negative bacteria are placed into nonnutritive, nontoxic water of varying salinity and temperature, the cells, over a several-day time period, become smaller and rounded and lose the ability to grow on, or in, routine laboratory culture media (Fig. 2). However, such nonculturable cells remain viable, demonstrated by the direct viable count (DVC) method of Kogure et al. [*19*] and by a combination of the DVC and uptake of [^{3}H]–thymidine and [^{14}C]–glutamic acid [*17,18*]. Thus, when monitoring bacteria by conventional methods, that is, by the use of culture media, viviforms and somnicells go undetected. Grimes and Colwell [*20*] have recently demonstrated that somnicells retain most, if not all, of their genetic information, including plasmids. Such bacteria, including GEMs, are, therefore, potentially capable of supplying and/or acquiring genetic information even though they may be considered nonviable by classical bacteriological culturing methods.

Another area of active research in our laboratory deals with chitin degradation by marine bacteria, notably by *Vibrio* spp. The chitinase gene of *V. vulnificus* has been cloned into *E. coli*

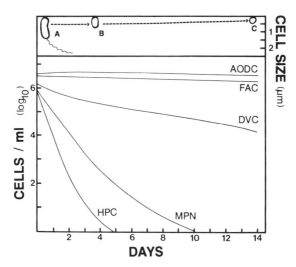

FIG. 2—*Transition of living, culturable cells* (A) *to somnicells* (B *and* C), *showing the various "fates," as measured by acridine orange direct count (AODC), fluorescent antibody count (FAC), direct viable count (DVC), most probable number (MPN), and heterotrophic plate count (HPC).*

[*21*] and the gene is now being sequenced. A gene probe has also been developed, so that chromosomal DNA and plasmids from other marine bacteria, including deep sea bacteria, can be examined for the presence of the chitinase genes. Chitin is one of the few biomolecules that survives descent to the abyssal plain without substantive degradation, and it is utilized by a variety of marine bacteria.

Genetic exchange between deep sea bacteria under pressure at low temperature (about 430 atm and 3°C) is also being actively investigated. Using the approach outlined in Fig. 3, deep sea isolates with known plasmid profiles and unique colonial morphologies are mated in pressure chambers [*22*] and examined for plasmid exchange. Suspected transconjugates/transformants are subjected to further study, for example, restriction enzyme analysis of plasmids, to confirm acquisition of new plasmid bands. Acquisition of plasmids under conditions of high pressure and low temperature will be useful information for predicting the fate of plasmids in GEMs released to the deep sea, whether accidental or to accomplish a specific purpose, for example, to degrade toxic chemicals discharged at deep ocean dump sites.

In early stages of development is the production of a probe specific for unique sequences in 16S ribosomal DNA. The "ribosomal RNA transcription unit," which contains DNA coding for 16S rRNA, is present in all bacteria, including GEMs, in multiple copies, and is highly conserved. Despite its conservation, however, certain "signature sequences" in the DNA coding for 16S rRNA make the region highly specific, and therefore an appropriate biotarget for tracking GEMs released to the open ocean. We have begun development of a ^{32}P-based probe and plan to develop a nonradioactive probe for this application (B.A. Ortiz-Conde, M.T. MacDonell, and R.R. Colwell, work in progress).

Conclusion

There exists an increasingly significant body of evidence which suggests that genetic transfer occurs readily among bacteria in natural systems, including the ocean. Freshwater bacteria discharged to the sea may appear to become nonviable in the conventional sense. However, viability is not a necessary prerequisite to the propagation of some portions of bacterial genetic

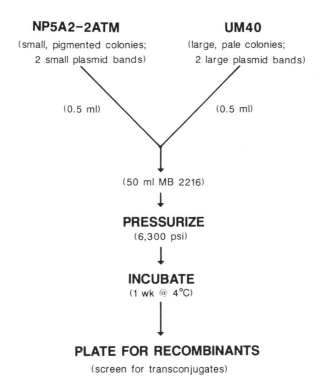

FIG. 3—*Pressure mating protocol for two bacterial strains (NP5A2-2ATM and UM40) isolated from deep seawater samples and mated in Marine Broth 2216 (MB 2216) under high pressure [about 443 kg/cm^2] and low temperature (4°C).*

information. In addition, the release of GEMs to the ocean to solve site-specific problems, for example, degrade toxic waste, must be considered in the context of natural processes. Eventual selection of the best-fit gene pool for a given habitat will occur, regardless of whether GEMs are present or absent. Therefore, GEMs modified only in expression or specificity of genes already known to exist in that specific environment, or reasonably assumed to exist, given the known input of allochthonous microbiota, may be expected to have a minimal effect on the habitat, as measured by genetic exchange. The impact of GEMs carrying novel genes, however, would be difficult to predict and would require more information than is available at the present time.

A frequently considered approach to ensure safety of experiments with GEMs is to construct novel genetic elements in debilitated hosts. Experiments in our laboratory, however, have shown that organisms concluded by employment of classical culturing methods to be nonviable under a specified set of environmental conditions may not only be intact, but capable of nutrient uptake and nucleic acid metabolism, hence "viviform." It would be imprudent to adopt strategies requiring a large initial input of GEMs and to depend upon natural attrition for their removal. If transformation of autochthonous bacteria by free and sediment-adsorbed nucleic acids occurs at a significant rate in nature, then not even lysis of allochthonous GEMs will assure the desired minimal environmental impact.

Release of GEMs to address site-specific environmental problems may prove to be a very important tool for ecologists. However, guidelines for the use of GEMs for this purpose must acknowledge the scope of problems that can arise. Amplification of genes already existing in a given environment constitutes a reasonable use of molecular genetic technology. In fact, we would argue that there is already a great deal more genetic exchange and modification in the marine environment than heretofore recognized. The impact of release of novel genes into the marine environment is much more difficult to predict, given the myriad of selective pressures on a new competitor in such an ecosystem. Perhaps in those cases where there is doubt as to the safety of the release of GEMs, the focus can be placed on release of GEM gene products instead.

Acknowledgments

Work described in this paper was supported, in part, by the Microbial Ecology Branch of the U.S. Environmental Protection Agency (Cooperative Agreement Number CR812246-01-0) and the Molecular Biology Program of the Office of Naval Research (Contract N00014-86-K-0197).

References

[1] Colwell, R. R. and Grimes, D. J., in *Aquatic Toxicology and Environmental Fate: Ninth Volume, ASTM STP 921*, T. M. Poston and R. Purdy, Eds., ASTM, Philadelphia, 1986, pp. 222–230.
[2] Guerry, P., Ph.D. thesis, University of Maryland, Aug. 1975.
[3] Sizemore, R., Ph.D. thesis, University of Maryland, Aug. 1975.
[4] Hada, H. S. and Sizemore, R. K., *Applied and Environmental Microbiology*, Vol. 41, No. 1, January 1981, pp. 199–202.
[5] Cavari, B., Roshanski, N., and Colwell, R. R., *VIes Journées Étud Pollutions*, Cannes, 1982, pp. 615–618.
[6] Wickham, G. S. and Atlas, R. M., *Abstracts of the Annual Meeting of the American Society for Microbiology*, Q147, 1986, p. 308.
[7] Baya, A. M., Brayton, P. R., Brown, V. L., Grimes, D. J., Russek-Cohen, E., and Colwell, R. R., *Applied and Environmental Microbiology*, Vol. 51, No. 6, June 1986, pp. 1285–1292.
[8] Lorenz, M. G., Aardema, B. W. and Krumbein, W. E., *Marine Biology*, Vol. 64, 1981, pp. 225–230.
[9] Aardema, B. W., Lorenz, M. G., and Krumbein, W. E., *Applied and Environmental Microbiology*, Vol. 46, No. 2, August 1983, pp. 417–420.
[10] Deflaun, M. F., Paul, J. H., and Jeffrey, W. H., *Abstracts of the Annual Meeting of the American Society for Microbiology*, Q145, 1986, p. 308.
[11] Winn, C. D. and Karl, D. M., *Limnology and Oceanography*, Vol. 31, No. 3, May 1986, pp. 637–645.
[12] Wortman, A. T., and Colwell, R. R., "Occurrence of Plasmids in Bacteria Isolated from the Deep Sea," *Applied and Environmental Microbiology*, Vol. 54, 1988, in press.
[13] Baross, J. A., Liston, J., and Morita, R. Y. in *International Symposium on Vibrio parahaemolyticus*, T. Fujina et al., Eds., Saikon Publishing Co., Ltd., Tokyo, 1974, pp. 129–137.
[14] Morrison, W. D., Miller, R. V., and Sayler, G. S., *Applied and Environmental Microbiology*, Vol. 36, No. 5, Nov. 1978, pp. 724–730.
[15] Mach, P. A. and Grimes, D. J., *Applied and Environmental Microbiology*, Vol. 44, No. 6, Dec. 1982, pp. 1395–1403.
[16] Gealt, M. A., Chai, M. D., Alper, K. B. and Boyer, J. C., *Applied and Environmental Microbiology*, Vol. 49, No. 4, April 1985, pp. 836–841.
[17] Roszak, D., Ph.D. thesis, University of Maryland, April 1986, pp. 118–126.
[18] Roszak, D. B. and Colwell, R. A., *Applied and Environmental Microbiology*, Vol. 53, Nov. 12, December 1987, pp. 2889–2983.
[19] Kogure, K., Simidu, U., and Taga, N., *Canadian Journal of Microbiology*, Vol. 25, No. 3, 1979, pp. 415–420.
[20] Grimes, D. J. and Colwell, R. R., *FEMS Microbiology Letters*, Vol. 34, 1986, pp. 161–165
[21] Wortman, A. T., Somerville, C. C., and Colwell, R. R., *Applied and Environmental Microbiology*, Vol. 52, No. 1, July 1986, pp. 142–145.
[22] Deming, J. W., Tabor, P. S., and Colwell, R. R., *Microbial Ecology*, Vol. 7, 1981, pp. 85–94.

Vlasta Molak[1] *and Jerry F. Stara*[2]

Genetically Engineered Microorganisms in the Aquatic Environment: Environmental Safety Assessment

REFERENCE: Molak, V. and Stara, J. F., **"Genetically Engineered Microorganisms in the Aquatic Environment: Environmental Safety Assessment,"** *Aquatic Toxicology and Hazard Assessment: 10th Volume, ASTM STP 971*, W. J. Adams, G. A. Chapman, and W. G. Landis, Eds., American Society for Testing and Materials, Philadelphia, 1988, pp. 43–50.

ABSTRACT: The deliberate release of genetically engineered microorganisms (GEMs) into the environment is a promising approach to the amelioration of environmental problems such as pollution and pest control.

GEMs may find their way into the aquatic environment, (oceans, streams, groundwaters, lakes, wastewaters, and sludges) either by direct release into that environment or, after uses on land, via surface runoff and soil seepage into groundwaters, streams, lakes, and estuaries. Since experience with the deliberate release of GEMs is very limited, it is necessary to develop approaches for risk assessment of the potential adverse effects that could be caused by the introduction of a large number of novel microorganisms into the aquatic environment. The accepted approach to risk assessment (hazard identification, exposure, and dose-response assessment, leading to risk characterization) has been used to develop models for risk assessment of the deliberate release of GEMs into the aquatic environment. The new concept introduced in this risk assessment is that of "critical mass," defined as the number of GEMs applied at a specific site that would result in replacement of endogenous microorganisms. This "critical mass" would depend both on the properties of GEMs and on the number and properties of endogenous microorganisms at the site of GEMs application. An experimental assessment of critical mass may be possible by using aquatic microcosms corresponding to the aquatic environment into which GEMs would be released.

KEYWORDS: genetic engineering, microorganisms, water, risk assessment, modeling

Deliberately released, genetically engineered microorganisms (GEMs) may ameliorate some pressing environmental problems such as pollution and pest control. Some of these GEMs will ultimately end up in the aquatic environment, such as surface waters (waste water, streams, lakes, seas) and groundwater. Although the potential benefits of the deliberate release of GEMs are high, the introduction of large numbers of such GEMs into the aquatic environment may pose hazards. This paper will present an approach to the evaluation of risk of deliberate release of GEMs in the aquatic environment, both to human health and to ecosystems. The assumption will be that the engineered genes will not be able to "travel" from GEMs to other bacteria in the aquatic ecosystem.

Elements of Risk Assessment

A scheme for the risk assessment for chemicals [*1*] can, with some modifications, be applied to the risk assessment of a deliberate release of GEMs [*2*]. Experimental evidence substantiates

---

[1] Biochemist and Environmental Criteria and Assessment Office (ECAO), Environmental Protection Agency, Cincinnati, OH 45268.

[2] Deceased, formerly director of ECAO.

the premise that GEMs do not have properties that are basically different from those of naturally occurring microorganisms. Nutritional requirements, growth rate, and the dieoff rate of a GEM may be different from those of the parent microorganism, but no new properties are observed that do not already exist in nature. The only difference between a particular GEM and a naturally occurring microorganism is that the GEM has been developed to express a specific feature that makes it useful for a particular task. For example, *Pseudomanas fluorescens* with the inserted toxin gene from *Bacillus thuringiensis* has the ability to produce a toxin that can destroy certain undesirable insects [3]. Therefore, risk assessment of a deliberate release of large quantities of GEMs into the environment can be performed similar to the risk assessment of naturally occurring microorganisms.

Three basic elements of risk assessment need to be considered (Fig. 1): hazard identification, dose response, and exposure assessment [1].

With respect to hazard assessment, microorganisms can be divided into two groups: pathogens and nonpathogens. Although pathogens (human, animal, and plant) clearly present a hazard, nonpathogens are not necessarily safe. Depending on their fate and transport in the environment, nonpathogens may replace endogenous populations or alter the ecosystem in some undesirable way. Therefore, GEMs that are not pathogenic to humans or to useful species cannot automatically be excluded from the risk assessment.

The dose response for pathogens has features in common with dose response relationships for chemicals. The minimal infectious dose (ID_{50}) can be established in similar fashion to the LD_{50} for chemicals [4]. For nonpathogens, we suggest that a new dose response concept, that of "critical mass," be considered. "Critical mass" is the number of GEMs applied at a specific site that would cause replacement of the endogenous microorganisms and continuous persistence of GEMs in that site. This critical mass would depend both on the properties of GEMs and on the number and properties of endogenous microorganisms. The aquatic environment into which GEMs may be released is already occupied by numerous other species of microorganisms. Bacterial concen-

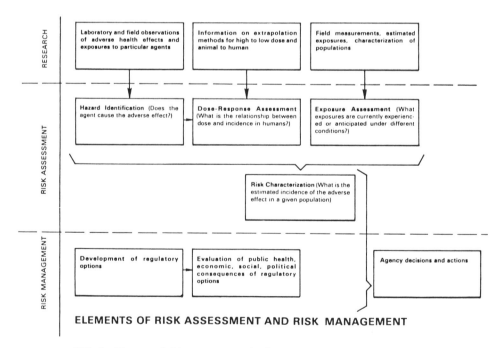

FIG. 1—*Elements of risk assessment and risk management (adapted from Ref. 1).*

trations in the water column range from 10^{-3} to 10^4 bacteria mL^{-1} in oligotrophic waters (organic carbon level 1 to 15 mg/L) to 10^7 to 10^8 bacteria mL^{-1} in eutrophic waters (organic carbon level > 15 mg/L) [5]. The air-water interface in oceans and lakes is covered with a film of hydrophobic hydrocarbons, fatty acids, and esters, overlaying a layer of polysaccharide protein complexes. This layer, rich in nutrients, attracts bacteria, resulting in a concentration of 10^8 bacteria mL^{-1} [5].

Benthic concentrations of microorganisms are generally higher than concentrations in the water column by 1 to 4 orders of magnitude [5]. The ratio of the concentration of GEMs to the concentration of endogenous microorganisms may determine the ultimate fate of GEMs and, consequently, the risk they pose. However, one also has to consider other factors which could affect the fate of GEMs. For example, if a GEM is not equipped to live in an aquatic environment, even if a large number are introduced in a particular aquatic environment they will not survive. Therefore, a concept of "critical mass" may be applicable only when GEMs are introduced into an environment suitable for their survival and they can compete with endogenous microorganisms.

The last element in risk assessment, exposure assessment, is a function of environmental fate and transport for both pathogenic and nonpathogenic GEMs. This, in turn, is determined by the "dose response of the ecosystem," or initial number and properties of GEMs applied relative to the number and properties of endogenous microorganisms.

Fate and Transport of GEMs

GEMs can appear in the aquatic environment as a result of:

1. Direct application in the aquatic environment, such as GEMs used in oil spills and in pollutant degradation in wastewaters and sludges.

2. Transport of GEMs into the aquatic environment from other environments. For example, microorganisms used as pesticides in soil or plant leaves may be transported to surface waters and seawaters as the result of surface runoff or into groundwater as a result of soil seepage. Also, GEMs used in soil for degradation of pollutants can ultimately end up in surface waters, seawater, or groundwater.

A survey of national water quality in the United States [6] revealed that bacterial pollution presents a serious problem in maintaining water quality. This bacterial pollution is believed to be due to releases from municipal wastewater treatment plants, combined sewers, urban runoff, feedlots, pastures and rangeland, and septic systems. Runoff from agricultural lands is one of the main sources of nonpoint pollution, both chemical and bacterial [7]. Therefore, bacteria originating in agricultural lands have the potential to gain access to and survive in aquatic environments. Excessive top soil loss from agricultural lands due to surface runoff is a serious problem in this country and around the world. The estimated yearly top soil loss from U.S. cropland is 1680 million tons [8]. If the use of GEMs on agricultural lands were to become widespread, transport of GEMs with the top soil due to surface runoff might become an important source of GEMs in the aquatic environment. On a smaller scale, the use of GEMs on a field of limited size would have an impact only on surface waters near the application site.

Ocean dumping of municipal sludge is another potential source of GEMs in the aquatic environment, since GEMs may be used in the degradation of toxic contaminants in municipal sludge.

Modeling of Fate and Transport

Modeling of the fate and transport of GEMs is similar to that for naturally occurring microorganisms, based on the assumption that the two categories of microorganisms are not basically different.

In general, modeling can be qualitative or quantitative. Qualitative modeling is performed in cases where little knowledge of quantitative relationships in the system exists. Quantitative modeling is applied to systems where sufficient quantitative data about the system exist for formulation of mathematical relationships between various components.

In analytical quantitative modeling, enough is known about the system that various pathways for microorganisms' transport and fate can be described by mathematical equations. Properly chosen parameters in these equations should lead to realistic estimates of the final concentrations of GEMs in the environment. In black box quantitative modeling, an empirical relationship between the input and output is experimentally determined. Details of the ecological system are treated as a black box. In order to obtain the proper output function, the inside of the black box should mimic the real ecosystem into which the GEMs will be released. A properly designed aquatic microcosm that would mimic the aquatic environment into which GEMs would be released would help to determine the mathematical relationship between output and input [9,10].

Current Applicable Models

Recently, the U.S. Environmental Protection Agency (EPA) has initiated an effort to develop methodology for pathogen risk assessment for various disposal options for municipal sludge [11]. Pathogen risk assessment was performed for ocean disposal, landfilling, and land application. Although GEMs are not necessarily pathogenic, their fate and transport can be treated similarly to the fate and transport of pathogens. In the application of these models, the assumption is that no gene exchange occurs between GEM and endogenous species. This is a reasonable assumption since companies who are planning the release of GEMs are aware of the potential for more difficulty in obtaining EPA clearance if the engineered genes can "travel." In addition, the level of complexity that would be required for modeling such a system is enormous.

The model used for ocean disposal is qualitative, since it was felt that insufficient data existed for quantitative risk assessment. The relationship among various components of the ocean ecosystem are presented in Fig. 2. Concentrations of various bacteria and viruses in sediments are 10 to 10^4 times higher than in the water column [5,12,13]. If the GEMs can survive and replace endogenous populations in sediments and the water column where they are applied, there may be a reason for concern. The survival of a particular GEM in the water column and in sediments may depend on the numerical relationship between the introduced GEM and endogenous species at a particular marine site.

A more quantitative approach could be taken by finding the relationship of concentration of a

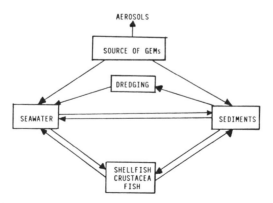

FIG. 2—*The relationship among components of the ocean ecosystem.*

GEM at a particular marine site as a function of the initial concentration of the GEM released. This is described by the following equation

$$n_{GEM}(x,t) = n_{GEM}(x,t_o)f(t - t_o) \quad (1)$$

where $n_{GEM}(x,t)$ = number of GEMs at Site x per unit volume at the time $(t - t_o)$ after application, and
$f(t - t_o)$ = empirically derived function of time for a particular ecological system.

Function $f(t - t_o)$ would have to be empirically derived in aquatic microcosms that mimic a particular site of GEMs application [9,10].

An example of a semiquantitative model is DRASTIC, designed to follow the fate and transport of microorganisms from landfill into groundwater [14]. In this model, a relative rank is assigned to various factors describing the site, including depth to water table, net recharge, aquifer media, soil media, topography impact of the vadose zone, and hydraulic conductivity of the aquifer. The relative ranking scheme is based on the "contamination potential index" that is calculated by an equation combining the various factors just listed [15]. If GEMs were applied to degrade toxic contaminants in landfills, this model might give the relative potential for the contamination of groundwater at a particular landfill site compared to other sites. This type of model does not take into account characteristics of microorganisms such as growth and survival.

An example of an analytical model is the "Sandia model" [16], a computer model that attempts to evaluate and quantify the hazard posed by enteric pathogens in sludge. Although this model primarily describes pathogen fate and transport in soil and through the food chain, it also includes surface runoff and groundwater contamination. This type of model could be used for risk assessment of GEMs that are initially applied to soil, either as pesticides or for pollution control. Because of surface runoff and soil seepage, these GEMs could conceivably also affect groundwater and surface waters. Figure 3 describes a simplified version of this model.

Since the Sandia model has yet to be validated experimentally, another approach to the evaluation of GEMs in the aquatic environment due to surface runoff can be taken using the universal soil loss equation (USLE). The USLE is an empirical equation describing the loss of top soil due to runoff [17].

$$L = R \times K \times LS \times C \times P \quad (2)$$

where L = computed average annual soil loss,
R = rainfall erosion index,
K = soil erodibility factor,
LS = topographic factor,
C = cover and management factor, and
P = support practice factor.

The average annual number of GEMs that would end up in the aquatic environment can be estimated from the following equation

$$N_{GEM} = a \times m \times L \quad (3)$$

where N_{GEM} = number of GEMs lost from the application site,
a = adsorption factor for GEM to soil particles,
m = average number of GEMs per unit mass of soil at the application site, and
L = computed average annual soil loss.

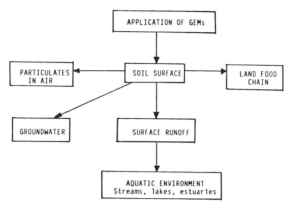

FIG. 3—*Transport of GEMs from site of application in soil.*

If one assumes that GEMs associate with soil particles, the average concentration of GEMs in receiving waters will be determined by the amount of topsoil lost according to the USLE and the volume of receiving waters.

Since fluctuations in m are on time scales much smaller than a year, it may be more appropriate to consider surface runoff in a storm event. In that case a modified universal soil loss equation (MUSLE) is used [*18*]

$$L_e = 11.8 \times (vq_p)^{0.56} \times K \times LS \times C \times P \qquad (4)$$

where L_e = soil loss due to storm event,
v = volume of runoff, and
q_p = the peak runoff rate.

The number of GEMs that would end up in the aquatic environment as a consequence of surface runoff during a storm is

$$N_{\text{GEM}(t_o)} = a \times m(t_o) \times L_e \qquad (5)$$

where $N_{\text{GEM}(t_o)}$ = total number of GEMs lost from the application site due to surface runoff during the storm at time t_o,
a = adsorption factor for GEM to soil particles,
$m(t_o)$ = density of GEMs at the application site at time t_o, and
L_e = soil loss due to storm event.

Equation 5 can be combined with Eq 1 to determine the density of GEMs in the aquatic system due to runoff

$$n_{\text{GEM}(x,t)} = N_{\text{GEM}(t_o)} \times f(t - t_o) = L_e \times a \times m(t_o) \times f(t - t_o) \qquad (6)$$

where L_e = soil loss due to storm event,
V = volume of aquatic environment into which GEMs were introduced,

a = adsorption factor for GEM to soil particles,
$m(t_o)$ = density of GEMs at the application site at time t_o, and
$f(t - t_o)$ = empirically determined function of time for a particular GEM and aquatic environment of release.

Conclusion

Several models applicable to the risk assessment of GEMs in aquatic environment exist. They vary in their predictive value from purely qualitative to quantitative. None of the quantitative models has yet been validated. Therefore, the exact prediction of the fate and transport of GEMs is not yet possible. However, semiquantitative estimates based on characteristics that can be determined in a contained environment, such as a greenhouse or aquatic microcosm, can be made. These would include the survival function in the aquatic environment $[f(t)]$, survival in the soil environment $[m(t)]$, and adsorption to soil particles (a). Each GEM has to be evaluated for a particular aquatic environment of deliberate release, since $f(t)$ may depend on the properties of the GEM, the initial concentration of GEM at the application site, and the properties of the aquatic environment.

References

[1] Ruckelshaus, W. D., "Risk Assessment and Management: Framework for Decision Making," U.S. EPA 600/9-85-002, U.S. Environmental Protection Agency, Washington, DC, December 1984.
[2] Molak, V. and Stara, J., "Risk Assessment of Deliberate Release of Genetically-Engineered Microorganisms," in *Proceedings of 1985 Meeting of the Society for Risk Analysis*, L. Lave, Ed., 1986.
[3] Lythy, P., Cordiev, J. H. L., and Fisher, H. M., Bacillus Thuringiensis as a Bacterial Insecticide: Basic Considerations and Application," in *Microbial and Viral Pesticides*, E. Kurstak, Ed., Marcel Dekker, Inc., New York and Basel, Switzerland, 1982, pp. 35–74.
[4] Ward, R. L. and Akin, E., "Minimum Infective Dose of Animal Viruses," *CRC Critical Reviews in Environmental Control*, Vol. 14, 1984, pp. 197–310.
[5] Grant, W. D. and Long, P. E., *Environmental Microbiology*, Halstead Press, New York and Toronto, 1981.
[6] "National Water Quality Inventory 1984," report to Congress, EPA 440/4-85-029, Environmental Protection Agency, Washington, DC, August 1985.
[7] "Nonpoint Source Pollution in the U.S.," report to Congress, U.S. Environmental Protection Agency, Office of Water Program Operations, Washington, DC, January 1984.
[8] Brown, L., "The Global Loss of Topsoil," *Journal of Soil and Water Conservation*, 1984.
[9] Bretthauer, R., "Laboratory Aquatic Microcosms," in *Microcosms in Ecological Research*, J. P. Giesy, Jr., Ed., Technical Information Center, U.S. Dept. of Energy, Washington, DC, 1980, pp. 416–445.
[10] Giddings, J. M., "Types of Aquatic Microcosms and Their Research Applications," in *Microcosms in Ecological Research*, J. P. Giesy, Jr., Ed., Technical Information Center, U.S. Dept. of Energy, Washington, DC, 1980, pp. 248–266.
[11] Battelle, Inc., "Pathogen Risk Assessment Feasibility Study," U.S. Environmental Protection Agency, Washington, DC, November 1985.
[12] Goyal, S. M., Gerba, C. P. and Melnick, J. L., "Occurrence and Distribution of Bacterial Indicators and Pathogens in Canal Communities Along the Texas Coast," *Applied Environmental Microbiology*, Vol. 34, 1977, pp. 139–149.
[13] Goyal, S. M., "Viral Pollution of the Marine Environment," *C.R.C. Critical Review in Environmental Control*, Vol. 14, 1984, pp. 1–32.
[14] Aller, L., Bennet, T., Lehr, J. H., et al., DRASTIC: A Standardized System for Evaluating Groundwater Pollution Potential Using Hydrogeologic Settings," EPA 600/2-85/018, Environmental Protection Agency, Washington, DC, 1985.
[15] Yates, M. V., "Septic Tank Siting to Minimize the Contamination of Groundwater by Microorganisms," American Association for Advancement of Science, Washington, DC, 1985.
[16] BDM Corp., "Sewage Sludge Pathogen Transport Model Project," Contract No. IAG-78-8-X0226, U.S. EPA Request No. 224, Environmental Protection Agency, Washington, DC, 1980.

[17] Wischmeier, W. H. and Smith, D. D., "Predicting Rainfall Erosion Losses—A Guide to Conservation Planning," Handbook No. 527, U.S. Department of Agriculture, Washington, DC, 1978.
[18] Williams, J. R., "Sediment and Yield Prediction with Universal Equation Using Runoff Energy Factor," in *Present and Prospective Technology for Predicting Sediment Yields and Sources,* USDA ARS-5-40, U.S. Department of Agriculture, Washington, DC, 1975.

Application of Paleoecological Methodology to the Analysis of Environmental Problems

U. M. Cowgill[1]

Paleoecology and Environmental Analysis

REFERENCE: Cowgill, U. M., **"Paleoecology and Environmental Analysis,"** *Aquatic Toxicology and Hazard Assessment: 10th Volume, ASTM STP 971*, W. J. Adams, G. A. Chapman, and W. G. Landis, Eds., American Society for Testing and Materials, Philadelphia, 1988, pp. 53–62.

ABSTRACT: A discussion is presented which shows the types of information that may be gleaned from the chemical, mineralogical, and biological analyses of lake muds. Illustrations are taken from a variety of studies carried out in various countries.

Substantial rises in the concentration of sulfur in the recently deposited mud as well as in the surface water of Linsley Pond (North Branford, Connecticut) are indicative of industrial expansion in the region over the past five decades. Similarly, stack emissions produced by the combustion of fossil fuels have caused rises in sulfur concentrations and subsequent changes in pH in unbuffered waters of Southern and Southwestern Scandinavia. These phenonmena have brought about shifts in population size and distribution among the lake plankton.

Recent technological growth has resulted in a progressive increase in mercury in the sediments of Lake Windermere, United Kingdom, Lake Huleh, Israel, and Lake Biwa, Japan. No doubt reflecting man's global activities, similar rises in concentrations of lead, cadmium, and silver have been noted in recently deposited muds in various sections of the world. The steady rise of phthalates noted in recently deposited sediments is indicative of the advent of the use of plastics in the many areas of human endeavor.

To ascertain which of the various hypotheses advanced to explain the sudden collapse of the ancient Mayan culture in Central America were valid, a series of lake cores were taken in Guatemala and subjected to chemical and biological analyses. The ecologically oriented hypothesis appears to be invalid. Furthermore, these studies showed that a high agricultural civilization developed and that under modern conditions a larger agricultural population employing indigenous methods can be supported. In addition, these investigations showed that early man changed a savanna to a tropical forest and that his activities brought about only a small increase in the erosion rate.

Additionally, paleoecological methods may be used to elucidate past climates, the advent of agriculture in a region, processes of plant selection and domestication, effects of slash and burn agriculture, the size of one time plankton populations, the chemical concentration of the atmosphere in ancient times, the history of a landscape, and the effect of road building on nearby lakes.

KEYWORDS: uses of the past, sulfur, mercury, cadmium, silver, ancient Maya, seasonal swamps, Laguna de Petenxil, phthalates, ice cores, slash and burn agriculture

Sometimes it is important to discover when a particular measurable ecological phenomenon began. One discipline that is devoted to such discovery is called paleoecology and one of its subfields is paleolimnology. The most common material from which paleoecological information may be gathered is lake mud. Cores of mud are extracted from the bottom of lakes with piston borers. This mud core is then extruded from the casing in which it was collected, the stratigraphy is examined, and samples are removed at regular intervals. Usually, the first group of such samples is subjected to dating analyses since knowledge of sedimentation rates is an important aspect of any landscape under study. Carbon-14 (^{14}C) analysis will provide reasonably accurate dates (the accuracy depending upon the amount of organic carbon and the length of counting time) from

---

[1] Associate environmental consultant, Health and Environmental Sciences, Mammalian and Environmental Toxicology, 1702 Bldg., The Dow Chemical Co., Midland, MI 48674.

about 35 000 B.P. (Before Present) to relatively recent time, although recent time is better identified by the use of ^{210}Pb and ^{137}Cs. Simultaneously, samples are removed from the mud for moisture analysis. All biological, chemical, and mineralogical data are then calculated on the basis of moisture so that comparison among the data from various disciplines is possible.

This introductory discussion is devoted to illustrating the types of information that may be gleaned from the chemical, mineralogical, and biological analyses of lake muds. Examples are taken from a variety of investigations carried out in various parts of the world.

Chemistry

Sulfur

The chemical composition of a lake is governed by the action of three significant processes: precipitation, watershed drainage, and the oxygen status of the mud [1]. The granitic regions of the world have lakes which are particularly susceptible to acidic precipitation in the sense that they are weakly acidic and impoverished in their ionic content. Thus they exhibit a low buffering capacity [1]. This situation is largely due to the fact that granite weathers slowly. Had man not populated such regions, the lakes contained therein would have waters that reflect the pH of the precipitation and drainage they receive. If their mud-water interfaces were anoxic in the summer, released sulfides from the mud would also affect the final pH of the water.

In recent years lakes situated near heavily industrialized regions have received precipitation and drainage containing increased sulfur content [1]. When such lakes are soft water lakes, located chiefly though not exclusively in granite regions, detrimental effects to biota have been noted [1].

Well-buffered lakes with neutral to alkaline pH ranges have maintained their hydrogen ion concentrations at stable levels for the past century, but the sulfate content of the surface water has increased severalfold [2].

Linsley Pond (North Branford, Connecticut) is a small kettle hole approximately 14.5 m deep and about 9.4 ha in surface area. It was formed about 13 000 years ago in a valley initially filled with glacial drift containing stagnant ice blocks. On melting, the ice produced a lake. Figure 1 shows the sulfur content of the water column of Linsley Pond in 1937 (20 Oct. 1937) [3]. At that time the oxygen was depleted, though not to the extent it is now, there was no hydrogen sulfide being produced, and sulfur as sulfate was accumulating in the lower waters during the height of summer stagnation [4]. In 1980, from a depth of 4 to 14 m the water was devoid of dissolved oxygen during the summer months, hydrogen sulfide was present throughout this zone, and the sulfur was depleted in the lower waters owing to the precipitation of ferrous sulfide. The total sulfur content of the surface waters has increased fourteenfold in a little more than four decades. However, the pH of the water has remained stable [5,6].

Table 1 [5–7] shows the chemical composition of the surface mud in the deepest portion of the lake. Only manganese remained stable. The sulfur content has increased 2.7 fold in 43 years. The lake has been losing iron over the past four decades [6]. It has been noted elsewhere [5–6] that though pyrite or iron sulfide is formed and precipitates out of solution, there is usually an excess of sulfur which is greater than the quantity utilized in pyrite formation. An excess of sulfur is prevalent below 5 m. It should not be assumed that well-buffered lakes can continue to receive sulfur from atmospheric precipitation with no long-term detrimental effects. It is clear that Linsley Pond is losing iron, not only from its waters but from the mud as well [2,5,6]. In addition, only a portion of the iron precipitated as pyrite in any given year is released as ferrous iron during the following eutrophic period [6]. As more sulfur is added to the lake system, more iron is precipitated. When the system runs out of iron there will be no cation to precipitate out the phosphorus which diffuses from the anoxic mud. The phosphorus will accumulate and a permanent algal bloom will result.

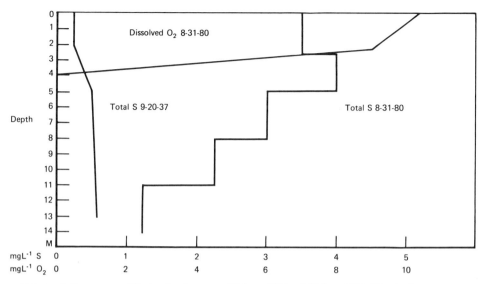

FIG. 1—*Sulfur content of Linsley Pond water on 20 Sept. 1937 and 31 Aug. 1980. Dissolved oxygen content is shown for 31 Aug. 1980.*

Mercury

Recent technological growth has resulted in a progressive increase in mercury in the sediments of Lake Windermere [8], Lake Huleh [9], and Lake Biwa [10]. Figure 2 shows the distribution of mercury [8] in sediment samples from Lake Windermere and Lake Huleh [9] since 500 B.C. Unfortunately, data from Lake Biwa [10] is reported on a wet weight basis and the dry weights are not available, thus these data cannot be compared with those from the other two lakes, which show mercury (μg/kg) on a dry weight basis.

Table 2 illustrates the mercury content of some of the core samples grouped according to mercury concentration. A rise in mercury in the Lake Huleh sediments and Lake Windermere sediments begins about 1200 A.D. (Fig. 2). Mercury was used in religious ceremonies in the fourth century B.C. according to Aristotle [11], and according to Kaufman [12] was common in Egyptian tombs as early as 1500 B.C. Until the sixteenth century [13] consumption was slight and mostly for cosmetic and medicinal purposes. Prior to 1870 [14] the Almaden Mine in Spain, which began production in 400 B.C., the Idria Mine in Yugoslavia, which initiated production in 1470 A.D., and Monte Amiata in Italy, which began in 1868, completely dominated European

TABLE 1—*Chemical composition (%) of Linsley Pond mud at 14.5 m.*

Element	1940[a]	1972[b]	1980
Fe	8.46	8.10	7.95
Mn	0.178	0.179	0.179
S	0.400	0.940	1.07
P	0.111	0.254	0.306

[a] See Ref. 7.
[b] See Ref. 5.

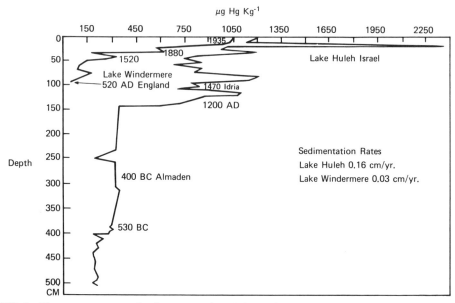

FIG. 2—*The mercury content of sediment cores from Lake Windermere, England, and Lake Huleh, Israel.*

production. In addition to such activities, the burning of fossil fuels [15], heavy industrial activity, denudation of land surfaces, and sewage disposal have all contributed to the global mercury distribution which eventually finds its way to the atmosphere [8,9].

The Lake Huleh study consisted of a 54-m core [16]. This core spans a period that exceeds the limits of ^{14}C dating, that is, in excess of 35 000 years. The early sediments are indicative of a swamp in which peat was deposited [16]. This swamp was converted to a lake, presumably by the damming of the Jordan River by the late Pleistocene Yarda Basalt some 30 000 years ago. Subsequent to this first lacustrine period, water levels permitted peat to be again deposited. About 23 000 years ago a second lacustrine episode began. This too was interrupted with low water levels, which permitted the deposition of peat. The modern lacustrine period began about 21 000 years ago. Thus, this core spans three lakes. Man was not believed to have been present in this region until Natufian man [17] came about 10 000 years ago. Only the recently deposited sediment contains very high quantities of mercury. These observations suggest that upward postdepositional

TABLE 2—*Mercury content of lake mud samples.*

	Lake Windermere[a]				Lake Huleh[b]		
N	Depth, cm	µg kg^{-1}	%CV	N	Depth, cm	µg kg^{-1}	%CV
5	50 to 70	122 ± 36	29.1	81	142 to 792	271 ± 43	15.9
4	30 to 44	287 ± 82	28.6	20	37 to 142	888 ± 165	18.6
3	20 to 26	608 ± 26	4.2	3	22 to 32	1137 ± 76	6.7
3	5 to 15	1030 ± 23	2.2	4	2 to 17	1428 ± 627	43.9

NOTE: N = number of samples; %CV = % coefficient of variation.
[a] See Ref. 8.
[b] See Ref. 9.

TABLE 3—*Cadmium, lead, and silver in Lake Huleh mud samples[a].*

Depth, cm	Cd, µg kg^{-1}	Pb, mg kg^{-1}	Ag, µg kg^{-1}	Stratigraphy
0 to 77	1660	94.9	69	Modern Lake Mud Surface
82 to 792	580	97.2	55	Grey Lake Mud
4245 to 4425	780	113.5	54	Second Lacustrine Period
4890 to 4941	2490	111.5	110	First Lacustrine Period

[a] Ref. 9.

migration of mercury by organisms, or through interstitial waters by ionic or molecular diffusion, or by compaction of the sediment, are not responsible for the rise in mercury concentration observed in recent sediments in England [8], Israel [9], and Japan [10]. Since the surface muds of earlier lacustrine periods fail to show a rise in mercury concentration so obvious in recent sediments, such upward mobility does not seem to have been significant in the cores studied.

Cadmium, Lead, and Silver

The average concentration of cadmium, lead, and silver in Lake Huleh sediments is shown in Table 3 [9]. Both silver and cadmium are higher in the surface sediment than in the grey lake mud below, although the rise in silver is hardly significant. The slight automobile traffic in this region is confirmed by the low lead concentration, which reflects only the natural erosion processes in the Huleh Basin.

According to Nriagu [18] average lead levels in river sediments that are believed to show evidence of pollution are estimated to be near 98 mg kg^{-1} (range 3.9 to 3700 mg kg^{-1}), which is not far from the amount observed in the surface sediments of Lake Huleh. It is believed that one third of this amount is anthropogenic in origin [19]. Lake sediments are a primary sink for lead in aquatic environments. It is interesting to note that the beginning of lead smelting in the Neolithic in Yugoslavia was recorded in the sediments of Lake Biwa in Japan (personal communication, S. Horie, Institute of Paleolimnology and Paleoenvironment on Lake Biwa, Japan). There is little doubt that atmospheric input is the main source of lead in recent sediments as well as the rise in concentration coincident with the initiation of smelting.

According to Forstner [20], core sediment investigations indicate that next to mercury, cadmium tends to be the most important metal enrichment of all heavy metals studied. This certainly appears to be true of the Huleh sediments where the surface sediments contain 2.9 times as much as the grey lake mud below. It has been suggested [21] that much of the cadmium found in lake sediments originates from the burning of fossil fuels.

The silver concentration in air resulting from cloud seeding with silver iodide crystals is of the order of 0.1 ng m^{-3} [22–24], which results in a concentration range of silver in rain water as a result of this process of 0.04 ng L^{-1} to 15 µg L^{-1} [22–23]. There is little doubt that this process is a major contributor to silver concentrations in recent sediments. In fact a silver cycle now exists in Linsley Pond as a result of cloud seeding to encourage snow fall in Montreal [5,25].

Phthalates

Peterson and Freeman [26] studied the distribution of a number of phthalate esters in dated (^{210}Pb) sediment cores to determine whether an environmental accumulation had occurred during the past century. Bis(2-ethylhexyl) phthalate is the most highly produced phthalate ester. A comparison of production of this compound with quantities measured in a dated sediment core taken from the Chesapeake Bay produced a linear correlation coefficient of 0.96. This coefficient is statistically significant beyond the 0.1% level.

Polychlorinated biphenyls

Not all substances that accumulate in surface sediments of anthropogenic origin remain deposited in the mud. Polychlorinated biphenyls (PCB), it was once thought, became inactivated through deposition in sediments as a result of their lipophilic properties. Recently Larsson [27] showed that PCBs are transported from the sediment into the contiguous water and that this process follows a seasonal cycle, the highest concentrations in water being recorded in the summer and the lowest in the winter. Larsson showed that PCB concentrations in the air over water were positively correlated with levels found in the water. Thus, sediments may act as a source for chlorinated hydrocarbons and may be instrumental in maintaining a continuous cycle such that they are transported from sediment to water to air and eventually returned through airborne fallout and resedimentation.

Before completing the chemical discussion on paleoecology and the kinds of information that may be gleaned from such studies, it should be pointed out that much may be learned from the analysis of ice cores. For example Stauffer et al. [28] point out that air entrapped in bubbles of cold ice has essentially the same chemical composition as that of the atmosphere at the time of formation. Analyses of air extracted from these bubbles show a steady rise in atmospheric methane during the past two centuries. The analyses were carried out in an Antarctic ice core. The amount detected is now about twice that of three centuries ago. The source of this increase is thought to be the result of greater emissions of methane and less removal of methane by hydroxyl radicals. In addition, the authors believe that the methane rise may be due in part to a rise in carbon monoxide concentration.

Interdisciplinary Studies

Guatemala

The southernmost portion of the Yucatan Peninsula was occupied by the Mayan peoples during the first millenium A.D. These people developed one of the highest known indigenous cultures of the New World. This culture collapsed before 1000 A.D., and the Peten, the northernmost state of Guatemala, never regained the population it once had supported, nor the degree of civilization it had once attained. As a result of this state of affairs, a series of hypotheses were advanced over the years. The subsequent discussion will show how these hypotheses were examined in an ecological light and which were either accepted or refuted.

The first major hypothesis that was proposed to explain the decline and fall of Mayan civilization was that of O. F. Cook [29]. He suggested that the agricultural practices of this indigenous population were detrimental to the soil and that burning of brush would encourage the encroachment of weeds. Further, he proposed an abnormal rise in population size and that this phenomenon forced the indigenous farmer to forego the usual rest period common to slash and burn agriculture. The loss of the rest period, so necessary for the accumulation of nutrients to support the next crop, caused a decline in yield culminating eventually in a food shortage.

To examine Cook's [29] hypothesis, a number of different types of studies were carried out. The first of these was a modern agricultural study [30]. In the Central Peten it is possible to sow four corn crops annually. Two of these are emergency crops, one for the dry season and one for the wet season. The main crop is planted two years in succession while the emergency crops are only sown once on any particular parcel of land. The agricultural methods have not changed, with the exception of the introduction of the machete, since the description of Bishop Landa [31], published in the sixteenth century.

Soil samples [30] were gathered from cornfields that had rested for variable periods of time since the last planting, fields that had supported one crop, fields that had supported two or three crops, and finally, fields that had supported each of the two emergency crops. In addition, a sampling program was carried out to ascertain the effect of burning. The results [30] were illuminating. Cultivation caused a decline in total nitrogen; organic matter; available phosphorus; and exchangeable sodium, potassium, calcium, magnesium, and pH. Rest or fallow, on the other hand, brought about an increase in the concentration of all substances studied except pH. Burning brought about an increase in potassium and magnesium and no change in sodium or pH. All other items measured declined.

The yields obtained by this indigenous population, utilizing only a dibble stick and a machete, are comparable to yields obtained by farmers of the American Midwest prior to the advent of hybrid corn.

Current agricultural practices in the Peten have persisted for the past four centuries with no discernible decline in yields [30,31]. To discover how long these agricultural practices had existed, cores were taken from Laguna de Petenxi [32], a small lake west of Lago de Peten Itza. The main archaeological sites in the region were probably not established much before the second half of the first millenium.

Analyses of samples from the lake cores showed that calcium, strontium, potassium, and sometimes sodium, were indicative of agricultural events in the basin [32]. Five periods, when some or all of these events are maximally enriched, are recognizable:

1. *Circa* 2000 B.C.
2. *Circa* 100 B.C.
3. *Circa* 400 A.D.
4. *Circa* 900 A.D.
5. *Circa* 1260 A.D.

These five periods correlate fairly well with the incidence of burnt grass fragments which were found embedded in the mud [33]. Doubtless, these observations are indicative of intermittent agricultural activity, quite reminiscent of modern practices [30,31]. The last epoch, around 1260 A.D., may be considered an intense period of agricultural activity, though the rate of erosion is only slightly higher than that of the oldest vegetational zone. Thus, it would appear that our studies [30,32,33] have failed to uncover much evidence to substantiate the hypotheses advanced by Cook [29]. Furthermore, it is quite clear that slash and burn agriculture has persisted in this region since 2000 B.C.

The pollen sequence [34] indicates the presence of agricultural man for the entire period that the cores represent, roughly four thousand years. The basin was never without some agricultural activity as evidenced in both cores by the presence of charred grass fragments at all levels studied [33]. The most important observation revealed by the pollen analyses of these cores is that man came to a region that was a grassland and turned the savannah into a high forest rather than the reverse, which has been noted elsewhere in the world. Thus, it would appear that the hypotheses advanced by Cook [29] have not been substantiated by our studies [30,32–34].

C. W. Cook [35,36] and Ricketson and Ricketson [37] during the 1930s proposed an extension of O. F. Cook's [29] hypotheses. These workers suggested that population pressure had caused an intensive and extensive use of the land, bringing about excessive erosion and filling up all the lakes which are now bajos or seasonal swamps.

The Cook-Ricketson hypothesis was examined by digging a pit in El Bajo de Santa Fé [38], a large seasonal swamp east of the ancient city of Tikal, which is still considered the largest Maya city in Guatemala. This pit (339 cm long, 122 cm wide, and 511 cm deep) apparently had been a seasonal swamp for the past 11 560 years as evidenced by the discovery of amorphous carbonized organic matter of an unknown source which gave a ^{14}C date of 11 560 ± 360 years B.P. [*39*]. The presence of root casts were encountered the entire extent of the dug pit. The clay, which comprised the majority of the sediment, was montmorillonite which had been formed by the decomposition of limestone from the surrounding scarp. The decomposition could not have been formed in place since datable organic carbon was found at 511 cm and root casts were present through the pit. The limestone at the top of the surrounding scarp when treated with hydrochloric acid contains 1 to 1.07% residue. This residue is largely montmorillonite and on examination with an X-ray diffractometer produced a diffractogram that was similar to that provided by the clay of the dug pit [*38*].

It may be concluded therefore that this bajo deposit has been a seasonal swamp for the past 11 560 years. Further critical evidence to support this conclusion is the absence of pollen in the sediments, the absence of diatoms in the sediments, and the irregular occurrence of isolated unwaterworn flints embedded in the clay without any indication of sorting or the formation of littoral or shorelike deposits. The petrology of these flints indicates that they were formed by replacement in partially dolomitized marine limestone [*40*]. It is believed that they arrived at the site of the pit by a process of mass watering [*38,40*].

Thus, it would appear that our investigations indicate that this swamp was never a lake. The evidence for this may be enumerated below:

1. Absence of pollen in the sediments.
2. Absence of diatoms in the sediments.
3. Presence of hematite [ferric oxide (Fe_2O_3)] the entire extent of the pit, indicating continuous presence of oxygen.
4. Presence of live roots at the top of the pit, followed by dead ones beneath this live root layer, followed by fossilized roots filled with gypsum [calcium sulfate ($CaSO_4$)]
5. Carbonized organic matter at 5.11 m, giving a ^{14}C age of 11 560 ± 360 years B.P.
6. Unwaterworn unsorted presence of flints formed prior to their entry into the swamp.

All these observations are indicative of a seasonal swamp and demonstrate the absence of a lacustrine deposit.

Final Comments

In addition to the above discussion it should be noted that paleoecological methods have been used to elucidate past climates [*41–47*]; the advent of agriculture in a region [*34,41,44–45,47*]; processes of plant selection and domestication [*34,41,48*]; the effects of slash and burn agriculture [*30,32–34*]; the size of past plankton populations [*49–50*]; the chemical concentration of the atmosphere in ancient times [*28*]; the history of a landscape [*51*]; and the effect of road building on nearby lakes [*51*].

Thus, many problems confronting the modern investigator may be elucidated by the study of the past through interdisciplinary research on lake muds, ocean sediments, and polar ice cores.

References

[*1*] Cowgill, U. M., "Acid Precipitation: A Review," *Proceedings of the Miami International Symposium on the Biosphere, Studies in Environmental Science,* Vol. 25, Elsevier, Amsterdam, 1984, pp. 233–259.
[*2*] Cowgill, U. M., *Developments in Applied Spectroscopy,* Vol. 6, Plenum Press, New York, 1968, pp. 299–321.
[*3*] Hutchinson, G. E., *Ecological Monographs,* Vol. 11, 1941, pp. 21–60.
[*4*] Hutchinson, G. E., *A Treatise on Limnology,* Vol. 1, John Wiley & Sons, Inc., New York, 1957, pp. 766–774.

[5] Cowgill, U. M., *Archiv für Hydrobiologie,* Suppl., Vol. 45, 1974, pp. 1–119.
[6] Cowgill, U. M., *Archiv für Hydrobiologie,* Suppl., Vol. 50, 1976, pp. 401–438.
[7] Hutchinson, G. E. and Wollack, A., *American Journal of Science,* Vol. 238, 1940, pp. 493–517.
[8] Aston, S. R., Bruty, D., Chester, R., and Padgham, R. C., *Nature,* Vol. 241, 1973, pp. 450–451.
[9] Cowgill, U. M., *Nature,* Vol. 256, 1975, pp. 476–478.
[10] Fuwa, K., "Paleogeochemistry of Mercury," *Lake Biwa,* Monographiae Biologicae, S. Hovie, Ed., Vol. 54, Dr. W. Junk Publishers, Dordrecht, The Netherlands, 1984, pp. 579–585.
[11] Agricula, G., *De Re Metallica* (1556), translation by Hoover, H. C. and Hoover, L. H., Dover, New York, 1950, p. 432.
[12] Kaufman, C., "Mercury Profile," *Metal Statistics,* Vol. 1977, American Metal Market, New York, 1977, pp. 149.
[13] *Mineral Facts and Problems,* Bureau of Mines, Bulletin 639, Government Printing Office, Washington, DC, 1970.
[14] *Mineral Facts and Problems,* Bureau of Mines, Bulletin 671, Government Printing Office, Washington, DC, 1980.
[15] Joensuu, O. I., *Science,* Vol. 172, 1971, pp. 1027–1028.
[16] Cowgill, U. M., *Archiv für Hydrobiologie,* Vol. 66, 1969, pp. 249–272.
[17] Perrot, J., *L'Anthropologie,* Vol. 70, 1966, pp. 438–483.
[18] *Biogeochemistry of Lead in the Environment,* Vol. 1A and 1B, J. R. Nriagu, Ed., Elsevier-North Holland, Amsterdam, 1978.
[19] Craig, P. J., in *The Handbook of Environmental Chemistry,* Vol. 1, Part A, O. Hutzinger, Ed., Springer-Verlag, Berlin, 1980, pp. 69–128.
[20] Forstner, U. in *The Handbook of Environmental Chemistry,* Vol. 3, Part A, O. Hutzinger, Ed., Springer-Verlag, Berlin, 1980, pp. 59–107.
[21] Suess, E. and Erlenkeuser, H., *Meyniana (Kiel),* Vol. 27, 1975, pp. 63–69.
[22] Sargent, *Technology Review,* Vol. 71, 1969, pp. 42–47.
[23] Standler, R. B. and Vonnegut, B. J., *Applied Meteorology,* Vol. 11, 1972, pp. 1398–1401.
[24] Hodge, V. F. and Folsom, T. R., *Nature,* Vol. 237, 1972, pp. 98–99.
[25] Cowgill, U. M., *Developments in Applied Spectroscopy,* Vol. 10, Plenum Press, New York, 1973, pp. 331–352.
[26] Peterson, J. C. and Freeman, D. H., *Environmental Science and Technology,* Vol. 16, 1982, pp. 464–469.
[27] Larsson, P., *Nature,* Vol. 317, 1985, pp. 347–349.
[28] Stauffer, B, Fischer, G., Neftel, A., and Oeschger, H., *Science,* Vol. 229, 1985, pp. 1386–1388.
[29] Cook, O. F., "Annual Report," Vol. 1919, Smithsonian Institution, Washington, DC, 1919, pp. 303–326.
[30] Cowgill, U. M., *Transactions, Connecticut Academy of Arts & Sciences,* Vol. 42, New Haven, CT, 1961, pp. 1–56.
[31] Tozzer, A. M., "Landa's Relacion de las cosas de Yucatan," Peabody Museum Archaeology and Ethnology Papers No. 18, Harvard University, Cambridge, MA, 1941.
[32] Cowgill, U. M. and Hutchinson, G. E., *Memoirs Connecticut Academy of Arts & Sciences,* Vol. 17, 1966, pp. 7–63.
[33] Hutchinson, G. E. and Goulden, C. E., *Memoirs Connecticut Academy of Arts & Sciences,* Vol. 17, 1966, pp. 67–73.
[34] Tsukada, M., *Memoirs Connecticut Academy of Arts & Sciences,* Vol. 17, 1966, pp. 63–66.
[35] Cook, C. W., *Journal of the Washington Academy of Sciences,* Vol. 21, 1931, pp. 283–287.
[36] Cook, C. W., "A Possible Solution of Maya Mystery," sci. service radio talks presented over CBS, 1933, pp. 362–365.
[37] Ricketson, O. G. and Ricketson, E. B., Publication 477, Carnegie Institution of Washington, Washington, DC, 1937.
[38] Cowgill, U. M. and Hutchinson, G. E., *Transactions of the American Philosophical Society,* Vol. 53, 1963, pp. 1–44.
[39] Stuiver, M. and Deevey, E. S., *Radiocarbon,* Vol. 4, 1962, pp. 250–262.
[40] Sanders, J. E., *Transactions of the American Philosophical Society,* Vol. 53, 1963, pp. 45–49.
[41] Tsukada, M., *Paleoecology II: Synthesis* [In Japanese], Kyoritsu Publishing Co., Tokyo, 1974.
[42] Birks, H. H. in *Lake Sediments and Environmental History,* Chap. 14, University of Minnesota Press, Minneapolis, 1984.
[43] Horie, S., *Lake Biwa,* Monographic Biologicae, Vol. 54, Dr. W. Junk Publishers, Dordrecht, Netherlands, 1984, pp. 423–474.
[44] Brugam, R. B., *Quaternary Research,* Vol. 9, 1978, pp. 349–362.

[45] Hammen, Th. van der and Gonzalez, E., *Leidse Geologische Mededelingen,* Vol. 25, 1960, pp. 261–315.
[46] Tsukada, M., *American Journal of Science,* Vol. 265, 1967, pp. 562–585.
[47] Bonatti, E., *Transactions of the American Philosophical Society,* Vol. 60, Pt. 4, pp. 26–31.
[48] Iverson, J., Danmarks Geologiske Undersoegelse II RK., Vol. 66, 1941, pp. 7–68.
[49] Brugam, R. B., *Ecology,* Vol. 59, 1978, pp. 19–36.
[50] Engstrom, D. R., Swain, E. B., and Kingston, J. C. *Freshwater Biology,* Vol. 15, 1985, pp. 261–288.
[51] Hutchinson, G. E., Bonatti, E., Cowgill, U. M., Goulden, C. E., Leventhal, E. A., Mallett, M. E., Margaritora, F., Patrick, R., Racek, A., Roback, S. A., Stella, E., Ward-Perkins, J. B., and Wellman, T. R., "Ianula: An Account of the History and Development of the Lago Di Monterosi, Latium, Italy," *Transactions of the American Philosophical Society,* Vol. 60, Pt. 4, 1970, pp. 1–178.

Richard B. Brugam[1]

Long-Term History of Eutrophication in Washington Lakes

REFERENCE: Brugam, R. B., **"Long-Term History of Eutrophication in Washington Lakes,"** *Aquatic Toxicology and Hazard Assessment: 10th Volume, ASTM STP 971,* W. J. Adams, G. A. Chapman, and W. G. Landis, Eds., American Society for Testing and Materials, Philadelphia, 1988, pp. 63–70.

ABSTRACT: Paleolimnological techniques were used to reconstruct past changes in lake water quality at Meridian Lake, Washington. A core was taken from the lake and dated using ^{210}Pb and pollen analysis. Diatom microfossils showed two periods of increasing trophic status. The first came in the 1880s when the watershed was deforested, and the second occurred in the late 1940s when the watershed was developed for suburban housing. In the Meridian core, eutrophication was indicated by increasing percentages of *Asterionella formosa* in the fossil diatom assemblage. In 15 other western Washington lakes where core tops and bottoms were examined, *Asterionella* also increased with human disturbance. On the basis of previous laboratory measurements of nutrient-limited growth, *Asterionella* seems to be adapted to lakes with high phosphorus levels but low silica levels—characteristics of lakes which have eutrophied.

KEYWORDS: paleolimnology, diatoms, pollen, Washington, eutrophication, *Asterionella formosa,* sediment

There has been an increased interest in the use of engineering methods to reverse cultural eutrophication in lakes [1]. When engineering modifications of lakes and watersheds are planned under the Federal Clean Lakes Program, diagnostic studies are required. These usually involve constructions of phosphorus budgets for the lake using data gained in a year or two of field study. Historical information about lake water quality other than anecdotal evidence that the "lake has been getting bad" is seldom available.

Paleolimnological studies offer a solution to the problem of poor historical data on lakes. It is possible to reconstruct the history of a lake using the fossil remains of organisms from the sediment. With this information, one can answer questions about the timing of lake deterioration. It is also possible to compare present-day conditions with those prevailing before human settlement in the watershed.

In only a few cases have paleolimnological studies been included in lake restoration programs. These include work by Smeltzer and Swain in Vermont [1] and by Barnes et al. [2] in Washington.

A number of different kinds of fossil remains are good indicators of past lake trophic status. I have, however, chosen to use diatoms. They are algae with shells of silicon dioxide which are well preserved in lake sediment. Diatoms are also sensitive to water quality changes, with different species preferring different environmental conditions [3]. While taxonomy of diatoms is sometimes a problem, the various species can be easily recognized from their fossilized silica shells.

In addition to analysis of microfossils, a sediment core must be dated. Good dating allows the

---

[1] Associate professor, Department of Biological Sciences, Southern Illinois University, Edwardsville, IL 62026.

comparison of changing lake conditions with historically documented changes in human activities in the lake watershed. Radionuclides (such as ^{210}Pb, ^{137}Cs, and ^{14}C) are most commonly used to date sediment cores. Of these radionuclides, ^{210}Pb is best suited for studies of recent human disturbances because its half-life is 22.3 years, making it a useful dating tool over the last 100 to 120 years. ^{137}Cs is less useful because it yields only one date—1954, the beginning of atmospheric nuclear testing.

Pollen analysis is a useful supplement to radionuclides for dating lake sediment. The first colonization of lake watersheds in North America by European farmers has normally resulted in catastrophic deforestation. Pollen of forest trees and agricultural weeds preserved in lake sediment records this deforestation. In most cases this event is well dated for a particular lake.

This paper will explore some of the applications of paleolimnological methods to the reconstruction of eutrophication in western Washington lakes. Of 15 lakes cored, Meridian Lake was chosen for more thorough examination because its history was already well known [2].

Description of Meridian Lake

The history of human disturbance in Meridian Lake is representative of many lakes in western Washington. The lake is located 6.5 km east of Kent, Washington on a small plateau. Maximum lake depth is 27 m. The area of the lake is 60 ha, and the area of the drainage basin is 289 ha. Development of the plateau was begun because of the discovery of coal at Black Diamond, 10 km southeast of the lake [2]. The United States Geological Survey (USGS) Land Survey map of 1900 lists the lake as Cow Lake and the watershed as burnt-over and restocking. According to Barnes et al [2], deforestation of the watershed occurred between 1880 and 1890 as a result of fires set by farmers to clear land. Between the time of land clearance and the mid-1940s, the lake watershed was probably pasture and cultivated land. From the 1940s to the present the watershed has been rapidly urbanized. According to Barnes et al [2], 75% of the buildings in the watershed in 1973 have been built since 1943.

Until the 1973–74 installation of sewers, dwellings in the Meridian Lake watershed were served by septic tanks. Mean total phosphorus concentrations in 1972–73 were 23 µg/L and the mean Secchi disk transparency was 4.2 m. In 1976–77, mean total phosphorus had improved to 18µg/L and mean Secchi disk transparency was 5.1 m [4]. Davis et al. [4] classified the trophic status of the lake as "early mesotrophy." My own measurement of total phosphorus on 10 Sept. 1984 was 7 µg/L. The Secchi disk transparency at that time was 6.8 m.

According to Davis et al. [4], the lake water quality is good. However, the major environmental problem in the lake is nuisance growth of *Myriophyllum spicatum*, the Eurasian milfoil. This luxuriantly growing rooted aquatic plant lives in the littoral zone of the lake.

Methods

Sediment cores were taken with a gravity corer during the summer and fall of 1984 from the deepest parts of 15 lakes from King and Snohomish counties in western Washington. Care was taken to minimize disturbance of the sediment-water interface. Cores were carried upright to the lab, extruded, sliced at 1-cm intervals, and stored in plastic bags at 4°C. Only the core top and core bottom were analyzed for diatoms. It was assumed that core tops represent recently deposited sediment and that core bottoms represent material deposited before the arrival of European farmers. The core from Meridian Lake, near Kent, Washington, was more closely examined.

Microscope slides of diatoms were prepared from 0.1 cm^3 aliquots of lake sediment by treating with hot nitric acid (HNO$_3$) and potassium dichromate (K$_2$Cr$_2$O$_7$). After washing, the samples were quantitatively deposited on microscope slides using Battarbee's technique [5]. Final mounting was in hyrax, a high refractive index mounting medium that is especially suited for diatoms.

Pollen preparations were made using the acetolysis technique of Faegri and Iversen [6]. Final mounting was in silicone oil.

^{210}Pb analysis of sediment followed methods used by Kharkar et al. [7] for marine plankton with appropriate modifications for sediment. Samples from contiguous depth lamina were ashed, spiked with a known amount of stable lead, and leached in 8 N hydrochloric acid (HCl). Lead was recovered from the leachate using ion exchange, precipitated as lead chromate, and spread on a counting planchette. Ingrowth of ^{210}Bi was followed using a Beckman-Sharp Low Beta II anticoincidence counter. Dates were calculated using the constant rate of supply (crs) method of Appleby and Oldfield [8].

Total phosphorus levels of the lakes from which core tops and core bottoms were analyzed for diatoms were measured in early fall 1984 or late spring 1985. Water samples were taken about 0.3 m below the lake surface. In the lab, unfiltered aliquots of lake water were digested with persulfate according to *Standard Methods* [9]. After digestion, phosphorus was measured spectrophotometrically (with a 10-cm cell) using the molybdophosphoric acid blue method with stannous chloride as a reducing agent [9].

Results

In the 15 lakes from which core tops and bottoms were examined, *Asterionella formosa* or *Fragilaria crotonensis* increase in the tops (Fig. 1). Many lakes had no detectable numbers of

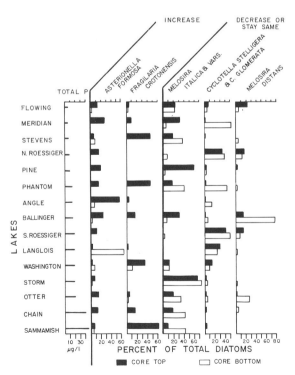

FIG. 1—*A comparison of core bottom (open rectangles) with core top (darkened rectangle) samples from western Washington lakes. Left-most column shows the late summer total phosphorus concentrations for the lakes. Lakes are arranged in order of ascending total phosphorus concentration.*

these species before human disturbance began. Only Langlois Lake had a significant decrease in *A. formosa* since European settlement. The species that decrease with human disturbance are different from lake to lake and include *Melosira distans, Cyclotella glomerata, Cyclotella stelligera, Melosira italica,* and *Melosira italica tennuissima.*

There seems to be no consistent relationship between the variations in fossil diatom assemblages and total phosphorus levels in the lakes. But there is a strong relationship between the onset of human disturbance and increases in *A. formosa* and *F. crotonensis.*

The Meridian Lake core was examined to determine the chronology of man's impact on a particular lake. There are two periods of change in the diatom assemblages from Meridian Lake (Fig. 2). The earliest occurs at 14 cm in the core, where *Asterionella formosa* first appears in significant numbers. A pair of small *Stephanodiscus* species (*S. hantzschii* and *S. minutus*) also appears for the first time at 14 cm. Low numbers of *Melosira distans* and *Cyclotella comta* are also present at this depth. The next change occurs at 8 cm in the core where *Asterionella* increases. With each increase in *A. formosa* there is a decrease in *Cyclotella stelligera* and *Cyclotella glomerata*. At the top of the core *A. formosa* decreases a bit and *Melosira italica* increases.

The date of these changes is important because it allows comparisons with watershed events. In the Meridian Lake core, ^{210}Pb analysis yields a period between 1877 and 1885 for the first *Asterionella* rise and between 1930 and 1946 for the second increase (Table 1).

Pollen analysis provides an independent check on the ^{210}Pb results (Fig. 3). In western Washington the presettlement forest was composed of huge Douglas fir (*Pseudotsuga menziesii*), cedar (*Thuja plicata*), and western hemlock (*Tsuga heterophylla*). Around the turn of the century much of the western Washington lowlands was logged and converted to pasture or burnt-over lands with the reseeding ("restocking") of Douglas fir or red alder (*Alnus rubra*) [10]. This change is evident from the pollen diagram from Meridian Lake. At 14 cm in the core, pollen from pasture weeds first appears. Alder pollen increases somewhat later (11 cm). These changes mark the time of deforestation of the lake watershed and are coincident with the first rise in *Asterionella formosa*

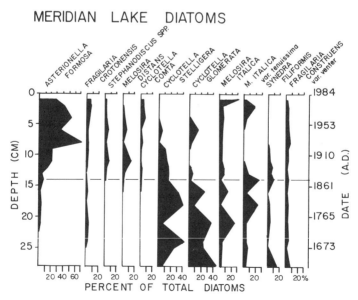

FIG. 2—*Fossil diatom assemblages from the Meridian Lake core. The horizontal line denotes the deforestation horizon.*

TABLE 1—*Meridian Lake, Washington* ^{210}Pb *results. The date listed in the column on the far right refers to the date of the* top *of each depth interval.*

Depth Range, cm	^{210}Pb Activity, DPM/g Ash	Ash Weight, g	Unsupported ^{210}Pb, DPM/gm Ash	Age, Years	Date, A.D.	
0–2	29.16	0.0816	28.94	0	1984	
2–4	34.31	0.1168	33.34	11	1983	
4–6	20.24	0.1094	19.27	26	1958	
6–8	15.95	0.2376	14.98	37	1947	⎫ *Asterionella* rise
8–10	19.49	0.1420	18.51	53	1931	⎭
10–12	15.10	0.0653	14.13	74	1910	
12–13	14.31	0.0410	13.33	87	1897	
13–14	10.67	0.0473	9.70	99	1885	⎫ Deforestation
14–15	22.82	0.0179	21.85	106	1877	⎭
15–17	7.96	0.0459	6.98	123	1861	
17–19	3.57	0.0774	2.60	152	1832	
19–21	0.00					
21–23	2.52	Supported ^{210}Pb				
23–25	0.00					
25–27	1.37					
	0.97 Average supported ^{210}Pb					

NOTE: DPM = decays per minute.

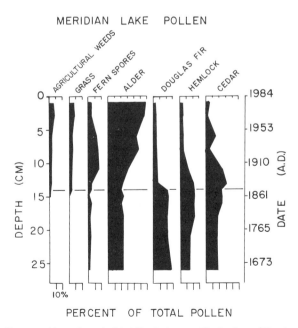

FIG. 3—*Fossil pollen assemblages from the Meridian Lake core. The horizontal line denotes the deforestation horizon.*

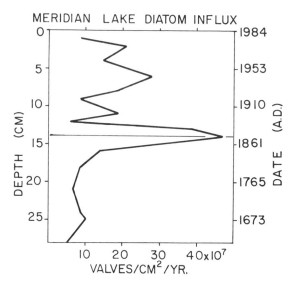

FIG. 4—*Diatom influx to the Meridian Lake core. The horizontal line denotes the deforestation horizon.*

in the core. ^{210}Pb dates these changes between 1877 and 1885, a date range consistent with the USGS survey of forest resources in the area conducted between 1894 and 1896. This survey found that the watershed of Meridian Lake was "burnt and restocking" (presumably with Douglas fir) at that time.

The 1930 to 1946 increase in *A. formosa* is not related to any change in the forest around the lake. It is more likely associated with the increase in the number of people living around the lake. At the time of the 1894 USGS forest survey, the lake had about 12 houses in its watershed, and by 1973 there were about 390 [2].

Because the Meridian Lake core is well dated, it is possible to determine influxes of diatoms to the lake sediment. Influx is a measure of the rate of sedimentation of diatom microfossils at the core site (Fig. 4). Changes in influx of diatoms may reflect changes in diatom production or changes in the sedimentation pattern of the lake [11]. Diatom influx to the Meridian Lake core is low before deforestation, reaches a peak during the deforestation period, and then declines to a level higher than during the predeforestation period.

Discussion

The Meridian Lake core shows major changes in diatom assemblages that are associated with increasing human disturbance in the lake watershed. The increase in *A. formosa* from 1877 to 1885 is associated with the conversion of the lake watershed from forest to pasture. The second increase in the same species during 1930 to 1947 is attributed to urbanization.

The dating of the Meridian Lake sediment record from this study does not agree with earlier work of Barnes et al. [2]. They find a major change in sediment chemistry at 14 cm in their sediment core and attribute that to urbanization of the lake watershed in the 1940s. They base their dating on a constant initial concentration (cic) analysis of ^{210}Pb data but do not verify their conclusions with pollen analysis. The pollen analysis presented here confirms that the 14-cm level is from the time of deforestation—not the post-war urbanization of the watershed.

To interpret the changes in the Meridian Lake core it is first necessary to understand the ecology of *Asterionella formosa*. The core top and core bottom data from lakes other than Meridian shows that *A. formosa* responds positively to human disturbance. Stockner and Benson [12,13] found an increase in the diatom family Araphidineae in Lake Washington and in the English Lake District. This family contains *Asterionella formosa* and *Fragilaria crotonensis*, both of which have increased in the lakes of western Washington. Although this family does not seem to respond positively to human activity in Minnesota [14,15], it does seem to increase with human disturbance in western Washington. Stockner [13] related the abundance of that family to lake trophic status.

The core data collected from various lakes in Washington suggest that the interpretation of *A. formosa* as a direct indicator of lake phosphorus levels is too simple. There is no clear relationship between the abundance of *A. formosa* and total phosphorus in the lakes sampled (Fig. 1). The strongest relationship that the data display is between human disturbance and *A. formosa*. It is unclear from the paleolimnological data what aspect of human disturbance encourages the growth of this diatom.

Tilman et al. [16] find that *A. formosa* is able to use silica at low concentrations, but that it is less efficient in using phosphate. Unfortunately, there are strong variations in responses to phosphorus and silica limitation between *Asterionella formosa* clones. Tilman et al.'s results do, however, suggest a reason for the association between *A. formosa* and human disturbance. As a lake eutrophies, phosphorus levels increase because of pollution from human activities in the lake watershed. At the same time silica levels seem to decrease. This change occurs because the greater diatom growth allowed by high phosphorus levels causes faster and more efficient removal of silica from the water column (the Schelske-Stoermer hypothesis [17]). Some clones of *Asterionella formosa* have the characteristics (relatively efficient silica utilization, but inefficient phosphorus utilization) that would fit them to eutrophic lakes with low silica levels. Because the important factor in controlling the abundance of particular diatoms is the relative supply rates of silica and phosphorus [18], one might expect that relationships between particular diatom species and total phosphorus would be poor. Diatom species abundances should, however, be correlated with silicon (Si) to phosphorus (P) supply ratios [16].

In midwestern lakes another diatom species, *Stephanodiscus minutus*, behaves like *Asterionella* does in western Washington [19]. Tilman et al. [16] show that *Stephanodiscus minutus* is an even more efficient silica user and less efficient phosphorus user than *A. formosa*. It would seem that *S. minutus* would be even more suitable for eutrophic environments than *A. formosa*. The distribution of *S. minutus* suggests, however, that it is limited to high alkalinity lakes [14,15]. Gensemer and Kilham's experiments show that *Stephanodiscus minutus* does not grow at pH < 7.0, but that *Asterionella* does. [20]. The lakes sampled in this study all have low alkalinity and soft water, but pHs are nearly all above 7.0. Gensemer and Kilham's experimental protocol did not differentiate between the effects of low pH and of low concentrations of dissolved inorganic carbon that are a consequence of low pH. It may be that *Asterionella* is a more efficient user of dissolved inorganic carbon than *Stephanodiscus*. At any rate, the field data show that *Stephanodiscus* increases in hard-water lakes (like those of Minnesota) as a result of human disturbance and that *Asterionella* responds similarly in soft-water lakes (like those of western Washington).

The influx data for Meridian Lake support the argument that the increase in *Asterionella formosa* represents a decrease in the Si:P ratio in the lake. This change would occur because of increasing phosphorus loading rates. At each increase in *Asterionella* there is an increase in diatom influx. If the Schelske-Stoermer hypothesis is applicable to Meridian Lake, one would expect an increase in the loss of silica to the sediment. These increased losses would be seen as an increase of the diatom influx to the sediment. The fossil diatoms in the sediment represent the storage of biogenic silica in the lake sediment that would normally be available as dissolved silica for the support of diatom production. Figure 4 shows the expected increase in diatom influx to the sediment.

It seems that the trophic status of Meridian Lake increased in two increments. The first resulted from deforestation in the 1880s. The second came in the 1940s and probably can be attributed to greater residential development of the lake watershed. Any "reclamation" of the lake can certainly be expected to reverse eutrophication to the levels seen after deforestation. It would be difficult to imagine any engineering change that would return the lake to its pristine, predeforestation condition.

Acknowledgments

I thank W. T. Edmondson of the Zoology Department, University of Washington for providing lab space and encouragement. I also thank Sally Abella, Arni Litt, Katie Frevert, and Mardi Varela for help in the lab and field. I am grateful to Joanne I. Davis from Municipality of Metropolitan Seattle Water Quality Division for providing a copy of the management report on Meridian Lake. Support was provided by the graduate school of Southern Illinois University at Edwardsville, by the Andrew Mellon Foundation, and by the U.S. Department of Energy through grants to W. T. Edmondson (DE-AM06-76R102225).

References

[1] Smeltzer, E. and Swain, E. B., *Proceedings of the Fourth Annual Conference on Lake and Reservoir Management*, North American Lake Management Society, Merrifield, VA, 1985, pp. 268–274.
[2] Barnes, R. S., Lazoff, S., and Spyridakis, D. E. "Appendix 1, Phosphorus Loading to Lake Meridian," in "A Study of the Trophic Status and Recommendations for the Management of Lake Meridian," J. I. Davis, J. M. Buffo, D. S. Sturgil, and R. I. Matsuda, Municipality of Metropolitan Seattle Water Quality Division, Seattle, WA, Feb. 1978.
[3] Lowe, R., "Environmental Requirements and Pollution Tolerance of Freshwater Diatoms," Environmental Monitoring Series, No. EPA-67014-74-005. U.S. Environmental Protection Agency, Washington, DC, 1974.
[4] Davis, J. I., Buffo, J. M., Sturgil, D. S., and Matsuda, R. I., "A Study of the Trophic Status and Recommendations for the Management of Lake Meridian," METRO, Seattle, Washington, Feb. 1978.
[5] Battarbee, R. W., *Limnology and Oceanography*, Vol. 18, 1974, pp. 647–653.
[6] Faegri, K. and Iversen, J., *Textbook of Pollen Analysis*, Hafner, New York, 1975, pp. 1–237.
[7] Kharkar, D. P., Thomson, J., Turekian, K. K., and Forster, W. O., *Limnology and Oceanography*, Vol. 21, 1976, pp. 407–414.
[8] Appleby, P. G. and Oldfield, R., *Catena*, Vol. 5, 1978, pp. 1–8.
[9] *Standard Methods for the Examination of Water and Wastewater*, American Public Health Association, New York, 1985.
[10] Davis, M. B., *Northwest Science*, Vol. 47, 1973, pp. 133–137.
[11] Brugam, R. B. and Speziale, B. J., *Ecology*, Vol. 64, 1983, pp. 578–591.
[12] Stockner, J. and Benson, W. W., *Limnology and Oceanography*, Vol. 12, 1967, pp. 513–532.
[13] Stockner, J., *Verhandlungen der Internationalen Vereinigung für Theoretische und Angewandte Limnology*, Vol. 13, 1982, pp. 349–372.
[14] Brugam, R. B., *Freshwater Biology*, Vol. 9, 1979, pp. 451–460.
[15] Brugam, R. B. and Patterson, C., *Freshwater Biology*, Vol. 13, 1983, pp. 47–55.
[16] Tilman, D., Kilham, S. S., Kilham, P., and Dressler, R. L., *Annual Review of Ecology and Systematics*, Vol. 13, 1982, pp. 349–372.
[17] Schelske, C. L. and Stoermer, E. F., *Science*, Vol. 173, 1972, pp. 423–424.
[18] Titman, D., *Science*, Vol. 192, 1976, pp. 463–465.
[19] Bradbury, J. P., *Geological Society of America, Special Paper* No. 171, Geological Society of America, Boulder, CO, 1975, pp. 1–74.
[20] Gensemer, R. W. and Kilham, S. S. *Canadian Journal of Fisheries and Aquatic Science*, Vol. 41, 1984, pp. 1240–1243.

Thomas L. Crisman[1]

The Use of Subfossil Benthic Invertebrates in Aquatic Resource Management

REFERENCE: Crisman, T. L., "**The Use of Subfossil Benthic Invertebrates in Aquatic Resource Management,**" *Aquatic Toxicology and Hazard Assessment: 10th Volume, ASTM STP 971,* W. J. Adams, G. A. Chapman, and W. G. Landis, Eds., American Society for Testing and Materials, Philadelphia, 1988, pp. 71–88.

ABSTRACT: Since the beginning of this century, benthic invertebrates have been used as biotic indicators of water quality. The objective of any lake management plan is both to identify the principal watershed disturbances/practices contributing to a lake problem and to rank their importance as part of a cost-effective abatement strategy. In most cases we see only the end result of the disturbance process without having a sufficient historical data base on either the predisturbance condition or how the biota responded to the type and intensity of disturbance. Without such a data base, the paleolimnological approach is the only way to delineate past lake conditions and the response to disturbance events. Although remains of several faunal groups are preserved in lacustrine sediments, bryozoans, chaoborids, and chironomids show the greatest promise for paleolimnological reconstructions. These groups can be used to make qualitative statements on water quality, but their use in quantitative reconstructions must await construction of calibration models relating subfossil assemblages to known water chemistry and trophic state, and questions on the relationship between producing population and observed subfossil assemblage, postmortem redeposition, differential preservation, and basic ecology are addressed. It is suggested that paleolimnological reconstructions couple detailed land-use histories with multiparameter analyses from dated cores.

KEYWORDS: paleolimnology, bryozoa, chaoborus, chironomids, benthos

The biological and chemical historical data base for most lakes is usually sparse. When an extensive record does exist, interpretations based on such data are often approached cautiously due to potential error arising from changes in methodology and analysts. In addition to characterizing the current condition of an impacted lake, limnologists need a historical perspective on both the preimpact lake condition and how the lake responded to a given level of disturbance. Without such a perspective, effective management and/or restoration plans are not possible.

Lacking a reliable and extensive historical data base, the only alternative for estimating past lake conditions is through paleolimnological analyses. An excellent overview of the nature of the paleolimnological record is provided by Frey [1], and detailed reviews of the chemical [2] and invertebrate [3–5] records exist. The biotic record is extremely diverse. Autotrophic communities are represented by both macrophytes (seeds, tissues) and algae (primarily diatoms, chrysophytes, and other silaceous taxa). Although incidental reports of subfossil remains exist for most invertebrate groups [3], the greatest attention has been given to cladocerans (littoral and planktonic) and dipterans (*Chaoborus* and chironomid midges).

The biotic groups that are of the greatest value in paleolimnological reconstructions are those displaying clear interspecific ecological differences and whose subfossils are both reasonably

---
[1] Associate professor, Department of Environmental Engineering Sciences, University of Florida, Gainesville, FL 32611.

abundant in lacustrine sediments and retain key features needed for taxonomic identification. A majority of the past paleolimnological research effort has centered on the use of diatoms and cladocerans (littoral and planktonic) with advances in the use of benthic invertebrates lagging behind. In part this oversight of the benthos is due to the fact that their remains are not overly abundant in sediments and the general lack of detailed ecological studies on individual taxonomic groups. Three invertebrate groups (bryozoans, chaoborids, chironomid midges) do show promise for use in paleolimnological interpretations as a result of recent ecological studies of extant communities and increased use in multiparameter paleolimnological studies. The purpose of the present paper is to review recent advances in our understanding of these three components of the benthos and to demonstrate their general applicability for reconstructing past lake conditions.

Examples of the most frequently encountered remains of bryozoans, chaoborids, and chironomids are presented in Fig. 1. Bryozoans (Ectoprocta) are represented in the paleolimnological record

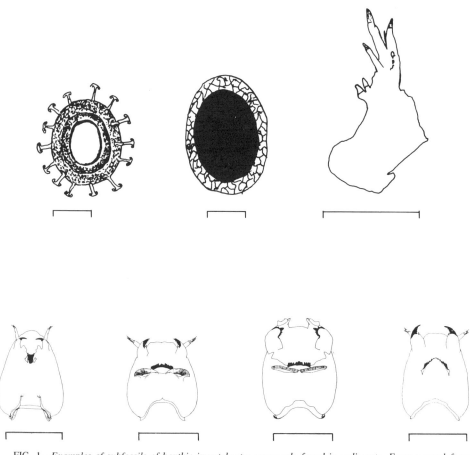

FIG. 1—*Examples of subfossils of benthic invertebrates commonly found in sediments. From upper left to lower right: spinoblast of the bryozoan* Pectinatella, *floatoblast of the bryozoan* Plumatella, *mandible of* Chaoborus, *and representative head capsules of the chironomids Tanypodinae, Chironomini, Tanytarsini, and Orthocladiinae.*

by their statoblasts, two-valved chitinous "overwintering" reproductive structures containing cells and yolky nutrient that are formed from the funiculus of the polypide [6]. A majority of bryozoans belong to the class Phylactolaemata, all four families of which produce statoblasts. The second class of bryozoans found in freshwater, Gymnolaemata, lack statoblasts, producing instead winter buds of sclerotized incipient zooids (hibernacula) that are not as likely to be preserved in the sediments [7].

Four major types of statoblasts are recognized based on external morphology and mode of dispersal [6,8]. Both sessoblasts and piptoblasts lack peripheral float cells and thus are poorly dispersed from the producing colony. Sessoblasts are always attached to a substrate. Spinoblasts are the largest statoblasts (70 to 175 μm) and are characterized by peripheral float cells fringed with hooks or spines. The final statoblast type, floatoblasts, also have peripheral float cells but lack marginal armament. These are by far the most common type of statoblasts found in sediments and reach their greatest diversity in the Plumatellidae. Owing to the low number of species and the often pronounced differences in structure between species, statoblasts are often sufficient to identify individual species of bryozoans.

Chaoborus is represented in the sediment record by mandibles, premandibular fans, antennal segments, prelabral scales, and pupal respiratory horns [3]. By far the most common subfossil remains encountered are mandibles. These often have the premandibular fans attached. Although most taxonomic keys rely on a variety of features including antennae and prelabral appendages, individual species can usually be identified in the sediment record solely by the arrangement of their mandibular teeth.

The most common remains of chironomids found in sediments are the chitinous head capsules. Although complete taxonomy is dependent upon a variety of features of the larvae and adult males, head capsules can be identified to the generic level relatively easily for most groups except the Tanytarsini. Three subfamilies of the Chironomidae are common in freshwater. The Tanypodinae are distinguished from all other chironomids by the presence of a forked-shaped ligula. The remaining subfamilies possess a broad-toothed labial plate (mentum) instead of a ligula that is usually subtended by a paralabial (ventromental) plate. Subfossil taxonomy of these groups is based primarily upon the number, size, and arrangement of the labial teeth and the size, shape, and ornamentation of the paralabial plate. Mandibles are often attached to the head capsules but are only infrequently used in taxonomic keys.

The subfamily Chironominae is comprised of two tribes. The Chironomini are characterized by possessing striated paralabial plates that in most taxa are widely separated. Tanytarsini also possess striated paralabial plates, but in most taxa these nearly touch at the center of the labial plate. This tribe is further distinguished by having long tubercles upon which the antennae attach. The paralabial plates in the final two common subfamilies of freshwaster chironomids, Orthocladiinae and Diamesinae, are either greatly reduced and unstriated or totally absent.

Sediments are prepared for invertebrate analysis by first deflocculating in a weak (10%) potassium hydroxide (KOH) solution, then filtering through a screen. The choice of screen size depends upon the invertebrate group of interest. Bryozoan statoblasts are quite large (70 to 175 μm), but chaoborid mandibles and chironomid head capsules are often less than 100 μm. Through a comparative study Walker and Paterson [9] demonstrated that screens larger than 100 μm lose small head capsules, but they found no differences in the retention efficiency of screens below this size. While some workers use screens as small as 37 μm, most prefer a size of approximately 80 μm. The 37-μm size should be used if the sample is to be subsampled also for cladoceran remains.

The screened fraction is then examined under a dissecting microscope, and individual remains are isolated with wire loops or a micropipet. Each remain is then mounted in a medium such as polyvinyl lactophenol or silicone oil and identified under a binocular microscope.

Bryozoa

Pennak [10] noted that bryozoans are most common along the shoreline in water less than 1 m deep but are frequently encountered up to a depth of 5 to 10 m. Colonies are found attached to hard substrates and macrophytes and generally avoid bright light. They tend to be sporadically located within a lake, and individual colonies are usually not temporally stable.

Frey [3] noted that fossil bryozoan statoblasts were identified from lacustrine sediments by a few workers in the 1880s and 1890s, and Wesenberg-Lund [11] provided illustrations of the statoblasts of several species found in sediments. Since that time it has been common for palynologists to include stratigraphic profiles of statoblasts in addition to pollen, but interpretations of such data are almost always lacking.

Paleolimnological interpretation of bryozoan data began with the work of Deevey [12] at Linsley Pond. The postglacial maximum of *Plumatella repens* statoblasts coincided with the period of maximum organic matter in the sediments. Deevey attributed the bryozoan increase to an expansion of their substrate, littoral macrophytes, coincident with a period of increasing trophic state. He suggested further that the later disappearance of bryozoans in the lake resulted from a progressive increase in trophic state to favor phytoplankton and a loss of littoral vegetation due to algal shading. Although Harmsworth [13] gave no interpretations of the postglacial bryozoan profile from Blelham Tarn, Tsukada [14] attributed periods of reduced *Plumatella* abundance coincident with an increase in *Bosmina* in a core from Lake Nojiri, Japan to a reduction in aquatic macrophytes associated with increased water depth.

The occurrence of individual species in long cores may be of importance in climatic and biogeographic interpretations. Deevey [12] felt that the presence of *Pectinatella magnifica*, a warm-water stenotherm, in the spruce-fir pollen zone of Linsley Pond indicated that summer temperature of surface water reached 18 to 20°C during this early postglacial period. Fredskild et al. [15] found numerous statoblasts of *Cristatella mucedo* in sediments of a Greenland lake deposited 2600 to 3600 years ago and noted that extant specimens of this species have never been collected in Greenland. The biogeographic and climatic significance of this find have not been resolved.

Several authors have noted parallel trends between the stratigraphy of statoblasts and those of other invertebrate remains. For chironomids, Stahl [16] noted that increased *Plumatella* abundance corresponded with periods of increased importance of *Sergentia* in the midge fauna of Myers Lake, Indiana, while Deevey [12] reported the same relationship between statoblasts and *Chironomus* in Linsley Pond. An interpretation of these apparent relationships was not given. Finally, Roback [17] observed that peaks in *Plumatella* statoblasts and turbellarian cocoons both occurred in Lago di Monterosi, Italy approximately 1200 years ago and attributed this to water level fluctuations.

Few studies have attempted to use bryozoan statoblasts in reconstructions of recent lake histories. Birks et al. [18] attributed increased abundance of *Plumatella* and *Cristatella mucedo* in recent sediments of Elk Lake and Lake Sallie, Minnesota to historical expansions in the extent of macrophytic vegetation. Prat and Daroca [19] had hoped to use statoblasts in a similar fashion for interpreting the eutrophication history of four Spanish reservoirs but found too few remains to be of value. Most recently, Crisman [20] used annual accumulation rates of bryozoan statoblasts in a core from Lake Maxinkuckee, Indiana as suggestive evidence for an expansion in aquatic macrophytes associated with land clearance by European settlers in the mid-1800s.

Deevey's [12] original assertion that increased statoblast abundance in cores is associated with periods of littoral zone expansion was based on casual observations of bryozoan distributions and has been applied by subsequent researchers in paleolimnological reconstructions without testing its validity. Deevey's bryozoan interpretations for Lindsley Pond were later challenged by Vallentyne and Swabey [21], who suggested that the increased statoblast abundance was the result of redeposition of sediments from the littoral zone rather than an expansion of the littoral zone. Their argument was based on increased breakage of pollen, diatoms, and statoblasts in this core section and the ratio of *Bosmina* heads to shells.

The first attempt to provide a calibration model for interpreting bryozoan statoblasts in sediment cores was the work of Crisman et al. [22]. Statoblasts were isolated from the surface sediments of 30 Florida lakes of known water chemistry, trophic state, and macrophyte cover. Annual accumulation rates of statoblasts were calculated from 210-Pb analyses and regressed against individual chemical and biological parameters. The results of this study showed that while the abundance of filter feeding bryozoans tended to increase with increasing trophic state, the extent of the vegetated littoral zone appeared to be a much more important controlling factor. Statoblast accumulation rates increased linearly with increasing macrophyte cover up to approximatley 50% bottom coverage, above which value they decreased progressively. The results of this initial investigation suggested that while bryozoans are both food (phytoplankton) and substrate (macrophyte) limited in oligotrophic lakes and respond favorably to moderate increases in both parameters, they can become severely food limited in excessively macrophyte-dominated lakes.

Bryozoan statoblasts show great potential for use in paleolimnological reconstructions, but they should be used cautiously until a number of problems are addressed. The relationship between accumulation rates of statoblasts in sediments and the actual producing population needs to be quantified for individual species. The production of statoblasts displays a great deal of interspecific variability ranging from 1 to 2 statoblasts/polypide for *Fredericella sultana* to 15 to 20/polypide for *Plumatella repens* [23]. Additionally, some species produce more than one type of statoblast. There is great need for additional work on individual species. Fortunately, the paleolimnological approach taken by Crisman et al. [22] provides a novel way to approach the ecology of such a spatially and temporally heterogeneous group. When these questions are tackled, then bryozoans will be an extremely informative paleolimnological tool.

Chaoborus

Chaoborus larvae are routinely sampled as part of both the benthos and plankton. This apparent discrepancy in habitat preference arises from the fact that chaoborids remain at or near the sediment/water interface during the day and vertically migrate in the water column at night to feed on zooplankton. Such a diurnal cycle minimizes the impact of sight predators such as fish. The maximum abundance of *Chaoborus* has long been assumed to be in stratified eutrophic lakes, where in addition to darkness, larvae minimize predation pressure by remaining in the anoxic hypolimnion during the day.

Deevey [12], in his postglacial history of Linsley Pond, was the first to identify *Chaoborus* subfossils from lake sediments. He attributed their increased abundance in the core to eutrophication of the basin. Subsequent investigations have included *Chaoborus* as a routinely analyzed animal microfossil. Goulden [24], Marland [25], Harmsworth [13], Deevey [26], and Roback [17] presented chaoborid data as part of their postglacial core investigations but did not provide any detailed interpretation of the data.

Following the lead of Deevey [12], most workers have attributed increased *Chaoborus* abundance in postglacial cores to basin eutrophication and the development of an anoxic hypolimnion [16,27,28]. In his literature review, Stahl [29] noted that *Chaoborus* abundance generally paralleled that of *Chironomus* in paleolimnological investigations. Hofmann [30] attributed the postglacial shift in chironomid dominance to *Chironomus* coincident with the establishment of maximum *Chaoborus* abundance in cores from both Schohsee and Grosser Plöner See to progressive lake eutrophication and the development of an anoxic hypolimnion. Subsequently, in a late and postglacial paleolimnological reconstruction for Grosser Seeberger See [31], he cautioned against using chaoborid data alone in making interpretations because some species such as the core dominant in this lake, *Chaoborus flavicans,* are found in both oligotrophic and eutrophic lakes.

Instead, he suggested that an increase in the ratio of *Chironomus/Chaoborus* may be a better indicator of hypolimnetic deoxygenation. Stahl [16] noted the coexistence of two species of *Chaoborus* in Myers Lake throughout most of the postglacial but was unable to use stratigraphic changes in their relative importance in his paleolimnological reconstruction due to a paucity of ecological data on each species.

In marked contrast to the long core studies just described, chaoborids rarely have been used in reconstructions of recent lake history. In the few studies that have been completed, they have proven to be a useful component of multiparameter investigations examining the development of culturally related eutrophication and meromixis.

Brugam [32] noted that the cultural eutrophication of Linsley Pond accelerated following both a change in agricultural practices (approximately 1915) and increased residential development in the watershed (post-1960). *Chaoborus* accumulation rates increased in response to both events with the largest increase (300%) associated with recent urbanization. The profile of *Chironomus* paralleled that of *Chaoborus* thus providing, as proposed by Hofmann [30], further evidence of the development of an anoxic hypolimnion as a consequence of progressive eutrophication.

Crisman [20] observed that accumulation rates of chaoborid mandibles in Lake Maxinkuckee, Indiana were higher prior to progressive cultural eutrophication during the past 30 years, while rates for chironomid genera characteristic of eutrophic lakes increased sharply post-1950. He suggested that increased chaoborid abundance during the preimpact period reflected the fact that the lake had always experienced morphometrically induced hypolimnetic anoxia during summer. Limnological observations from the 1890s and the failure of four lake trout stockings between 1890 and 1894 were cited as additional evidence for this interpretation. The lack of a response by *Chaoborus* to recent lake eutrophication was attributed to increased predation pressure following recent fisheries management including introduction of walleye pike.

Frey [33] noted that *Chaoborus* has occurred in Langsee, Austria principally since the lake became meromictic approximately 2000 years ago due to increased imput of silt and clay associated with human land clearance. Unlike Frey [33], Crisman and Swain (unpublished) observed that *Chaoborus* essentially disappeared in Brownie Lake, Minnesota once the lake became meromictic in the early 1920s associated with dredging and water level manipulations. Discrepancies such as these underscore the need to include additional factors including trophic state and fish community structure in addition to the profundal oxygen regime in all interpretations of chaoborid core data.

Although past investigators have assumed that *Chaoborus* abundance is positively related to increasing trophic state and/or hypolimnetic deoxygenation, Eriksson et al. [34] among others have observed that chaoborid abundance actually increases in oligotrophic lakes that have lost their fish populations as a result of recent progressive acidification. Freed from fish predation, chaoborids in such lakes become the dominant planktonic predators. To date, no one has applied the paleolimnological approach to this interesting biotic response to anthropogenic lake acidification.

Although individual chaoborid species can be identified by their subfossil mandibles, paleolimnological interpretations are often hindered by a paucity of ecological data. Often the data that do exist are confusing. For example, Alhonen [35] stated that chaoborid abundance increases with increasing humic acid content of the water, but Walker et al. [36], in a survey of 29 Canadian lakes, never found *Chaoborus* in peat pools. Finally, past paleolimnological interpretations of chaoborid data have been restricted to deep stratified lakes. Crisman is currently completing a calibration model for *Chaoborus* in shallow lakes representing a gradient of trophic conditions based on accumulation rates of mandibles in the surface sediments of the same 30 Florida lakes used by Crisman et al. [22] for a bryozoan calibration model. Initial analysis of the data suggests that the accumulation rate of chaoborid mandibles in the sediments is not strongly related to lake trophic state as measured by chlorophyll concentrations.

A number of problems need to be addressed by future investigations. The relationship between accumulation rates of chaoborid mandibles in sediments and the level of the extant population that produced them is unknown. While this in part is a function of difficulties in sampling an organism such as *Chaoborus* that is both benthic and planktonic, Deevey [*12*] suggested that *Chaoborus* may be underrepresented in the sediment record because their exuviae float on the water surface and therefore can be wind dispersed. This discussion has demonstrated the need for separating the influence of trophic state/oxygen from that of alterations in predation pressure from fish as factors controlling chaoborid abundance in sediments. Until this distinction can be made, chaoborid data are best interpreted as part of multiparameter paleolimnological investigations.

Chironomids

Compared to bryozoans and chaoborids, a great deal of attention has been given to the modern ecology and paleolimnology of chironomid midges. Early in the twentieth century, Thienemann [*37,38*] noted distinct compositional differences in profundal chironomid communities of deep stratified German lakes based on summer hypolimnetic oxygen concentrations. Using this relationship, he proposed a lake typology scheme based on chironomid dominance along an oxygen (and trophic state) gradient from *Tanytarsus* lakes (oligotrophic) through *Stictochironomus/Sergentia* lakes (mesotrophic) to *Chironomus* lakes (eutrophic). Refinements to this basic classification system were made by several workers including Lundbeck [*39*] and Brundin [*40,41*] for Europe and Deevey [*42*] and Saether [*43,44*] for North America. An excellent review of the use of such lake typologies worldwide including their historical development is provided by Brinkhurst [*45*].

Frey [*3*] noted that Andersson [*46*] and Ekman [*47*] were the first to recognize the subfossil remains of chironomids in lake sediments, while Gams [*48*], in his investigation at Lunzer Obersee, Austria, was the first to use subfossils to delineate the postglacial history of chironomids in an individual lake. The first use of chironomid subfossils in a multiparameter paleolimnological investigation was the excellent work of Deevey [*12*] at Lindsley Pond. Since that time, detailed histories of chironomids in long cores representing the entire period since formation have been constructed for lakes in North America [*16,25,49–54*], Central America [*24*], Europe [*13,17,27,28,30,31,33,55–59*], Iceland [*60*], Japan [*14*], and New Zealand [*62*].

Several workers have interpreted chironomid profiles in light of the typological schemes discussed earlier as a means of delineating the history of lake trophic state. The "classical" succession from oligotrophy (*Tanytarsus*) to eutrophy (*Chironomus*) during the course of basin history has been documented for numerous lakes [*12–14,27,30,56,57*]. Tsukada [*14,61*] at Lake Nojiri, Japan and Goulden [*56*] at Esthwaite Water, England attributed much of the increased trophic state to initial land clearance by humans.

One must not assume that eutrophication is an unavoidable undirectional process consisting of distinct stages that are correlated with advancing lake age. In both Schohsee and Grosser Plöner See, Hofmann [*30*] noted that the progression from oligotrophy to eutrophy lacked a distinct intermediate mesotrophic stage. Instead, the fauna in these lakes was defined as being transitional to the higher trophic state. In addition, the trophic state of several lakes based on chironomid interpretations has not changed appreciably since lake formation [*16,28,31,49,50,54,55,59*], and Tarn Moss, England [*55*] appears to have actually declined in productivity.

Chironomid interpretations of long cores have not been restricted to general lake trophic changes. Dickman et al. [*51*] and Cheek [*53*] used chironomid subfossils as evidence that two Canadian lakes, Pinks and Crawford, had been meromictic since formation, and Frey [*33*] used this approach as one line of evidence for the human-induced meromixis of Langsee, Austria.

Several of the investigations just mentioned that interpreted chironomid changes from a trophic state perspective also included additional factors including climate, water level, and sediment composition in their interpretations. The orthoclad *Heterotrissocladius* was a common faunal element in many lakes during the late glacial and early postglacial and either declined drastically or disappeared shortly thereafter [*30,50,57–59*]. In addition, Hofmann [*58*] suggested that the

Tanytarsus lugens community is extremely cold stenothermal and restricted to the late glacial in most lakes. Hofmann [30], on the basis of cores taken in shallow and deep sections of Schohsee and Grosser Plöner See, documented that those stenothermal species (including *Sergentia* and *Lauterbornia*) common in late glacial and early postglacial times and still surviving in the lake following climatic amelioration do so by changing their habitat from littoral to the profundal.

Several authors have examined the influence of water level fluctuations on both trophic state and the contribution of the littoral fauna to the total subfossil assemblage [31,4,50,55,57]. With regard to the latter, Megard [49] felt that Dead Man Lake, New Mexico (currently 1 m deep) has been so shallow throughout its history that the littoral chironomid fauna cannot be used to assess past oscillations in system trophic state. The most interesting study on the effect of water level on chironomids is that of Lawrenz [50], who showed that although Green Lake, Michigan had always been oligotrophic, four distinct chironomid zones could be identified from the core based primarily on past climatically induced water level fluctuations.

Finally, both Lawrenz [50] and Bryce [55] suggested that changes in sediment composition in association with water level fluctuations were at least partially responsible for observed chironomid changes in their cores. In fact, the reversal of the classical succession pattern of *Chironomus* replacement of *Tanytarsus* in Tarn Moss was attributed by Bryce [55] to a period of increased marl deposition during a higher water stage.

Only within the past decade have chironomids been used in reconstructions of recent lake history. A majority of studies have concentrated on the chironomid response to cultural eutrophication. The midge community of several European lakes [63–66] changed dramatically during the late 1940s as a result of a postwar acceleration in cultural eutrophication attributed to increased sewage loading from expanding resident and tourist populations, phosphorus-rich detergents, agricultural fertilizers, and feed lots. Brugam [32] noted a chironomid response at Linsley Pond to changing agricultural practices about 1915, but the greatest response was to residential development following World War II. Recently, Crisman [20] noted a major chironomid response in Lake Maxinkuckee, Indiana to both a postwar population expansion and the increased input of agriculturally derived erosion products. The entry of the inorganic sediment into the lake was facilitated by the destruction of a marsh at the mouth of a stream draining the major portion of the catchment devoted to agriculture, thereby losing its ability to trap inorganic sediment before entry into the lake. In most of the just-cited studies, lake eutrophication was accompanied by a shift in chironomid dominance to *Chironomus* with *Procladius* and *Dicrotendipes* as the principal subdominants.

In some cases, the chironomid response to cultural eutrophication has been either temporally delayed or absent. Warwick [67] noted that although the Bay of Quinte began to eutrophy following European settlement, the chironomid assemblage remained characteristic of oligotrophic lakes due to maintenance of inorganic-dominated lacustrine sediments from erosion products derived from land clearance. Only in later years when the catchment soils were stabilized and nutrient loading continued to increase with expanding populations, did the chironomids adequately reflect the progressive eutrophication of the basin. Wiederholm [68] failed to document a pronounced response of profundal midges to the well-known cultural eutrophication of Lake Washington and attributed this to the fact that the eutrophication period was too short for development of an anoxic hypolimnion in this deep lake. Finally, Crisman [20], through an analysis of historical limnological data, documented that the lack of a pronounced chironomid response in Lake Maxinkuckee to land clearance and population expansion pre-1900 reflected the fact that the lake had been characterized by midsummer morphometrically derived hypolimnetic deoxygenation even during periods of lower trophic state.

Two recent investigations have examined the chironomid response to both historical cultural eutrophication and recent attempts at lake restoration. Kansanen [66] presented a detailed account of alterations in the chironomid community in a Finnish lake associated with expanding population and industrial development post-1870. While the midge community was seriously altered by an expanding paper industry from 1870 to 1940, the fauna collapse was associated with post-World War II industrial expansion and diversification and expanding human population. Following implementation of pollution control measures in 1975, some positive response in the littoral fauna of the lake has been noted. Brodin [69] also noted an increase in chironomid diversity following installation of a sewage treatment plant at Lake Vaxjosjon, Sweden in the 1930s.

The application of subfossil chironomids as a tool for reconstructing lake responses to recent perturbations has not been limited solely to eutrophication. Traunsee, a 189-m-deep lake in Austria, was holomictic until industrial discharges from alkali works began in the 1880s [70]. Oxygen levels in the monimolimnion have been maintained by input of riverine water, but the profundal midge community has been completely eliminated due to the buildup of an insoluble alkaline sludge of calcium carbonate ($CaCO_3$), magnesium carbonate ($MgCO_3$), and calcium sulfate ($CaSO_4$) from the industrial discharge. Establishment of meromixis at Brownie Lake, Minnesota in the 1920s associated with artificial water level manipulations and dredging resulted in complete elimination of profundal chironomids due to permanent anoxia (Crisman and Swain, unpublished).

Warwick [67] noted that many of the subfossil head capsules deposited post-1950 displayed serious deformations, including asymmetrical teeth in both mentum and mandibles and cuticular thickening of the head capsule wall. He attributed such abnormalities to industrial and/or agricultural contaminants and supported his interpretations with published data on the response of extant communities to such perturbations. Although no one has applied the paleolimnological approach for delineating the response of aquatic biota to specific contaminants, the initial results of Warwick [67] suggest the potential value of subfossil chironomids in this expanding area of applied research.

The rapidity of the response of chironomid communities to environmental improvement was demonstrated by Clair and Paterson [71] for a coastal lake in Canada where the chironomid fauna was completely destroyed as the result of salt water intrusion during a single storm in 1869 but quickly reestablished itself following lake freshening. Prat and Daroca [19] used subfossil chironomids as part of a multiparameter investigation of the post-filling trophic history of four Spanish reservoirs. They noted that following an initial productivity pulse, trophic state declined sharply in those reservoirs receiving low nutrient loading. During low water periods the trophic state of all reservoirs increased regardless of nutrient loading.

Most recently, subfossil chironomids have been used to document the response of benthic invertebrates to progressive lake acidification. In an analysis of cores from two Swedish lakes (pH 4.3 to 4.7) currently receiving acidic deposition, Henrikson et al. [72] documented a reduction in Tanytarsini and an increase in *Phaenopsectra* and *Psectrocladius* as contributors to the chironomid assemblage of surface sediments relative to their importance in sediments presumably deposited during the preacidification period. Although *Chironomus* is also a major component of the extant fauna, it was not reported in the sediment record. Emeis-Schwarz and Kohmann [73] analyzed chironomid remains from a Bavarian lake (pH 3.6 to 4.8) core and recorded a reduction in the number of taxa, an increase in the importance of *Chironomus,* and a reduction and eventual elimination of *Tanytarsus* in the upper 40 cm of the core that they attributed to progressive lake acidification. Hofmann [74] reviewed an unpublished study of Bordin at Lake Flarken, Sweden, who found a reduction in the number of taxa and increased importance of two species of *Chironomus* in the most recently deposited sediments that he related to a pH shift from 6.5 (1974) to 5.0 to 5.5 (1979 to 1982). Hofmann [74] also reported a reduced number of taxa and faunal dominance (75%) by *Chironomus* in a core from acidic Reinbeker Tonteich and noted good agreement between extant and subfossil faunas for the two years he examined, 1933 (pH 3.2) and 1950 (pH 3.3). Finally, as part of a reconstruction for the entire postglacial, Walker and Paterson [54] documented a historical faunal shift in humically colored Woods Pond (pH 4.0 to 4.8). Increased importance of *Chironomus* and *Psectrocladius* and a reduction in *Tanytarsus* in the upper part of the sediment record were attributed to encroachment of a *Sphagnum* mat associated with fire and land clearance.

The fact that *Chironomus* can assume dominance in both eutrophic and oligotrophic (acidic) lakes underscores the need for a calibration model for subfossil chironomids that covers the whole trophic spectrum. Raddum and Saether [75] examined extant chironomid communities in five acidic Norwegian lakes (pH 4.47 to 6.25), and Walker et al. [36] published a similar survey on 29 clear and colored acidic lakes (pH 4.0 to 7.3) in eastern Canada. Based on comparable data from numerous temperate lakes, Saether [44] divided the oligotrophic range into six distinct chironomid indicator groups. Finally, the only published investigation specifically examining subfossil chironomids relative to known lake conditions is that of Crisman [76], who related the distribution of subfossil head capsules in surface sediments from 22 acidic Florida lakes (pH 4.01 to 7.15) to detailed data on current water chemistry and trophic state for each lake. The results of this initial study demonstrated the potential of chironomids as acidity indicators, but faunal differences between these subtropical lakes and comparable temperate systems, especially regarding the importance of Orthocladinae, underlined the importance of constructing similar calibration models for any lake region where paleolimnological reconstructions are to be made.

In addition to dividing the oligotrophic range into six divisions, Saether [44] also defined three and six divisions for the mesotrophic and eutrophic ranges, respectively. He concluded that profundal chironomid communities in oligotrophic and mesotrophic lakes are controlled more by food availability than by hypolimnetic oxygen levels, while in eutrophic lakes, organic matter accumulation and oxygen levels are so linked as to preclude separation of the importance of each as controlling factors for faunal distributions. Currently, there are no published calibration models for subfossil chironomid taxa relative to a broad range of trophic conditions, but a nearly completed investigation by Crisman on 160 Florida lakes will be a start, at least for the subtropics.

Based on an extensive data base for Swedish lakes, Wiederholm [77] devised the benthic quality index (BCI) relating the importance of five indicator chironomid species to total phosphorus concentrations/mean lake depth. Wiederholm and Eriksson [64] applied the BCI in reverse, whereby subfossil chironomid assemblages for select levels in a core from Ekoln Bay of Lake Malaren, Sweden were used to estimate past phosphorus and chlorophyll levels. Where historical limnologial data existed, there was reasonable agreement between predicted and actual values. This initial investigation demonstrates that until detailed calibration models specifically relating subfossil chironomids to lake parameters are developed, existing water quality indices may be of value in paleolimnological reconstructions.

While the original lake typologies stressed the relationship between profundal chironomid communities and hypolimnetic oxygen concentrations, subsequent investigators have questioned such a simplistic interpretation and stressed other environmental variables including water level fluctuations, sediment composition, extent of macrophytes, food availability, predation intensity, and climate as equally important controlling factors. Future interpretations of core profiles of subfossil chironomids must consider possible temporal changes in each of these factors. Incorporation of the habit and functional group categories of Merritt and Cummins [78] may be of value in this endeavor.

Megard [49] noted that Dead Man Lake, New Mexico (currently 1 m deep) has always been dominated by Tanytarsini and orthoclads in spite of its eutrophic status and suggested that typological schemes based on profundal chironomids are of no value in interpreting cores from shallow lakes. The latter point was reiterated by Hofmann [30] and Alhonen [35]. Carter [63] and Devai and Moldovan [65] noted the persistence of taxa characteristic of lower trophic state in the recent sediments of shallow eutrophic Lough Neagh and Lake Balaton, respectively, that they attributed to the lack of oxygen depletion in these unstratified systems. Unlike Megard [49] these authors as well as Brodin [69] noted major stratigraphic changes in the chironomid assemblages of their short cores that they attributed to recent changes in substrate, food, and, secondarily, oxygen resulting from cultural eutrophication. While such investigations demonstrated the utility of subfossil chironomids in paleolimnological reconstructions for shallow lakes, their interpretations would have been greatly aided by detailed ecological data on littoral zone chironomids, especially their relationship to macrophyte species composition, water depth, and intensity of fish predation.

Deep water sediment cores are thought to provide an integrated record of all microhabitats within a lake basin. Given the fact that chironomid head capsules in most cases can be identified only to the generic level and that within many genera littoral and profundal species frequently coexist within a single lake, interpretation of the subfossil chironomid record from deep water cores would be seriously hindered if there had been pronounced redeposition of littoral remains into profundal areas. Although Carter [63], Wiederholm and Eriksson [64], and Warwick [67] found no evidence for redeposition of littoral chironomid remains, several authors [51,53,66,68,69] have suggested that redeposition has had a major influence on their sediment records. Wiederholm [68] attributed the apparent minor response of the chironomid community to the cultural eutrophication of Lake Washington to the fact that only 5 of the 51 taxa that he encountered in his core were truly profundal, and these accounted for only 20% of the total remains. Kansanen [66] reported that the importance of the littoral fauna in the sediment record is in part related to the distance of the coring site from the shore. He found that subfossil assemblages were comprised of approximately 36% littoral remains at sites 120 to 240 m from shore, but this contribution was reduced to 6% at sites 900 to 1500 m from shore.

Stahl [29] suggested that while littoral remains are likely to be redeposited in deeper water, remains of the actual profundal community should not be subject to much postmortem transport. Additionally, Hofmann [30] suggested possible differential susceptability of taxa to redeposition of remains. Iovino [79] investigated the importance of intrabasin lateral transport of chironomid remains by comparing living and subfossil assemblages isolated from surficial sediments along depth profiles of Pretty Lake, Indiana. His results demonstrated good agreement between living and subfossil distributions but indicated that lateral transport of remains may be important in that portion of the littoral zone above the base of the epilimnion (effective wave depth). Remains of littoral midges were often concentrated at the epilimnion-metalimnion boundary, but below this point subfossil assemblages reflected the distribution of the living species, with little evidence of contamination from littoral remains. Additional investigations by Stark [52] at Elk Lake, Minnesota and Walker et al. [80] in Canada led to the same conclusions.

Iovino [79] in his survey observed that differential preservation of both instars and taxa contribute to the general underrepresentation of the extant fauna by the subfossil assemblage. He attributed the general paucity of first and second instars in the sediment record to the fact that chironomids passing through these developmental stages dissolve the chitinous head capsule before molting. Both Wiederholm and Eriksson [64] and Devai and Moldovan [65] felt that these early instars were simply too weakly sclerotized to be preserved. Further, it is felt [30,66] that the fourth instar of *Chironomus* is underrepresented in the subfossil record because the head capsule of this final larval instar adheres to the pupa upon molting. During adult emergence, the pupal exuvia with its attached larval exuvia floats on the water surface where it can be dispersed by wind and water currents.

It has been suggested that *Glyptotendipes* [63], in addition to *Chironomus* in the Chironomini and many of the Tanypodinae [17,55], may be underrepresented in the subfossil record. Regarding the latter, Iovino [79] did not find *Procladius* underrepresented in the sediment record of his stratified alkaline lakes, but Walker et al. [80] noted that this genus was so underrepresented in their shallow humically colored lakes as to be of little value in paleolimnological reconstructions for this lake type. These conflicting data raise more questions than they answer and underscore the need for further research on chironomid preservation under a variety of lake conditions.

In addition to postmortem redeposition and differential preservation, interpretation of the chironomid record is hindered by problems relating to voltinism, incomplete subfossil taxonomy, and bioturbation of the sediment record. The potential accumulation rate of head capsules in sediments for a given species in part depends on the number of generations (univoltine to multivoltine) completed per year. In addition, chironomid taxonomy is based on nonpreservable larval anatomy and adult males in addition to preservable features of the head capsule. While most subfossil specimens can be readily identified to genus, there are few taxonomic aids [16,81] for specific identifications. Finally, the living chironomid fauna can disturb the stratigraphic record by reworking the upper 15 [65] to 50 cm [66] of the sediment profile. Kansanen [66] also suggested that molting of instars deep in the sediments could introduce error in core interpretations by providing records for taxa for periods predating either their establishment or dominance in a lake.

Multiparameter Studies

It is apparent that the species composition and abundance of the three principal components of the benthos found in the sediment record are controlled by a complex set of environmental parameters, the rank ordering of which is group dependent. Bryozoans appear to be controlled by substrate and food availability, chaoborids principally by fish predation, and chironomids by an interaction of oxygen, substrate, and food. Because these parameters do not necessarily display a strong covariant linkage, it is conceivable that interpretations of the timing and magnitude of the lake response to a given environmental perturbation based separately on bryozoans, chaoborids, and chironomids could vary greatly.

Four studies have been selected to evaluate the correspondence of environmental interpretations based on individual invertebrate groups (Fig. 2). These were selected from the rather limited number of multiparameter paleolimnological studies that have sufficient time resolution to permit both correlation of watershed events with lake responses and calculation of annual accumulation rates for each chemical and biological parameter. Each study is limited to the last few decades of lake history based on 210-Pb dated core profiles.

Lake Maxinkuckee, Indiana is currently mesotrophic but displays a number of qualities characteristic of eutrophic lakes [20]. The vegetated littoral zone is restricted aerially, and the phytoplankton assemblage has been dominated by blue-green algae since at least the 1880s. Summertime hypolimnetic deoxygenation also has been common for at least 100 years.

Consistent with available monitoring data, phosphorus accumulation rates in the sediments tripled after 1960 coincident with destruction of a marsh at the mouth of an inlet stream to increase drainage from grain fields. Thus, although the lake appears to have been moderately productive throughout the period of European occupation, it has experienced accelerated cultural eutrophication for at least the past 25 years.

In spite of persistence of an anoxic deep water refuge, *Chaoborus* was extremely rare in the sediment record, and thus of no value in indicating anoxia or trophic state oscillations. Chironomids, on the other hand, more adequately reflect known historical trends. The accumulation rates of *Chironomus* increased progressively throughout the postsettlement period and displayed a ten-fold increase coincident with the post-1960 phosphorus enhancement. With the exception of a pulse during the 1950s, the abundance of Tanytarsini remained relatively constant throughout the core. Relatively constant bryozoan accumulation rates since at least 1890 suggest that the extent of the vegetated littoral zone has changed little in the past 95 years. This conclusion is supported by the rather limited historical data base on the macrophyte community.

Brownie Lake, Minnesota is a meromictic eutrophic lake in Minneapolis [82]. The area around the lake was settled in the 1850s and remained rural until the beginning of the twentieth century. Meromixis was established in the early 1920s associated with dredging and water level manipulations.

Phosphorus accumulation rates began to rise shortly before 1920 and increased progressively until the early 1950s after which they remained relatively constant. *Chaoborus* was abundant before 1920 but disappeared following initiation of meromixis. The same trend is also displayed by chironomids and bryozoans. The one recent subfossil Tanytarsini record is attributed to redeposition from the littoral zone. Bryozoan statoblasts were found at only one core level and thus are of no value in reconstructing lake history. While the subfossil benthos were extremely useful at identifying when meromixis began, their elimination from profundal areas precluded their use in delineating subsequent basin eutrophication.

Lake Vanajanselka, Finland is culturally eutrophic with human disturbance beginning about 1800 [66]. Although paper mills were established in the last quarter of the nineteenth century, progressive industrial expansion and diversification began in the 1940s, especially after World War II. Population trends for the area paralleled those of industrial development.

Phosphorus accumulation rates remained relatively constant from 1880 to 1950 after which they increased progressively. The phosphorus trend thus matches that of industrial development in the area. The chaoborid data gave no indication of the recent cultural eutrophication of the basin. In fact, accumulation rates were slightly lower post-1960 when phosphorus accumulation was maximal. *Chironomus* accumulation rates began to increase earlier than those of phosphorus, 1920s versus 1950, but unlike the latter declined sharply in the 1960s. Tanytarsini accumulation rates were greatest prior to 1920 when they were equal to or slightly greater than those of *Chironomus*. By the 1940s accumulation rates reached minimum levels at which they were maintained. The chironomid succession in Vanajanselka provides a good record of the cultural eutrophication of the basin. Prior to 1920, a lower trophic state is indicated by the shared dominance of *Chironomus* and Tanytarsini. For the next four decades *Chironomus* increased progressively while the Tanytarsini decreased slightly in response to increased anthropogenic phosphorus loading to the system. The one discrepancy is the pronounced decline in *Chironomus* accumulation rates in the 1960s at a time when phosphorus sediment concentrations continued to increase progressively. Kansanen [66] noted that the biological oxygen demand (BOD) of sewage discharged to the lake increased dramatically during this period and that the chironomid community of the lake collapsed as a result of associated permanent hypolimnetic anoxia. Thus, in spite of continuing basin eutrophication, one of the principal eutrophic indicators, *Chironomus,* was severely reduced as a result of habitat loss.

Linsley Pond, Connecticut currently is eutrophic with a residentially developed watershed [32]. The area was cleared for agriculture by European settlers in 1700, but the lake changed little until about 1915 when Brugam [32] suggested that establishment of a dairy herd and other changes in agricultural practices enhanced lake productivity. Basin eutrophication has continued since then and greatly accelerated following residential development of the watershed in the 1960s.

Phosphorus accumulation rates were not altered as a result of changing agricultural practices at the beginning of the twentieth century [32]. Values did begin to increase following World War II and peaked during the 1960s coincident with urbanization of the watershed. *Chaoborus* accumulation rates increased shortly after 1915 and displayed maximum core values associated with the urbanization of the 1960s. *Chironomus* accumulation rates did not respond to the agricultural changes after 1915, but *Chironomus* did replace *Dicrotendipes* as the dominant faunal element. As with phosphorus and *Chaoborus, Chironomus* accumulation rates increased sharply post-1960, reflecting watershed urbanization. Although all parameters changed markedly as a result of post-1960 urbanization, evidence of the earlier (1915) watershed disturbance was provided by only the two biological parameters.

The results of the four investigations just discussed caution against basing paleolimnological interpretations on single parameters. These investigations demonstrate that the best approach for reconstructing lake histories is to couple analyses of several biological and chemical parameters from dated cores with detailed information on past changes in watershed practices. Because of the often weak covariance linkage of biotic and abiotic controlling factors, individual benthic faunal elements can be sensitive indicators of one level of disturbance while appearing to be unresponsive to either subsequent disturbances in the same lake or comparable disturbances in other lakes.

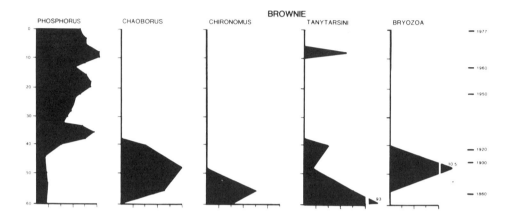
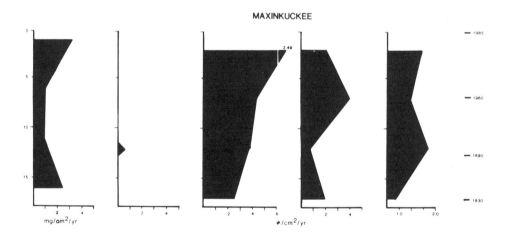

FIG. 2—*Profiles of 210-Pb calculated accumulation rates of phosphorus and selected subfossil invertebrates from Lakes Brownie, Minnesota (Crisman and Swain, unpublished), Vanajanselkä, Finland [66], Maxinkuckee, Indiana [20], and Linsley Pond, Connecticut [32]. Data for the latter three sites were taken from the original references.*

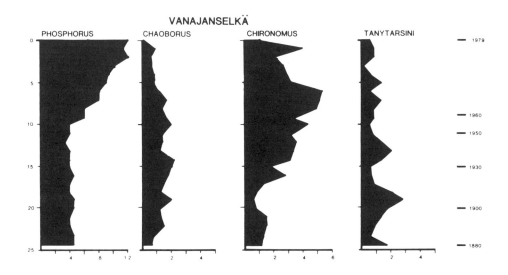

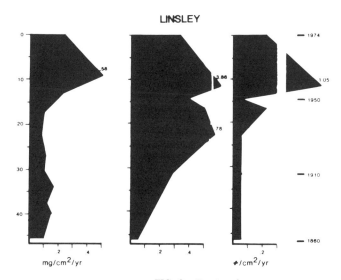

FIG. 2—*Continued*.

Conclusions

Since the beginning of this century, benthic invertebrates have been used as biotic indicators of water quality. Although remains of several faunal groups are preserved in lacustrine sediments, bryozoans, chaoborids, and chironomids show the greatest promise for paleolimnological reconstructions. While subfossils of these three faunal elements can be used to make qualitative statements on water quality, their use in quantitative reconstructions of past lake conditions must wait until detailed calibration models have been constructed for subfossil assemblages relative to water chemistry and trophic state and questions regarding the relationship between subfossil assemblages and the populations that produced them, postmortem redeposition, differential preservation, and basic ecology are addressed.

The paleolimnological approach has great potential application to problems regarding lake and watershed management. In most cases we see only the end result of the disturbance process without having a sufficient historical data base on either the predisturbance conditions of the lake or how the biota responded to changes in the type and intensity of watershed disturbance. The latter point is important in that it is indeed rare to find that the perceived or actual demise of a lake system is the result of a single isolated watershed disturbance. As amply demonstrated by many of the case histories reviewed here, the process in most cases is additive.

The objective of any lake management or restoration plan is not only to identify the principal watershed disturbances/practices that are contributing to a perceived lake problem but to rank them so that a management plan can be devised whereby the most important contributers can be controlled at the least cost. Without a historical data base the paleolimnological approach offers the only way of evaluating how a lake responded to individual watershed events. Such paleolimnological reconstructions must take a multiparameter approach. Because water quality is but one of a complex set of abiotic and biotic factors controlling the composition and abundance of each faunal group and because individual factors do not necessarily exhibit a strong covariance linkage during the disturbance process, it is essential to base reconstructions on as many subfossil parameters as possible. With such an approach, the potential of the paleolimnological approach as a tool in applied research is assured.

References

[1] Frey, D. G., *Mitteilungen Internationale Vereinigung Limnologie,* Vol. 20, 1974, pp. 95–123.
[2] Engstrom, D. R. and Wright, H. E., Jr. in *Lake Sediments and Environmental History,* E. Y. Haworth and J. W. G. Lund, Eds., Minnesota, Minneapolis, MN, 1984, pp. 11–68.
[3] Frey, D. G., *Archiv fur Hydrobiologie Ergebnisse der Limnologie,* Vol. 2, 1964, pp. 1–114.
[4] Frey, D. G., *Canadian Journal of Zoology,* Vol. 54, No. 12, 1976, pp. 2208–2226.
[5] Crisman, T. L. in *Biology and Quaternary Environments,* D. Walker and J. C. Guppy, Eds., Australian Academy of Sciences, Canberra, 1978, pp. 69–102.
[6] Bushnell, J. H., *Ecological Monographs,* Vol. 36, No. 3, 1966, pp. 95–123.
[7] Bushnell, J. H. and Rao, K. S., *Transactions of the American Microscopical Society,* Vol. 93, No. 5, 1974, pp. 524–543.
[8] Rogick, M. D., *Annales of the New York Academy of Sciences,* Vol. 45, No. 3, 1943, pp. 163–178.
[9] Walker, I. R. and Paterson, C. G., *Hydrobiologia,* Vol. 122, 1985, pp. 189–198.
[10] Pennak, R. W., *Fresh-Water Invertebrates of the United States,* Ronald Press, New York, 1953.
[11] Wesenberg-Lund, H. C., *Meddelungen Dansk Geologiske Foreningen,* Vol. 3, 1896, pp. 51–85.
[12] Deevey, E. S., Jr., *American Journal of Science,* Vol. 240, No. 4, 1942, pp. 233–264.
[13] Harmsworth, R. V., *Ecological Monographs,* Vol. 38, No. 2, 1968, pp. 223–241.
[14] Tsukada, M., *Transactions of the Connecticut Academy of Arts and Sciences,* Vol. 44, No. 6, 1972, pp. 339–365.
[15] Fredskild, B., Jacobsen, N., and Roen, U., *Meddelelser om Gronland,* Vol. 198, No. 5, 1975, pp. 1–44.
[16] Stahl, J. B., *Investigations on Indiana Lakes and Streams,* Vol. 5, 1959, pp. 47–102.
[17] Roback, S. A., *Transactions of the American Philosophical Society,* Vol. 60, No. 4, 1970, pp. 150–162.

[18] Birks, H. H., Whiteside, M. C., Stark, D. M., and Bright, R. C., *Quaternary Research,* Vol. 6, 1976, pp. 249–272.
[19] Prat, N. and Daroca, M. V., *Hydrobiologia,* Vol. 103, 1983, pp. 153–158.
[20] Crisman, T. L., "Historical Analysis of the Cultural Eutrophication of Lake Maxinkuckee, Indiana," Lake Maxinkuckee Association, Culver, IN, 1986.
[21] Vallentyne, J. R. and Swabey, Y. S., *American Journal of Science,* Vol. 253, 1955, pp. 313–340.
[22] Crisman, T. L., Crisman, U. A. M., and Binford, M. W., *Hydrobiologia,* Vol. 143, 1986, pp. 113–118.
[23] Bushnell, J. H. in *Living and Fossil Bryozoa,* G. P. Larwood, Ed., Academic, London, 1973, pp. 503–520.
[24] Goulden, C. E., *Memoirs of the Connecticut Academy of Arts and Sciences,* Vol. 17, 1966, pp. 84–120.
[25] Marland, F. C., "The History of Mountain Lake, Giles County, Virginia: An Interpretation Based on Paleolimnology," Ph.D. dissertation, Virginia Polytechnic Institute and State University, Blacksburg, VA, 1967.
[26] Deevey, E. S., Jr., *Mitteilungen Internationale Vereinigung Limnologie,* Vol. 17, 1969, pp. 56–63.
[27] Alhonen, P. and Haavisto, M. L., *Bulletin of the Geological Society of Finland,* Vol. 41, 1969, pp. 157–164.
[28] Czeczuga, B., Kossacka, W., and Niedzwiecki, E., *Polskie Archiwum Hydrobiologii,* Vol. 26, No. 3, 1979, pp. 351–369.
[29] Stahl, J. B., *Mitteilungen Internationale Vereinigung Limnologie,* Vol. 17, 1969, pp. 111–125.
[30] Hofmann, W., *Archiv fur Hydrobiologie Ergebnisse der Limnologie,* Vol. 40, 1971, pp. 1–74.
[31] Hofmann, W., *Archiv fur Hydrobiologie,* Vol. 82, 1978, pp. 316–346.
[32] Brugam, R. B., *Ecology,* Vol. 59, No. 1, 1978, pp. 19–36.
[33] Frey, D. G., *Memorie dell Istituto Italiano di Idrobiologia,* Suppl. Vol. 13, 1955, pp. 143–161.
[34] Eriksson, M. O. G., Henrikson, L., Nilsson, B. I., Nyman, G., Oscarson, H. G., Stenson, A. E., and Larsson, K., *Ambio,* Vol. 9, 1980, pp. 248–249.
[35] Alhonen, P., *Archiv fur Hydrobiologie,* Vol. 86, No. 1, 1979, pp. 13–26.
[36] Walker, I. R., Fernando, C. H., and Paterson, C. G., *Hydrobiologia,* Vol. 120, 1985, pp. 11–22.
[37] Thienemann, A., *Internationale Revue Hydrobiologie,* Vol. 6, 1913, pp. 243–249.
[38] Thienemann, A., *Archiv fur Hydrobiologie,* Vol. 13, 1922, pp. 609–646.
[39] Lundbeck, J., *Archiv fur Hydrobiologie,* Suppl. Vol. 7, 1926, pp. 1–473.
[40] Brundin, L., *Report of the Institute of Freshwater Research of Drottningholm,* Vol. 30, 1949, pp. 1–914.
[41] Brundin, L., *Report of the Institute of Freshwater Research of Drottningholm,* Vol. 37, 1956, pp. 185–235.
[42] Deevey, E. S., Jr., *Ecological Monographs,* Vol. 11, No. 2, 1941, pp. 413–455.
[43] Saether, O. A., *Verhandlungen Internationale Vereinigung Limnologie,* Vol. 19, 1975, pp. 3127–3133.
[44] Saether, O. A., *Holarctic Ecology,* Vol. 2, 1979, pp. 65–74.
[45] Brinkhurst, R. O., *Benthos of Lakes,* St. Martins Press, New York, 1974.
[46] Andersson, G., *Bulletin of the Commission on the Geology of Finland,* Vol. 8, 1898, pp. 1–210.
[47] Ekman, S., *Internationale Revue Hydrobiologie,* Vol. 7, 1915, pp. 146–204 and 275–425.
[48] Gams, H., *Internationale Revue Hydrobiologie,* Vol. 18, 1927, pp. 305–387.
[49] Megard, R. O., *Ecology,* Vol. 45, No. 3, 1964, pp. 529–546.
[50] Lawrenz, R. W., "The Developmental Paleoecology of Green Lake, Antrim County, Michigan," M.S. thesis, Central Michigan University, Mt. Pleasant, MI, 1975.
[51] Dickman, M., Krelina, E., and Mott, R., *Verhandlungen Internationale Vereinigung Limnologie,* Vol. 19, 1975, pp. 2259–2266.
[52] Stark, D. M., *Archiv für Hydrobiologie,* Suppl. Vol. 50, 1976, pp. 208–274.
[53] Cheek, M. R., "Paleoindicators of Meromixis," M.S. thesis, Brock University, St. Catharines, Ontario, Canada, 1979.
[54] Walker, I. R. and Paterson, C. G., *Freshwater Invertebrate Biology,* Vol. 2, No. 2, 1983, pp. 61–73.
[55] Bryce, D., *Transactions of the Society of British Entomology,* Vol. 15, 1962, pp. 41–54.
[56] Goulden, C. E., *Verhandlungen Internationale Vereinigung Limnologie,* Vol. 15, 1964, pp. 1000–1005.
[57] Gunter, J., *Hydrobiologia,* Vol. 103, 1983, pp. 231–234.
[58] Hofmann, W., *Hydrobiologia,* Vol. 103, 1983, pp. 235–239.
[59] Schakau, B. and Frank, C., *Verhandlungen der Gesellshaft fur Okologie,* Vol. 12, 1984, pp. 375–382.
[60] Einarsson, A., *Freshwater Biology,* Vol. 12, 1982, pp. 63–82.
[61] Kadota, S., *Paleolimnology of Lake Biwa and the Japanese Pleistocene,* Vol. 4, 1976, pp. 297–307.
[62] Deevey, E. S., Jr., *Records of the Canterbury Museum,* Vol. 6, No. 4, 1955, pp. 291–344.
[63] Carter, C. E., *Freshwater Biology,* Vol. 7, 1977, pp. 415–423.

[64] Wiederholm, T. and Eriksson, L., *Hydrobiologia,* Vol. 62, 1979, pp. 195–208.
[65] Devai, G. and Moldovan, J., *Hydrobiologia,* Vol. 103, 1983, pp. 169–175.
[66] Kansanen, P. H., *Annales Zoologici Fennici,* Vol. 22, 1985, pp. 71–104.
[67] Warwick, W. F., *Canadian Entomologist,* Vol. 112, 1980, pp. 1193–1238.
[68] Wiederholm, T., *Northwest Science,* Vol. 53, 1979, pp. 251–256.
[69] Brodin, Y., *Archiv fur Hydrobiologie,* Vol. 93, No. 3, 1982, pp. 313–326.
[70] Löffler, H., *Hydrobiologia,* Vol. 103, 1983, pp. 135–139.
[71] Clair, T. and Paterson, C. G., *Hydrobiologia,* Vol. 48, 1976, pp. 131–135.
[72] Henrikson, L., Olofsson, J. B., and Oscarson, H. G., *Hydrobiologia,* Vol. 86, 1982, pp. 223–229.
[73] Emeis-Schwarz, H. and Kohmann, F. in *Versauerung in der BRD,* B. Lenhart, Ed., Verlag E. Schmidt, Stuttgart, West Germany, 1984, pp. 135–142.
[74] Hofmann, W. in *Proceedings of a Workshop on Paleolimnological Studies of the History and Effects of Acidic Precipitation,* S. Norton, Ed., United States Environmental Protection Agency, Washington, DC, 1984, pp. 328–344.
[75] Raddum, G. C. and Saether, O. A., *Verhandlungen Internationale Vereinigung Limnologie,* Vol. 21, 1981, pp. 399–405.
[76] Crisman, T. L. in *Paleoecological Investigation of Recent Lake Acidification (PIRLA): Interim Report,* D. F. Charles and D. R. Whitehead, Eds., Electric Power Research Institute, Palo Alto, CA, 1986, in press.
[77] Wiederholm, T., *Naturvadsverkets Limnologie Undersucken,* Vol. 10, 1976, pp. 1–17.
[78] Merritt, R. W. and Cummins, K. W., *An Introduction to the Aquatic Insects of North America,* Kendall-Hunt Publishing, Dubuque, IA, 1984.
[79] Iovino, A. J., "Extant Chironomid Larval Populations and the Representativeness and Nature of Their Remains in Lake Sediments," Ph.D. dissertation, Indiana University, Bloomington, 1975.
[80] Walker, I. R., Fernando, C. H., and Paterson, C. G., *Hydrobiologia,* Vol. 112, 1984, pp. 61–67.
[81] Hofmann, W., *Archiv fur Hydrobiologic Ergebnisse der Limnologie,* Vol. 6, 1971, pp. 1–50.
[82] Swain, E. B., "The Paucity of Blue-Green Algae in Meromictic Brownie Lake: Iron Limitation or Heavy-Metal Toxicity?," Ph.D. dissertation, University of Minnesota, Mineapolis, 1984.

Ronald B. Davis,[1] Dennis S. Anderson,[1] Donald F. Charles,[2] and James N. Galloway[3]

Two-Hundred-Year pH History of Woods, Sagamore, and Panther Lakes in the Adirondack Mountains, New York State

REFERENCE Davis, R. B., Anderson, D. S., Charles, D. F., and Galloway, J. N., "Two-Hundred-Year pH History of Woods, Sagamore, and Panther Lakes in the Adirondack Mountains, New York State," *Aquatic Toxicology and Hazard Assessment: 10th Volume, ASTM STP 971*, W. J. Adams, G. A. Chapman, and W. G Landis, Eds., American Society for Testing and Materials, Philadelphia, 1988, pp. 89–111.

ABSTRACT: Diatom remains in sediment cores from Woods Lake (L) (pH 4.9), Sagamore L (6.0), and Panther L (7.2) in the Adirondack Mountains, N.Y. were studied to infer lake water pH for the past 200 years. Historical studies and analyses of sedimentary ^{210}Pb, Ca, Mg, K, pollen, and charcoal provided data for chronostratigraphy and for distinguishing the effects on inferred pH (IpH) of watershed disturbance versus acid deposition. Indications of watershed logging were found in post-1890 sediment at the three lakes, but the pH responded (increase 0.5–1.0 IpH unit) only in poorly buffered Woods L. Mean IpHs prior to anthropogenic disturbance and acid deposition were 5.1–5.2 (Woods), 6.7–7.0 (Sagamore), and 7.4–7.6 (Panther) compared to IpHs for surface-sediment of 4.8, 6.3, and 6.9, respectively, indicating an acidification but also that edaphically induced pH differences between the lakes have remained about the same. By 1940, the pH of Sagamore and Woods L had started to decrease. Despite watershed disturbances (which would tend to raise lake pH) at Woods L since 1950, the lake has continued to acidify. Acid deposition is the likely cause of acidification of the three lakes. At Panther L, however, the decrease in IpH since the 1960s is unlikely to reflect a decrease in the lake's summer pH due to the well-buffered water, but may indicate intensified depression of lake water pH restricted to periods of snowmelt.

KEYWORDS: lake acidification, diatoms, pH, acid rain, paleolimnology

The effects of anthropogenic acid deposition on lakes are imperfectly understood, in part because of an absence of comparative data on the chemical (including pH) and biological conditions of lakes before they were exposed to, and during the early stages of exposure to, acid deposition. The absence of these early data can be offset in part by inferences on past conditions based on indicators in the sediment [1,2], an approach we apply to three lakes in the southwestern Adirondack Mountains, N.Y.

Adjacent lakes in the same region often differ greatly in their responses to acid deposition. These responses are controlled by lake morphology and hydrology, and the natural geological/biological and anthropogenic conditions of the watershed. Differences in these factors and their

---

[1] Professor and assistant scientist, respectively, Department of Botany and Plant Pathology and Institute for Quaternary Studies, University of Maine, Orono, ME 04469.
[2] Research associate, Department of Biology, Indiana University, Bloomington, IN 47405. (Present address: U.S. EPA, Corvallis, OR 97333.)
[3] Associate professor, Department of Environmental Sciences, University of Virginia, Charlottesville, VA 22903.

effects at Woods, Sagamore, and Panther Lakes (L) have been intensively studied in the Integrated Lake–Watershed Acidification Study (ILWAS) [3,4]. Our objective has been to carry out sedimentary diatom studies at these lakes to reconstruct the water pH of the past 200 years and to combine insights from this paleoecological work with insights gained from ILWAS. In addition to diatoms, other sedimentary parameters were studied in combination with land use history to help distinguish strictly terrestrial effects on lake pH from effects of atmospheric inputs.

Diatom taxa vary widely in pH distribution; therefore, potentially they are good indicators of pH. Their taxonomically identifiable frustules are well preserved in the sediment of most small, soft-water lakes. Calibrational equations relating diatom assemblages in surface sediments to contemporary water pH have been developed for several regions including the Adirondacks [5], and these equations have been applied to down-core diatom assemblages to infer past water pH [6,7].

Study Sites

The study lakes are small, relatively oligotrophic and with soft water, and have watersheds that are 92 to 99% forested (Table 1). The bedrock and overlying deposits are granitic [3]. Although the three lakes and their watersheds receive similar acidic precipitation [8], the lakes differ in chemistry, including pH (Table 1) [9]. Predominant air-equilibrated pHs at the outlets from 1 May to 30 November (1977–1981) were 4.8–5.0 at Woods L, 5.5–6.5 with central tendency of 6.0 at Sagamore L, and 7.0–7.4 at Panther L, based for the most part on weekly or biweekly samples. Late winter and early spring (snowmelt periods) air-equilibrated pHs at the outlets were lower: 4.5–4.8 at Woods, 5.0–5.6 at Sagamore, and 4.8–6.5 at Panther L [Electric Power Research Institute (EPRI)/University of Virginia (U.Va.) data].

We characterize the pH of the lakes (Table 1) by May through November averages, rather than by annual averages as in Ref. 9 for two reasons: (1) May through November largely spans the period when the lakes are free of ice, and this is the major period of diatom productivity [10]; and (2) the diatom/pH calibrations used in this paper are based on Charles [5], who used pHs

TABLE 1—*Selected characteristics of the study lakes. Values of chemical parameters are means for the lake outlet. Alkalinity and Al are based on data from 1 Jan. 1978 to 31 Dec. 1980 [9], and organic carbon from April to Nov. 1980 and Feb. to July 1981 [63]. pH is the predominant range of air-equilibrated readings for the 1 May to 30 Nov. periods from 1 Nov. 1977 to 30 June 1981; specific conductance is the mean for the same period [EPRI/U.Va. data]. Physical features are from Ref. 64, except for lake flushing time from Ref. 65. Forest cover is from Ref. 4.*

	Woods	Sagamore	Panther
pH	4.8–5.0	5.5–6.0[a]–6.5	7.0–7.4
Alkalinity, µeq/L	−10	31	147
Dissolved organic carbon (DOC), mg/L	2	7	4
Conductance, µmhos/cm at 25°C	22	30	35
Al, monomeric; µeq/L	13	1.5	0.5
Lake surface area, km^2	0.26	0.66	0.18
Lake mean depth, m	4.0	8.8	3.5
Lake max depth, m	12.0	23.0	7.0
Lake flushing time, days	221	50	208
Watershed area, km^2	2.1	48.9	1.2
Forest cover, % of watershed	96	92	99
Bedrock outcrop and soils <6 cm, % of watershed	80	40	40
Groundwater, estimated mean residence time	8–12 months	...	8–12 years

[a] Central tendency.

from the ice-free seasons at 37 other Adirondack lakes. We and Charles [5] use air-equilibrated pHs because of the closer statistical relationships (than for unequilibrated pHs) with diatom distributions in the Adirondacks [5].

Methods

Panther and Woods L sediment was cored at water depths of 6 and 12 m, respectively, in February 1980 using a 10-cm-diameter modified Davis-Doyle [11] stationary-piston corer. Cores were sectioned in the field. The top few most-watery increments were removed by aspiration with a wide-mouth pipet. Deeper increments were sectioned by upward extrusion and spatula. Sagamore L was cored at a 13-m depth in March 1978 by Woods Hole Oceanographic Institution personnel using a 21-cm-diameter gravity corer with sphincter core retainer [22].

Potassium, Ca, and Mg concentrations were determined by dissolving dried sediment in a mixture of HNO_3, $HClO_4$, HF, and HCl [13]. After dilution, samples were analyzed by flame atomic absorption on an Instrumental Laboratories 751 Spectrophotometer. The matrices of standards and samples were matched by making up standards in each extracting solution, acidified to the same degree as the sample solutions. The determinations had a coefficient of variation of ±10%.

Dating of sediment was accomplished by a combination of ^{210}Pb analysis and chronostratigraphic markers (pollen, charcoal, geochemistry). ^{210}Pb activities were determined by J. D. Eakins at Atomic Energy Research Establishment at Harwell, England. Computation of dates back to ca. 1850 were based on the constant rate of supply (CRS) model [14]. Older dates to ca. 1800 were estimated by extrapolation from the oldest CRS date using the sediment accumulation rate for that part of the core computed by the constant initial concentration (CIC) model [15].

Samples for pollen and charcoal analyses were processed by standard methods [16]. Three hundred to 400 pollen grains were identified per sample. Pollen percentages of *Ambrosia*, Gramineae, *Rumex*, *Polygonum*, *Plantago*, and certain heavily logged tree taxa, for example, *Pinus strobus*, were used as chronostratigraphic markers of historically dated forest clearance, agriculture, and logging in the region (≤50 km from lake, in Adirondacks) [17–30]. Charcoal was counted on the pollen slides [31] and related to large forest fires of known date in the region. Major events of known date in the watersheds themselves (windthrow, logging, fire) in some cases also provided chronostatigraphic information.

Samples for diatom analysis were digested in chromic acid and the rinsed residues used for slide preparation by the settling method [32]. Six hundred diatom valves were identified for each sample [7]. Duplicate slides are in the diatom herbarium at the Philadelphia Academy of Natural Sciences, and complete diatom data are in Ref. 33. Three regression equations relating sedimentary diatom assemblages to lake water pH were used (Table 2): (1) multiple linear regression of Hustedt pH groups [5]; (2) multiple linear regression of the first principal component (1PC) of 22 selected taxa [7] (Table 3); and (3) stepwise, multiple linear regression of taxon clusters [5]. The equations represent three widely different approaches to expressing pH-related variance in the diatom data. Additional methodological details are given in Ref. 33.

Results and Interpretations

Woods Lake

Pollen and Charcoal Stratigraphy (Fig. 1)—We interpret the small increases from 21 to 16 cm in Gramineae, *Ambrosia*, and *Rumex* pollen to indicate initial forest clearance in the region from 1794–1832 [22–26] and estimate that 18.5 cm dates to the midpoint of this period: 1813. The substantial increase in these pollen types from 10.5 to 8 cm probably is due to the intensification of regional settlement from 1880 to 1900 [29] and therefore we estimate that 9.25 cm dates 1890.

TABLE 2—*Regressions used for pH inference, and regression statistics, based on calibrational data set from 37 Adirondack lakes [5]. Equations (1) and (3) are from Table 5 in Ref. 5. p = probability that H_o is true (F-test). AcB = acidobiontic, Ac = acidophilous, Ind = "indifferent" (circumneutral), Al = alkaliphilous.*

(1) Four Hustedt pH groups, multiple regression

pH = 8.14 − 0.041 AcB − 0.034 Ac − 0.0098 Ind − 0.0034 Al
r^2 = 0.94; S_e ± 0.28 pH unit; p = 0.001

(2) First principal component of 22 taxas, multiple regression

See Table 3 for taxa and their coefficients
Y intercept = pH 5.995
r^2 = 0.85; S_e ± 0.43 pH unit; p = 0.001

(3) Six taxon clusters, stepwise multiple regression

pH = 5.30 − 0.033 Cluster B + 0.0049 Cluster C + 0.053 Cluster E + 0.019 Cluster F
r^2 = 0.89; S_e ± 0.38 pH unit; p = 0.001

In 1908 an intense fire burned about 5 km² of forest northeast of the lake, including the northeast half of the watershed [20; W. R. Marleau, personal communication]. The sharp, up-core increase in charcoal between 8.5 and 6.0 cm (Fig. 1) may be from this fire, and so we estimate that 7.25 cm dates 1908. The continued increase in charcoal above 6 cm (see also charcoal stratigraphy for Sagamore and Panther L, Figs. 5 and 9) could be a function of continued erosion of the charcoal that had been produced in 1908, especially during the logging from 1914 to 1927 and the blowdown and logging in 1950 and 1951 [W. R. Marleau, personal communication]. The charcoal peak at 2.25 cm suggests intense erosion, perhaps due to the mechanized logging in the 1960s and 1970s [W. R. Marleau, personal communication; C. Schofield, personal communication]. Linear

TABLE 3—*The 22 taxa from the Charles [5] data set used for the multiple regression of the first principal component (of the taxa) on pH (Table 2). For each taxon, the regression coefficient is given. E = euplanktonic.*

Cyclotella comta and *C. stelligtera*   0.004   (E)
All *Melosira distans* (sensu Hustedt, 1927–1966)   −0.003
Melosira italica subsp. *subarctica*   0.018   (E)
Asterionella formosa   0.012   (E)
Fragilaria construens (all varieties) and *F. pinnata*   0.009
Fragilaria virescens   0.045
Fragilaria virescens v. 1 (=*F. acidobiontica* sensu [38])   −0.010
All *Synedra*   0.032
Tabellaria flocculosa v. *flocculosa* (excludes strain 3p)   −0.007
Tabellaria flocculosa v. *flocculosa* strain 3p (sensu Koppen)   0.003   (E)
Achnanthes marginulata   −0.025
Other *Achnanthes*   0.045
Anomoeoneis serians v. *brachysira*   −0.015
Anomoeoneis vitrea   0.154
All *Eunotia*   −0.011
Frustulia rhomboides (all varieties)   −0.036
Navicula subtilissima and *N. mediocris*   −0.035
Other *Navicula*   0.048
All *Nitzschia*   0.054
All *Pinnularia*   −0.011
Stauroneis gracillima   −0.048
Surirella delicatissima   −0.042

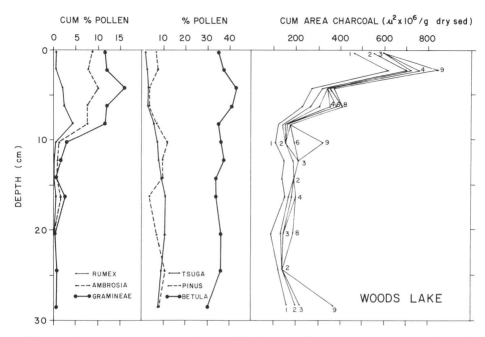

FIG. 1—*Pollen and charcoal stratigraphy in the Woods L core. The charcoal areas are given in particle area (μ^2) classes [31], viz., 1 = 72–576, 2 = 576–1296, 3 = 1296–2016, 4 = 2016–2880, 5 = 2880–3600, 6 = 3600–4320, 7 = 4320–5040, 8 = 5040–6480, and 9 = greater than 6480. Cum = each class is additive on the prior one.*

interpolation between the 1890 pollen date at 9.25 cm and 1980 at the surface results in a date of 1909 for 7.25 cm, in good agreement with the 1908 charcoal date.

The opposing trends of *Pinus* and *Betula* pollen above 10 cm (the up-core increase in *Betula* is slightly delayed) could reflect the logging of *P. strobus* that increased greatly in the late 1800s and early 1900s [*18–25,26*], and the successional response of *B. lutea* in the logged tracts. Alternately, or in addition, we may be seeing a vegetational response to the 1908 fire (there were numerous other large fires in the region between 1899 and 1908) [*18–21*].

^{210}Pb Stratigraphy and Dating, and Related Geochemical Stratigraphy (Fig. 2)—Extrapolated ^{210}Pb dates and pollen chronostratigraphy indicate that 20 cm dates about 1800. ^{210}Pb CRS dates and pollen and charcoal chronostratigraphies indicate that 10 cm dates 1880–1900. Sediment profiles of K and Mg are nearly constant below 9 cm; above this they change. Assuming that these changes are due to erosion caused by the first logging of the watershed from 1896 to 1900 [*18*; W. R. Marleau, personal communication], we assign a "geochemical date" of 1898 to 9 cm.

The departure of the ^{210}Pb profile from exponentiality above 4.5 cm indicates that dilution by old soils or sediments eroded to the coring site, an increased rate of sediment accumulation, bioturbation (0 to 4.5 cm depth of partial mixing), or some combination of these processes has been occurring. The near-vertical part of the ^{210}Pb profile from 2.5 to 4.5 cm correlates with renewed logging and postulated erosion in the 1960s, according to the ^{210}Pb CRS model. The few chironomid tubes and red chironomid larvae observed in the top few cm of the core provide indisputable evidence of bioturbation (partial mixing), which reduces temporal resolution of our interpretations.

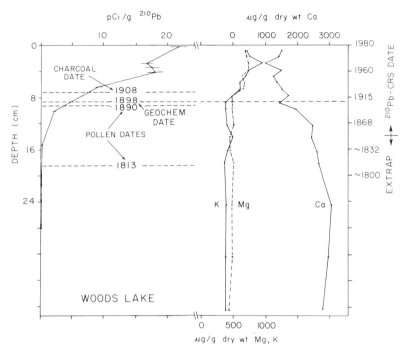

FIG. 2—^{210}Pb *stratigraphy and CRS dates, extrapolated (extrap.)* ^{210}Pb *dates as explained in text, chronostratigraphic markers, and K, Mg, and Ca in the Woods L core. Error bars on* ^{210}Pb *activities are standard count errors based on binomial distributions. See text for explanation of chronostratigraphic markers (charcoal, geochemistry, and pollen).*

Other Geochemical Changes—The concentration of Ca in the sediment appears to have been decreasing since the early 1800s while K and Mg have remained essentially unchanged. The Ca decrease becomes more rapid at ca. 1875, but when K and Mg start increasing around 1900 (ca. 9 cm; beginnings of logging) the decrease returns to its former, slower rate (Fig. 2). Calcium accumulation rates computed on the basis of ^{210}Pb dates parallel the decreases in concentration to the early 1900s and then the rate becomes essentially constant. Decreasing Ca accumulation rates since the mid- to late-1800s have also been seen in sediment cores from several acidic lakes in northern New England, and it has been proposed that this is a result of increased leaching of Ca (the same is observed for Mn) from sediment precursors (soils) in the terrestrial watershed, from sediment in the lake shallows, *in situ* at the site of coring, or at more than one of these locations by anthropogenically acidified waters [2,34–36]. Much of the Ca in lake sediment resides in organic matter, but as in Woods L (EPRI/U.Va. data) the decreases of Ca are not usually paralleled by decreases in organics. The 1800s beginnings of decreases in Ca (and Mn) precede the major increase in sulfur emission in the Northeast beginning around 1900 [37]. *In situ* leaching of sediment already deposited would render the chronology of the beginning of Ca decrease erroneously old.

Stratigraphy of Diatom Taxa (Fig. 3)—Below 12.5 cm (ca. 1863) there is little change, but between that depth and 10 cm (ca. 1890) the euplanktonic, "pH indifferent" (really circumneutral) *Cyclotella stelligera* and the euplanktonic, alkaliphilous *Melosira italica* subsp. *subarctica* begin to increase, while certain benthic, acidophilous taxa decrease, viz., *Melosira* spp., *Anomoeoneis*

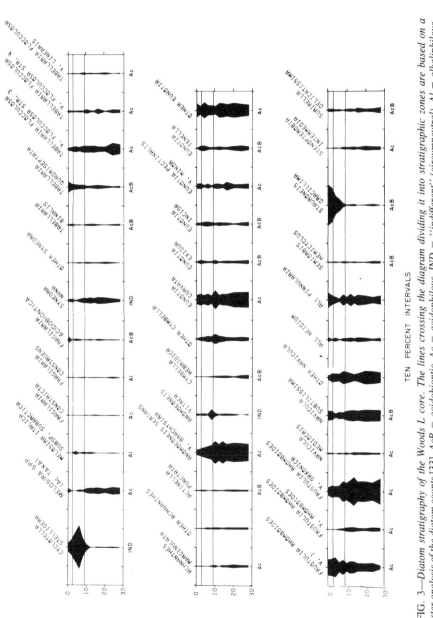

FIG. 3—Diatom stratigraphy of the Woods L core. The lines crossing the diagram dividing it into stratigraphic zones are based on a cluster analysis of the diatom counts [33]. AcB = acidobiontic, Ac = acidophilous, IND = "indifferent" (circumneutral), Al = alkaliphilous. "Melosira spp." (Ac taxa) below 12.5 cm consist largely, in order of abundance, of M. perglabra (6.15), M. distans v. nivalis (5.38), and M. distans v. nivaloides (5.92). In parentheses we give abundance-weighted mean pHs of these taxa in the Adirondack calibration data set. At the sediment surface, only the most "acidic" taxon, M. distans v. nivalis remains.

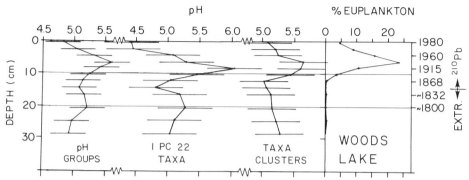

FIG. 4—*pH reconstructions with S_e bars based on three regressions, and profile of total euplanktonic diatoms for the Woods L core.*

serians v. *brachysira, Frustulia rhomboides* (except v. *saxonica*), and certain *Eunotia* species. The overall stratigraphy of pH groups [*33*] indicates that acidobionts and acidophils reach minimal percentages at 8.25 and 6.25 cm, respectively. Indifferents peak at 6.25 cm. The least acidic period, then, is between 9.25 and 5.25 cm, covering the time from the first logging in the watershed from 1896–1900 to ca. 1946, and including the fire in 1908 and logging episodes that followed and continued to 1927 in unburned parts of the watershed. Given a small amount of sediment mixing, the period of implied higher pH correlates well with this series of disturbances. The trend of increasing, inferred pH (IpH) reverses above 6 cm (ca. 1940): the benthic, acidobiontic taxa *Fragilaria acidobiontica* [*38*], *Tabellaria quadriseptata,* and especially *Stauroneis gracillima* (= *Navicula tenuicephala*) increase, while *C. stelligera* and the benthic, acidophilous *T. flocculosa* v. *flocculosa* str. III decrease. By 4 cm (ca. 1960) the total percentage of acidobionts becomes greater than at any prior depth, and by the time the surface is reached the indifferents have returned to low percentages similar to those deep in the core [*33*]. The two samples above 2.5 cm are in a separate cluster (Fig. 3) suggesting, together with the aforementioned results, that the most recent two decades have been a uniquely acidic period of lake history.

pH Reconstruction (Fig. 4)—Below 12.5 cm (pre-1860, prior to heavy regional air pollution) there are only minor trends in IpH, and the mean IpHs are 5.1 (pH groups), 5.1 (1PC), and 5.2 (clusters). Above 12 cm, starting about 1875, the three reconstructions indicate increasing IpH; the total increase above pre-1860 means are 0.5, 1.0, and 0.5 unit, respectively. This is followed by a decrease in IpH starting about 1930 for the pH group and cluster versions, and starting about a decade earlier for the 1PC version. Inferred pHs for the surface are 4.8, 5.1, and 4.4, respectively.

The regression standard errors (S_e) for prediction of pH (Y) of individual samples are 0.28 to 0.43 pH unit (Table 2).[4] Even for clear trends in IpH, as in the top four samples (above 6.5 cm) of the reconstructions, there is much overlap of the S_e bars between adjacent samples. When these four IpHs are viewed as a linear regression versus depth, F-tests (H_o implies no significant trend) indicate significant trends of decreasing IpH in the pH group ($p = 0.003$) and 1PC ($p = 0.009$) reconstructions, but only marginal significance for the cluster version ($p = 0.090$). (Similar results are obtained for the increasing IpH from 12.5 to 8 cm.) One-tail t-tests (unequal variance of

---

[4] There is reason to believe [*7*] that S_e is misleadingly large for assessing changes in temporal sequences of samples at the same lake because it incorporates sources of variance (from the calibrational set of lakes) that are constant within the time period covered at the individual lake.

groups) to determine if recent IpH (top two samples; ≳1970) is the same (H_o) or less than early IpH (bottom three samples; ≲1800) indicate rejection of H_o at $p = 0.004$ for 1PC version but not for the pH group ($p = 0.267$) and cluster ($p = 0.218$) versions. In summary, evidence is sufficient for stating that the pH of Woods L has been decreasing since 1920–1930, but the contention that the recent pH is lower than the pH prior to anthropogenic acid deposition (AAD) is tenuous except for the 1PC reconstruction.

The sedimentary profile of euplanktonic diatoms (Fig. 4) suggests an increase in pH starting just above 12 cm at ca. 1875, peaking at 6.25 cm (ca. 1930), and not quite returning to pre-1860 values at the surface. Euplanktonic diatoms are rare in surface sediments of Adirondack lakes with water pH below 6.0 [5] and of northern New England lakes below pH 5.7 [7]. This supports the trends in IpH and strongly suggests that during the first few decades of this century the pHs in Woods L rose to 5.5–6.0 or slightly higher, at least seasonally when diatom production was taking place.

The 4% euplanktonic diatoms in the surface sediment seems high considering: (1) the fact that the pH at the lake in 1977–1981 rarely rose as high as 5.5 (EPRI/U.Va. data); and (2) the scarcity of euplanktonic diatoms in the plankton in 1979 [39]. This suggests an enrichment of the surface by older sediments due to bioturbation, physical mixing of the sediment, or delayed focusing and deposition, or some combination of these processes, as argued earlier on other grounds. It is reasonable to assume that mixing has been occurring during the entire period represented by the core, and that each sample represents some unknown span of years greater than indicated by the net rate of sediment accumulation. This also means that any trends (for example, indications of decreasing pH) approaching the sediment surface within the mixing zone or within the range of depth spanning the period of delayed deposition, or both, are diminished [40]. Thus, at Woods L IpHs for most recent times would not indicate as much acidification as has actually occurred. The contention that the pH of Woods L is now lower than its pre-AAD pH may be somewhat less tenuous than the t-tests indicate. Sediment mixing and/or delayed deposition may also explain why diatom IpH values for surface sediment in lakes with low measured pH (lakes thought to have been acidified recently) are often higher than the measured pH [5,7].

While the Woods L core is not ideal for sharp chronological resolution, three major periods of differing IpH can be distinguished: (1) pre-1860, IpH 5.1–5.2; (2) 1860–1960, higher IpH, peaking at 5.6–6.1 between 1900 and 1930; and (3) post-1960, IpH 4.4–5.2 depending on the method of inference.

Sagamore Lake

Pollen and Charcoal Stratigraphy (Fig. 5)—The consistent occurrence of pollen associated with forest clearance and agriculture begins between 47 and 43 cm. This probably corresponds to the period of regional settlement from 1795 to 1820 with the most significant settlement centered about 1815 [22,28], the date we assign to the 45-cm level. The increase between 30 and 25 cm of pollen indicators of landscape disturbance likely is due to intensified regional settlement 1880–1900 [24–28], and, accordingly, we assign the date 1890 to 27.5 cm. The increase in charcoal between 24 and 20 cm may be due to extensive fires in the region [18–21] and increased use of fuel wood at the lake around the turn of the century [30].

^{210}Pb Stratigraphy and Dating, and Related Geochemical Stratigraphy (Fig. 6)—The pollen date of 1815 for 45 cm is only seven years older than the date obtained for that level by ^{210}Pb date extrapolation. The pollen date of 1890 for 27.5 cm is only one to four years older than the ^{210}Pb date for that level. The changes in K, Mg, and Ca at 8–10 cm, and the peaks in these constituents at 3.5 cm, could be due to erosion that resulted, respectively, from renewed human activities in the watershed in the 1950s (see below) and the first logging in 1975–1976 [30; H.

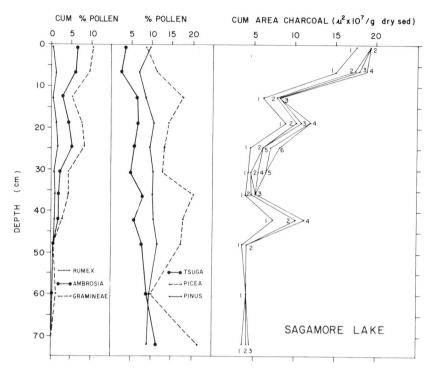

FIG. 5—*Pollen and charcoal stratigraphy in the Sagamore L core. See Fig. 1 caption.*

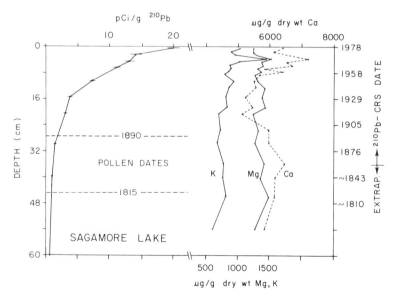

FIG. 6—*^{210}Pb stratigraphy and CRS dates, extrapolated (extrap.) ^{210}Pb dates as explained in text, chronostratigraphic markers, and K, Mg, and Ca in the Sagamore L core. See Fig. 2 caption.*

Kirschenbaum, personal communication]. The departure from exponentiality in the ^{210}Pb curve around 4 cm is consistent with the postulated erosion in the 1970s. Erosion would have increased the flux of old materials entering the lake. No herpobenthos were observed in the core.

Stratigraphy of Diatom Taxa (Fig. 7)—Starting at the bottom of the core, there is steady change in diatom assemblages driven by the increase in *Tabellaria flocculosa* v. *flocculosa* (the euplanktonic strain IIIp *sensu* Koppen) from 8.8% at 61–59 cm (roughly mid-1700s) to 35.4% at 30–29 cm (ca. 1880–1890). Above 29 cm, *T. flocculosa* v. *flocculosa* str. IIIp starts decreasing, to 3.8% at the surface. Starting 35 to 30 cm (ca. 1860–1880) there is a series of minor but correlated and therefore probably significant increases in additional euplanktonic forms, viz., *Cyclotella comta, C. stelligera*, and *Asterionellaz ralfsii* v. *americana* (long form). These three peak at 25–24 cm (ca. 1900–1905) with a total of 13.3% and then reverse, reaching less than 5% by 17–16 cm (ca. 1925–1930) and maintaining less than 5% to 0 cm. Maximum percentages of all euplanktonic diatom taxa combined occur 32.5–22.5 cm or ca. 1875–1910 (Fig. 8).

Several small changes consistently indicating decreasing pH begin above 20 cm (after ca. 1920). Along with the decreasing plankters are increases in the acidophilous *Achnanthes marginulata, Eunotia curvata*, and other *E*. spp., and the acidobiontic *E. exigua*. Above 12 cm (ca. 1943), there are decreases in the "indifferent" *Synedra rumpens* and the alkaliphilous *Fragilaria pinnata*, and an increase in the acidophilous *Tabellaria flocculosa* v. *flocculosa* str. IV.

History of human uses of the watershed was studied [28,30] to try to explain the diatom sequence. A hunting cabin existed at the lake ca. 1877, but in 1890 the watershed still had forests essentially undisturbed by humans and with no sign of recent forest fire. In the 1890s several buildings were erected near the lake shore as part of a summer estate and forest preserve that surrounded the lake by 2–5 km. Sporadic building continued at the same site at least through the 1920s, including in 1901 a prototype flowing water sewage system whose outfall probably went into the lake. Since 1954, the facilities have been used as an educational center and now include about 20 buildings, in all accommodating ca. 100 persons. Although these activities (and any associated nutrient inputs to the lake) have been at the southwest end and near the outlet of the lake, winds during the season of minimal flushing could have distributed nutrients sufficiently to account for some of the planktonic productivity revealed by the core. However, this history is not consistent with the 1700s beginnings of the increase of planktonic diatoms and the reversal since the early 1900s.

In 1979 diatoms were only sporadically present in plankton samples [39]. For a short period in the spring and one in early winter, diatoms constituted up to 12% of the "plankton" biomass, but, apart from *Asterionella*, the taxa were not euplanktonic [39]. As at Woods L the surface sediment appears to contain an admixture of old diatom frustules.

pH Reconstruction (Fig. 8)—There is a slight increase in IpH from the 1700s to the mid-1800s followed (in the 1PC and cluster versions) by a decrease with minima coinciding with the period of maximum percentages of plankters centered around 1900. Just above 20 cm (ca. 1920) in the 1PC and cluster versions, an IpH decrease starts and continues to 0 cm.[5] In the pH group version, an IpH decrease starts above 12.5 cm (ca. 1945). F-tests for the regression of IpH versus depth for the top four samples (above 12.5 cm) in each reconstruction indicate significnt ($p = 0.028$–0.036 that H_o true) trends of decreasing IpH. The mean IpHs below 20 cm are: pH groups 6.8, 1PC 7.2 and clusters 6.9. The respective IpHs for the surface are lower: 6.4, 6.6, and 6.0. One-tail t-tests (unequal variance) to determine if recent IpH (top two samples; ≳1970) is the same (H_o) or less than early IpH (bottom three samples; ≲1830) indicate rejection of H_o at $p = 0.043$

---

[5] The colorimetrically measured pH of 7.0 for surface water in 1933 [41] is close to the IpHs for the period.

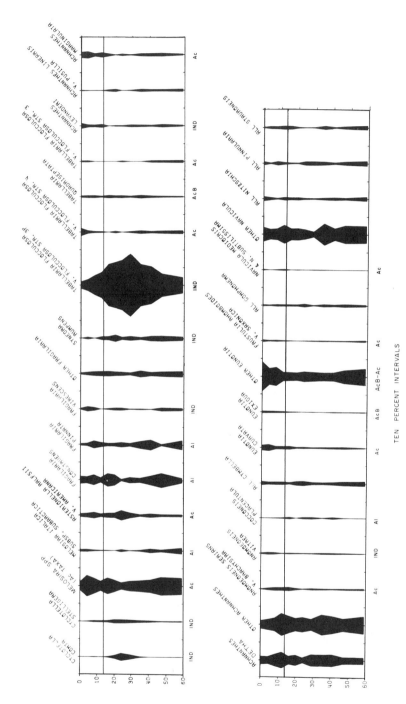

FIG. 7—Diatom stratigraphy of the Sagamore L core. See Fig. 3 caption except for Melosira specific to Woods L. In Sagamore L, "Melosira spp. (Ac taxa)" consist largely of M. distans v. nivaloides, M.d. v. tenella, M. lirata, and M. perglabra v. floriniae.

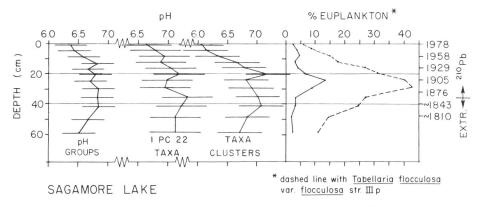

FIG. 8—*pH reconstructions with S_e bars based on three regressions, and profiles of total euplanktonic diatoms for the Sagamore L core.*

(pH group), 0.037 (1PC), and 0.057 (cluster). Overall, the results indicate that Sagamore L has been acidified in recent decades.

Panther Lake

Pollen and Charcoal Stratigraphy (Fig. 9)—Nonarboreal pollen types associated with settlement, forest clearance, and agriculture appear at 48–46 cm. We estimate that 47 cm dates ca. 1800 when such activities were beginning in the region [*21–24*]. The decreases in pollen of *Pinus*,

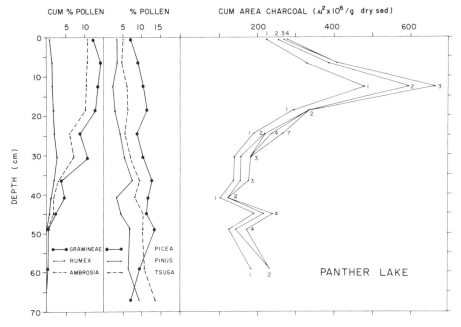

FIG. 9—*Pollen and charcoal stratigraphy in the Panther L core. See Fig. 1 caption.*

Picea, and *Tsuga* starting 36–31 cm, and the substantial increase in nonarboreal pollen types at this depth, probably result from intensification of logging and settlement in the southwestern Adirondacks in the final third of the 19th century [21–28]; the depth and date midpoints are 33.5 cm and 1883. Charcoal begins a marked increase at 30–25 cm, probably due to the high frequency and large size of forest fires around the turn of the century [18–21]. The pronounced charcoal peak at 12–12.5 cm is unlikely to be due to a forest fire in the watershed or region. According to the ^{210}Pb-CRS chronology, this depth dates to 1950. No forest fires occurred in the watershed during this century; the few forest fires that occurred in the region ($\gtrsim 50$ km from lake) since 1930 were all small [New York State Department of Environmental Conservation (P. Hartman, personal communication) and Adirondack League Club records (M. Hanna, personal communication)]. We suggest, therefore, that the charcoal peak is due to a pulse in erosion of old soils containing charcoal from earlier fires. This is correlated with a blowdown in 1950 and salvage logging of affected spruces near the lake [D. Webster, personal communication].

^{210}Pb Stratigraphy and Dating, and Related Geochemical Stratigraphy (Fig. 10)—The pollen date of 1883 at 33.5 cm agrees closely with the ^{210}Pb date, but the pollen date at 47 cm is 24 years older than the extrapolated ^{210}Pb date of 1824. The geochemical profiles are fairly smooth in the lower four fifths of the core except for a single notch at 18.25 cm. The watershed of Panther Lake remained unlogged until the 1890s when large pines and spruces were removed. Then essentially no disturbance took place until 1950–1951 [D. Webster, personal communication]. No trace of the 1890s logging is seen in the core, but the 1950 event may be indicated by the departure of the ^{210}Pb profile from exponentiality at 11–6 cm when old materials would have been eroded from the watershed. Bioturbation is unlikely to be involved. No herpobenthos were seen in the core. The jet blackness (FeS) of the upper sediment in February 1980 indicated that bottom waters were anoxic, which would exclude herpobenthos.

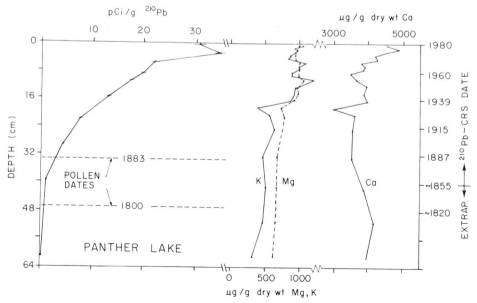

FIG. 10—*^{210}Pb stratigraphy and CRS dates, extrapolated (extrap.) ^{210}Pb dates as explained in text, chronostratigraphic markers, and K, Mg, and Ca in the Panther L core. See Fig. 2 caption.*

Stratigraphy of Diatom Taxa (Fig. 11)—Below 40 cm (pre–mid-1800s), there is little change that is interpretable in terms of pH despite the overall increase in planktonic forms from 27 to 40% (Fig. 12). Above 40 cm, planktonic *Cyclotella stelligera* continues to increase and peaks at 18–18.5 cm (ca. 1930), as does the total of euplanktonic taxa (57%). Above 16 cm (since ca. 1940), planktonic forms and the benthic alkaliphil *Fragilaria construens* nom. var. decrease. The most marked change occurs between 6 and 4.5 cm where *Melosira* spp. (especially the acidophilous *M. distans* v. *nivaloides*) and *Nitzschia* spp. (mostly *N. fonticola*) increase. The "indifferent" *Tabellaria flocculosa* v. *flocculosa* str. IIIp, *C. stelligera*, and *C. comta* decrease. Overall, the shifts above 6 cm are toward a more acidophilous flora. These shifts and the rapid decrease in planktonic forms indicate that an acidification started at ca. 1965–70. The cluster of three counts above 6 cm (Fig. 11) is most different from the other clusters, and its counts are individually most unlike each other [*33*], suggesting that conditions in the lake since 1965–70 have been changing rapidly. No disturbance has occurred in the watershed during this period [D. Webster, personal communication].

pH Reconstruction (Fig. 12)—Except for the top two levels of the core, there is no sustained change in IpH that is consistent between the three reconstructions. The mean IpHs below the top two levels are 7.1, 7.8, and 7.7 for the pH group, 1PC, and cluster versions, respectively. The IpHs for the surface sediment are 0.6, 0.3, and 0.7 unit lower. F-tests for the regression of IpH versus depth for the top three samples (above 4.5 cm) indicate a significant ($p = 0.034$ that H_o true) trend of decreasing IpH for the 1PC version but questionable or no trend for the pH group ($p = 0.097$) and cluster ($p = 0.144$) versions. One-tail t-tests (unequal variance) to determine if recent IpH (top two samples; ≳1970) is the same (H_o) or less than early IpH (bottom three samples; ≲1820) give mixed results. H_o is rejected at $p = 0.002$ and 0.021 for the pH group and cluster versions, respectively, but not for 1PC whose mean IpH of recent samples is greater than early samples.

The pH group and cluster diatom IpHs of 6.5 and 6.9 for the sediment surface are lower than the predominant range of variation of measured water pH in the ice-free season (Table 1). The *annual* range of measured pH of surface waters in Panther L is great: 4.8 to 7.4 (even slightly lower or higher on rare occasions), and the low part of this range results from acidic snowmelt (EPRI/U.Va. data). Although most diatom productivity takes place in the ice-free seasons (the period on which our diatom/pH calibrations are based) and outside the period of snowmelt, some planktonic diatom taxa bloom late in the period of ice cover and continue into spring overturn [*10,42*]. Anthropogenic depression of the pH of spring meltwaters [*43*] may account for the observed decreases in IpH for Panther L. Although decreases in IpH don't begin until ca. 1970, influences of acidified snowmelt could have begun by 1940 as suggested by decreases in planktonic taxa. Apart from the spring flush of meltwater, this lake is relatively well buffered. Its summer pH is unlikely to have been significantly affected by acid deposition [*9*].

Summary pH Reconstructions, and Concluding Interpretations

The summary pH reconstruction for each lake (Fig. 13) is a weighted mean of the IpHs by the three different equations. The weighting was in inverse proportion to the S_e (Table 2) so that the pHs by the equation with the smallest S_e contributed the most to the mean. The weighting factors were also utilized in the computation of the pooled S_e for the means.

The lakes appear to have maintained their pH differences back at least to the mid-1700s (Fig. 13). In the period around 1800 and before, prior to any significant industrialization upwind and prior to any direct effects of nonaboriginal Americans on the watersheds, the mean IpHs (mIpH) were 5.1–5.2, 6.8, and 7.4–7.6 in Woods, Sagamore, and Panther L, respectively. That is, Sagamore was ca. 1.6 units higher than Woods, and Panther was 0.7 unit higher than Sagamore.

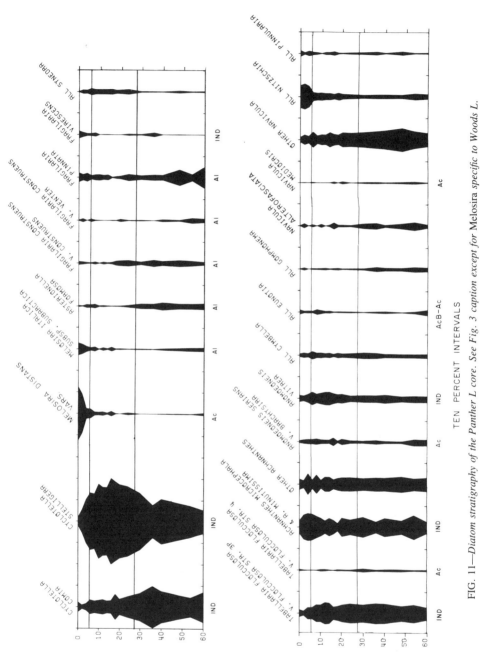

FIG. 11—Diatom stratigraphy of the Panther L core. See Fig. 3 caption except for Melosira specific to Woods L.

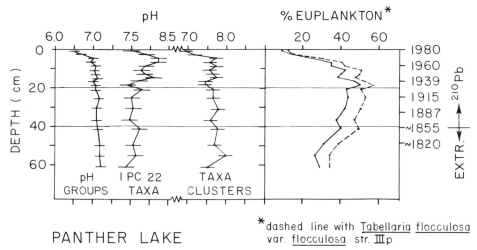

PANTHER LAKE

*dashed line with <u>Tabellaria flocculosa</u> var. flocculosa str. IIIp

FIG. 12—*pH reconstructions with S_e bars based on three regressions, and profile of total euplanktonic diatoms for the Panther L core.*

Similar differences in mIpH are indicated for ca. 1980: Sagamore is 1.6 units higher than Woods, and Panther is 0.6 unit higher than Sagamore. Despite the effects of acid deposition, other factors continue their major influences on lake pH. These paleoecological results support the ILWAS model [44].

The structure and depths of unconsolidated surficial deposits and soils, hydrologic flow paths, and soil-water contact areas and contact times in watersheds are key factors accounting for pH differences between lakes, for example, the extreme difference between Woods and Panther L [3,4,45,46]. The Woods L watershed has thin deposits (Table 1), minimal contact with minerals for water flowing to the lake, and therefore minimal generation of alkalinity in that water. Panther L watershed has deep deposits (Table 1) and long flow paths (except in rapid snowmelt episodes) and substantial generation of alkalinity in the ground water before it enters the lake [3,4,45,46].

Our mIpHs for Woods L for the period before anthropogenic acid deposition (AAD) are in agreement with previous workers [9] who, by assuming an 80% reduction in $SO_4^=$ deposition and applying principles of watershed mass balance and electroneutrality of solutions, inferred that the upper limit of lake pH before the onset of AAD would have been 5.8.[6] However, our mIpH of 7.5 for Panther L in the pre-AAD period differs from the estimate of 6.4 in Ref. 9. The estimate of pH 6.4 was derived by imposing the $SO_4^=$ reduction on present *year-round* lake chemistry, that is, on the annual mean non–air-equilibrated pH 6.2 [9]. Had the authors [9] used the present air-equilibrated pH from only the ice-free period as the starting point for their calculation, as we did for diatom/pH calibrations and inferences, then their pre-AAD pH would have been much closer to our diatom-inferred value.

The cores from the three lakes have major increases of planktonic diatoms reaching maxima in the late 1800s (Sagamore) or early 1900s. At Woods L, the increase didn't begin until the late-1800s, but at Sagamore and Panther L it began in the late-1700s (or possibly earlier, prior to coverage by our cores). We have no definite explanation for these early increases. While there are no *known* correlated changes in the watersheds, certain important changes could have taken place,

---

[6] The ILWAS model [44] predicts that a halving of atmospheric sulfur (S) inputs would result in May–Nov. pHs of 5.5–6.2 in Woods Lake.

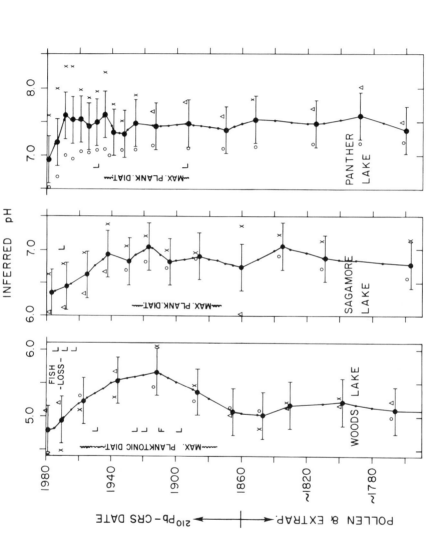

FIG. 13—Summary pH reconstructions for the three lakes. The curves are based on weighted $(1/S_e)$ means of lpHs by three regressions. The small black dots represent the points, mostly at 5 or 10-year intervals along the original curves (Figs. 4, 8, 12), where the means were computed. The large black circles are where diatom counts were made. The error bars are the pooled S_e, incorporating the $1/S_e$ weighting in the computation. The extreme lpH values for each diatom count are given: open circle = pH group, triangle = cluster, X = 1PC, L = logging in watershed, F = forest fire in watershed. Fishery information is from C. Schofield (personal communication).

including: (1) natural disturbance, succession of vegetation, and effects on soils and runoff; and (2) beaver activity and effects on hydrology and chemical outputs from wetlands (Sagamore L watershed is 9% wetland [4]). In the lake, food chains and grazing on diatom plankton could have changed. Possible causes originating outside the watersheds include: (1) change in climate affecting hydrology, water temperature, and chemistry (for example, end of the Little Ice Age [47]; and (2) change in inputs of dust from upwind areas.

The increases of diatom plankton at Sagamore and Panther L parallel the great expansion of agriculture between Albany and Rochester [48]. At Sagamore L the decrease in diatom plankton parallels the major decline of this agriculture and the stabilization of the fields by native vegetation. The type of agriculture practiced in the region would have released large amounts of dust [48; D. Smith, personal communication], increasing fluxes of P, Si, and other nutrients in downwind fallout areas [49] such as the nutrient-poor Adirondacks. This hypothesis is weakened by the absence of chronologically similar responses at Woods L and other lakes studied paleolimnologically in the Adirondacks [2; D. Charles, unpublished data]. However, lake/watershed systems undoubtedly differ in their responsiveness to changing inputs of nutrients from dust. By further study of the sediment cores from Sagamore and Panther L for minerals foreign to the watersheds and traceable to distant upwind areas [50], correlations with land uses may be determined more precisely.

At Woods L the period of highest IpH is centered on a series of disturbances in the watershed, including the first logging, an intense forest fire, and additional logging in the unburned area (Fig. 13). In the same period, major changes in beaver populations were taking place. There are several ways in which these disturbances could have led to higher pH of the lake water: (1) the fire would have increased base cation and micronutrient concentrations in runoff [51]; (2) the logging would have increased dissolved nutrients and inorganic and organic particulates in runoff [52]; (3) the early successional growth following these disturbances could have taken up NO_3^- in preference to NH_4^+ [53], thereby generating alkalinity in soils and runoff; and (4) increased beaver activity would have led to expanded areas of standing water and saturated soil, possibly increasing denitrification and $SO_4^=$ reduction and generating alkalinity in substrates and runoff. Beaver activity greatly increased in the southwestern Adirondacks in 1910–1930 [D. Webster, personal communication], including at the "beaver meadow" [54] around the major inlet to Woods L. Lake water pH would have increased as a result of intensified algal photosynthesis resulting from (a) enhancement of micronutrients in inflows, and (b) increase in particulates from the watershed. These particulates would have yielded micronutrients and base cations by decomposition and, in an acidic, oligotrophic lake, by exchange for protons [55]. Positive responses of lake water pH to historic disturbances in the watershed have also been inferred by the diatom method for certain New England lakes with water chemistries and surrounding terrain similar to Woods' [56].

Woods is the only one of the three study lakes with a positive response of IpH correlated with watershed disturbance. Woods is the most oligotrophic and poorly buffered of the three lakes, so small changes in chemical fluxes can most easily affect its character. The 1890s to 1920s appears to have been the only lengthy period since our record began when the lake had positive (slight) bicarbonate alkalinities. While alkalinity and/or photosynthesis in the other two lakes also may have been elevated during periods of watershed disturbance, those lakes' buffering capacities would have dampened pH changes, explaining why we do not detect correlated increases in IpH there.

More difficult to explain for Woods L is the decrease in inferred pH for the top ca. 6 cm of the core which covers the period of 1950 blowdown and the logging that immediately followed and recurred intermittently into the 1970s. We suggest that opposing forcing functions have been in operation in this period: first, the likely positive influence on lake pH of watershed disturbance and, second, the negative effect of acid deposition. The second appears to have exerted the dominant influence. Unfortunately, quantification of these complex influences at Woods L is impossible for the period in question.

If we knew the ecological and physiological mechanisms of diatom response to pH, our environmental interpretations could become more secure, but this is not the case [57,58]. Does water pH have a direct effect on diatoms or is the effect indirect via environmental factors that are altered by or correlated with pH? pH covaries with a number of other water chemistry factors [59].[7] As pH changes so does speciation of inorganic carbon ($CO_3^=$, HCO_3^-, CO_2), and this may alter the competitive balance among algae with differing inorganic carbon requirements [58]. Increased ionic activity and toxicity of metals at lowered pH can reduce growth rates differentially for algal taxa, and the metals may influence silica acid uptake [58]. A Si response mediated by pH is a prime candidate mechanism, but published research on this subject is sparse and inconclusive [57–61]. Recent analysis of a calibrational data set of 63 lakes in northern New England indicates a significant correlation between pH and dissolved Si (R. B. Davis and D. S. Anderson, unpublished data). A further problem is the difference in water chemistry between sites of measurements (open water) and sites of growth of benthic diatoms. Despite these unknowns and problems, empirical relationships between the remains of diatoms in midlake sediments and open water pH are strong [57]. However, until more information is available on relevant aspects of diatom ecology and physiology, pH inferences based on sedimentary diatoms will continue to be strictly correlational and empirical [57].

Inferred pH has decreased in recent decades in all three lakes, and the mIpH for 1980 is lower than at any time back at least to the late 1700s. F-tests for the regression of mIpH versus depth for the top five samples (since ca. 1920) in Woods, top four samples (since ca. 1940) in Sagamore, and top three samples (since ca. 1970) in Panther L indicate rejection of H_o at P = 0.001, 0.029, and 0.078, respectively. One-tail t-tests (unequal variance) to determine if recent mIpH (top two samples; ≳1970) is the same or less than early mIpH (bottom three samples; ≲1830) indicate marginal rejection of H_o for Woods (p = 0.053) and Panther L (0.071), and strong rejection for Sagamore L (0.008). These results, taken together with (1) the qualitatively unique changes in diatom assemblages near the top of the cores, and (2) the likelihood of some redeposition of old diatom frustules at the top, lead us to conclude that the three lakes are significantly more acidic now than in the past. We have been unable to explain the unprecedentedly low mIpHs for the top samples in terms of recent changes originating in the watersheds, and therefore propose that acid deposition is the likely cause. We do not mean to imply, however, that the chronologies of lake and atmospheric acidification have been the same. Sulfur emissions in the Northeast rose sharply from 1880 to 1920, and by 1920 were comparable to those in most recent decades [37]. It takes time for acid deposition to overcome the buffering capacities of lake/watershed systems [62].

Acknowledgments

Persons who helped in the laboratory include R. S. Anderson (pollen), A. Zlotsky (history and charcoal), and D. Banning (sediment chemistry). Frode Berge assisted in sediment coring at Woods and Panther Lakes. We appreciate the historical and land use information provided by M. Hanna, P. Hartman, H. Kirschenbaum, W. Marleau, C. Schofield, E. Skidmore. D. Webster, and D. Smith. Insightful discussion of results was provided by C. Carlson, C. Cronan, R. Goldstein, S. Norton, D. Whitehead, R. Wright, and three anonymous reviewers. The project was funded by The Electric Power Research Institute and by the Maine Agricultural Experiment Station (U.S.D.A./Hatch).

---

[7] At this stage of research, we must make the simplest assumption, viz., that the covariant chemical relationships remain constant through time, in which case pH can be inferred even if one or the other chemical factors is causing the diatom response.

References

[1] *Diatoms and Lake Acidity*, J. P. Smol, R. W. Battarbee, R. B. Davis, and J. Merilainen, Eds., Junk, Dordrecht, The Netherlands, 1986.
[2] Charles, D. F. and Norton, S. A. in *Acid Deposition: Long Term Trends*, National Research Council, Committee on Monitoring and Assessment of Trends in Acid Deposition, National Academy Press, Washington, DC, 1986, Chap. 9, pp. 335–431.
[3] April, R. H. and Newton, R. M., *Water, Air, and Soil Pollution*, Vol. 26, 1985, pp. 373–386.
[4] Cronan, C. S., *Water, Air, and Soil Pollution*, Vol. 26, 1985, pp. 355–371.
[5] Charles, D. F., *Ecology*, Vol. 66, 1985, pp. 994–1011.
[6] Battarbee, R. W., *Philosophical Transactions of the Royal Society of London*, Vol. 305, 1984, pp. 451–477.
[7] Davis, R. B. and Anderson, D. S., *Hyrobiologia*, Vol. 120, 1985, pp. 69–87.
[8] Johannes, A. H., Altwicker, E. R., and Clesceri, N. L., *Water, Air, and Soil Pollution*, Vol. 26, 1985, pp. 339–353.
[9] Galloway, J. N., Schofield, C. L., Peters, N. E., Hendrey, G. R., and Altwicker, E. R., *Canadian Journal of Fisheries Aquatic Sciences*, Vol. 40, 1983, pp. 799–806.
[10] DeNicola, D. M. in *Diatoms and Lake Acidity*, J. P. Smol et al., Eds., Junk, Dordrecht, The Netherlands, 1986, pp. 73–85.
[11] Davis, R. B. and Doyle, R. W., *Limnology and Oceanography*, Vol. 14, 1969, pp. 643–648.
[12] Burke, J. C., *Limnology and Oceanography*, Vol. 13, 1968, pp. 714–718.
[13] Pratt, P. F. in *Methods of Soil Analysis*, C. A. Black, Ed., American Society of Agronomy, Madison, WI, 1965, pp. 1019–1021.
[14] Appleby, P. G. and Oldfield, F., *Catena*, Vol. 5, 1978, pp. 1–8.
[15] Robbins, J. A. in *Biogeochemistry of Lead in the Environment*, J. O. Nriagu, Ed., Elsevier, Amsterdam, 1978, pp. 285–393.
[16] Faegri, K. and Iverson, J., *Textbook of Pollen Analysis*, Hafner Press, NY, 1975.
[17] Colvin, V., "New York (State of) Seventh Annual Report on the Progress of the Topographical Survey of the Adirondack Region of New York," Weed, Parsons and Co., Albany, 1880.
[18] "Annual Reports of the Forest Commission," New York (State of) Forest Commission, State of New York, Albany, 1890–1899.
[19] "Annual Reports of the Forest, Fish, and Game Commission," New York (State of) Forest, Fish, and Game Commission, State of New York, Albany, 1907–1909.
[20] New York (State of) Conservation Commission, "Fire Protection Map of the Adirondack Forest," State of New York, Albany, 1916.
[21] Miller, W. J., "Adirondack Mountains," New York State Museum Bulletin 193, 1917.
[22] Higgins, R. L., "Expansion in New York with Especial Reference to the 18th Century," Ohio State University Studies, Contributions in History and Political Science No. 14, Ohio State University, 1931.
[23] Larson, C. C., "Forest Economy of the Adirondack Region," State University of New York, College of Forestry Bull. 19, Syracuse, NY, 1956.
[24] Ellis, D. M., Frost, J. A., Syrett, H. C., and Carmen, H. J., *A Short History of New York State*, Cornell University Press, Ithaca, NY, 1957.
[25] Ketchledge, E. H., "Changes in the Forests of New York," The N.Y. State Conservationist, N.Y. State Conservation Dept., Albany, NY, 1965.
[26] Marleau, W. R., *Big Moose Station*, Marleau Family Press (Delta Lithograph, Van Nuys, CA), 1986.
[27] *Geography of New York State*, J. H. Thompson, Ed., Syracuse University Press, NY, 1966.
[28] Adirondack Mountain Club, "Adirondack Bibliography Supplement 1956–1965," The Adirondack Museum, Blue Mountain Lake, 1973.
[29] New York State College of Agriculture, "The People of Herkimer County, New York: Trends in Human Resources and Their Characteristics 1900–1960," Agricultural Experiment Station of Cornell University Bulletin 62-21, Ithaca, N.Y., 1963.
[30] Applegate, H. L., *The Story of Sagamore*, University College of Syracuse University, Syracuse, NY, 1975.
[31] Anderson, R. S., Davis, R. B., Miller, N. G., and Stuckenrath, R., Jr., *Canadian Journal of Botany*, Vol. 64, 1986, pp. 1977–1986.
[32] Battarbee, R. W., *Limnology and Oceanography*, Vol. 18, 1973, pp. 647–653.
[33] Davis, R. B., Anderson, D. S., Charles, D. F., Galloway, J. N., "Two-hundred Year pH History of Woods, Sagamore and Panther Lakes in the Adirondack Mountains, New York State, USA," Completion Report, Electric Power Research Institute, Palo Alto, CA, 1984.

[34] Kahl, J. S., Norton, S. A., and Williams, J. S. in *Geological Aspects of Acid Deposition*, O. P. Bricker, Ed., Butterworth Publishers, Boston, MA, 1984, pp. 23–36.
[35] Norton, S. A., Davis, R. B., and Anderson, D. S. "The Distribution and Extent of Acid and Metal Precipitation in Northern New England," Completion Report, U.S.F.W.S., Dept. of Interior, Project 14-16-0009-79-040, Washington, DC, 1985.
[36] Davis, R. B., et al., "A Comparative Paleolimnological Study of the Impacts of Air Pollution on Three Northern New England Lakes, Preliminary Results," Interim Report, Electric Power Research Institute, Palo Alto, CA, 1986.
[37] Husar, R. B. in *Acid Deposition: Long Term Trends*, National Research Council, Committee on Monitoring and Assessment of Trends in Acid Deposition, National Academy Press, Washington, DC, 1986, Chap. 2, pp. 48–92.
[38] Charles, D. F. in *Diatoms and Lake Acidity*, J. P. Smol et al, Eds., Junk, Dordrecht, The Netherlands, 1986, pp. 35–44.
[39] Baumgartner, K. J., "A Quantitative Study of the Phytoplankton of Three Adirondack Lakes with Differing pH Values," M.S. thesis, Cornell University, Ithaca, NY, 1981.
[40] Davis, R. B., *Limnology and Oceanography*, Vol. 19, 1974, pp. 466–488.
[41] New York (State of) Conservation Department, "A Biological Survey of the Raquette Watershed," Bull. 8, State of New York, Albany, NY, 1933.
[42] Davis, R. B., Bailey, J. H., Scott, M., Hunt, G., and Norton, S. A., "Descriptive and Comparative Studies of Maine Lakes," Technical Bulletin 88, Maine Life Sciences and Agriculture Experiment Station, Orono, Maine, 1978.
[43] Galloway, J. N., and Baker, J. P. in *The Acidic Deposition Phenomenon and its Effects: Critical Review Papers, Vol. II—Effects Sciences*, R. A. Linthurst, Ed., U.S.E.P.A., Washington, DC, 1984, pp. 4-45–4-48.
[44] Gherini, S. A., Mok, L., Hudson, R. J. M., Davis, G. F., Chen, C. W., and Goldstein, R. A., *Water, Air, and Soil Pollution*, Vol. 26, 1985, pp. 425–459.
[45] April, R., Newton, R., and Coles, L. T., *Geological Society of America Bulletin*, Vol. 97, 1986, pp. 1232–1238.
[46] Peters, N. E. and Murdoch, P. S., *Water, Air, and Soil Pollution*, Vol. 26, 1985, pp. 387–402.
[47] Baron, W. R., Gordon, G. A., Borns, H. W., and Smith, D. C., *Journal of Climate Applied Meteorology*, Vol. 23, 1985, pp. 317–319.
[48] Smith, D. C., "Agriculture in New York State to 1880: The Impact of Public Funds," in *Agricultural Artifacts in Museums: Historical Context and Curatorial Practice*, The Farmer's Museum, Cooperstown, NY, 1984, pp. 14–27.
[49] Dillon, P. J. and Reid, R. A. in *Atmospheric Input of Pollutants to Natural Waters*, S. Eisenreich, Ed., Ann Arbor Science, Ann Arbor, MI, 1981, pp. 183–198.
[50] Smith, R. M., Twiss, P. C., Krauss, R. K., and Brown, M. J., *Soil Science Society America, Proceedings*, Vol. 34, 1970, pp. 112–117.
[51] Tiedemann, A. R., Helvey, J. D., and Anderson, T. D., *Journal of Environ. Quality*, Vol. 7, 1978, pp. 580–588.
[52] Bormann, F. H., and Likens, G. E., *Pattern and Process in a Forested Ecosystem*, Springer-Verlag, NY, 1979.
[53] Robertson, G. P., *Ecology*, Vol. 63, 1982, pp. 1561–1573.
[54] Cronan, C. S., and DesMeules, M. C., *Canadian Journal of Forest Res.*, Vol. 15, 1985, pp. 881–889.
[55] Stumm, W., and Baccini, P. in *Lakes—Chemistry, Geology, Physics*, A. Lerman, Ed., Springer-Verlag, NY, 1978, pp. 91–126.
[56] Norton, S. A., Davis, R. B., and Anderson, D. S., "The Distribution and Extent of Acid and Metal Precipitation in Northern New England, Part 2, Sediment Analyses," final report to U.S. Fish and Wildlife Service, Columbus, OH, Grant 14-16-0009-79-040, 1985.
[57] Davis, R. B. and Smol, J. P., in *Diatoms and Lake Acidity*, J. P. Smol et al., Eds., Junk, Dordrecht, The Netherlands, 1986, pp. 291–300.
[58] Gensemer, R. W. and Kilham, S. S., *Canadian Journal of Fisheries Aquatic Sciences*, Vol. 41, 1984, pp. 1240–1243.
[59] Anderson, D. S., Davis, R. B., and Berge F. in *Diatoms and Lake Acidity*, J. P. Smol et al., Eds., Junk, Dordrecht, The Netherlands, 1986, pp. 97–113.
[60] Riedel, G. F., and Nelson, D. M., *Journal of Phycology*, Vol. 21, 1985, pp. 168–171.
[61] Stokes, P. M., *Water, Air and Soil Pollution*, Vol. 30, 1986, pp. 421–438.
[62] Wright, R. F., Cosby, B. J., Hornberger, G. M., and Galloway, J. N., *Water, Air and Soil Pollution*, Vol. 30, 1986, pp. 367–380.
[63] Cronan C. S. and Aiken, G. R., *Geochim, Cosmochim. Acta*, Vol. 49, 1985, pp. 1697–1705.

[64] Chen, C. W., Gherini, S. and Summers, K. in *Electric Power Research Institute, The Integrated Lake-Watershed Acidification Study*, Electric Power Research Institute, Palo Alto, CA, 1981, pp. 2-1–2-22.
[65] Hendrey, G. R., Galloway, J. N., and Schofield, C. L. in *Electric Power Research Institute, The Integrated Lake-Watershed Acidification Study*, Electric Power Research Institute, Palo Alto, CA, 1981, pp. 7-1–7-5.

Sediment Assessment and Toxicity Testing

Jerry M. Neff,[1] Barny W. Cornaby,[2] Ralph M. Vaga,[2] Thomas C. Gulbransen,[1] Judith A. Scanlon,[1] and David J. Bean[2]

An Evaluation of the Screening Level Concentration Approach for Validation of Sediment Quality Criteria for Freshwater and Saltwater Ecosystems

REFERENCES: Neff, J. M., Cornaby, B. W., Vaga, R. M., Gulbransen, T. C., Scanlon, J. A., and Bean, D. J., **"An Evaluation of the Screening Level Concentration Approach for Validation of Sediment Quality Criteria for Freshwater and Saltwater Ecosystems,"** *Aquatic Toxicology and Hazard Assessment: 10th Volume, ASTM STP 971,* W. J. Adams, G. A. Chapman, and W. G. Landis, Eds., American Society for Testing and Materials, Philadelphia, 1988, pp. 115–127.

ABSTRACT: The U.S. Environmental Protection Agency (EPA) has initiated an effort to develop sediment quality criteria for both freshwater and marine ecosystems. The Screening Level Concentration (SLC) approach is one of several methods EPA is evaluating for calculating sediment quality criteria. The SLC approach uses field data on the cooccurrence in sediments of benthic infaunal invertebrates and different concentrations of the nonpolar organic contaminant of interest. The SLC method is designed to estimate the highest concentration (normalized to sediment organic carbon concentration) of a particular nonpolar organic contaminant in sediments that can be tolerated by approximately 95% of benthic infauna. As such, the SLC value could be used in a regulatory context as the concentration of a contaminant in sediment that, if exceeded, could lead to environmental degradation.

This paper describes the method for calculating the SLC and evaluates the SLC approach empirically for nonpolar organic contaminants in freshwater and marine sediments in terms of its statistical properties and its dependence on the characteristics of the data base. SLCs are calculated for five contaminants in freshwater sediments (total polychlorinated biphenyls (PCBs), dichlorodiphenyl trichloroethane (DDT), dieldrin, chlordane, and heptachlor epoxide) and nine contaminants in saltwater sediments (total PCBs, DDT, naphthalene, phenanthrene, fluoranthene, benz(a)anthracene, chrysene, pyrene, and benzo(a)pyrene). The method used to calculate SLCs is illustrated for total PCBs in freshwater and saltwater sediments. Differences in SLC values for PCBs and DDT between freshwater and saltwater sediments are discussed. The SLC approach demonstrates sufficient merit to warrant further evaluation and elaboration. Given a large enough data base and minor modifications of the methods for calculating an SLC for a specific contaminant, the approach can provide a conservative estimate of the highest concentration, normalized to sediment organic carbon, that 95% of the benthic infauna can tolerate in sediment.

KEYWORDS: sediment, nonpolar organic contaminant, benthic fauna, sediment organic carbon concentration, screening level concentration, sediment quality criteria, freshwater, saltwater

Background

The U.S. Environmental Protection Agency, Criteria and Standards Division (EPA-CSD) has initiated an effort to develop sediment quality criteria for nonpolar organic contaminants and

---

[1] Battelle, Department of Ocean Sciences, Duxbury, MA 02332.
[2] Battelle, Department of Environmental Sciences, Columbus, OH 43201.

metals. Sediment quality criteria are to be used in conjunction with water quality criteria to protect freshwater and saltwater bodies and their uses, including fisheries, recreation, and drinking water.

The development of technically sound sediment quality criteria that can be applied widely to sediments from different sources is a difficult task. Chemical contaminants interact in complex, often poorly understood ways with sediment particles and may be present in sediments in a variety of adsorbed or solid forms. As a general rule, chemical pollutants associated with sediments are much less bioavailable and toxic to aquatic organisms than the same pollutants in solution in the water [1,2]. However, there is no known simple relationship between the total concentration of a contaminant in sediment and its toxicity to aquatic organisms in contact with that sediment.

The screening level concentration approach answers part of this need. The approach uses field data on the concentration of specific nonpolar organic contaminants in sediments and the presence of specific taxa of benthic infauna in that sediment to calculate screening level concentrations (SLCs). The SLC is defined here as the concentration of a nonpolar organic contaminant in sediment that, if exceeded, could lead to environmental degradation and therefore would warrant further investigation. It is an estimate of the highest concentration of a particular nonpolar organic pollutant in sediment that can be tolerated by approximately 95% of benthic infauna. The SLC is a means of determining if a particular chemical is present in sediments at a high enough concentration to be of concern. It can not be used alone to determine whether a particular sediment is or is not hazardous, because the sediment may contain many contaminants for which no SLC is available. The SLC approach is complementary to and can be used to verify the results of the triad approach to sediment quality determination in which assessments of sediment quality involve, at a minimum, measurements of concentrations of toxic chemicals in the sediments, toxicity of the sediments to representative infauna, and evidence of modified resident infaunal community structure in the contaminated sediments [3,4].

SLCs are calculated from contaminant concentrations, normalized to sediment organic carbon concentration, rather than concentrations in bulk sediment. This normalization is based on the premise, supported by much theory and experimental data, that bioavailability of nonpolar organic pollutants from sediments is dependent upon the organic carbon content of the sediment, the lipid content of the organism, and the relative affinities of the chemical for sediment organic carbon versus animal lipid [5,6]. A nonpolar organic contaminant will be distributed among three phases, the sediment organic fraction, the tissue organic fraction, and the sediment pore water, in proportion to the respective sediment organic carbon/water and tissue lipid/water partition coefficients of the contaminant. Thus, the bioavailability of a nonpolar organic pollutant in sediment will be proportional to the ratio of the partition coefficient of the pollutant in the tissue organic fraction of the animal to the partition coefficient of the pollutant in the sediment organic fraction, and will be inversely proportional to the sediment organic carbon concentration. Abernethy et al. [7] showed that the acute toxicities of 38 hydrocarbons and chlorinated hydrocarbons to freshwater and saltwater microcrustaceans is controlled by organism-water partitioning, which for nonpolar organic compounds is a reflection of aqueous solubility.

General Data Requirements

Large data bases containing information on the biology and chemistry of surficial sediments from freshwater and saltwater ecosystems are required to calculate screening level concentrations. The calculation of an SLC for a given nonpolar organic contaminant requires data bases containing matched (synoptic, if possible) observations of species composition of benthic infauna, concentration of the organic contaminant of interest in the sediment, and concentration of total organic carbon in the sediment. Data on sediment grain size are also useful, but not essential, because the relationship between sediment organic carbon and contaminant bioavailability does not hold for very coarse sediments. At a minimum, 20 observations of the presence of a particular species in sediments containing different concentrations of the contaminant of interest are required for calculation of a species screening level concentration (SSLC). A minimum of ten SSLCs are required to calculate an SLC. These numbers were chosen somewhat arbitrarily for the initial evaluation of the SLC approach.

Due to the preliminary nature of this approach, data bases were sought that fulfilled the aforementioned minimum criteria. These data bases were not subjected to any extensive quality assurance review, nor were the QA/QC backgrounds of the data bases evaluated. Lacking this more extensive review, SLCs developed using these data sources will be illustrative of the mechanics and conceptual utility of the approach, but are not proposed at this stage of development for regulatory purposes. Before SLCs could be used in a regulatory context, the methodolgy used to collect and assess geological, chemical, and biological data would require a comparability assessment. Inconsistencies in taxonomic identifications, for instance, may affect SLC values, yet only a superficial review of taxonomic criteria has been conducted in this study. A more thorough review of the biological, chemical, and geological data may also result in refinements to and improvements in the sensitivity of the SLC methodology.

Materials and Methods

Data Sources

The freshwater data sets used in this study were located by systematically contacting various government agencies, private consulting firms, and universities, and by searching the open literature. The data base compiled for calculating freshwater SLCs consisted of 80 individual data sets representing 323 separate sampling stations. Sampling stations were located in six states: Illinois, Michigan, Indiana, New York, Ohio, and Wisconsin. Data from both lotic and lentic ecosystems were included in the analysis.

Potentially useful saltwater data sets were identified by searching a computerized inventory of marine pollution monitoring programs. In addition, several consulting firms, universities, or individual investigators that were known to have performed or participated in marine benthic monitoring and assessment programs were contacted. From these saltwater data bases, a total of 19 field surveys or monitoring cruises were identified that contained data suitable for derivation of SLCs. The 19 data sets contained data from 293 sampling stations. Nearly equal numbers of stations were located in each of three major regions; the New York Bight, the Southern California Bight, and Puget Sound.

Calculation of Screening Level Concentrations

Separate SLCs were derived for freshwater and saltwater sediments and were based exclusively on the respective freshwater and saltwater data bases. However, the procedures used to calculate freshwater and saltwater SLCs were the same.

First, we identified all the stations in the data base at which the contaminant of interest was analyzed in the sediments. For each of these stations, we prepared a list of all species of benthic infauna that were present at that station. We then normalized contaminant concentrations to the total organic carbon concentration of the sediment at each station by the simple formula:

TOC-normalized contaminant concentration = X/TOC (µg contaminant/g organic carbon)

where X is the contaminant concentration in the bulk sediment (µg contaminant/kg sediment dry weight), and TOC is the concentration of total organic carbon in the sediment (g organic carbon/kg sediment dry weight).

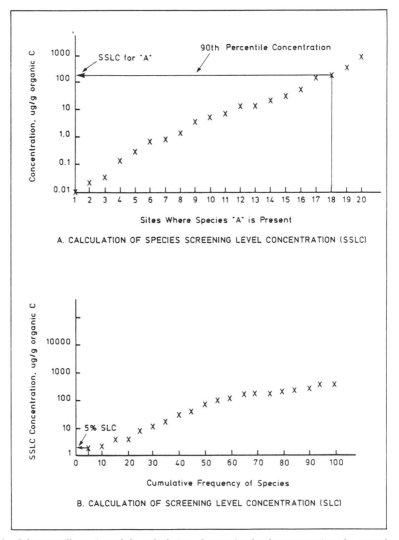

FIG. 1—*Schematic illustration of the calculation of screening level concentrations for nonpolar organic contaminants in sediments.*

For each species that was present at 20 or more stations, we plotted the organic carbon normalized concentration of the chemical in the sediment for all samples (or stations) in which the species was present, versus the cumulative frequency of such cooccurrences, proceeding from the least to the most contaminated station (Fig. 1a). From this plot, we estimated the sediment contaminant concentration below which 90% of the samples containing the species occurred. This concentration was defined as the species screening level concentration (SSLC) of the contaminant. The 90% cutoff was chosen arbitrarily at the beginning of this exercise to give a conservative estimate of the SSLC. This procedure was repeated for each benthic species present in the data set at 20 or more stations, thereby generating a number of SSLCs for a given contaminant.

We then constructed a cumulative frequency distribution (based on rank, which in turn was based on the SSLC values) of all SSLCs for the contaminant (Fig. 1b) and calculated the 5th percentile (the SSLC value above which 95% of all SSLCs fall) of that distribution by linear interpolation between the two nearest quantiles. This interpolated value was designated as the screening level concentration (SLC) of the contaminant.

Because the SSLCs for each contaminant were not normally distributed (Kolmogorov D-Statistic, $\alpha = 0.05$) [8], standard statistical (distribution-free) techniques were used to calculate a confidence interval for the SLCs. Order statistics were employed to set a confidence interval for the 5th percentile of the SSLC cumulative frequency distribution for each contaminant. Confidence intervals were set using the binomial distribution as described by Mood et al. [9]. The interval that provided a confidence coefficient greater than or equal to 95% was chosen.

Results

The freshwater data base contained data for the presence of a total of 103 different infaunal invertebrate taxa. However, only a total of 23 species, representing seven families, orders, or classes, were present at a sufficient number of sampling stations to be included in the analysis. The freshwater benthic species used in the analysis included eight oligochaetes, five ephemeropterans, three trichopterans, one chironomid, one isopod, two amphipods, and one gastropod. For all five contaminants for which freshwater SLCs were calculated, the taxa found most frequently in the sample were Oligochaeta and Ephemeroptera.

The saltwater data base contained data for the presence of a total of 117 species of marine benthic invertebrates. Of these, only 60 species were present at a sufficient number of sampling stations to be included in the analysis. The most abundant saltwater taxa used to calculate SLCs were the Polychaeta, followed by the Crustacea and Mollusca.

The distribution of total organic carbon concentrations in freshwater sediments ranged from 5.0 to 366 g/kg dry weight. The distribution of total organic carbon concentrations in saltwater sediments ranged from 0.31 to 303 g/kg dry weight. The highest value for saltwater sediments was somewhat anomalous, in that the second highest value was 160 g/kg. The distribution of total organic carbon, bulk contaminant concentrations, and organic carbon normalized contaminant concentrations were not normally distributed (Kolmogorov D-Statistic, with $\alpha = 0.05$) in either freshwater or saltwater sediments. The range of organic carbon normalized concentrations of contaminants spanned at least one order of magnitude in freshwater sediments and at least two orders of magnitude in saltwater sediments.

As an example of the application for the SLC approach, we will use the calculation of SLCs for total PCBs in freshwater and saltwater sediments.

A species screening level concentration was calculated for each species of infaunal invertebrate that was present at a minimum of 20 stations where PCBs were analyzed in the sediments. SSLCs for total PCBs were calculated for 21 species of freshwater invertebrates (Table 1). The freshwater SSLCs ranged from 0.286 to 103.4 µg PCBs/g total organic carbon. SSLCs for total PCBs were calculated for 51 species of saltwater invertebrates (Table 2). The saltwater SSLCs ranged from 3.394 to 71.32 µg PCBs/g total organic carbon. Examples of cumulative frequency distribution curves of organic carbon normalized PCB concentrations in sediments for three species each of freshwater and saltwater infaunal invertebrates are presented in Figs. 2a and 3a, respectively. The SSLCs for some representative freshwater animals were 0.379 µg/g for *Stenonema terminatum*, 7.44 µg/g for *Hyalella azteca*, and 103.45 µg/g for *Peloscolex multisetosus*. The SSLCs for some representative saltwater animals were 3.87 µg/g for *Nephtys ferruginea*, 9.14 µg/g for *Pholoe minuta*, and 58.77 µg/g for *Spiophanes berkeleyorum*. Similarly, SSLCs were calculated for the other species in the freshwater and saltwater data bases.

To calculate an SLC for total PCBs, a cumulative frequency distribution was constructed using the SSLC values for the freshwater or the saltwater sediments, respectively (Figs. 2b and 3b). The

TABLE 1—*Cumulative frequency and values for species screening level concentrations for total polychlorinated biphenyls in freshwater sediments. The number of observations used to calculate each SSLC also is given.*

Rank	Cumulative Frequency, %	SSLC, μg/g Organic Carbon	No. of Observations	Organism
1	4.8	0.286	25	Cyrnellus fraternus
2	9.5	0.379	35	Stenonema terminatum
3	14.3	0.606	28	Stenonema pulchellum
4	19.0	0.650	34	Hydropsyche orris
5	23.8	0.722	37	Hydropsyche frisoni
6	28.6	0.722	42	Stenonema integrum
7	33.3	0.949	20	Stenonema exiquum
8	38.1	1.905	54	Stenacron interpunctatum
9	42.9	3.137	20	Pentaneura mallochi
10	47.6	4.655	25	Asellus intermedius
11	52.4	7.442	36	Hyalella azteca
12	57.1	9.318	26	Potamothrix vejdovskyi
13	61.9	24.260	26	Valvata sincera
14	66.7	29.259	23	Limnodrilus claparedeianus
15	71.4	29.600	56	Tubifex tubifex
16	76.2	34.286	43	Peloscolex ferox
17	81.0	45.714	20	Limnodrilus udekemianus
18	85.7	52.778	20	Linmodrilus cervix
19	90.5	52.778	55	Limnodrilus hoffmeisteri
20	95.2	56.338	56	Gammarus fasciatus
21	100.0	103.448	31	Peloscolex multisetosus

shape of both distributions was nearly linear, with PCB concentrations evenly distributed over the entire range of observed values. The 5th percentile of these distributions (equivalent to the SLC value) was calculated by interpolation. The SLC values were 0.290 and 4.26 μg total PCBs/g total organic carbon for freshwater and saltwater sediments, respectively.

Screening level concentrations also were calculated for an additional four nonpolar organic contaminants in freshwater sediments and an additional eight nonpolar organic contaminants in saltwater sediments (Table 3). The SLC values ranged over five orders of magnitude (0.008 to 43.4 μg/g). The SLCs for nonpolar contaminants in saltwater sediments were much higher than those for nonpolar contaminants in freshwater sediments. Direct comparison of freshwater and saltwater SLCs was possible for total PCBs and DDT. The saltwater sediment SLCs for these compounds were at least an order of magnitude higher than the corresponding freshwater sediment SLCs.

Discussion

The procedure for calculating a screening level concentration for a nonpolar organic contaminant generates from an existing data base distributions of the contaminant concentrations, normalized to sediment organic carbon concentration, in sediments. Calculation of an SSLC is based upon a subsample from this parent distribution. The nature of the subsample is determined by the presence of a given species of infaunal invertebrate at the sampling stations. The value of the SLC is calculated from the distribution of the SSLCs for the contaminant.

This approach to developing sediment quality criteria has several intuitively appealing attributes. It makes use of field data on the coexistence of specific levels of sediment contamination and a resident infauna, making extrapolations from laboratory to field conditions unnecessary. It utilizes

TABLE 2—*Cumulative frequency and values for species screening level concentrations for total polychlorinated biphenyls in saltwater sediments. The number of observations used to calculate each SSLC also is given.*

Rank	Cumulative Frequency, %	SSLC, µg/g Organic Carbon	No. of Observations	Organism
1	2.0	3.394	21	Spiochaetopterus costarum
2	3.9	3.871	32	Nephtys ferruginea
3	5.9	4.583	24	Harmothoe extenuata
4	7.8	4.634	22	Euchone elegans
5	9.8	4.634	22	Scalibregma inflatum
6	11.8	4.714	24	Drilonereis longa
7	13.7	4.714	27	Spiophanes bombyx
8	15.7	4.841	29	Anobothrus gracilis
9	17.6	4.841	27	Arctica islandica
10	19.6	4.841	30	Euchone incolor
11	21.6	4.841	26	Ninoe nigripes
12	23.5	6.000	23	Nephtys incisa
13	25.5	6.000	33	Nucula proxima
14	27.5	7.500	25	Mediomastus ambiseta
15	29.4	7.500	33	Tharyx acutus
16	31.4	8.000	39	Aricidea catherinae
17	33.3	8.000	22	Caulleriella of killariensis
18	35.3	8.000	24	Goniadella gracilis
19	37.3	8.000	24	Unciola irrorata
20	39.2	8.854	25	Lumbrinereis hebes
21	41.2	9.143	54	Pholoe minuta
22	43.1	10.000	23	Paraonis gracilis
23	45.1	10.000	27	Pherusa affinis
24	47.1	10.000	26	Phyllodoce mucosa
25	49.0	10.000	33	Tharyx annulosus
26	51.0	10.625	30	Lumbrinereis acicularum
27	52.9	10.625	29	Pitar morrhuanus
28	54.9	10.941	32	Tellina agilis
29	56.9	11.417	24	Glycera dibranchiata
30	58.8	11.731	37	Amphiodia (amphispina) urtica
31	60.8	13.769	25	Heterophoxus oculatus
32	62.7	16.935	55	Euphilomedes carcharodonta
33	64.7	18.644	21	Prionospio cirrifera
34	66.7	27.736	28	Cossura longocirrata
35	68.6	30.118	21	Ampelisca brevisimulata
36	70.6	33.103	26	Compsomyax subdiaphana
37	72.5	33.905	20	Sthenelanella uniformis
38	74.5	39.683	20	Armandia brevis
39	76.5	40.017	23	Glycinde armigera
40	78.4	40.017	56	Pectinaria californiensis
41	80.4	41.143	109	Prionospio steenstrupi
42	82.4	42.765	38	Nephtys cornuta franciscana
43	84.3	45.045	74	Capitella capitata
44	86.3	46.025	90	Axinopsida sericata
45	88.2	46.307	20	Chloeia pinnata
46	90.2	47.817	56	Prionospio pinnata
47	92.2	52.058	100	Glycera capitata
48	94.1	52.058	67	Macoma carlottensis
49	96.1	56.307	89	Parvilucina tenuisculpta
50	98.0	58.774	42	Spiophanes berkeleyorum
51	100.0	71.315	40	Tellina carpenteri

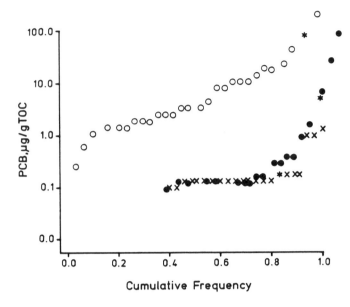

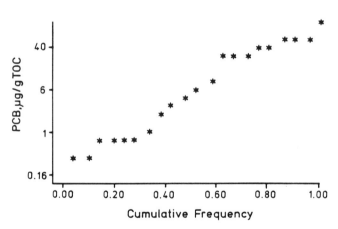

FIG. 2—(A) *Cumulative frequency distribution of total PCBs in freshwater sediments for* Stenonema treminatum *(X),* Hyalella azteca *(●), and* Peloscolex multisetosus *(○). (B) Cumulative frequency distribution of species screening level concentrations (SSLCs) for total PCSs in freshwater sediments. SSLC values are in* µg *PCB/g sediment organic carbon.*

data on only the *presence* of species in sediments containing given concentrations of contaminants. Thus, no *a priori* assumptions are made about a causal relationship between levels of sediment contamination and the distribution of infaunal populations. Because no causal relationship is assumed, it is not necessary to take into account the wide variety of natural environmental factors, such as water depth, sediment texture, and salinity, that affect the composition and distribution of

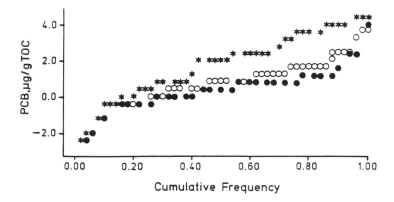

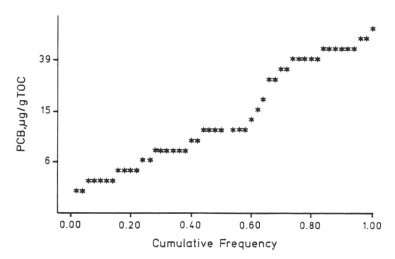

FIG. 3—(A) *Cumulative frequency distribution of total PCBs in saltwater sediments for* Nephtyhs ferruginea (●), Pholoe minuta (○), *and* Spiophanes berkeleyorum (*). (B) *Cumulative frequency distribution of species screening level concentrations (SSLCs) for total PCBs in saltwater sediments. SSLC values are in* µg *PCBs/g sediment organic carbon.*

benthic infaunal communities. However, because the method uses actual observations from the field of the cooccurrence in the sediments of multiple species of benthic infauna and concentrations of contaminants, valid *a posteriori* inferences can be made about the range of contaminant concentrations that the benthic infauna can tolerate in sediment.

The value of the SLC for any contaminant depends on the characteristics of the data base from which it was calculated. The range and distribution of contaminant concentrations in the data base used to calculate an SLC will have a marked effect on the value of the SSLCs, and therefore the SLCs generated. The SLC calculation process, by its very nature, makes no *a priori* assumptions about a causal relationship between a given concentration of the contaminant of interest in sediments and the presence or absence of a particular species of benthic infauna in those sediments. Therefore,

TABLE 3—*Summary of screening level concentrations (SLCs) for freshwater and saltwater sediments. Values in micrograms of contaminant per grams of sediment organic carbon (parts per million).*

	SLC [Confidence Interval (CI) and α]	
Compound	Freshwater	Saltwater
Heptachlor Epoxide	0.008 (CI = 0.0 to 0.029, α = 0.02)	...
Chlordane	0.098 (CI = 0.0 to 0.136, α = 0.04)	...
Dieldrin	0.021 (CI = 0.0 to 0.084, α = 0.04)	...
Polychlorinated Biphenyls	0.290 (CI = 0.0 to 0.65, α = 0.02)	4.26 (CI = 0.0 to 4.63, α = 0.03)
DDT	0.190 (CI = 0.0 to 0.283, α = 0.02)	42.8 (CI = 0.0 to 113.7, α = 0.03)
Naphthalene	...	36.7 (CI = 0.0 to 41.4, α = 0.03)
Phenanthrene	...	25.9 (CI = 0.0 to 38.4, α = 0.03)
Fluoranthene	...	43.2 (CI = 0.0 to 64.3, α = 0.04)
Benz(a)anthracene	...	26.1 (CI = 0.0 to 41.0, α = 0.03)
Chrysene	...	38.4 (CI = 0.0 to 60.5, α = 0.03)
Pyrene	...	43.4 (CI = 0.0 to 74.4, α = 0.06)
Benzo(a)pyrene	...	39.6 (CI = 0.0 to 46.8, α = 0.03)

it is possible to have a data set in which all concentrations of the contaminant of interest are well below the concentration in sediments that would adversely affect the distribution of benthic infauna. SLCs calculated with such a data set would be conservative and the SLC would have little regulatory relevance. On the other hand, if most observations are from a heavily contaminated area, most of the pollutant-sensitive species would be absent and the SLC would be based primarily on pollutant-tolerant species. In such a case, the SLC would be too high. As the range of contaminant concentrations upon which the SLC is based increases, the likelihood of these types of biases in the SLC decreases.

In many instances, concentrations of different contaminants were determined at the same sampling sites. A high correlation between the concentrations of two or more contaminants in the parent distribution can lead to problems in interpreting the SLC values. For example, a correlation analysis of the concentration of pesticides in the freshwater data base showed that there was a high degree of correlation among dieldrin, chlordane, and heptachlor epoxide (dieldrin compared to chlordane: $r = 0.84$, $p = 0.001$, $n = 324$; chlordane compared to heptachlor epoxide: $r = 0.89$, $p = 0.001$, $n = 304$). This correlation leads to the possibility that any two subsamples drawn from the parent distribution to calculate SSLCs for two different contaminants will also be correlated. When this is the case, it is not possible to ascertain to which of the two contaminants the benthic community is responding. Such distinctions can be made by increasing the size of the data base to include enough sampling sites to decrease the correlation among contaminants of interest.

Nearly always, contaminated sediments contain more than one contaminant at an elevated concentration. The infauna resident in the contaminated sediments, as well as the populations that have been eliminated from the contaminated sediments, are responding to the multiple contaminants present and not just to the contaminants of interest. The SLC approach can not take into account multiple contaminant interactions in sediments. As a result, the SLC value for a particular contaminant may be conservative (for example, lower than the benthic infauna could tolerate if the contaminant of interest was the only contaminant in the sediment). Because the mix and relative proportions of different contaminants present in the sediments will vary substantially from location to location, this conservative bias in the SLC will tend to decrease as the number of observations upon which SSLCs are based is increased.

All SLCs for freshwater sediments are lower than all SLCs for saltwater sediments by at least one order of magnitude (Table 3). This pattern is exemplified best by the two contaminants for which we have comparative freshwater and saltwater sediment SLCs: PCBs and DDT. The SLC for PCBs in saltwater sediments is 15 times higher than the corresponding value for freshwater sediments. There is a 225-fold difference in the SLCs for DDT in freshwater and saltwater sediments. There are several possible reasons for these differences. The most important are the following: (1) differences in the range and distribution of values of organic carbon normalized contaminant concentrations for freshwater and saltwater sediments in the two data bases; (2) differences in the relative sensitivity of the freshwater and saltwater benthic infauna used in this analysis; and (3) differences in the solubility of the nonpolar organic contaminants in freshwater and saltwater.

In the freshwater and saltwater data sets used to calculate SLCs, the observed organic carbon normalized concentrations of the contaminants in sediments were distributed quite differently. This could account for much of the difference in the SLC values between freshwater and saltwater sediments. For example, in the freshwater data set, approximately 10% of the observations of the organic carbon normalized concentration of DDT in sediments were below 0.5 $\mu g/g$, and only 10% of observations were above 30 $\mu g/g$. However, in the corresponding saltwater data set, approximately 10% of observations were below 1.0 $\mu g/g$, and approximately 75% of observations were above 30 $\mu g/g$. As a result, 47.6% of the SSLCs for DDT in freshwater sediments were below 0.35 $\mu g/g$ organic carbon; whereas 47.4% of the SSLCs for DDT in saltwater sediments were at or below 208 $\mu g/g$ organic carbon. Because the saltwater data set for DDT contained a large number of observations at stations heavily contaminated with DDT, and possibly other contaminants, it is likely that many pollutant-sensitive species of benthic infauna were not present at the stations sampled, and therefore were not used for the SLC calculation. Absence of SSLCs for pollutant-sensitive species would yield an SLC value higher than it should be.

The differences between freshwater and saltwater data sets for PCBs are similar to but not as large as those described above for DDT.

Although differences in the sensitivity of freshwater and saltwater benthic invertebrates to sediment-associated nonpolar contaminants could result in some differences in the SLC values, it is unlikely that such differences would be large enough to account for more than a fraction of the differences in SLC values observed here for freshwater and saltwater sediments. Current water quality criteria for DDT and PCBs indicate that there are only small differences in the apparent sensitivity of freshwater and saltwater animals to these two chemicals.

Within any freshwater or saltwater benthic community, different species may vary substantially from one another in relative sensitivity to sediment-bound contaminants. Attempts have been made to relate at least part of this differential sensitivity to feeding habits of the different species [10]. Theoretically, benthic animals that ingest sediments as a source of nutrition (deposit feeders) should be more sensitive than filter feeders (that filter particulate food from the overlying water column), suspension feeders (that feed on the organic floc at the sediment-water interface), or carnivores to sediment-bound contaminants. Therefore, the relative proportion of species characteristic of the different feeding types in the benthic communities sampled will influence the shape of the SLC curves generated and the values of the SLCs obtained.

Salinity of the ambient medium does affect the physical and chemical behavior of many chemicals. Kadeg et al. [11] reviewed the literature, discussing effects of salinity on the behavior of nonpolar organic chemicals in aqueous media. The aqueous solubility of PCBs, DDT, and polycyclic aromatic hydrocarbons decreases with increasing salinity. As a result, the presence of electrolytes (salts) in solution increases the sorption of nonpolar organic chemicals by sediments. Therefore, it is reasonable to infer that nonpolar organic chemicals adsorbed to sediments will be less bioavailable in saltwater than in freshwater. There are relatively few data available that are suitable for testing this inference [1]. Boehm [12] measured the concentration of several nonpolar organic pollutants in sediments and resident infaunal polychaetes and bivalves from the New York

Bight. Bioaccumulation factors for the contaminants from the sediments (concentration in animal tissues/concentration in sediment) ranged from 0.001 to 0.7 in the polychaetes *Nephtys* sp. and *Pherusa affinis* and from 0.002 to 4.46 in the bivalve *Nucula proxima*. Bioaccumulation factors for several polycyclic aromatic hydrocarbons (PAH) ranged from 0.01 to 0.24 in the polychaetes and 0.002 to 3.20 in the bivalve. Eadie et al. [*13–15*] studied the concentrations of several PAHs in sediments and benthic oligochaetes and arthropods from the Great Lakes. Bioaccumulation factors from sediments for individual PAHs in the amphipod *Pontoporeia hoyi* ranged from 1 to 45. Bioaccumulation factors from sediments for different PAHs in the oligochaete *Limnodrilus hoffmeisteri* ranged from 0.1 to 2.3. This limited comparison supports the inference that bioavailability of nonpolar organic contaminants from sediments will be inversely related to salinity of the overlying water. Because bioavailability and toxicity of a nonpolar organic chemical are directly related, we can infer that there may be a tendency for freshwater organisms to be more sensitive than saltwater organisms to sediment-adsorbed contaminants. This inference is consistent with our analysis and may account for a small part of the difference in SLCs for freshwater and saltwater sediments. This conjecture is very preliminary and requires further experimental verification.

Of the three possible reasons for the differences between the freshwater and saltwater SLC values, the most important probably is that the two data bases had different ranges of contaminant concentrations, normalized to sediment organic carbon concentration, in sediments. The freshwater concentrations tended to be low, as evidenced by the many zero contaminant values. The saltwater data base tended toward the more highly polluted sediments. Based on these observations, the freshwater SLC values may be conservative and the saltwater SLC values may be too high.

The SLC approach has demonstrated sufficient merit to warrant further evaluation and elaboration. Given a large enough data base and minor modifications to the methods for calculating SSLCs and SLCs, the approach will provide a conservative estimate of the highest organic carbon normalized concentrations of individual contaminants in sediments that can be tolerated by approximately 95% of benthic infauna. As the number and range of observations contributing to the calculation of the SLC for a contaminant increases, one would expect the SLC values calculated to asymptotically approach some ideal "true" SLC values for freshwater and saltwater sediments. It is essential that the data base contain organic carbon normalized concentrations of the sediment contaminants of interest that span a wide range (preferably two orders of magnitude or more) and include values from locations known to be heavily contaminated as well as locations known to be relatively clean. Low and intermediate sediment contaminant concentrations are needed to ensure that pollutant-sensitive species are not excluded from the analysis. High values are needed to ensure that benthic communities are in fact being adversely affected at some stations by the contaminant of interest. Data from areas containing clearly defined gradients of concentrations of the contaminant of interest in the sediments would be ideal for use in calculating an SLC. In the present investigation, the freshwater data base was dominated by low contaminant concentrations and the saltwater data base was dominated by high contaminant concentrations. The result was that freshwater SLCs tended to be low and saltwater SLCs tended to be high. As the number of observations in the data base increases, the magnitude of this bias toward high or low values will decrease.

Acknowledgments

This work was supported by contract No. 68-01-6986 from the Environmental Protection Agency, Criteria and Standards Division.

References

[1] Neff, J. M., *Fresenius Zeitschrift fur Analitische Chemie,* Vol. 319, 1985, pp. 132–136.
[2] Lake, J., Hoffman, G. L., and Schimmel, S. C., "Bioaccumulation of Contaminants from Black Rock Harbor Dredged Material by Mussels and Polychaetes," Technical Report D-85-2, U.S. Army Corps of Engineers and U.S. Environmental Protection Agency, Washington, DC, 1985.
[3] Chapman, P. M. and Long, E. R., *Marine Pollution Bulletin,* Vol. 14, 1983, pp. 81–84.
[4] Long, E. R. and Chapman, P. M., *Marine Pollution Bulletin,* Vol. 16, 1985, pp. 405–515.
[5] Karickhoff, S. W. and Morris, K. R. in *Fate and Effects of Sediment-Bound Chemicals in Aquatic Systems,* Proceedings of the Sixth Pellston Workshop, 1986, K. L. Dickson, A. W. Maki, and W. Brungs, Eds., Society of Environmental Toxicology and Chemistry, Rockville, MD, in press.
[6] Adams, W. J. in *Fate and Effects of Sediment-Bound Chemicals in Aquatic Systems,* Proceedings of the Sixth Pellston Workshop, 1986, K. L. Dickson, A. W. Maki, and W. Brungs, Eds., Society of Environmental Toxicology and Chemistry, Rockville, MD, in press.
[7] Abernethy, S., Bobra, A. M., Shiu, W. Y., Wells, P. G., and Mackay, D., *Aquatic Toxicology,* Vol. 8, 1986, pp. 163–174.
[8] Sokol, R. R. and Rohlf, F. J., *Biometry,* W. H. Freeman, San Francisco, 1969.
[9] Mood, A. M., Graybill, F. A., and Boes, D. C., *Introduction to the Theory of Statistics,* McGraw-Hill, New York, 1974.
[10] Maurer, D., Leathem, W., and Menzie, C., *Marine Pollution Bulletin,* Vol. 12, 1981, pp. 342–347.
[11] Kadeg, R. D., Pavlou, S. P., and Duxbury, A. S., "Sediment Criteria Methodology Validation. Work Assignment 37, Task II. Elaboration of Sediment Normalization Theory for Nonpolar Organic Chemicals," report to U.S. Environmental Protection Agency, Criteria and Standards Division, Washington, DC, 1986.
[12] Boehm, P. D., "Organic Pollutant Transforms and Bioaccumulation of Pollutants in the Benthos from Waste Disposal-Associated Sediments," technical report submitted to U.S. Dept. of Commerce, National Oceanic and Atmospheric Administration, Rockville, MD, 1982.
[13] Eadie, B. J., Faust, W., Gardner, W. S., and Nalepa, T., *Chemosphere,* Vol. 11, 1982a, pp. 185–191.
[14] Eadie, B. J., Faust, W. R., Landrum, P. F., Moorehead, N. R., Gardner, W. S., and Nalepa, T. in *Polynuclear Aromatic Hydrocarbons: Formation, Metabolism, and Measurement,* M. Cooke and A. J. Dennis, Eds., Battelle Press, Columbus, OH, 1983, pp. 437–449.
[15] Eadie, B. J., Landrum, P. F., and Faust, W., *Chemosphere,* Vol. 11, 1982b, pp. 847–849.

Kimberly A. Knapp[1] *and Benjamin C. Dysart*[2]

A Technique for Better Sediment Characterization

REFERENCE: Knapp, K. A. and Dysart, B. C., **"A Technique for Better Sediment Characterization,"** *Aquatic Toxicology and Hazard Assessment: 10th Volume, ASTM STP 971*, W. J. Adams, G. A. Chapman, and W. G. Landis, Eds., American Society for Testing and Materials, Philadelphia, 1988, pp. 128–137.

ABSTRACT: A two-phase technique for determining the particle size distribution (PSD) of a sediment sample was developed. The hydrometer method of PSD analysis was selected for this research. For each sample, standard preparation for PSD analysis, that is, chemical dispersion of soil aggregates, was preceded by nondispersed PSD analysis. Resulting data were plotted as percentage finer by weight (arithmetic scale) versus grain size in millimeters (log scale). Each sediment sample's curve comprised a dispersed portion and a nondispersed portion. These curves can be used to describe the transport behavior of the soil, information that is necessary for the rational choice and design of effective erosion control measures. Sensitivity analysis, including investigation of the effects of sampling approach, temperature, dispersing time, meniscus, and concentration, showed this technique to be a simple, quick, reproducible, and inexpensive means of characterizing an eroded material.

KEYWORDS: sediment, hazard assessment, erosion, flocculation

Sediment's many detrimental effects on water quality and aquatic habitat are well documented. Plainly, not all sizes of sediment are equally damaging to environmental quality or damaging in the same ways. Because the surface area of colloidal clay is so much greater than that of sand, the adverse effects of sediment which are influenced by surface area, such as adsorption of dangerous toxics, are far greater in the presence of a substantial percentage of fines.

Therefore, it is widely recognized that the particle size distribution (PSD) of a soil is a critical indicator of its potential for causing environmental damage. The standard preparation of samples for PSD analysis [1] involves chemically dispersing the soil sample in order to break natural, aggregate particles down into primary particles (that is, individual sand, silt, and clay particles) and to neutralize the charges that frequently cause extensive flocculation. However, adsorption is just one stage of damage to environmental quality; another, and perhaps more important, stage is the transport of the sediment overland and instream. Unfortunately, the dispersed PSD frequently provides a very poor description of the particle sizes of the sediment which are actually transported through the environment. Because sediment transport is governed by the degree to which the particles are flocculating and settling, a nondispersed PSD would be a much more valuable indicator of a sediment's potential hazard to the quality of receiving waters and aquatic ecosystems in general.

The purpose of this research was to develop and evaluate a two-phase (dispersed and nondispersed) PSD analysis technique for use in determining a sediment's potential for effecting environmental damage and as a guide for the design of effective sediment controls.

---

[1] Environmental engineer, Georgia Department of Natural Resources, Atlanta, GA 30334.
[2] Professor, Clemson University, Clemson, SC 29631.

Procedure

Sampling Approach

Sediment samples were collected over a two-year period from the site of a 1000-MW pumped-storage hydroelectric facility now under construction in northwest South Carolina. The site's extremely steep slopes and high annual rainfall, combined with the extensive land-surface disturbances underway, produced an abundance of eroded material and required the installation of a wide variety of sediment control measures by Duke Power Co., the project owner. Soil samples were collected so as to represent the path of the eroded material from its original source to and/or through various sediment control measures and into the site's mountain streams. Samples were taken under dry conditions, during runoff-producing precipitation events, and immediately following such storm events.

Laboratory Methodology

The hydrometer, bottom withdrawal tube and the pipet are the most commonly used methods for determining the PSD of the fine fraction of a material [2]. The latter two are more time-consuming, more complicated, and more expensive than the first method [2–4] but are often used because of the large quantity of sediment residue required for the hydrometer test. For this research, a large number of samples were to be processed and adequate quantities of sediment were available; hence, the hydrometer method was judged to be the most appropriate.

Analysis began with a nondispersed hydrometer test (that is, the sample was not oxidized and no dispersing agent was added, so the flocs and other aggregates remained intact). After each representative sample of approximately 40 to 100 g was oven-dried and weighed, it was transferred to distilled water to bring the total volume of the soil-water mixture to 1000 mL; the sample was then allowed to soak overnight. A plunger was used to agitate the mixture, the hydrometer was inserted, and the time at which the hydrometer's every 0.001 g/cm^3 increment reached the water surface was recorded. Nondispersed hydrometer tests typically ran 3 to 10 min before the last reading at 1.001 g/cm^3 was taken.

Next, the sample was allowed to settle, the clear supernatant was decanted, and the remaining slurry was treated with hydrogen peroxide to oxidize any organics present. Dispersion of the soil sample followed standard procedures [5], with a dispersing agent of sodium hexametaphosphate and sodium carbonate. The sample was placed on a horizontal reciprocating shaker at 120 oscillations/min for at least 8 h to ensure total dispersion, but less than 20 h to avoid abrasion or shearing of the discrete particles. After dispersion, a dispersed hydrometer test was run in the same manner as the nondispersed test, except the end of the test was defined as the elapsed time of 24 h if that occurred prior to reaching a 1.001 g/cm^3 reading as in the nondispersed test. The sand fraction was then run through 1.000, 0.500, 0.250, 0.125, and 0.062-mm U.S. standard 3-in.-diameter sieves, following standard wet-sieving procedures [2].

Data Analysis

Calculations were based on Stokes' law for the velocity of a freely falling sphere, as described in the ASTM procedure [1]. Results were plotted as percentage finer than by weight (arithmetic scale) versus grain size in millimeters (log scale). Each sample's curve contains both nondispersed (lower curve) and dispersed (upper curve) portions for the fines and a sieved portion for the coarse material.

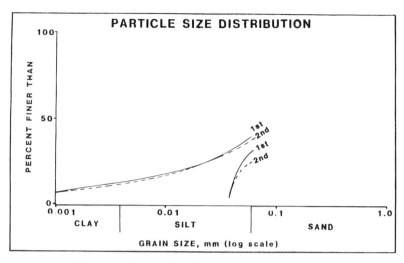

FIG. 1—*Sensitivity analysis: Two portions from same sample.*

Results

Sensitivity Analysis

In order to determine which parameters influenced the reproducibility of the resulting curves and the magnitude of this influence, sensitivity analysis was performed.

First, nondispersed and dispersed tests were run on two 40-g portions randomly chosen from the same sample. Figure 1 shows the results of the first run versus the second run. The nondispersed curves track each other fairly closely; the dispersed curves are nearly identical.

One 40-g portion was then tested twice under the same conditions. The results (Fig. 2) show virtually no difference between the first and second run.

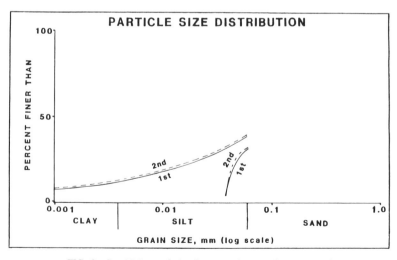

FIG. 2—*Sensitivity analysis: Same portion tested consecutively.*

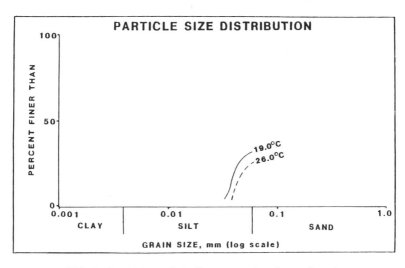

FIG. 3—*Sensitivity analysis: Temperature (nondispersed tests).*

Next, temperature effects were investigated. Nondispersed tests were run on a sample at 19.0 and 26.0°C. Dispersed tests were run at 21.0 and 23.0°C. See Figs. 3 and 4. It should be noted that temperature corrections have been made in the calculations. However, since over 85% of the samples were run at temperatures between 20.0 and 23.0°C, once the temperature corrections had been made temperature had little effect on the resulting curves.

Standard procedures call for dispersing a sample "overnight" on a horizontal shaker [1]. As reported earlier, this dispersing time was limited to between 8 and 20 h. To determine the effect of dispersing at these two time limits, one 40-g portion was dispersed for 8 h, a hydrometer analysis made, and then the portion was returned to the shaker for an additional 12 h. Figure 5 shows that the effect of dispersing time is negligible between 8 and 20 h.

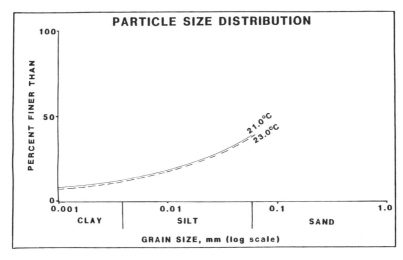

FIG. 4—*Sensitivity analysis: Temperature (dispersed tests).*

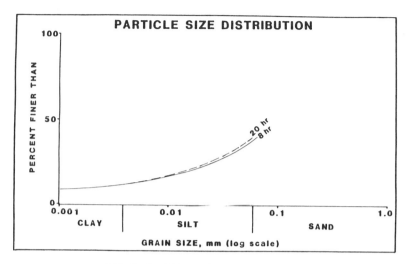

FIG. 5—*Sensitivity analysis: Dispersing time.*

ASTM procedures also recommend making a correction to account for the meniscus which forms around the hydrometer stem. Nondispersed and dispersed curves were calculated with and without this correction. As can be seen in Fig. 6, this correction appears to be negligible.

Finally, the effects of suspension concentration were considered. Because natural suspensions have not been dispersed, this effort was confined to nondispersed analyses. Both a coarse sample (with approximately 20% fines) and a fine sample (with approximately 75% fines) were analyzed. The resulting curves for the coarse sample are shown in Fig. 7. Concentration was varied between 20 and 500 g/L or 20 000 and 500 000 mg/L. Figure 8 shows the curves for the fine material, where concentration was varied from 10 to 130 g/L or 10 000 to 130 000 mg/L. These figures demonstrate that, while the general shape of the curves does not change with concentration, the

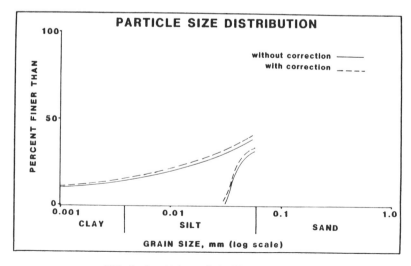

FIG. 6—*Sensitivity analysis: Meniscus correction.*

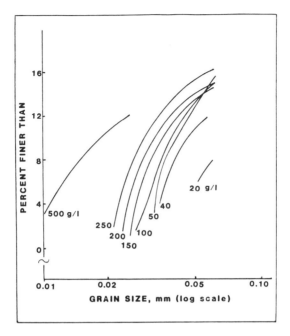

FIG. 7—*Sensitivity analysis: Concentration (coarse sample).*

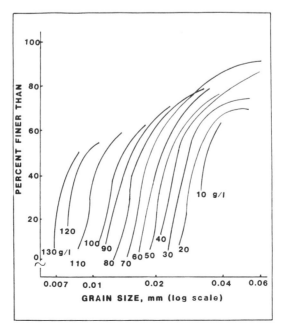

FIG. 8—*Sensitivity analysis: Concentration (fine sample).*

percentage finer than a given diameter increases as concentration increases. This result suggests that variations in dry weights of samples should be held to a minimum, and certainly upper and lower limits should be imposed. Recall that upper and lower limits of 100 and 40 g, respectively, were used throughout this research.

In conclusion, the PSD results from the hydrometer method appear to be reproducible. Sampling technique, temperature (after correction had been made), dispersing time, and meniscus proved to have negligible effects on results. Concentration had a substantial effect; large variations in concentration should be avoided.

PSD Analysis

The sampling effort was concentrated in the area of the site where construction of a project access highway was underway. The area included cleared and grubbed right-of-way, cut and fill slopes, and roadway in various degrees of completion. The route of the transported soil began at a fill section of the highway and continued down a cut section with a slope in excess of 10%. The soil then encountered a drainage divide, where its path bifurcated and passed over two separate lines of multiple silt fences. It was transported through small dug basins, brush barriers, and paved ditches. The considerable material not contained in these sediment controls then passed into high-quality mountain streams.

Soil samples were classified as in situ or disturbed area samples (referred to as "source samples"), material deposited or in transit between sources and in-place sediment control measures (referred to as "intermediate samples"), samples from in and around in-place sediment control measures (referred to as "control measure samples"), and elevated streamflow suspended sediment samples (referred to as "stormflow samples"). For each of the four sample types, every curve calculated was given equal weight in an averaging procedure to obtain a composite curve. These composites, constructed from 12, 12, 20, and 4 PSD samples, respectively, are presented in Figs. 9–12. The curves provided several significant insights into the nature of the eroded material at the construction site sampled.

Without exception, the nondispersed portion of the PSD curve falls off to 0% finer than in the silt range of diameters. However, no dispersed curve shows less than 6% finer than 0.001 mm

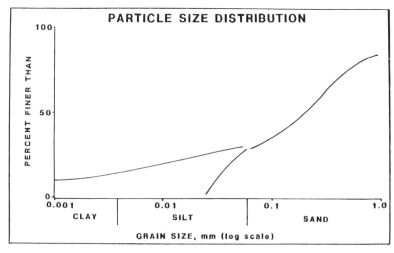

FIG. 9—*Composite PSD curve for source samples (N = 12).*

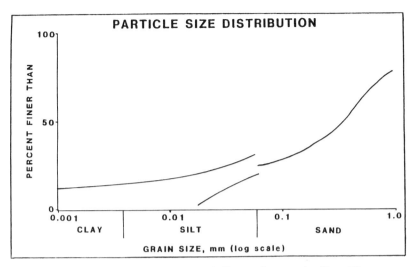

FIG. 10—*Composite PSD curve for intermediate samples (N = 12).*

(well into the clay range). This result suggests that, while the samples contain a substantial fraction of primary clay particles, in the field, where erosion, transport, and impact take place, these particles flocculate to form sand- or silt-size aggregates which behave and settle like much larger particles.

Were a designer to use a dispersed PSD curve in his selection and planning of erosion control systems at this site, he would see that the eroded material contains a sizable fines fraction. From this, the designer may assume that it is economically infeasible to control this fraction. Even worse, though less likely, he may implement very expensive control systems in order to capture

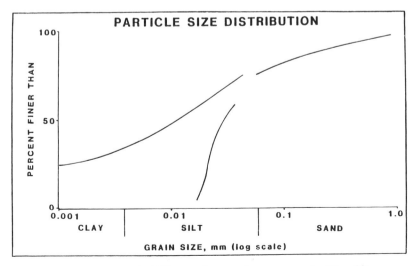

FIG. 11—*Composite PSD curve for control measure samples (N = 20).*

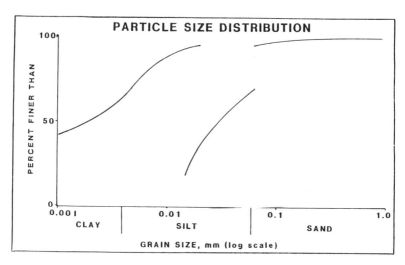

FIG. 12—*Composite PSD curve for stormflow samples (N = 4).*

this clay fraction. In actuality, however, the clay fraction is flocculating and behaving as silt-sized particles, and an entirely different set of control measures would be indicated.

Each of the four types of samples exhibits PSD curves which track fairly closely (within 20% for each of the four groups.). Hence, the use of a composite curve for each of the sample groups seems valid and appropriate for design purposes.

Finally, a definite trend in PSD is exhibited as one moves from source to intermediate samples downslope to control measures and finally to stormflow material leaving the project area and entering off-site streams. The composite curves shift distinctly upward from the abscissa, proving that, for a given grain size, percentage finer than increased with a material's distance from its erosion source. This result is summarized in Table 1. This increase is to be expected since, as material is transported from its source to its stream destination, gravity removes heavier particles from suspension.

Conclusions

PSD analysis by the hydrometer method provides a simple, quick, reproducible, and inexpensive means of characterizing eroded material. While a dispersed PSD is valuable in determining primary particle sizes, its worth in characterizing a sample as it actually exists in the field, frequently in

TABLE 1—*Percent finer than: three grain sizes at distances from source.*

Sample Type	% < 0.062 mm for Sieved Curve	% < 0.02 mm for Nondispersed Curve	% < 0.004 mm for Dispersed Curve
Source	28	0	13
Intermediate	26	3	14
Control measure	77	15	35
Stormflow	96	37	64

aggregated form, is very questionable at best. A nondispersed PSD provides a very different and much more appropriate and useful picture of in-field particle size and particle behavior.

The dual dispersed and nondispersed PSD analysis presented here provides a tool by which the environmental hazard posed by an eroded material can be assessed and mitigated. The characterization of the material eroded at the particular site studied in this research determined that the material contains a large percentage of primary clay and poses inherent adsorption/absorption dangers to downstream water quality. It further determined that this clay is flocculating to form aggregates which behave and settle like much large particles, a determination necessary for the rational choice and design of effective sediment control measures.

Acknowledgments

This research was performed in support of a Master of Science thesis at Clemson University, Clemson, South Carolina. It was funded by the construction site owner Duke Power Co., by the National Wildlife Federation, and by the U.S. Geological Survey.

References

[1] ASTM Method for Particle-Size Analysis of Soils [D 422–63 (1972)], *1986 Annual Book of ASTM Standards,* Vol. 04.08, American Society for Testing and Materials, Philadelphia, 1986.
[2] *National Handbook for Recommended Methods for Water-Data Acquisition,* U.S. Geological Survey, Reston, VA., 1977.
[3] Day, P. R., "Particle Fractionation and Particle Size Analysis," in *Methods of Soil Analysis,* C. A. Black, Ed., American Society of Agronomy, Madison, WI, 1965, pp. 545–567.
[4] Kaddah, M. T., "The Hydrometer Method for Detailed Particle Size Analysis: Graphical Interpretation of Hydrometer Readings and Test of Method," *Soil Science,* 118, Vol. 2, pp. 103–107, 1974.
[5] *Field Manual for Research in Agricultural Hydrology,* Brakensiek, D. L., Osborn, H. B., and Rawls, W. J., Coordinators, Handbook No. 224, U.S. Department of Agriculture, Washington, DC, 1979.

Charles A. Pittinger,[1] Vincent C. Hand,[2] Julie A. Masters,[2] and
Linda F. Davidson[3]

Interstitial Water Sampling in Ecotoxicological Testing: Partitioning of a Cationic Surfactant

REFERENCE: Pittinger, C. A., Hand, V. C., Masters, J. A., and Davidson, L. F., "**Interstitial Water Sampling in Ecotoxicological Testing: Partitioning of a Cationic Surfactant,**" *Aquatic Toxicology and Hazard Assessment: 10th Volume, ASTM STP 971*, W. J. Adams, G. A. Chapman, and W. G. Landis, Eds., American Society for Testing and Materials, Philadelphia, 1988, pp. 138–148.

ABSTRACT: Methods for collecting interstitial water (IW) in laboratory ecotoxicological testing were compared and used to characterize the partitioning behavior of a cationic surfactant in batch sediment/water systems. IW sampling methods included a conventional centrifugation technique, dialysis through a rayon membrane, and a novel fritted glass diffusion sampler. Dodecyl trimethyl ammonium chloride (TMAC) was the test compound monitored radiometrically in each static-renewal experiment. A 10-mg/L TMAC solution was delivered to test systems as overlying water, replenished daily; TMAC levels in sediment, interstitial water, and overlying water were monitored throughout the 20-day experiments. IW samples collected by the centrifugation procedure contained greater TMAC concentrations than samples collected by either dialysis or the fritted glass sampler. Equilibration of TMAC concentrations across the dialysis membrane was slow (>5 days), resulting in a substantial lag. Advantages of using the fritted glass sampler for ecotoxicological testing are discussed. Following rapid changes in TMAC levels in each compartment during the initial five days of each experiment, order of magnitude differences were observed in TMAC concentrations among sediments, overlying water, and interstitial waters, in decreasing order. The partition coefficient measured in separate shake-flask experiments performed at a high solids concentration was similar to the observed ratio between TMAC levels in sediments and interstitial water. Sorption of TMAC onto sediments was consistent with a first-order kinetics process.

KEYWORDS: interstitial water, pore water, sampling methods, ecotoxicology, cationic surfactant, partitioning, sediment, sorption

Accurate characterization of chemical exposure during ecotoxicological testing is critical to sound environmental risk assessment. Dose levels in bioassays with aquatic species are usually not known as precisely as with mammalian species, owing to differences in exposure techniques (that is, submersion versus injection or ingestion). Toxicity tests with benthic organisms in sediment/water systems offer unique challenges in dose quantitation, as multiple routes of exposure to test substances are possible. Depending upon the ecology of the test species and the design of the test, these routes may include uptake of soluble chemical fractions in overlying or interstitial

---

[1] Environmental toxicologist, Packaged Soap and Detergent Division, The Procter and Gamble Co., Ivorydale Technical Center, Cincinnati, OH 45217.

[2] Research chemist and technician, respectively, Enviornmental Safety Department, The Procter and Gamble Co., Ivorydale Technical Center, Cincinnati, OH 45217.

[3] Information specialist, Human and Environmental Safety Division, The Procter & Gamble Co., Ivorydale Technical Center, Cincinnati, OH 45217.

(pore) waters and/or ingestion of sorbed fractions associated with particulates. The route of exposure may affect a chemical's absorption, distribution, biotransformation, and excretion within an organism and thus may influence the magnitude and mode of the toxic response.

While dose-response relationships may be inferred from chemical levels measured in sediments, interstitial water, or overlying water (OW), only that relationship based upon the principal route of exposure is most realistic and relevant in describing an organism's sensitivity. Knowledge of the exposure route may be critical in effectively regulating environmental exposures to protect aquatic communities. Interstitial water (IW) may be a principal route of exposure for organic chemicals to benthic organisms [1,2], but can be particularly difficult to monitor. A variety of IW collection techniques, as reviewed by Bricker [3] and by Kriukov and Manheim [4], have been developed for particular sampling applications. These include collection by centrifugation [5,6], porous samples [7], dialysis [8,9], pressure filtration [10], and solvent displacement [11,12]. The majority of these, however, are designed for *in situ* use and are not well-suited to studies in confined laboratory test systems.

Monitoring of IW concentrations of test chemicals in laboratory bioassays should ideally: (1) provide representative samples with sufficient frequency to track the dynamics of the test system; (2) offer minimal disturbance to sediments and benthic organisms; (3) provide adequate sample volume for analytical characterization; and (4) occupy relatively little space in test chambers. The characterization of partitioning patterns among sediments, IW, and OW assists in understanding of the fate and dynamics of test compounds in simulated sediment/water systems. Such information is useful in determining routes of exposure, accurately defining dose-response relationships, and evaluating chemical hazards to aquatic organisms. This paper assesses the applicability of three IW collection techniques to laboratory ecotoxicological testing and demonstrates partitioning of a sorptive test chemical among sediments, IW, and OW in batch laboratory systems.

Materials and Methods

Two *in situ* methods for collecting IW samples were each compared to IW collections by conventional sediment centrifugation in 20-day, static-renewal experiments. The methods compared were dialysis through a regenerated cellulose (rayon) membrane and collection by filtration through a fritted glass sampler. Dodecyl trimethyl ammonium chloride (TMAC), a cationic surfactant representative of those used in commercial detergent formulations, was the test compound monitored in these studies. TMAC has a molecular weight of 263.5 g/mole and is highly soluble in water (>1000 mg/L). Radiolabelled TMAC prepared by Amersham (Arlington, IL) was labelled with carbon-14 on the alpha carbon of the alkyl chain; its specific activity and pruity were 95 μCi/mg and 94.9%, respectively. It was diluted before use with nonlabelled TMAC from Eastman Kodak (Rochester, NY), with a purity greater than 99%.

In two experiments, a sediment/water slurry was siphoned into a series of beakers or centrifuge bottles and allowed to settle for 24 h. Dialysis bags or fritted glass samplers were positioned in the sediments prior to settling. The ratio of sediment to water was 1:1.2 (w/v) in the first (dialysis) experiment and 1:1.4 (w/v) in the second (fritted glass sampler) experiment. TMAC was introduced into each system by replenishing the OW daily with a 10-mg/L solution. Care was taken to minimize disturbance of the sediment when replenishing test systems. Test systems remained at an ambient temperature of 24 ± 2°C throughout the experiments. Test systems were sacrificed in triplicate on Days 1, 5, 10, 15, and 20, and samples of OW, IW, and sediments were collected for TMAC analyses. Sacrificed test systems were not replenished on the day of collection, so that TMAC values in OW represented solutions that had equilibrated in test systems for 24 h.

Natural stream sediments used in these experiments were collected from the Little Miami River near Xenia, Ohio, and from Rapid Creek near Rapid Creek, South Dakota. Neither sediment contained detectable levels of TMAC at the start of the experiments. The sediments were sieved before use and stored wet at 4°C. Particle sizes favored the fine clay and silt fractions over the sand fraction. The Total Organic Carbon content of the sediments was 1.8%. High-quality water from a reverse osmosis filtration system (hardness = 0 meq/mL; pH = 7.0) was used in these experiments.

Sample Collection

The following were performed on each sampling day, after removing OW from test systems:

1. *IW Sample Collection by Centrifugation.* After dialysis bags or fritted glass samplers were sampled and removed from the test systems, the entire volume of sediment from each of three test systems was immediately centrifuged (10 000 g, 40 min). Supernatants of centrifuged samples were analyzed for TMAC concentration and compared with IW values obtained by dialysis or the samplers from respective test systems.

2. *IW Sample Collection by Dialysis.* Rayon dialysis bags (Collodion Bags SM1320-00, Sartorius Filters Inc., Haywood, California) with an internal capacity of approximately 5 mL were used in order to minimize membrane deterioration during the experiments. After soaking in distilled water for 1 to 2 h, the water-filled bags were sealed with polypropylene closures (Fisher No. 08-670-11AA) and buried approximately 2 cm below the sediment surface (one per system). Following removal of overlying water, bags were retrieved from sacrificed test systems and the contents analyzed for TMAC. A separate experiment was performed to determine rates of diffusion across the dialysis membrane. Twelve bags were submersed in 500 mL of a 10-mg/L TMAC solution without sediments. The solution was not replenished during the 5-day diffusion experiments. Samples of the external solution and the bag contents were collected in triplicate on each sampling day and analyzed for TMAC.

3. *IW Sample Collection by the Fritted Glass Sampler.* Fritted glass IW samplers were prepared in-house from modified gas dispersion tubes (Fisher No. 28630). The cylindrical frit section (10 to 15 μm pore size) was removed from each dispersion tube and attached 2.5 cm beneath a glass sphere with an internal capacity of 10 to 15 mL. An open-ended glass tube (7 cm length; 11 mm I.D.) was attached to the top of the sphere. Prior to use, the frit section was wrapped with a strip of 0.7-μm glass fiber filter paper (Whatman GF/F) to prevent clogging. The sampler was fixed firmly against the side of a test chamber, positioned with the frit buried approximately 2.5 cm beneath the sediment/water surface (Fig. 1). IW was sampled by pipetting through the open end of the tube extending above the water surfaces. Following evacuation, the sampler refilled by hydrostatic pressure in 1 to 2 h.

Channeling Experiments with the Fritted Glass Sampler

An experiment was initially conducted with the fritted glass sampler to determine whether OW was drawn into the sampler by channeling across the external surface, thereby interfering with the integrity of IW samples. Six samplers were embedded in approximately 125-mL sediment in 250-mL beakers, and the OW and IW in the samplers were removed by pipetting. OW was replaced with 80 mL of (0.001% v/v Formsilabs Red Water Soluble) dye solution. OW and IW from two beakers were successively collected after 3, 6, and 24 h; after removing OW and the IW samplers, each set of beakers was sacrificed to collect IW samples by the centrifugation technique. OW samples were similarly centrifuged. Following preparation of a standard curve, dye concentrations in OW and IW samples were measured by absorbance at 553 nm using a Beckman 26 spectrophotometer. Dye concentrations were expressed as a percentage of the concentration of the dye solution placed in the beakers.

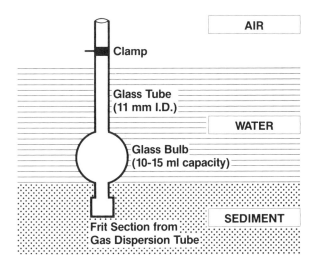

FIG. 1—*Fritted glass sampler used to collect interstitial water. Frit section was wrapped with Whatman glass fiber filter.*

Analytical

TMAC levels in sediment and water samples were determined by direct liquid scintillation counting techniques. To remove CO_2 generated during TMAC biodegradation, aqueous samples were acidified with 1.2 N HCL and purged with nitrogen; sediment samples were acidified and stirred mechanically. Acidified sediments were diluted to a final concentration of 30 to 50 mg per 10-mL sample. Percent solids was measured to determine the actual mass of sediment analyzed in each sample. Aqueous and sediment samples in Fisher Scintillation Surfactant cocktail were counted for 2 min on a Tracor Analytic Mark III Liquid Scintillation Counter. Results reported in dpm's were converted to TMAC concentrations; standard curves were developed to confirm a linear response of the counter.

Triplicate IW and OW samples collected on each sampling day were also analyzed by direct counting procedures. The volume of IW samples collected by dialysis was measured and diluted by water to 10 mL (if necessary). Ten millilitres of the cocktail were added, and the samples were mixed and counted as above. TMAC concentrations were corrected for the amount of water added to each sample, as necessary.

K_d Determination

A separate study was conducted to determine the partition coefficient (K_d) for TMAC in the same sediment slurry used in the dialysis experiment. A modification of the U.S. Environmental Protection Agency's (EPA) shake-flask procedure [*13*] was employed. The slurry was spiked with 10 mg/L TMAC and allowed to equilibrate 24 h at 24 ± 3°C by continuous shaking. Sediment and water were separated by centrifugation, and TMAC analyses were performed on the two phases as described above. K_d was calculated as the ratio of sediment concentration (μg/g dry weight) to dissolved concentration (mg/L).

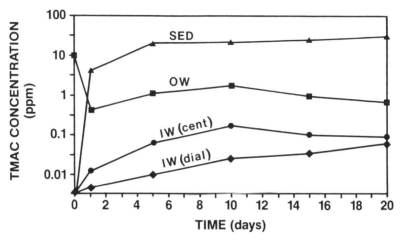

FIG. 2—*Partitioning of TMAC among sediments (▲, μg/g dry weight) overlying water (■, mg/L) and interstitial water (●, mg/L collected by centrifugation; ◆, mg/L collected by dialysis) in the first experiment.*

Results and Discussion

Comparison of IW Sampling Techniques

TMAC levels measured in sediment, OW, and IW in the first experiment are shown in Fig. 2. Differences between IW samples collected by centrifugation and by dialysis were apparent. TMAC concentrations in centrifuged IW samples increased from 0.011 mg/L on Day 1 to 0.168 mg/L on Day 10, then decreased by Day 15 to 0.098 mg/L. TMAC concentrations in dialyzed samples increased at a slower rate, from 0.005 on Day 1 to 0.097 on Day 20. Samples collected by dialysis contained significantly lower TMAC concentrations than centrifuged samples over the course of the experiment [p(t-test) = 0.0016]. The dialysis procedure was also found to be less precise than centrifugation in this experiment; coefficients of variation in TMAC concentrations averaged over the test period were 36.9 and 16.4%, respectively.

It was hypothesized that restricted diffusion of TMAC caused a lag in the equilibration of TMAC concentrations across the dialysis membrane and lower IW values. (Sorption of TMAC to the dialysis membrane, as determined by solvent extraction of the membrane, was not significant.) The diffusion rate across the membrane was estimated by submersing dialysis bags in a TMAC solution of 10 mg/L without sediment and by sampling internal and external solutions daily. TMAC concentrations in the bags were approximately half (47 to 62%) of those measured in solution, and the time for equilibration was estimated to exceed 120 h (Fig. 3). This finding is consistent with the low TMAC concentration (0.01 mg/L) observed on Day 5 in the dialyzed IW samples (Fig. 2). More rapid and complete equilibration was observed in a similar diffusion experiment (10 mg/L, without sediment) with the fritted glass sampler (Fig. 3).

TMAC concentrations in IW samples collected by the fritted glass sampler were compared to centrifuged samples in the second experiment (Fig. 4). Again, concentrations in IW collected by centrifugation were significantly higher [p(t-test) = 0.0037] over the course of the experiment. TMAC concentrations increased in both sample types during the initial 10 days and continued to increase in centrifuged IW samples. IW concentrations collected by the fritted sampler appeared to stabilize after 10 days. The fritted sampler provided slightly greater precision than centrifugation; average coefficients of variation were 23.6 and 27.9%, respectively.

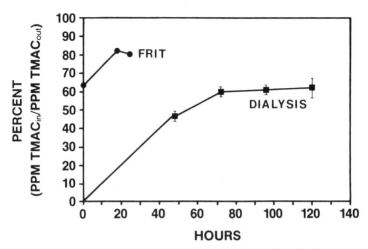

FIG. 3—*Uptake of TMAC from solution into two interstitial water samplers, the fritted glass sampler (●) and the dialysis bag (■). Results (mean ± SE) are expressed as the percent of TMAC concentration inside each sampler to the concentration surrounding the sampler at each time of interval. The fritted glass sampler was initially sampled after 1 h.*

The differences observed in IW concentrations among the techniques may be due to the unique nature of the samples collected by each. The centrifuged sample is an instantaneous measure of IW, whereas samples collected by dialysis and, to a lesser extent, by the fritted sampler represent integrated measures of conditions existing prior to the time of collection. Although centrifugation may be the most common method used to date, there is no assurance that it accurately reflects conditions in undisturbed systems. The severe disruption of sediments upon centrifugation could conceivably alter partitioning among compartments and could also change the oxidation state of certain chemicals. Oxidation during sample handling may reduce concentrations of inorganic

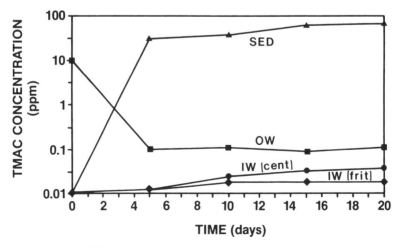

FIG. 4—*Partitioning of TMAC among sediments (▲, μg/g dry weight) overying water (■, mg/L) and interstitial water (●, mg/L collected by centrifugation; ◆, mg/L collected by the fritted glass sampler) in the second experiment.*

phosphate in interstitial water rich in ferrous iron [14]. Sampler composition can also affect the integrity of interstitial water samples [7]. The higher TMAC levels in the centrifuged samples versus those collected by dialysis or the fritted sampler may have been an artifactual result of the disruption occurring during centrifugation.

For ecotoxicological studies in the laboratory, the disruption of sediments associated with the dialysis and centrifugation procedures could have additional implications. Severe disruption of the substrate could injure vulnerable benthic organisms (for example, certain oligochaetes) and life stages (for example, midge larvae) or mask certain endpoints (for example, suspended chlorophyll concentrations). Sediment disruption might be minimized in the dialysis procedure by using a rigid and fixed sampler of the type described by Hesslein [15], modified to allow IW collections with the sampler in place. However, degradation of cellulitic membranes during use has been cited [7] as a further limitation of dialysis procedures.

In the IW channeling experiment with the fritted glass sampler, only low concentrations of dye (0.4 to 3.7% of the OW dye concentration) were measured in 3, 6, and 24-h IW samples collected by the sampler (Table 1). Dye concentrations did not appreciably increase in successive samples. This suggests that channelling of OW into the sampler is not a significant source of interference in these IW samples. By comparison, dye concentrations measured in respective IW samples collected by the centrifugation technique were similar or slightly higher (4.7 to 7.3% of OW). This was to be expected, as it is possible for small volumes of OW to be entrained in IW samples collected by this technique. The decrease in dye concentration in the OW observed over the 24-h test period was attributed to biodegradation and sorption of the dye to the sediment. When IW dye concentrations were normalized to the initial dye concentration to account for this activity, it again did not appear that the dye concentrations significantly increased in IW samples over time.

The fritted glass sampler represents an alternative for monitoring IW concentrations of cationic surfactants and may be useful for other chemical classes as well. The sampler does not require sacrificing of test systems or undue disturbance to the sediments and allows frequent and repeated sampling. The depth of sampling could be adjusted by varying the distance between the frit and the glass sphere. Provided that the volume of the sampler is small relative to the total IW volume in the system, it is unlikely that the collection of samples by this technique would appreciably alter the dynamics of the system. The fritted sampler is particularly well-suited to ecotoxicological studies with radiochemicals which require nondestructive, repeated sampling. We are currently using this method in research programs to evaluate sediment bioavailability of cationic materials and to validate the procedure with other classes of materials.

A fourth approach, solvent displacement, was also investigated as a possible technique for IW collection of TMAC samples. The technique is similar to the centrifugation procedure; an inert fluorocarbon solvent is added to sediment samples prior to centrifugation to displace the IW into

TABLE 1—*Dye concentrations in OW and IW samples collected in the channeling experiment with the fritted glass sampler.*

Sampling Interval	Percent of Initial Dye Concentration		
	OW	IW (Fritted Sampler)	IW (Centrifuged)
3	82.7	3.4	7.1
	82.9	1.9	7.2
6	70.8	0.4	7.3
	73.6	4.8	5.9
24	49.8	2.4	4.7
	44.4	3.7	4.7

a discrete layer. Batley and Giles [12] reported high recoveries of metals using this technique, without observing partitioning of the metals into the solvent. However, we found that the partitioning of TMAC into the recommended solvent, Fluorinert FC-72 (The 3M Corp., St. Paul, Minnesota), was significant (up to 15%), which precluded further use of the technique in these studies. The solvent displacement technique also necessitates disrupting sediments and so would not be ideally suited for ecotoxicological testing.

TMAC Partitioning in Sediment/Water Systems

Monitoring IW levels of chemicals, together with sediment and OW data, enables better characterization of test system dynamics during ecotoxicological testing. This is necessary to confirm that stable exposure concentrations are maintained within target ranges. Multicompartmental monitoring also aids in determining the equilibrium status of the system (for example, steady-state versus nonsteady-state) and the partitioning behavior of the chemical, variables which can have implications for toxicity testing. For example, the rapid changes in TMAC levels observed during the initial five days in both experiments (Figs. 2 and 4) indicate a need to allow sediment/water test systems to equilibrate prior to initiating benthic toxicity tests.

In the first experiment (Fig. 2), the range of TMAC levels measured in sediments, OW, and IW extended over two orders of magnitude, following rapid increases in sediment and IW levels during the first five days. TMAC was rapidly depleted from the OW within 24 h following daily replenishment. Levels of TMAC sorbed to sediment, expressed on a dry weight basis, were 20 to 30 µg/g between Days 5 and 20. TMAC levels in OW during this period were 0.6 to 1.7 mg/L, while those in IW were approximately one order of magnitude less, 0.07 to 0.17 mg/L.

In the second experiment (Fig. 4), TMAC levels at Day 20 in sediment were higher (68 µg/g) and OW and IW levels were lower (0.11 and 0.018 mg/L, respectively) than in the first experiment. These differences in magnitude of TMAC levels between the two experiments may be related to the higher sediment:water ratio in the second experiment; approximately 15% more TMAC was added to the test system in the second experiment, and higher sediment concentrations were observed. In both experiments, the relative stability of TMAC levels in each compartment during the later stages suggested that the systems were approaching equilibrium.

In both experiments, the partitioning patterns observed among compartments were likely the net result of two important removal processes for TMAC, sorption/desorption in sediments and biodegradation in water. TMAC concentrations in IW were low relative to OW in both experiments. It is speculated that the large surface area of the fine sediments provided a greater number of sorption sites in proximity to IW versus OW. In systems in which the rate of adsorption onto sediments exceeds the rate of desorption into IW, a net chemical flux would exist from the IW to the sediments. Replacement of TMAC into IW by diffusion from OW is likely to be slow relative to the rate of adsorption. This would promote the partitioning pattern observed in these experiments.

Prediction of chemical partitioning has potential value in designing toxicity studies (that is, selecting dose ranges and levels of exposure in a particular compartment). To determine whether the observed partitioning of TMAC between sediment and IW could be predicted from K_d measurements, conventional shake-flask experiments were conducted. A K_d value of 423 was determined in shake-flasks containing a high solids concentration (30%) equivalent to that in the static test system (Table 2). This value was compared to ratios of TMAC concentrations between sediment and OW and between sediment and IW on Day 20 in the dialysis experiment. The K_d value was similar to the ratio between sediment and IW levels and approximatley an order of magnitude greater than the ratio between sediment and OW levels. This demonstrates that K_d measurements have predictive value in estimating the partitioning of cationic surfactants between sediment and IW.

Determination of the kinetics of sorption/desorption processes can also assist in the design and

TABLE 2—*Partition coefficients (K_d) determined in shake-flask and beaker experiments.*

Experiment	TMAC Concentrations			K_d
	Sorbed, µg/g	OW Dissolved, mg/L	IW Dissolved, mg/L	
Shake-flask	33.5	7.9×10^{-2}	...	423
Static beaker[a]	30.25	6.7×10^{-1}	...	45
Static beaker[a]	30.25	...	9.0×10^{-2}	336

[a] Calculated from Day 20 results in the dialysis experiment.

conduct of toxicity studies. Kinetic parameters describing TMAC sorption in the experimental test systems were estimated by fitting the sediment data from the dialysis experiment to a first-order kinetic equation of the form

$$\text{TMAC} = \text{TMAC}_{eq}(1 - e^{-kt})$$

where TMAC = concentration (µg/g) on sediments at any time (t, days),
TMAC_{eq} = concentration on sediments at equilibrium,
e = 2.718
and k = sorption rate (days^{-1}).

The half-life to equilibrium, $t_{\frac{1}{2}}$, was calculated as $0.693/k$. Estimated values for k, TMAC_{eq}, and $t_{\frac{1}{2}}$ are listed in Table 3, and a comparison of observed and predicted sediment values is shown in Fig. 5. Close agreement was found between the predicted TMAC_{eq} value of 31.6 µg/g and the observed sediment concentration of 0.3 µg/g on Day 20, suggesting that the experimental test systems were approaching equilibrium. This information can be useful in ecotoxicological testing

TABLE 3—*Sorbed TMAC levels measured in the dialysis experiment and calculated first-order kinetic parameters.*

	Sorbed TMAC Levels	
	Sediment Concentration (µg/g)	
Day	Mean	S.E.
0	0	...
1	4.3	0.9
5	20.4	0.2
10	21.9	1.9
15	25.1	1.2
20	30.3	0.6
First-Order Kinetic Parameters[a]		
k (days^{-1})	0.201 (0.119, 0.282)	
TMAC_{eq} (µg/g)	28.2 (24.9, 31.6)	
$t_{1/2}$ (days)	3.45	

[a] Calculated from: $\text{TMAC} = \text{TMAC}_{eq}(1 - e^{-kt})$. Numbers in parentheses are lower and upper 95% confidence limits, respectively.

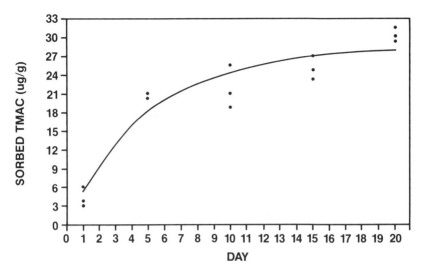

FIG. 5—*Comparison of measured and predicted TMAC levels from the dialysis experiment. Measured values are shown as points about the curve fit to a first-order kinetics equation.*

as a measure of test system stability and the time required to reach steady-state in a particular sediment/water system. The estimated half-life of 3.45 days indicates that 13.8 days (four half-lives) would be required for the sorption process to become 93.75% complete in the batch system used in the dialysis experiment.

Biodegradation kinetics can also influence partitioning in sediment/water systems, but are more difficult to measure *in situ* during ecotoxicological testing. Biodegradation was known to be a significant removal mechanism for TMAC in these experiments. In a preliminary study in which samples were not acidified and sparged, a 20-day mass balance of TMAC accounted for only 12.6% of the total amount added to the systems (adsorption to container walls was not significant). The remainder was presumed to have been volatilized as CO_2 from the open systems. This is consistent with the short (approximately 20 h) half-life for TMAC reported in preexposed lake water [16].

In addition, biodegradation may have contributed to the difference between IW and OW levels of TMAC observed in these experiments. There is evidence that sorbed TMAC in settled sediment systems is not directly available for degradation (R. J. Schimp, personal communication); desorption from sediments into IW and OW apparently increases TMAC availability for microbial attack. More rapid biodegradation of a cationic surfactant in a sediment/water slurry than in water alone (estimated half-lives were 4.9 and 13.8 days, respectively) has also been reported [17]. Thus it is possible that more rapid degradation of TMAC in IW than in OW served to enhance the difference in TMAC levels between these compartments.

Conclusions

It has been shown that the centrifugation, dialysis, and the fritted glass sampling methodologies yield significant differences in IW samples of a cationic surfactant. On the basis of criteria relevant to IW monitoring in laboratory toxicity tests, the fritted glass sampler was judged to be superior. Following an initial equilibration period characterized by substantial changes in TMAC levels in each compartment, order of magnitude differences in TMAC levels were observed among sediment,

OW, and IW, in decreasing order. It was postulated that these differences were primarily due to sorption/desorption processes occurring between sediments and IW and to metabolism of TMAC to CO_2 during biodegradation. Knowledge of the physical and biological kinetics of test materials in sediment/water systems were shown to have potential application in conducting ecotoxicological testing and in defining dose-response relationships.

References

[1] Adams, W. J., Kimerle, R. A. and Mosher, R. G. in *Aquatic Toxicology and Hazard Assessment (Seventh Symposium), ASTM STP 854*, R. D. Cardwell, R. Purdy, and R. C. Bahner, Eds., American Society of Testing and Materials, Philadelphia, 1985, pp. 429–453.

[2] Roesijadi, G., Anderson, J. W. and Blaylock, J. W., *Journal of the Fisheries Research Board of Canada*, Vol. 35, 1978, pp. 608–614.

[3] Bricker, O. P., "Sampling the Distribution and Speciation of Dissolved Chemical Substances in Interstitial Waters," *Biogeochemistry of Estuarine Sediments*, Proceedings of the 1976 UNESCO/SCOR Workshop, UNESCO, Paris, France, 1978, pp. 75–79.

[4] Kriukov, P. A. and Manheim, F. T. in *Dynamic Environment of the Ocean Floor*, K. A. Fanning and F. T. Manheim, Eds., Lexington Books, Lexington, MA, 1982, pp. 3–26.

[5] Van den Berg, C. M. G. and Dharmvanij, S., *Limnology and Oceanography*, Vol. 29, No. 5, 1984, pp. 1025–1036.

[6] Orem, W. H. and Gaudette, H. E., *Organic Geochemistry*, Vol. 5, No. 4, 1984, pp. 175–181.

[7] Zimmermann, C. F., Price, M. T. and Montgomery, J. R., *Estuarine and Coastal Marine Science*, Vol. 7, 1978, pp. 93–97.

[8] Hoepner, T., *Environmental Technology Letters*, Vol. 2, 1981, pp. 187–196.

[9] Mayer, L. M., *Limnology and Oceanography*, Vol. 21, No. 6, 1976, pp. 909–912.

[10] Hulbert, M. H. and Brindle, M. P., *Geological Society of America Bulletin*, Vol. 86, No. 1, 1975, pp. 109–110.

[11] Batley, G. E. and Giles, M. S., *Water Research*, Vol. 13, 1979, pp. 879–886.

[12] Batley, G. E. and Giles, M. S. in *Contaminants and Sediments 2*, R. A. Baker, Ed., Ann Arbor Science Publishers, Ann Arbor, MI, 1980, Chapter 7, pp. 101–17.

[13] U.S. Environmental Protection Agency, Office of Pesticides and Toxic Substances, *Federal Register*, Vol. 50, No. 188, Section 796.2750, 27 Sept. 1985, p. 39275.

[14] Bray, J. T., Bricker, O. P., and Troup, B. N., *Science*, Vol. 180, No. 4093, 1973, pp. 1362–1364.

[15] Hesslein, R. H., *Limnology and Oceanography*, Vol. 21, No. 6, 1976, pp. 912–914.

[16] Ventullo, R. M. and Larson, R. J., *Applied Environmental Microbiology*, Vol. 51, No. 2, 1986, pp. 356–361.

[17] Larson, R. J., *Residue Reviews*, Vol. 85, 1983, pp. 159–171.

Anne E. McElroy[1] and J. C. Means[2]

Factors Affecting the Bioavailability of Hexachlorobiphenyls to Benthic Organisms

REFERENCE: McElroy, A. E. and Means, J. C., "**Factors Affecting the Bioavailability of Hexachlorobiphenyls to Benthic Organisms,**" *Aquatic Toxicology and Hazard Assessment: 10th Volume, ASTM STP 971*, W. J. Adams, G. A. Chapman, and W. G. Landis, Eds., American Society for Testing and Materials, Philadelphia, 1988, pp. 149–158.

ABSTRACT: The purpose of this study was to simultaneously evaluate several key factors which potentially affect the accumulation of hydrophobic pollutant compounds in sediment by benthic organisms. These factors were: (1) organic carbon content of the sediment; (2) equilibration time of the pollutant with the sediment; and (3) feeding strategies of the organism. [^{14}C-U]2,4,5,2',4',5'-hexachlorobiphenyl (HCBP) was equilibrated with sediments from Narragansett Bay, RI containing either 4 or 2% total organic carbon for periods of either one or four weeks. Biological availability of the compound was assessed by following the kinetics of HCBP uptake into either the omnivorous deposit feeding polychaete *Nephtys incisa* or the deposit feeding bivalve *Yoldia limatula* exposed to bedded sediment.

In an attempt to normalize bioaccumulation data for differences in the lipophilic reservoirs between sediment and organism, an apparent preference factor (APF) was calculated comparing body burdens normalized to % lipid with sediment concentrations normalized to % total organic carbon (TOC)

$$APF = (HCBP_s/\%TOC)/(HCBP_o/\%lipid)$$

For both organisms, APF reached a constant value within 20 days. Length of isotope incubation with sediment had no observable effect on either the rate of approach to a constant APF or the value of APF reached. Even after normalizing with the preference factor calculation, significant differences were still evident between bioaccumulation from the two sediment types and between bioaccumulation from the same sediment by *Yoldia* and *Nephtys*. These data indicate that in addition to the influence of sediment TOC and lipid content of the organism, intrinsic properties of sediment and biological properties of the organism influence the transfer and bioaccumulation of hydrophobic compounds.

KEYWORDS: PCBs, polychaetes, bivalves, bioavailability, preference factors

A large number of physical, chemical, and biological factors have been shown to affect the bioaccumulation of organic compounds in the aquatic environment. These include the aqueous solubility of the compound; the presence or absence of polar substituent groups and heteroatoms; environmental factors such as temperature, salinity, and pH; and biological factors such as respiration, feeding, and excretion rates, intermediary metabolism, and trophic relationships. In the benthic environment, a number of questions concerning the bioavailability of organic compounds bound to sediment remain. What fraction of contaminants bound to sediment can be transferred to biota, what are the rate-limiting factors affecting this transfer, and what are the pathways through which transfer occurs are all unresolved questions. Since sediment serves as the primary reservoir for these compounds [1], it is important to develop a clear understanding of the factors affecting the transfer of contaminants in bedded sediment to the biota.

---

[1] Assistant professor, Environmental Sciences Program, University of Massachusetts/Boston, Boston, MA 02125.
[2] Professor, Environmental Studies, Louisiana State University, Baton Rouge, LA 70803.

Results from a number of studies on the partitioning behavior of nonpolar organic compounds to soils and sediment have demonstrated partitioning to be primarily controlled by physical/chemical properties of the compound. These factors include water solubility, the octanol-water partition coefficient (K_{ow}), and the sorbant's organic carbon content (f_{oc}) [2–4]. In a similar fashion, studies on the bioaccumulation of organic contaminants from the water column have also found bioaccumulation to be correlated with K_{ow} [5–9].

Using a thermodynamic approach, McFarland [10] combined two empirically derived equations for the correlation between K_{ow} and K_{oc} (the organic carbon normalized sediment partition coefficient) and K_{ow} and BCF (the lipid-normalized bioconcentration factor) to derive a "maximum bioaccumulation potential" (TBP) for neutral organic contaminants. TBP can be calculated using the following equation

$$TBP = (C_s/f_{oc})/0.52$$

where (C_s/f_{oc}) represents the chemical concentration in sediment normalized to the fraction of organic matter in the sediment. TBP is the maximum expected contaminant concentration in the organism normalized to lipid content of the organism in the same units as C_s. This approach relies on the premise that if partitioning of organics into or out of organisms is driven primarily by their lipophilicity, then at equilibrium the relative distribution of a compound between two phases (animal versus sediment, for example) should be related to the ratio of the lipophilic reservoirs in each.

As proposed by McFarland, the TBP could be used as a first tier in screening sediment for its potential to transfer contaminants to biota [10]. However, several assumptions are inherent in using this approach. It assumes that there are no physical or chemical barriers to transfer between phases and that the compound is not degraded. However, each of these potential interferences would tend to reduce bioaccumulation in the organism, so, as a maximum estimate, TBP should be conservative.

Lake et al. [11] used a similar approach to calculate a preference factor (PF) from empirical determination of organic carbon normalized sediment contaminant concentrations and contaminant body burdens normalized for lipid content using the following equation

$$PF = (C_s/f_{oc})/(C_o/\%lipid)$$

where C_s and C_o are in the same units. In this case, $C_o/\%lipid$ is equivalent to McFarland's TBP, so PF should approach a value of 0.5 as the organism comes to equilibrium with contaminants on sediment particles. It should be noted that uncertainties in the slopes and intercepts of the two empirical relationships used may lead to variations in the "optimum" value of PF.

Apart from strict partitioning controlling the transfer of contaminants between sediment and organism, other factors have been shown to influence bioaccumulation of organic contaminants. Bioaccumulation work is frequently done with "spiked" compounds, either radioactive or nonradioactive. This is done under the assumption that compounds experimentally added to sediment behave indistinguishably from chemicals "naturally" equilibrated with sediment. However, there is evidence suggesting that this assumption may not always be valid.

Differences observed between hydrocarbon patterns in benthic organisms and the sediment they live in led Farrington to postulate that petroleum hydrocarbons generated by fossil fuel combustion may be less bioavailable than those introduced by oil in the marine environment, and that not all compounds that are measured by chemical extraction are equally available to the biota [12]. Varanasi et al. [13] found that a ^{14}C-labeled polycyclic aromatic hydrocarbon spiked to a contaminated sediment was more available to benthic organisms than naturally occurring residues. Large differences between patterns of uptake and depuration of petroleum hydrocarbons have also been observed between organisms with different feeding strategies exposed to contaminated sediment [14–15]. In all cases, the differences observed may be due to differential availability, specialized uptake mechanisms, and/or metabolism of the compound.

It is necessary to understand these factors and how they interact if bioaccumulation data is to be accurately interpreted or if predictive estimates of the bioavailability of contaminants are to be made. In the present study, interactions between: (1) organic carbon content of the sediment; (2) equilibration time of the contaminant with the sediment; and (3) feeding strategies of the organism were simultaneously investigated. Apparent preference factors were calculated in an attempt to factor out the influence of strict partitioning behavior between lipophilic pools in sediments and organisms on bioaccumulation. A hexachlorobiphenyl was chosen as a model compound in this study because of its low water solubility and its resistance to metabolism.

Methods

Experimental Design

The experimental design is shown schematically in Fig. 1. Sediment was obtained by Smith-Mack grab from two locations (just north of Jamestown Island and off Sabin Point) in Narragansett Bay, RI, hand pressed through a 2-mm sieve and held at 4°C until use. Subsamples of sediment were analyzed for PCBs (using an acetonitrile/pentane extraction procedure now in use at the Environmental Protection Agency Narragansett Laboratory, R. Pruell, personal communication) and total organic carbon (TOC) content (Perkin Elmer H:C:N analyzer). The Jamestown (JT) and Sabin Point (SP) sediments were found to contain 50-ppb dry weight PCBs (calculated as Aroclor 1254) and $1.93 \pm 0.06\%$ TOC, and 903-ppb PCBs and $3.97 \pm 0.10\%$ TOC [mean $\pm$ standard deviation (SD)], respectively.

Nephtys incisa and *Yoldia limatula* were collected from the Jamestown site by scuba divers using an air lift which transported organisms and sediment to ship deck where the organisms were collected on a 2-mm sieve. Organisms were kept at ambient seawater temperatures until they were transported back to the laboratory, where they were acclimated to 20°C seawater and maintained in sediment from the Jamestown site until use.

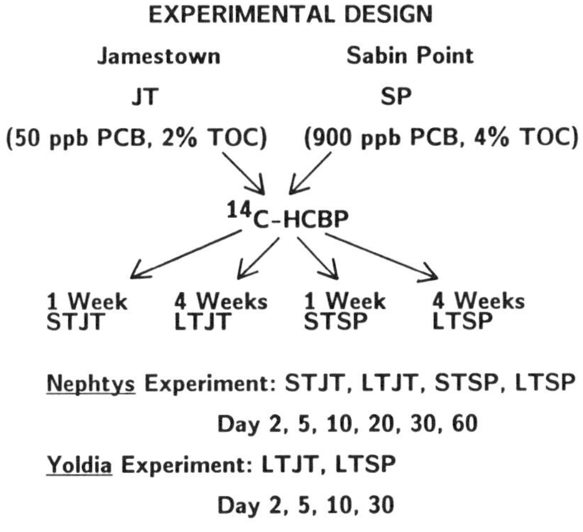

EXPERIMENTAL DESIGN

Jamestown   Sabin Point
JT   SP
(50 ppb PCB, 2% TOC)   (900 ppb PCB, 4% TOC)

^{14}C-HCBP

| 1 Week | 4 Weeks | 1 Week | 4 Weeks |
| STJT | LTJT | STSP | LTSP |

Nephtys Experiment: STJT, LTJT, STSP, LTSP
Day 2, 5, 10, 20, 30, 60

Yoldia Experiment: LTJT, LTSP
Day 2, 5, 10, 30

FIG. 1—*Experimental design.*

[^{14}C-U] – 2,4,5,2',4',5'-hexachlorobiphenyl (HCBP) obtained from Pathfinder Laboratories (specific activity = 12.5 mCi/mmol) was added to glass jars in a carrier solution of acetone. The jars were then carefully turned to evenly coat the interior of the jars with HCBP as the acetone evaporated. After all acetone had evaporated, wet Jamestown or Sabin Point sediment was added to each jar and seawater added to almost fill each jar, producing a slurry containing <10% sediment (dry weight). The sediment/seawater slurry was well mixed and the jars sealed and maintained on a roller mixer for a period of either one (ST) or four (LT) weeks. In this way the sediment was labeled via continuous exposure of the slurry to HCBP sorbed to the interior of the glass jars and the potential effects of the acetone carrier eliminated. After the labeling period, sediment of each of the four types (STJT, LTJT, STSP, LTSP) was split between a series of replicate chambers and allowed to settle. The small variation between HCBP content of sediment between replicate chambers and between the different labeling jars used to generate the ST and LT sediments (Table 1) indicates that this method reproducibly produced homogeneously labeled sediments.

For the *Nephtys* experiment, exposure chambers consisted of 150-mL beakers approximately half filled with bedded, labeled sediment (approximately 70 g wet weight), receiving filtered seawater at a rate of 50-mL/min and maintained in a constant temperature water bath at 20°C. Four worms (average wet weight 0.5 g each) were added to each chamber at the beginning of the experiment. The experiment was sampled on Days 0, 2, 5, 10, 20, 30, and 60, when one beaker of each sediment type was randomly collected.

Due to *Yoldia's* feeding activities, which resuspend large quantities of fine sediment [*16*], a somewhat different chamber design was required for the clams. *Yoldia* chambers were run as semiclosed systems to maintain a constant sediment reservoir of HCBP. The water column was mixed and aerated with a constant supply of filtered air but otherwise remained static. Chambers consisted of sealed pint jars containing approximately 30% sediment and 70% seawater and ten small clams. Clams had an average length of 17 mm and wet weight, including shell, of 0.40 g.

TABLE 1—*Bioaccumulation of hexachlorobiphenyl Day 30.*

	[Sediment], μg/gdw	[Organism]		PF
		μg/gww	μg/g Lipid	
		Nephtys		
Jamestown	1.03	0.397	23.4	2.36
	±0.09	±0.043	±5.2	±0.42
	(14)	(4)	(4)	(4)
Sabin Point	1.36	0.152	7.82	4.70
	±0.16	±0.021	±2.68	±1.25
	(14)	(4)	(4)	(4)
		Yoldia		
Jamestown	1.12	0.481	105.	0.595
	±0.03	±0.054	±80.	±0.454
	(7)	(2)	(2)	(2)
Sabin Point	1.44	0.157	29.3	1.16
	±0.05	±0.003	±0.9	±0.04
	(7)	(2)	(2)	(2)

NOTE: Values expressed as mean ±1 SD (n). Bioaccumulation data represent averages of accumulation from ST and LT sediments. Sediment concentrations represent averages of data from all samples. PF = ([sed]/%TOC)/([organism]/%lipid). gdw = gram dry weight; gww = gram wet weight.

Total wet weights are approximately 2 times the soft tissue wet weight. The vent from the chambers was fitted with a polyurethane foam plug to quantify any loss of HCBP due to volatilization. When chambers were sampled, the foam plugs were extracted three times each with 6 mL of acetonitrile (ACN). Total activity in the foam plug extracts was always less than 0.04% of total radioactivity in the chambers. On Day 20, all *Yoldia* in the chamber selected for sampling appeared to be dead or dying. This sample was not analyzed. The water column in the remaining *Yoldia* chambers was replaced on Day 20 with fresh filtered seawater and the chambers resealed. *Yoldia* harvested on Day 30 appeared to be in good health. Usable data from *Yoldia* chambers was therefore obtained as described above on Day 0, 2, 5, 10, and 30.

Analysis

When chambers were collected for sampling, organisms were removed from the sediment by sieving, rinsed, split into duplicate samples, and weighed prior to storage at $-20°C$ until analysis. Sediment was centrifuged at 10 400 g at 4°C for 20 min to remove pore water, homogenized and sonicated in ACN, and subsampled for analysis. Subsamples (of approximately 2.5-g dry weight) were extracted four times with 20 mL of ACN, the extracts combined, reduced by rotary evaporation, and subsampled in triplicate for analysis of radioactivity. The extracted sediment was stored 48 h in a dessicator under vacuum to remove residual ACN, and then dry weights were determined. Tissue samples (entire *Nephtys* and soft parts of *Yoldia*) were ground in anhydrous sodium sulfate (three or four times the wet weight of *Nephtys* or *Yoldia*, respectively), then extracted three times with 15 mL of ACN and the extracts combined. Tissue extracts (5 to 9 mL total volume) were then subsampled in duplicate for gravimetric analysis of lipid content (0.10 mL) and radioactivity (1.0 mL) as described below. Whole *Yoldia* were measured (long axis) prior to analysis. After removal of all soft tissue for extraction, *Yoldia* shells were extracted with acetone to assess residual activity. Activity extracted from shells ranged from 2.6 to 6.1% of total extracted activity and was not included in assessment of bioaccumulated HCBP.

Quantification of HCBP was done with liquid scintillation counting (LSC) on a Packard Tri-Carb 460-C, using Dimilume (Packard Instruments) as a scintillant. Foam plug and pore water extracts were analyzed whole, and subsamples of sediment and tissue extracts were analyzed in duplicate. Glass-distilled solvents obtained from Burdick and Jackson were used throughout. Subsamples of tissue and sediment extracts were also analyzed for total (radioisotope and nonradioactive) HCBP content. These data, the pore water data, and a more complete analysis of the bioaccumulation data will be discussed elsewhere.

Data Analysis

An apparent preference factor was calculated for each time point using the following equation

$$APF = (HCBP_s/\%TOC)/HCBP_o/\%lipid)$$

where $HCBP_s$ = HCBP concentration in sediment in $\mu g/g$ dry weight; %TOC = % total organic carbon in sediment in g/g dry weight; $HCBP_o$ = HCBP concentration in organism in $\mu g/g$ wet weight; and %lipid = % lipid in organism extracts in g/g wet weight. These were termed apparent preference factors because the concept of a "true" preference factor requires that the system be at thermodynamic equilibrium, which was not determined in this study. Since the concentration of HCBP in the sediment did not change over the course of the experiment, averages of all HCBP analyses for each sediment type (that is, STJT, LTJT, STSP, LTSP) for both *Nephtys* and *Yoldia* were used to calculate APF. TOC was determined on four individual samples each of Jamestown and Sabin Point sediment and the averages for each used in the PF calculation. $HCBP_o/\%lipid$

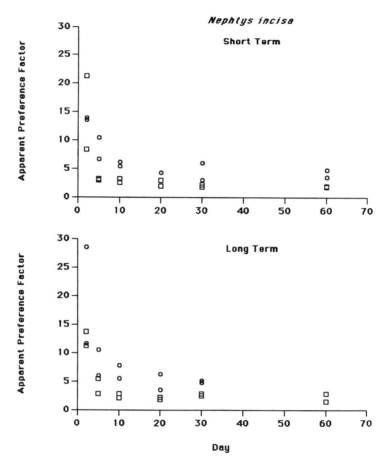

FIG. 2—*Apparent preference factor versus day of experiment for* Nephtys incisa. *Data from worms exposed to sediment equilibrated with isotope for one week (short term) are shown in the top panel. The bottom panel shows data from worms exposed to sediment equilibrated with isotope for four weeks. Circles represent exposure to Sabin Point sediment, and squares represent exposure to Jamestown sediment.*

was determined for each individual organism sample of 2 to 5 individual *Nephtys* or *Yoldia*. Means of the apparent preference factors calculated for duplicate organism samples from each chamber were analyzed for statistical differences between treatments using the parametric and nonparametric tests described below.

Results

Figure 2 shows the apparent preference factor (APF) calculated for *Nephtys* as a function of exposure period. The top and bottom panels show data for worms exposed to sediment equilibrated with isotope for one (short-term) of four (long-term) weeks, respectively. Individual values for the duplicate worm samples taken from each harvested chamber are shown to give an impression of variability in bioaccumulation between two pooled groups of organisms exposed to the same sediment. The long- and short-term equilibration data have been separated here for clarity, but it is clear from these two graphs that the STJT and LTJT and the STSP and LTSP curves are

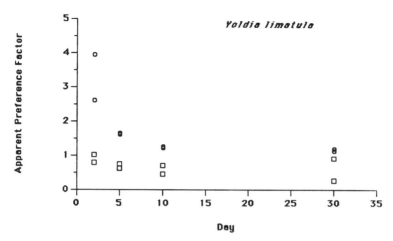

FIG. 3—*Apparent preference factor versus day of experiment for* Yoldia limatula. *All sediment was equilibrated with isotope for four weeks. Circles represent exposures to Sabin Point sediment, and squares represent exposures to Jamestown sediment.*

indistinguishable. However, there is a consistent difference between the Jamestown and Sabin Point sediment. For both the short- and long-term cases, APF approaches a smaller constant value more rapidly with the lower organic carbon, Jamestown sediment.

The observed trends were found to be statistically significant. A three-factor analysis of variance revealed no significant interactions between location (Jamestown versus Sabin Point), length of isotope equilibration (one versus four weeks), and sampling day (Days 2 to 60), but detected highly significant differences between the two locations ($P < 0.01$) and different sampling days ($P < 0.001$), but not between the one- and four-week labeling periods ($P < 0.4$). Multiple range testing indicated significant differences between Day 2, Days 5 and 10, and Days 10 to 60, indicating that by Day 20 *Nephtys* exposed to these sediments had reached a constant value of APF.

Figure 3 shows the results from *Yoldia* experiments expressed in the same way. With *Yoldia*, due to a shortage of experimental organisms, bioaccumulation was only assessed from sediments equilibrated with isotope for four weeks. The same trends observed with *Nephtys* are apparent in the *Yoldia* experiment, except that a constant APF value is reached more rapidly. The scales on this graph were expanded relative to those used with *Nephtys* to better visualize the kinetics. With the Jamestown sediment, *Yoldia* appears to be almost at a constant value by the first sampling period on Day 2. *Yoldia* exposed to Sabin Point sediment do not appear to reach a constant value until Day 10 and, as was seen with *Nephtys,* do so at a larger value of APF relative to *Yoldia* exposed to Jamestown sediment.

Because the same general trends were seen in both the *Yoldia* and *Nephtys* data, an analysis of variance was carried out on the combined data sets for Days 2, 5, 10, and 30, testing for differences between location, species, and sampling times. Species, location, and sampling day were all found to be significant ($P < 0.002, 0.03, 0.01$, respectively). A small but significant species/time interaction ($P < 0.03$) was observed due to *Yoldia* approaching a constant level of APF so quickly in the Jamestown sediment. Multiple range testing indicated that by Day 5, APF for *Yoldia* at both locations and reached a constant level. Due to the smaller sample size for *Yoldia* and unequal variances between the *Yoldia* and *Nephtys* data, the nonparametric Mann-Whitney U test was used to quantify the significant difference observed between the two species ($P < 0.025$).

By Day 30, it was possible to compare relative bioaccumulation of HCBP since both *Nephtys* and *Yoldia* had obtained a constant APF. Table 1 shows a comparison of bioaccumulation data between the two sediment types and two organisms on Day 30. Individual values for long and short isotope incubations have been pooled for the *Nephtys* data. Even though HCBP concentration in the Sabin Point sediment was 1.3 times that in the Jamestown sediment, concentration of HCBP in the organisms exposed to Sabin Point sediment accumulated on average only one third the amount of HCBP accumulated by organisms exposed to Jamestown sediment. Calculation of APF from these data normalizes these differences somewhat but does not alleviate them. Regardless of sediment type, *Yoldia* attains an APF value approximately one fourth that observed in *Nephtys*. It is also interesting to note that the difference between constant values of APF for the two sediment types is also a factor of 2 for either *Yoldia* or *Nephtys*.

Discussion

These results support the utility of using preference factors as predictors of maximum bioaccumulation potential for benthic organisms exposed to contaminants in sediment. In all cases APF approached a constant value within 20 days of exposure. However, these data also indicate that the absolute value of APF is still quite variable. Even at 30 days, significant differences between APF observed in the various treatments remained. In addition, the magnitude of the differences in APF between different treatments was just as large as observed when only comparing bioaccumulated concentrations normalized either to wet weight or lipid weight without even taking the sediment concentration or TOC into account (Table 1). These unresolved differences suggest that properties of sediment and organism affect both the rate and the amount of material transferred, above and beyond the influence of lipophilic reservoirs. Geometry of compound/particulate association, the type of organic carbon associated with a specific sediment, presence and levels of toxic contaminants, life-style and anatomy of the organism, and/or composition and availability of lipid pools within the organism may all contribute to the observed differences.

Life-style and/or feeding mode may control to what extent a benthic organism is exposed to sediment-sorbed contaminants. Although *Nephtys* is a soft-bodied organism living in close association with sediment, it irrigates its burrow for respiration and can secrete mucous. Both these processes may reduce its exposure to contaminants in the bulk sediment. *Yoldia*, on the other hand, processes large amounts of sediment during its feeding process [14]. Therefore, even though *Yoldia* lives in a shell, it may be exposed to contaminants in a much greater portion of the sediment reservoir. The special environment of the digestive tract may also be particularly efficient for uptake of compounds from sediment particles. The relative rate at which *Yoldia* accumulates HCBP relative to *Nephtys* suggests that its feeding mode facilitates accumulation. However, it is also possible that differences in the composition of lipid reserves between the two organisms contributed to the effect observed. Schneider [18] found that the PCBs in cod tissues were highly variable when normalized to wet weight, dry weight, or extractable lipids but were uniform when normalized to "fat" (nonpolar lipid) content.

The difference in HCBP bioaccumulation observed between the two sediment types may have also been influenced by the excessive amounts of contaminants (PCBs and other unknown contaminants) in the Sabin Point sediment relative to the Jamestown sediment. We have no way to assess this potential effect, but the similar survival and feeding patterns observed in both organisms in both sediment types suggest that toxicity due to contaminant concentration was not an overriding factor.

One surprising result was how fast both these organisms reached a constant level of HCBP body burden when normalized to lipid content. Previous work with PCBs has indicated that Nereid polychaetes can take more than 40 days to approach a constant residue level [19,20]. However, bioaccumulation in these studies was reported normalized only to either wet or dry weight. The results presented here indicate that even with a very hydrophobic compound like HCBP, organisms exposed to bedded sediments reach constant contaminant loads relative to their lipid pools relatively quickly. These findings suggest that extremely long exposures are not necessary to assess bioavailability. The similarity between HCBP bioaccumulation after either one or four weeks of isotope equilibration indicates that extremely long labeling periods are also unnecessary.

The APF values generated for HCBP can be compared with those obtained from field-collected organisms to see how well the "spiked" radio labeled HCBP mimics HCBP found in sediment from the field. PFs calculated for PCBs (as Aroclor 1254) for a number of different bivalves and polychaetes collected from Narragansett Bay and Long Island Sound by Lake et al. [11] averaged 0.35 and ranged from 0.24 to 0.50. Considering differences in methodology and the number of individual compounds lumped together in an Aroclor 1254 determination, the relatively good agreement between the APFs calculated for HCBP (averaging 2.0 and ranging from 0.6 to 4.7) and those generated by Lake et al. [11] indicates that the radioisotopes used in this laboratory experiment mimic naturally incurred residues reasonably well.

It is important to note that lipid content of the organism seems to be responsible for most of the variability in the APF observed. This is of consequence when comparing lipid normalized values from different laboratories. Slightly different, but generally accepted extraction methods for determining lipid content can drastically influence results obtained (J. Lake, unpublished results). If PF values are to be compared or used to regulate acceptable levels of contaminants in sediment, a single method should be used.

Acknowledgments

This work was done under the auspices of the National Academy of Sciences/National Academy of Engineering Senior Visiting Scientist Program with support from the United States Environmental Protection Agency. Although the research described in this article has been supported by the United States Environmental Protection Agency, it has not been subjected to Agency review and therefore does not necessarily reflect the views of the Agency, and no official endorsement should be inferred. Assistance from the staff at the Environmental Research Laboratory in Narragansett is gratefully acknowledged, with special thanks to Wayne Davis, Jim Heltsche, Norm Rubinstein, and John Sewal. Discussions with Jim Lake and Norm Rubinstein were influential in the development of this project.

References

[1] Wakeham, S. G. and Farrington, J. W. in *Contaminants and Sediments*, Vol. 1, R. A. Baker, Ed., Ann Arbor Science Publishers, Inc., Ann Arbor, MI, 1980, pp. 3–32.
[2] Karickhoff, S. W., Brown, D. S., and Scott, T. A., *Water Research*, Vol. 13, 1979, pp. 241–248.
[3] Means, J. C., Hassett, J. J., Woods, S. G., and Banwart, W. L. in *Polynuclear Aromatic Hydrocarbons*, P. W. Jones and P. Leber, Eds., Ann Arbor Science Publishing Inc., Ann Arbor, MI, 1979, pp. 327–340.
[4] Mackay, D., Mascarenhas, R., and Shiu, W. Y., *Chemosphere*, Vol. 9, 1980, pp. 257–264.
[5] Vieth, G. D., Defoe, D. L., and Bergstedt, B. V., *Journal of the Fisheries Research Board of Canada*, Vol. 36, 1979, pp. 1040–1048.
[6] Neely, W. B., Branson, D. R., and Blau, G. E., *Environmental Science and Technology*, Vol. 8, 1974, pp. 1113–1115.
[7] Chiou, C. T., Freed, V. H., Schmedding, D. W., and Kohnert, F. L., *Environmental Science and Technology*, Vol. 11, 1977, pp. 475–478.
[8] Southworth, G. W., Beauchamp, J. J., and Schmeider, P. K., *Water Research*, Vol. 12, 1978, pp. 973–977.
[9] Geyer, H. P., Sheehan, P., Kotzias, D., Freitag, D. and Korte, F., *Chemosphere*, Vol. 11, 1982, pp. 1121–1134.
[10] MacFarland, V. A. in *Dredging and Dredged Material Disposal*, Vol. 1, American Society of Civil Engineers, New York, 1984, pp. 461–467.

[11] Lake, J. L., Rubinstein, N., and Pavignano, S. in *Fate and Effect of Sediment-Bound Chemicals in Aquatic Systems,* K. L. Dixon, A. W. Maki, and W. A. Brungs, Eds., Society of Environmental Toxicology and Chemistry, in press.
[12] Farrington, J. W., Tripp, B. W., Teal, J. M., Mille, G., Tjessum, K., Davis, A. C., Livramento, J., Hayward, N. and Frew, N. M., *Toxicology and Environmental Chemistry,* Vol. 5, 1982, pp. 331–346.
[13] Varanasi, U., Reichert, W. L., Stein, J. E., Brown, D. W., and Sanborn, H. R., *Environmental Science and Technology,* Vol. 19, 1985, pp. 836–841.
[14] Shaw, D. G. and Wiggs, J. N., *Marine Pollution Bulletin,* Vol. 11, 1980, pp. 297–300.
[15] Boehm, P. D., Barak, J., Fiest, D., and Elskus, A., *Marine Environmental Research,* Vol. 6, 1982, pp. 157–188.
[16] Bender, K. and Davis, W. R., *Ophelia,* Vol. 23, No. 1, 1984, pp. 91–100.
[17] *SAS Users Guide Version 5,* SAS Institute, Cary, NC, 1985.
[18] Schneider, R., *Meeresforschung,* Vol. 29, 1982, pp. 69–79.
[19] Fowler, S., Polikarpov, G. G., Elder, D. L., and Villenuve, J.-P., *Marine Biology,* Vol. 48, No. 4, 1978, pp. 303–309.
[20] McLeese, D. W., Metcalfe, C. D., and Pezzack, D. S., *Archives of Environmental Contamination and Toxicology,* Vol. 9, No. 5, 1980, pp. 507–518.

Biomonitoring of Complex Effluents

James M. Marcus,[1] *Glenda R. Swearingen,*[2] *and Geoffrey I. Scott*[3]

Biomonitoring as an Integral Part of the NPDES Permitting Process: A Case Study

REFERENCES: Marcus, J. M., Swearingen, G. R., and Scott, G. I., **"Biomonitoring as an Integral Part of the NPDES Permitting Process: A Case Study,"** *Aquatic Toxicology and Hazard Assessment: 10th Volume, ASTM STP 971,* W. J. Adams, G. A. Chapman, and W. G. Landis, Eds., American Society for Testing and Materials, Philadelphia, 1988, pp. 161–176.

ABSTRACT: A case study of integration of extensive biomonitoring into the National Pollutant Discharge Elimination System (NPDES) permit of a synthetic organic chemical manufacturing firm in South Carolina to meet the objectives of the Clean Water Act is presented. NPDES monitoring requirements for specific chemicals and quarterly toxicity tests (*Mysidopsis bahia*) failed to predict severe water quality degradation in a small estuarine receiving stream. This severe disturbance was documented through benthic macroinvertebrate assessments and sediment elutriate bioassays. Additionally, chemical testing revealed the presence of a myriad of toxic, carcinogenic, and other organic chemicals in sediments and oyster tissue in the receiving stream and in outlying public shellfishing areas.

Implementation of monthly oyster-larvae bioassays, annual benthic macroinvertebrate community structure assessments, annual oyster recruitment/settlement assessments, and continuation of the mysid shrimp bioassays as NPDES permit requirements resulted in a comprehensive tool for actual determination of effluent toxicity as well as for guiding the in-facility reduction of toxicity. This comprehensive approach of laboratory toxicity testing of effluents and field monitoring of benthic macroinvertebrate communities provides a measure for assessing chronic or cumulative effects of complex effluents on aquatic communities, especially when those effluents are highly variable due to batch manufacturing processes. This case study suggests that biomonitoring is now developed to a level where it can be implemented to ensure directly that complex toxic effluents are regulated for optimum environmental protection.

KEYWORDS: Biomonitoring, toxicity, macroinvertebrates, priority pollutants, sediment, oysters

The National Pollutant Discharge Elimination System (NPDES) approach for point-source permitting and control has been primarily to monitor effluent quality via an array of physicochemical parameters. The selection of appropriate parameters usually follows United States Environmental Protection Agency (USEPA) guidance documents that are technology-based without water quality considerations for specific types of effluents. These guidelines impose categorical parameters and maximum allowable pollutant levels, suggest best professional judgment formulations, or allow combinations of both approaches [1]. Regardless of the guidance approach employed, the purpose of the NPDES permit strategy is to monitor and control specific pollutants at the "end of the pipe" as a tool for the protection of aquatic life. This strategy of protection is predictive in nature based on accumulated pollutant-specific data which may not be complete and is particularly limited

---
[1] Aquatic biologist, South Carolina Department of Health and Environmental Control, Bureau of Water Pollution Control, Columbia, SC 29201.
[2] Assistant professor, Shealy Environmental Services, Columbia, SC 29201.
[3] Project manager, University of South Carolina, School of Public Health—Department of Environmental Health Sciences, Columbia, SC 29208.

when assessing complex waste streams from chemical manufacturing industries. The detection, identification, and quantification of individual pollutants, intermediates, and reactive products present in complex effluents is essentially impossible due to both analytical and monetary limitations. The interaction of individual pollutants not only in the effluent but also in the receiving stream must be determined to assess thoroughly the biological effects of a complex effluent since chemical interactions may result in synergistic, antagonistic, and/or additive toxicological manifestations.

In-stream benthic macroinvertebrate assessments may provide clear evidence as to the existence of impact to aquatic communities especially when related to point-sources [2,3], given that the sources are discrete; there are no confounding nonpoint-sources; and, the natural substrate is amenable to qualitative or quantitative sampling. Even with this documentation, however, the lack of chemical-specific data from the effluent to accompany the in-stream assessment affords no opportunity for management of the effluent via process control of the wastewater treatment facility. Thus, both chemical and biological samplings and analyses are desirable to control and manage the effluent to protect aquatic life in the receiving water system [4].

A case study of an NPDES permit for a synthetic organic chemical manufacturing firm in coastal South Carolina is presented. NPDES monitoring requirements for a very limited number of specific chemicals per USEPA guidance/best professional judgment permit formulation and quarterly toxicity tests (*Mysidopsis bahia*) failed to predict severe water quality degradation in a small tidal receiving stream. Laboratory sediment elutriate toxicity tests and field biomonitoring techniques were integrated with an expanded physicochemical approach resulting in a more comprehensive NPDES permit. This approach allowed for determination of the cumulative impacts to the biota resulting from the accumulation of persistent chemicals in the sediments, for in-stream assessment of impacts related to current and recent discharges, for assessment of human health risks related to specific pollutants or pollutant groups in shellfish tissue from nearby public/private harvesting areas, and for development of a functional strategy for management and control of the effluent discharge relative to environmental impact and human health risk.

Project Background

The wastewater treatment facility serves a small chemical manufacturing firm specializing in dyes, dye intermediates, and other organic chemicals in Beaufort County, South Carolina. The firm has been in operation since 1967 with the production capability for 30 to 40 different organic products. These products are usually manufactured on a batch basis with production being changed frequently to meet new orders. Generally, about two weeks are required to stop production of one product and begin production of another. One result of this type of production scheme is an influent waste stream with the potential for highly variable characteristics. While the influent stream may be relatively consistent on a week(s)-to-month basis, a large variation occurs in the waste stream characteristics across an extended period of time (months to years).

The wastewater treatment facility consists initially of two separate waste streams for industrial and sanitary wastes. The industrial waste is routed to a pH neutralization basin while the sanitary waste flows to a small package plant. The effluents from the neutralization basin and package plant then flow together into a flow equilization tank and then into an aeration basin. After aeration, the waste is clarified, filtered by sand, chlorinated, and routed to polishing ponds. The final effluent is discharged to Campbell Creek at ebbing tide only.

Campbell Creek is a typically small (1.3 km in length) tidal creek along the South Carolina coast (Fig. 1). At low tide, the creek varies from approximately 0.3 m deep and 1.0 m wide at its upper reaches to 1.0 m deep and 10.0 m wide at its mouth. Campbell Creek lies in a salt marsh dominated by large stands of *Spartina alterniflora* and has a 1.5-m to 2.5-m tidal range with a daily tidal regime of two essentially equal high and low tides. With only minimal freshwater

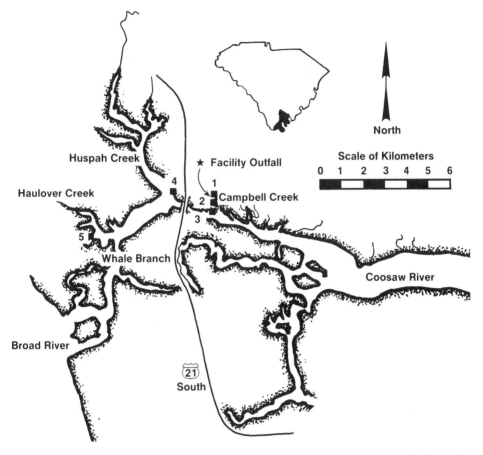

FIG. 1—*Sampling locations for Campbell Creek water quality assessment, Beaufort County, South Carolina.*

inflow, there is no vertical salinity stratification. These hydrological characteristics result in a well-defined intertidal zone of sandbars, mudflats, and oyster reefs. Public and privately leased shellfish harvesting grounds are located nearby in Whale Branch and Huspah Creek.

This facility had been operating under a NPDES permit generated, issued, and enforced by the South Carolina Department of Health and Environmental Control (SCDHEC). At the time of original permit issuance, technology-based guidelines established by USEPA for specialty chemical firms were followed in the derivation of permit conditions and limits. The resulting NPDES permit was principally oriented to control oxygen-demand impacts of the discharge as evidenced by permit conditions for flow, five-day biochemical oxygen demand (BOD_5), suspended solids, ammonia, and ultimate oxygen demand as a function of BOD_5 and ammonia discharge poundage over one tidal cycle. Conditions were also imposed for fecal coliform bacteria, lead, and toxicity through three 96-h flow-through toxicity tests per year using mysid shrimp (*Mysidopsis bahia*), although no permit limit was set for the toxicity condition. LC_{50}-values from this testing ranged from no significant mortality to 32%, while the predicted in-stream waste concentration was 0.66%. Other than lead and ammonia, there were no other specific inorganic or organic chemicals included in the permit. At the time of routine permit expiration and reissuance, concern was expressed within SCDHEC whether this permit approach was providing necessary protection of aquatic life in

Campbell Creek. To answer this concern, an ecological assessment was conducted in the Campbell Creek area to determine if this traditional NPDES permit approach had been successful in protecting the aquatic environment.

Assessment of the Conventional NPDES Permit Approach

This assessment consisted of chemical analyses of sediment and oyster tissue (*Crassostrea virginica*) for USEPA priority pollutants and biological analyses of the benthic communities including determination of hard and soft substrate community structures and oyster size-frequency distributions. Sediment elutriate toxicity testing was also conducted [5].

Sampling stations were established at various points in Campbell Creek and Whale Branch and at a control site in a tributary to Haulover Creek as follows (Fig. 1):

Station 1—Campbell Creek approximately 50 m north of existing treatment facility outfall.
Station 2—Campbell Creek approximately 300 m south of existing treatment facility outfall.
Station 3—Campbell Creek at confluence with Whale Branch.
Station 4—Whale Branch approximately 100 m west of US 21.
Station 5—Unnamed tributary to Haulover Creek approximately 25 m west of confluence with Haulover Creek and approximately 1.5 km west of confluence of Haulover Creek and Whale Branch. Established as a control station.

Sediment and oysters were collected from the midintertidal portion of the oyster reefs at each station for chemical analyses and for community structure determinations. The top 3 cm of sediment was collected at each station for analyses of USEPA priority pollutants. Three replicates of 30 oysters ($\geq$75 mm in height) per replicate were collected from the midintertidal region of the oyster reefs at each station for USEPA priority pollutant analyses. All nonpriority pollutant constituents detected were also identified and reported.

Analysis of the wastewater during initial stages of the assessment resulted in a mixture of organic chemicals, none of which were addressed either individually or as groups in the original NPDES permit. The effluent contained phenol, N-ethyl-benzenemethanamine, 3,3'-oxybis-propanitrile, 5-methyl-quinoline, 3-methyl-2(3H)-benzofuranone, tetramethyl pyrazine, 4-chloro-1,1'-sulfonyl bis-benzene, cyanide, and 3-(dimethylamino)-phenol during sampling in 1983. Some heavy metals were also detected (copper, iron, nickel, zinc) but at low levels considered typical for such effluents.

Chemical analyses from 1983 revealed a complex mixture of synthetic organics in both shellstock and sediments in Campbell Creek and Whale Branch (Table 1). Some compounds were identified at several stations over the area, while many were found only in Campbell Creek. In addition to the five stations depicted in Fig. 1, sediment was also collected from immediately at the facility outfall structure even though oysters were not present there. This outfall station and Station 2, at an oyster reef approximately 300 m south of the outfall site, exhibited the largest number of synthetic organics in any of the sediments (nine and twelve, respectively). Station 3 at the mouth of Campbell Creek exhibited the largest number of synthetic organics detected in shellstock (14). The control site (Station 5) exhibited the least number of synthetics in sediment (one) and shellstock (three) of all sites sampled.

The highly synthetic nature of the detected compounds and the use/generation of many in the manufacturing processes (benzene, toluene, 9,10-anthracendione, 6-oxabicyclo/3.1.0/hexane, chlorinated benzenes, phenols) implicated the wastewater treatment facility effluent as the source of these compounds. This implication was supported by this facility being the only point-source discharge in the area and by the absence of any nonpoint-source discharges there. The presence

TABLE 1—*Summary by groups of synthetic organic chemical compounds detected in sediments and oysters in the Campbell Creek/Whale Branch system.*

	Number of Compounds Detected Per Medium								
	Sediments			Oysters			Total		
Group	1983	1984	1985	1983	1984	1985	1983	1984	1985
Aliphatics	12	10	2	18	4	2	30	14	4
Paraffins	8	6	2	9	2	...	17	8	2
Olefins	...	1	...	3	...	2	3	1	2
Cyclics	4	3	...	6	2	...	10	5	...
Aromatics	16	11	2	7	0	1	23	11	3
Benzene	...	...	...	1	...	...	1	...	...
Other benzenes	9	11	2	2	...	1	11	11	3
PAH	7	...	...	4	...	...	11	...	...
Total	28	21	4	25	4	3	53	25	3
Number of priority pollutants	11	4	1	10	1	1	21	5	2

NOTE: PAH = polynuclear aromatic hydrocarbons.

of these organics, some of which were carcinogenic and mutagenic, in shellstock from harvesting areas was clearly unacceptable from a shellfish sanitation viewpoint.

Heavy metals levels from sediments and shellstock were within the expected ranges for the South Carolina coast. In general, the highest total metals burden in sediments occurred at the outfall site while the lowest burden was at Station 1 in Campbell Creek north of the outfall site. The total metals burden in shellstock was highest in Whale Branch near the mouth of Campbell Creek (Station 4) and lowest at the control site.

With these results communicated to the manufacturing firm and the existing NPDES permit still in force, USEPA priority pollutants analyses of sediments and shellstock were conducted again in December 1984 (Table 1)[6]. These data indicated the reduction not only in total numbers of synthetic organics in the aquatic system but also the reduction in priority pollutants and suspected carcinogens and mutagens. Shellstock from nearby harvesting grounds exhibited improved quality by the elimination of synthetic aromatics and significant reduction in synthetic aliphatic compounds. As in the 1983 sampling, there were no excursions of heavy metals in either sediments or shellstock above normal levels. Thus, from a chemical-specific viewpoint, the conditions had improved in Campbell Creek and outlying areas by December 1984.

This improvement in chemical contamination of the sediments and oyster tissue was likely driven by a concomitant improvement in the quality of the wastewater discharged to Campbell Creek. Given the previously noted numbers and types of organics present in the effluent during the initial assessment in 1983, each subsequent effluent sampling yielded fewer organics. A follow-up sampling in 1984 found three organics present in the wastewater: 1,1'-sulfonyl bis [4-chloro]-benzene, 2-(chloromethyl)-1,3-dixolane, and the ethyl ester of dipropyl carbamothoic acid. Likewise, two samplings in 1985 yielded only 1,1'-sulfonyl bis [4-chloro]-benzene.

This reduction in organics was due primarily to more active management of the waste treatment process by plant personnel. This was not the result of natural seasonal or temporal changes in the concentrations in sediments and oysters as evidenced by the concomitant reduction of organics in both media over time (Table 1). Pilot plant toxicity studies were conducted by plant personnel on various influent waste streams resulting in the complete removal of several highly toxic product wastes from the treatment facility altogether. Furthermore, SCDHEC required the Company to provide prior notice before any new product lines were added, thereby affording an assessment to be conducted regarding the treatability of the product wastes in the existing treatment facility. All effluent sampling conducted by SCDHEC or by the Company under the self-monitoring portion of the NPDES permit since 1983 showed a consistently reduced number of organics in the wastewater being discharged to Campbell Creek.

While the occurrence of these synthetic organics in sediments and shellstock indicated the influence of the wastewater discharge on the system, this presence did not necessarily equate with biological impact in the stream system. Sediment elutriate toxicity testing and benthic community structure analyses were used to assess the existence and degree of biological impact in the Campbell Creek system.

Results of the benthic macroinvertebrate community structure assessments in 1983 and 1984 clearly showed continued severe deleterious impact to the hard-substrate fauna in Campbell Creek. Even though the number of hard-substrate taxa was greater in Campbell Creek than elsewhere in 1983, this was primarily due to the lack of physical disturbance of the oyster reefs there via harvesting activities and to the apparent avoidance of Campbell Creek by key regulator and predator species (for example, mud crabs and blue crabs). The increased impact between 1983 and 1984 was evidenced by reductions in both mean species richness at Stations 1 and 2 and in abundance of organisms at Stations 1, 2, and 3 (Table 2). While, as previously noted, there had been improvements in the chemical contamination of sediment and tissue between 1983 and 1984, a similar positive response had not yet been observed in the hard-substrate biota. This was not surprising since biological effects, positive or negative, usually lag behind chemical exposure or exposure cessation, especially when length of reproductive cycles are also considered. This is significant because it suggests that the frequency with which field sampling should be conducted is specific to the ecosystem being studied.

Statistical comparisons (Tukey's Multiple Range Test: $p < 0.05$) between stations showed significantly fewer total numbers of oysters (Fig. 2) and juvenile oysters (≤ 11mm in height) in Campbell Creek than at the control station in 1983 and 1984 (Fig. 3). These oyster data indicated reduced oyster densities during 1983 and 1984 in Campbell Creek and to some extent in Whale Branch at the mouth of Campbell Creek. This continued diminution in oyster densities in 1984, given the improvements in chemical water quality by then, was due to the same lag in biological response as noted for the other hard-substrate fauna.

In 1983, the soft-substrate community in Campbell Creek reflected an unbalanced community dominated by large numbers of oligochaetes and opportunistic species of polydorid polychaetes. By 1984, some improvement was seen in the soft-substrate assemblages through increased species richness and a more-balanced community. This improvement in the soft-substrate community was due to the improved water quality conditions indicated by the chemical-specific data which would be most rapidly seen in the dominant annelids which have relatively short reproductive periods (on the order of a month).

Acute 96-h static toxicity tests were conducted on sediment elutriates [7] from several locations using adult grass shrimp (*Palaemonetes pugio*) and mummichogs (*Fundulus heteroclitus*) at sediment:water concentrations of 0.02, 0.002, 0.0006, and 0.00009%. Ninety-six-hour LC_{50}-values and fiducial limits were determined using the Trimmed Spearman-Karber method [8]. The 96-h static toxicity tests with grass shrimp and mummichogs resulted in no acute toxicity to the fish but significant toxicity to the shrimp at the outfall where the 96-h LC_{50} was 0.000168% (Table 3). Subsequent toxicity testing of downstream sediments at Stations 2 and 5 (control) resulted in 96-h LC_{50}-values of >0.2% and >2%, respectively. This suggested that the zone of acute toxic

TABLE 2—*Summary of the benthic macroinvertebrate taxa and specimens collected from the Campbell Creek/Whale Branch system.*

Sampling Date	Stations				
	1	2	3	4	5
	TOTAL NUMBER OF TAXA COLLECTED Hard Substrate, 0.01 m²				
1983	24	29	14	15	18
1984	21	24	22	26	30
1985	NS[a]	30	22	NS	31
	Soft Substrate, 23.2 m²				
1983	13	12	9	7	17
1984	14	19	16	15	15
1985	NS	15	19	NS	13
Sampling Date	MEAN NUMBER OF SPECIMENS COLLECTED ($n = 3$) Hard Substrate, 0.01 m²				
1983	330	386	310	271	451
1984	204	223	172	285	521
1985	NS	737	600	NS	776
	Soft Substrate, 23.2 m²				
1983	33	122	12	9	95
1984	65	77	50	29	122
1985	NS	503	83	NS	96

[a] NS = not sampled.

effects probably did not extend very far downstream from the outfall and that the role of sediment as a sink for the accumulation of persistant toxicants was important.

Observational data indicated more rapid toxicity at lower dilutions, suggesting that the aromatic organic compounds eluted from the sediment may have produced a narcotic effect on metabolism at higher concentrations. Behavioral observations tended to confirm this effect as grass shrimp were very lethargic, suggesting narcotic suppression at higher exposure concentrations. No mummichogs died during the 96 h of exposure; however, all exposed fish at all elutriate concentrations died within 24 h after return to clean seawater. These data also suggested the induction of a narcotic effect on metabolism due to the presence of aromatic organic compounds.

Adult grass shrimp were chronically exposed under flow-through conditions to a 0.00009% concentration of the sediment elutriate for 15 days, after which whole animal respiration rates (μL O_2/g/h) and allometric growth indices (dry weight/length × 1000) were determined. The allometric growth index addressed net growth effects based on changes in dry weight relative to length as a standardized component.

Chronic sublethal effects studies in grass shrimp indicated that exposure to the lower elutriate concentration of 0.00009% (approximately 53% of the 96-h LC_{50}-value) caused significant ($p < 0.05$) elevation in respiration rates and decreased allometric growth indices (Table 3) using Wilcoxon Rank Sum Scores [9]. Based on these data, it was estimated that the maximum allowable toxicant concentration (MATC) levels for the sediment at the outfall was greater than 1:10000 (sediment:water) and probably approached 1:100000. The likelihood of exceeding this low

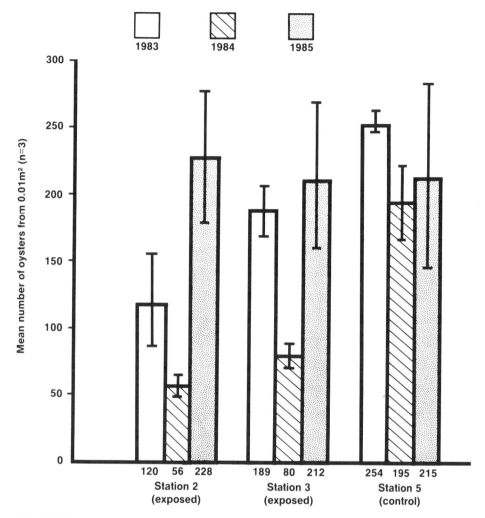

FIG. 2—*Three-year comparison of mean total numbers of oysters. Numbers below columns are mean values; bars represent ±1 standard deviation.*

concentration and minute exposure level in the stream was sufficient to suggest considerable risk of toxicity and/or sublethal effects to the aquatic biota, particularly crustaceans, in the area. When effects on larval or juvenile stages which may be more sensitive than adults are considered, the potential for impact seemed even greater.

The results of this assessment showed severe impact in the biota of Campbell Creek due to the wastewater treatment facility discharge as well as the presence of numerous synthetic organic chemicals, some of which were suspected carcinogens and/or mutagens in shellstock from nearby harvesting areas. All of the various components employed in the assessment were subsequently used to formulate and evaluate an NPDES permit approach that would allow for the restoration and protection of aquatic life in Campbell Creek.

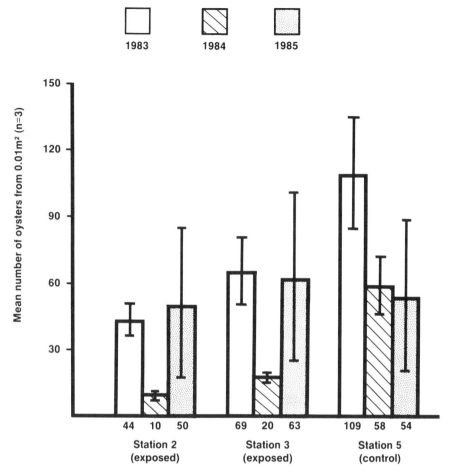

FIG. 3—*Three-year comparison of mean juvenile ($\leq 11mm$) numbers of oysters. Numbers below columns are mean values; bars represent ± 1 standard deviation.*

Formulation of a New NPDES Permit Approach

Neither the chemical-specific limitations on the existing permit nor the low-frequency mysid shrimp 96-h flow-through toxicity tests were able to predict and/or prevent the severe perturbations in the benthic macroinvertebrate communities in the receiving stream system or the presence of toxic pollutants in sediments and shellfish tissue. Therefore, an integrated biological and chemical-specific monitoring/assessment approach was adopted for the new NPDES permit tailored exclusively for particular characteristics of this project. The assessment component of this new approach was directly biologically based, while the monitoring component reflected increased levels of both biological and chemical-specific parameters. Emphasis was placed on monitoring the oyster population due to its ecological [10], recreational, and economical importance in the area and utility of the oyster as a sentinel organism [11]. Additionally, due to the relatively long life and sessile existence of oysters, age-class data could be obtained easily via size-frequency distributions, thereby affording a powerful technique for evaluating changes in water quality over time.

TABLE 3—*Summary of toxicology data from sediment elutriate toxicity testing in November 1983.*

Species	Station	Acute 96-h LC$_{50}$ (Lower, Upper 95% CI)	
Fundulus heteroclitus	Outfall 2 5	No mortality at all concentrations at all stations.	
Palaemonetes pugio	Outfall 2 5	0.000168% (0.000111 to 0.000284) >0.2% >2.0%	

Species	Station	Respiration Rate, µL O$_2$/g/h	Allometric Growth Index
Palaemonetes pugio	Outfall 5	^{a}3371 ± (402)^b 2436 ± (263)	^{a}1.47 ± (0.14)^b 1.91 ± (0.26)

NOTE: CI = confidence interval.
^{a} Significant difference between outfall and Station 5; $p < 0.05$.
^{b} Parenthetical values are ±1 standard error; $n = 11$ at outfall, $n = 9$ at Station 5.

A combination of monthly chronic bioassays using the oyster larvae shell development test [12] and seasonal acute toxicity testing using juvenile mysid shrimp was required to aid in predicting in-stream impact and in monitoring the variability of effluent toxicity. The mysid shrimp testing reflected only one more test per year than required in the original permit and was still low-frequency testing when considering the likely variable nature of the effluent. Conventional formulae for determining an allowable effective response level in the oyster larvae bioassay were not applied. However, an LC$_{50}$ of 66% was set as an NPDES permit condition for the flow-through mysid shrimp test based on a calculated in-stream waste concentration of 0.66% and the use of 100 as an acceptable application factor based on a recommended acute/chronic ratio [13,14]. The publication of additional toxics control guidance by USEPA [4] after this permit condition was derived and issued indicated that 10 is a more acceptable acute/chronic value, aside from the consideration of any uncertainty factors. The present concept behind the continued use of 100 as an application factor by SCDHEC is the allocation of 10 as the acute/chronic ratio component and 10 as a safety factor. Consideration of a safety factor is appropriate given the lack of multispecies testing to reduce uncertainties referred to in the later toxics guidance [4] and the intent to protect species more sensitive than that used in the test (mysid shrimp).

To insure protection of sensitive larval stages during the summer when peak spawning occurs in the oyster population, an oyster spat settlement study was required to determine if effluent quality would allow for recruitment and settlement of oysters in the receiving stream. Lack of recruitment or settlement in the summer would preclude establishment and maintenance of a healthy oyster reef community. Moreover, if recruitment/settlement occurred satisfactorily, then survival of the newly set spat would become the critical issue. This was addressed by using size-frequency distribution determinations. These were required along with monitoring of in-stream benthic macroinvertebrate community structure during the winter to assess overall impairment to the resident aquatic assemblages. Sampling was required in the winter since the community present would reflect animals which had recruited to the area during the summer and had survived. Also, these data should be less variable than in other seasons when population levels are fluctuating greatly due to reproduction of the animals.

Given the potential public and private harvesting of shellfish from areas in the zone of influence from the effluent discharge [15], analyses of oyster tissue from specified locations for USEPA priority pollutants were also required to be conducted at the beginning of the harvesting season as a new permit condition. These data, along with other data collected and analyzed by SCDHEC, are used by the shellfish program managers to assess the closure status of the grounds. Certain specific chemicals detected in oyster tissue and sediments during the initial assessments (for example, 9,10 anthracenedione, benzene, diphenylhydrazine, 1,2,3-trichlorobenzene, PCBs) were required in the permit to be routinely monitored in the wastewater treatment facility effluent. While evidence was clear that the number of synthetic organics in the effluent had been reduced significantly due to multiphasic management of the treatment system by the Company, these specific chemicals were included in the permit to ensure that this significant reduction remained in effect. Also, an inventory of raw products used and finished products produced by the manufacturing facility was required to be filed with SCDHEC. The addition of any new product lines or chemicals were required to be reported to SCDHEC for approval prior to the actual addition. An effluent toxicity reduction plan was required to be submitted to SCDHEC and, upon approval, implemented immediately.

Thus, the approach taken in development of the new NPDES permit for this facility integrated both chemical-specific and whole waste strategies to correct and monitor the in-stream impacts, to maintain satisfactory and safe water and shellfish quality and healthy aquatic communities, and to allow management of the manufacturing waste streams to reduce/eliminate toxicity in the effluent.

Validation of a New NPDES Permit Approach

After implementation of new permit conditions, various testing approaches were employed to determine whether or not the integrated permit concept had resulted in positive effects in the environment. Sediment and shellstock analyses from 1985 revealed even fewer numbers of synthetic organic constituents than either 1983 or 1984 (Table 1). Especially significant was the large reduction in aromatic compounds in both media. The elimination of some synthetic organics and significant reduction of other synthetic groups indicated improved chemical water quality due to more aggressive controls in the manufacturing facility of the types of waste products being sent to the treatment facility and more efficient treatment of the influent wastes once deposited there. This chemical-specific approach was necessary to address completely which components of the influent waste stream were either untreatable or poorly treated by the facility and should thus be eliminated from the manufacturing process or diverted from this treatment process to another alternative process. The chemical-specific approach was also necessary for evaluation of shellfish sanitation in the harvesting grounds. The presence/absence of known or suspected carcinogens over time in shellstock from those grounds were valuable data for decisions regarding closure of the grounds to harvesting.

Beyond these very real points of concern, however, remained the basic question of whether there had been any positive improvement in the severe ecological perturbation observed in 1983 and 1984. Chemical-specific data as just outlined could give only an indication of what would be expected. Effluent-toxicity data and in-stream biological data were employed again in the new integrated permit approach to address this fundamental question.

Immediately upon issuance of the new NPDES permit in May 1985, the oyster larvae bioassays were begun as required in the permit. Results of these tests were used to evaluate monthly variation in effluent toxicity. The EC_{50}-values from this test ranged from 13.2% to 40.5%, suggesting some swings in effluent toxicity at this facility, although this range cannot be considered conclusively broad given the inherent variability in toxicity testing. Overall, the monthly EC_{50} results indicated improving toxic potential in the effluent. It was not until June 1985, when the in-stream oyster spat recruitment study was conducted, that definitive evidence for incipient recovery of the oyster population in Campbell Creek was observed [16]. Results of that study found that oysters were recruiting to and settling in Campbell Creek. Differences in settlement of oysters during two exposure periods in June and July 1985 resulted in no significant differences (t-test; $p > 0.05$) between Station 2 in Campbell Creek and Station 5 at the control. Over five weeks exposure, the mean numbers of oysters ($\pm$ one standard deviation) settled were 43 $\pm$ 10 at the control and 10 $\pm$ 42 at Campbell Creek. The low number of oysters collected indicated that peak spawning had likely already occurred. Nevertheless, the data did clearly demonstrate the successful recruitment to and settlement of oysters in Campbell Creek.

In November 1985, the benthic macroinvertebrate assessment [17] clearly demonstrated that the oyster reef community had recovered substantially and that the soft-substrate community continued to show improvement (Table 2). A three-year comparison of the size-frequency distributions of the oyster population near the effluent discharge and at the control (Fig. 4) provided a clear record of improved oyster spat survival in Campbell Creek. The elimination/reduction of synthetic organics in the system, the recruitment/settlement of oyster spat in Campbell Creek, the survival of the spat and development into juvenile oysters, and the improvement in the hard and soft substrate community structures were observed as the more stringent NPDES permit was implemented. The significant increase (Tukey's Multiple Comparison Test; $p < 0.05$) in numbers of juvenile oysters in Campbell Creek from 1984 to 1985 clearly demonstrated that water quality had improved to allow the recruitment, settlement, survival, and growth of oysters to the same degree as the control. Consequently, the in-stream data provided direct assurance that aquatic life in the receiving stream was improving while the concomitant chemical-specific and effluent toxicity data allowed for better management and control of the effluent and assessment of the shellstock sanitation in the outlying areas.

Much of this improvement can be linked primarily to the concomitant reduction of toxicity and organic chemicals in the wastewater treatment facility effluent. Pilot plant toxicity and treatability studies on the influent waste streams and some finished products resulted in removal from the facility of highly toxic raw wastes as well as wastes which were shown to be treated poorly in the existing facility. More attention was devoted to operation and maintenance of the facility by Company personnel, allowing for the optimal operation of unit processes and better, more consistent treatment of the raw wastes that were input to the treatment facility. This focus on the types and amounts of influent wastes being routed to the facility was used in conjunction with the more frequent monthly oyster larvae toxicity tests to identify in-plant manufacturing process activities that could be modified to alleviate or reduce toxicity, synthetic organics, or both in the final effluent. The communication between the Company and SCDHEC prior to the addition of any new product lines allowed for careful consideration of possible consequences relative to toxicity and treatability in the facility. The success of these actions was manifested in improved effluent toxicity test results and in sustained reduction of synthetic organics in the effluent, sediments, and oyster tissue. This ultimately reduced the potential for environmental impact from the discharged effluent.

Conclusions

This case study has illustrated the utility of several biomonitoring requirements imposed in an actual NPDES permit in South Carolina and the necessity of those conditions to ensure completely the protection of the aquatic biota. Biomonitoring techniques allow determination of the actual, real-time condition of the biological communities as related to a point-source discharge. Additionally, they provide realistic tools for monitoring the environmental impact of changes in physical/chemical water quality over time. When used in conjunction with chemical-specific data, an integrated assessment of point-source impacts can be made and direct targeted steps for amelioration of those impacts can be planned and implemented. Paramount in this process is the identification and

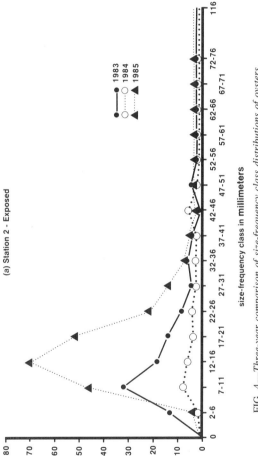

FIG. 4—*Three-year comparison of size-frequency class distributions of oysters.*

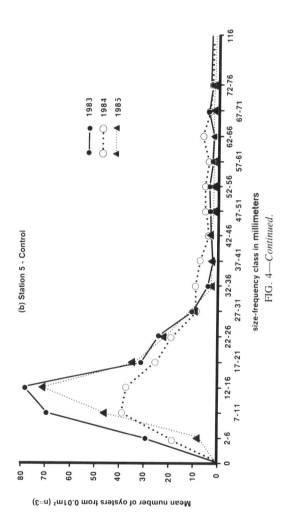

FIG. 4—*Continued.*

reduction of influent waste streams associated with various product lines that contribute potent toxicants or poorly-treatable materials to the wastewater treatment system.

In this project, effluent toxicity testing of oyster larvae combined with in-stream measurements of oyster spatfall for recruitment, of oyster size-frequency distributions for survival and population development, and of hard and soft substrate community structures for overall benthic community integrity were imposed as conditions in the NPDES permit. When taken in conjunction with expanded chemical-specific monitoring of the effluent and shellstock as permit conditions and with identification/reduction of toxic and nontoxic synthetic organics in the effluent by plant personnel, a comprehensive assessment and monitoring strategy was developed using the NPDES permitting process. The chemical-specific data component was necessary for process control management of the facility and for determination of shellfish sanitation relative to human health risk assessment. Concomitantly, the biomonitoring component of the new NPDES permit provided the basis for significant improvements in the quality of the aquatic biota in Campbell Creek and for field confirmation that those improvements had actually been manifested in the stream.

It should be noted that the biomonitoring techniques used in this assessment may not be universally applicable in all situations. The integration of these varied techniques for better NPDES permit conditions was designed to fit the type of treatment facility effluent and the site-specific discharge location characteristics. In other situations with other effluent characteristics and discharge locations, different biomonitoring techniques would likely be more appropriate.

Likewise, a distinction should be made between biomonitoring techniques for assessment versus those for compliance monitoring. Comprehensive biomonitoring was very useful in this project for the documentation and assessment of in-stream impacts and subsequent amelioration. Simultaneous chemical-specific data were useful in associating those impacts to the wastewater treatment facility effluent and in an active effort to reduce toxicity and synthetic organics discharge from the facility. Both testing components, when carried beyond the initial assessment stage, showed that the problem with the facility eliciting the impacts had been addressed by the industrial firm. Consequently, the focus of compliance monitoring could be concentrated on more frequent effluent synthetic organics measurements as NPDES permit conditions to ensure continued in-plant reduction of these compounds. Biomonitoring techniques would still remain useful but then would become less frequent given acceptable chemical-specific testing. Biomonitoring techniques would be used periodically to validate the chemical testing results so that environmental integrity is maintained in the receiving stream.

References

[1] Development Document for Effluent Limitations, "Guidelines and Standards for the Organic Chemicals and Plastics and Synthetic Fibers Industry: Proposed," EPA Publication 440/1-83/009-6, U.S. Environmental Protection Agency, Washington, DC, 1983.

[2] Gray, J. S., Boesch, D., Heip, C. Jones, A. M., Lassig, J., Vanderhurst, R., and Wolfe, D., *Rapports et Proces-Verbaux des Reunions, Conseil International pour Exploration de la Mer*. Vol. 179, 1980, pp. 237–252.

[3] Hart, C. W., Jr. and Fuller, S. L. H., *Pollution Ecology of Estuarine Invertebrates*, Academic Press, New York, 1979.

[4] "Technical Support Document for Water Quality-based Toxics Control," EPA Publication Unnumbered, U.S. Environmental Protection Agency, Washington, DC, 1985.

[5] Marcus, J. M. and Swearingen, G. R., "A Water Quality Assessment of Campbell Creek, Beaufort County, South Carolina," Technical Report 037-83, South Carolina Department of Health and Environmental Control, Columbia, SC, 1984.

[6] Marcus, J. M. and Swearingen, G. R., "A Summary of Water Quality Sampling Activities at Campbell Creek, Beaufort County, South Carolina," Technical Report 003-85, South Carolina Department of Health and Environmental Control, Columbia, SC, 1985.

[7] Tsai, C. F., Welch, J., Chang, K. Y., Schaeffer, J., and Cronin, L. E., *Estuaries*, Vol. 2, 1979, pp. 141–153.

[8] Hamilton, M. A., Russo, R. C., and Thurston, R. V., *Environmental Science and Technology*, Vol. 11, 1977, pp. 714–719; correction, Vol. 12, 1978, p. 417.
[9] Wilcoxon, F. and Wilcox, R. A., "Some Rapid Approximate Statistical Procedures," Lederle Laboratories, Pearl River, New York, 1964.
[10] Bahr, L. M. and Lanier, W. P., "The Ecology of Intertidal Oyster Reefs of the South Atlantic Coast: A Community Profile," U.S. Fish and Wildlife Service Publication FWS/OBS-81/15, Washington, DC, 1981.
[11] Farrington, J. W., *Oceanus*, Vol. 26, 1983, pp. 18–29.
[12] "Standard Methods for the Examination of Water and Wastewater," 15th ed., American Public Health Association, Washington, DC, 1980.
[13] "Bioassay in Water Quality Analysis and Effluent Monitoring," EPA Publication 430/1-74-014, U.S. Environmental Protection Agency, Washington, DC, 1974.
[14] "Methods for Measuring the Acute Toxicity of Effluents to Freshwater and Marine Organisms," EPA Publication 660/4-85/013, U.S. Environmental Protection Agency, Cincinnati, OH, 1985.
[15] "Dilution Study at Campbell Creek and Whale Branch, South Carolina," U.S. Environmental Protection Agency, Surveillance and Analysis Division Report, Athens, GA, 1978.
[16] "An Assessment of Oyster Spat Recruitment in Campbell Creek, Beaufort County, South Carolina," Aquatic Analysts, Columbia, SC, 1985.
[17] "A Macroinvertebrate Assessment of Campbell Creek, Beaufort County, South Carolina," Aquatic Analysts, Columbia, SC, 1986.

G. M. Degraeve,[1] W. H. Clement,[1] M. F. Arthur,[1] R. B. Gillespie,[1] and G. K. O'Brien[1]

Environmental Persistence/Degradation of Toxicity in Complex Effluents: Laboratory Simulations of Field Conditions

REFERENCE: Degraeve, G. M., Clement, W. H., Arthur, M. F., Gillespie, R. B., and O'Brien, G. K., **"Environmental Persistence/Degradation of Toxicity in Complex Effluents: Laboratory Simulations of Field Conditions,"** *Aquatic Toxicology and Hazard Assessment: 10th Volume, ASTM STP 971,* W. J. Adams, G. A. Chapman, and W. G. Landis, Eds., American Society for Testing and Materials, Philadelphia, 1988, pp. 177–189.

ABSTRACT: A laboratory method to monitor the persistence or degradation of toxicity in complex effluent/receiving water mixtures was developed and preliminarily field-validated. The objective of this research was to develop a method for possible future use in the National Pollutant Discharge Elimination System (NPDES) effluent permitting process that recognizes differential rates of toxicity degradation in effluent/receiving water mixtures.

Initial studies were made with the toxicant 2,4-dichlorophenol (DCP) added to a natural receiving water in the laboratory. After incubation under conditions similar to the field, toxicity (as measured periodically by Microtox®) paralleled the degradation (or persistence) of the toxicant as measured chemically. Loss of DCP in the laboratory was similar to predicted field-loss rate. Receiving waters and industrial and municipal effluents were then incubated in the laboratory in proportions similar to those observed in the field. Environmentally relevant conditions of temperature, photoperiod, light intensity, humidity, and aeration were maintained. Periodically, samples were analyzed for toxicity using the Microtox® system to monitor persistence (or degradation). Following development of test protocol with the DCP and effluents, the laboratory unit was field-validated using chlorine in Mississippi River water in the field and in the laboratory. Other field validations using complex effluents are underway.

This simple laboratory method can be used to predict the persistence (degradation) of effluent toxicity in natural receiving waters. Ultimately, one may be able to model effluent toxicity behavior in natural systems by tracking toxicity degradation in the laboratory, which could result in a cost-effective, toxicity-based method for use in the NPDES permitting process.

KEYWORDS: complex effluents, toxicity persistence, environmental simulation, Microtox®, field validation, NPDES permits, environmental toxicity persistence unit (ETPU)

On 9 March 1984, the U.S. Environmental Protection Agency (EPA) released a national policy statement which emphasizes an integrated approach which uses chemical analyses and biological testing to evaluate the effects of effluent discharges on receiving waters and to establish National Pollutant Discharge Elimination System (NPDES) permit limitations [1]. In September 1985, EPA issued a guidance document to support the national policy [2]. The "Technical Support Document for Water Quality-Based Toxics Control" emphasizes whole effluent biological testing and provides guidance for state and regional permit writers who may incorporate toxicity testing requirements into NPDES permits.

Water quality-based permits are being emphasized by EPA because the actual effects of complex

---

[1] Battelle, Columbus Division, Columbus, OH 43201.

effluents on aquatic communities are frequently different from effects predicted on the basis of single chemical laboratory toxicity tests [3,4]. A variety of chemical, physical, and biological fate processes in effluents and effluent/receiving water mixtures are responsible for differences between predicted effluent effects based upon individual contaminant toxicity tests and observed effects of effluents in receiving streams.

Although the practice of effluent toxicity testing has been gaining acceptance among the states, industry, and EPA since 1984, several issues remain. Effluent toxicity persistence/degradation in receiving waters has been poorly understood. The rate of degradation of instream effluent toxicity is highly variable among different effluent types and receiving waters, suggesting that the concept of environmental effluent toxicity persistence/degradation should be considered when evaluating the overall impact of an effluent on a water body [5]. Currently there is no mechanism in the NPDES permitting process to incorporate the rate of toxicity degradation into the permit requirements, which we feel should be changed to reflect differences in instream impact between complex effluents whose toxicities degrade at different rates.

The present cooperative research program between Battelle and U.S. EPA–Duluth was designed (1) to investigate toxicity persistence/degradation in effluent-receiving water mixtures and (2) to develop a field-validated, bench-scale laboratory procedure for predicting effluent toxicity persistence/degradation which can be utilized in the NPDES effluent permitting process.

Methods

Experimental Design

The parameter of most concern was the persistence/degradation of toxicity in effluent/receiving water mixtures, rather than the persistence/degradation of the contaminant(s). Therefore, our assumption was that toxicity persistence/degradation can be used to better understand the true instream impact of an effluent discharge, and, further, that modeling the kinetic behavior of toxicity as a single entity [analogous to biochemical oxygen demand (BOD)] may be achievable. However, unlike BOD, which is modeled on the assumption that the major controlling fate path is microbial activity, effluent toxicity persistence/degradation may be affected by chemical, physical, and microbial transformation/degradation mechanisms including photolysis, oxidation-reduction, hydrolysis, sorption, volatilization, ionization, and metabolism, all of which must be accounted for in order to simulate environmental toxicity persistence/degradation. This study was designed to: (1) examine individual fate processes for a suitable toxicant in the laboratory so that the most optimum set of conditions could be combined to realistically mimic environmental toxicity persistence/degradation; (2) develop a laboratory procedure, using the most optimum set of conditions, to simulate the necessary fate processes which might affect toxicity persistence/degradation; and (3) field validate that procedure using a simulated effluent in a natural receiving water. An overriding concern in this program is to develop a simple, workable system that can be easily duplicated in well-equipped environmental testing laboratories.

A five-tiered approach was used to accomplish these objectives:

1. The available literature was reviewed to identify the conditions and options affecting fate process simulations in the laboratory, laboratory fate process measurements, and appropriate toxicity testing systems.

2. Having selected the fate process simulation options which we felt were most important to duplicate in the laboratory, we conducted individual chemical fate screens with the toxicant 2,4-dichlorophenol (DCP) in a natural receiving water. DCP was chosen because of the information available on fate processes for this compound as well as its known toxicity to Microtox®.

3. Chemical and toxicological endpoints were measured for the individual fate processes, and

those values were compared to existing literature values to ensure that our laboratory fate process simulations were realistically simulating natural environmental conditions.

4. Having optimized individual fate process simulations in the laboratory, we developed our Environmental Toxicity Persistence Unit (ETPU), which was designed to mimic the environmental degradation of toxicity in a complex effluent/natural receiving stream mixture. We then compared the persistence of the model toxicant with the persistence of the toxicity in a natural receiving water with all fate and toxicity degradation processes in operation. We also compared the predicted (based upon individual fate measurements) degradation rate of toxicity with the measured rate of toxicity degradation in the ETPU.

5. Using a simulated effluent (chlorine as toxicant), we field-validated our laboratory ETPU by comparing toxicity persistence/degradation measured in the field with toxicity persistence/degradation simulated and measured in the laboratory.

Individual Fate Processes and Toxicity Screens

Prior to studying toxicity persistence of complex effluents, we examined the individual fate paths of DCP under controlled laboratory conditions. Fate processes examined included biodegradation, volatilization, photolysis, hydrolysis, and sorption, all in natural river water. The objective of fate screens with DCP was to measure individual fate processes, to compare our results with published literature values, and finally to determine whether toxicity persistence/degradation followed the persistence of the DCP. DCP concentration was determined by hexane extraction of the water samples followed by gas-liquid chromatographic (GLC) analysis of the hexane extracts. GLC analyses were conducted on a Varian 3700 using a 6-ft glass column packed with 1.5% SP-2250/1.95% SP-2401 on 100/120 Supelcoport, followed by electron capture detection. Toxicity was measured using the Microtox system, which measures reductions in light output by the luminescent marine bacterium *Photobacterium phosphoreum*. Two-mL samples were centrifuged (12000 rpm, 12 min, 20°C) and osmotically adjusted with 200 µL of Beckman osmotic adjusting solution. After calibrating the instrument and reconstituting the Microtox reagent according to the Beckman manual [6], background (undosed) bacterial light output was determined for all cuvettes. At this point, each cuvette contained 500 µL of Microtox diluent and 20 µL of reconstituted bacterial reagent. Immediately following light output determinations, 500 µL of the osmotically adjusted samples were added to appropriate duplicate cuvettes; 500 µL of Microtox diluent were also pipetted into triplicate controls. Light readings were then made at time-zero and again after 10 min of incubation. The 10-min reading for each cuvette was expressed as a percent of the light output relative to the undosed cuvettes, and these values were averaged (X_1). Similarly, the 10-min readings for the diluent controls were expressed relative to the undosed cuvettes, and the mean was determined (X_2). Finally, we divided X_2 by X_1, which produced toxicity of receiving water or receiving water/DCP mixture relative to controls. Consequently, using this system, 1.0 indicates no toxicity relative to controls, and 2.0 indicates a doubling of toxicity relative to controls.

Biodegradation—This process was examined using a river die-away method. Water was collected from Darby Creek, a moderately flowing stream in central Ohio that is largely removed from industrial effluents but is part of a wooded/agricultural watershed. Typical water characteristics of Darby Creek are shown in Table 1. DCP was added to duplicate 250-mL Erlenmeyer flasks at concentrations of 0, 5, or 10 mg/L in sterile and nonsterilized Darby Creek water. The tests were duplicated in both Darby Creek water containing suspended solids and in Darby Creek water from which solids had been removed by settling. The test concentrations were selected based on preliminary toxicity tests with the Microtox system, which showed that the EC_{50} for DCP was 4.65 mg/L. Biodegradation flasks were incubated at 22°C in the dark with constant shaking. Samples were removed periodically for Microtox and GLC analysis.

TABLE 1—*Typical water quality characteristics for Darby Creek water.*

Parameter	Settled Water	Nonsettled Water
Temperature	4.9°C	4.7°C
Dissolved oxygen	9.7 ppm	10.3 ppm
Conductivity	433 μmhos/cm	408 μmhos/cm
pH	7.8	7.8
Hardness	148 mg/L CaCO$_3$	208 mg/L CaCO$_3$
Alkalinity	172 mg/L CaCO$_3$	176 mg/L CaCO$_3$

Hydrolysis—The hydrolysis of DCP in Darby Creek water was evaluated based on the results of DCP measurements in the aseptic flasks of the biodegradation experiments. The flasks were protected from light (to avoid photolytic decomposition) in cotton-stoppered Erlenmeyer flasks (to minimize volatility) and thus represented a study of the hydrolytic path of DCP.

Sorption—Sorption in Darby Creek water was evaluated by comparing the results of the sterile biodegradation flasks containing suspended sediments with those containing minimal suspended sediments (after settling).

Photolysis—The photolysis of DCP was studied in 10 L of Darby Creek water in four 20-L glass aquaria dosed with 10 mg/L DCP. Two of the aquaria were exposed to a photoperiod of 16-h light/8-h dark while control aquaria were held entirely in the dark. Lighting conditions for the treatment group were a combination of incandescent, "grow light" fluorescent and black lights. Both groups were maintained at 22°C.

Volatilization—This process was examined in 20-L glass aquaria. In this experiment, 10 L of sterilized (autoclaved) Darby Creek water was dosed with 10 mg/L of DCP. The aquaria were mixed constantly with magnetic stirrers and were maintained in the dark in an environmental chamber with light surface air flow at 22°C. Samples were removed periodically over six days and analyzed by GLC to determine DCP concentrations.

Environmental Toxicity Persistence Unit Development and Calibration

After completing the individual fate screens, the selected fate processes were then combined in a single unit, and the degradation of toxicity and DCP was followed using Microtox and GLC. The environmental toxicity persistence unit (ETPU) consisted of an all glass 20-L tank containing 10 L of Darby Creek water. The decision to use a simple, static system was influenced by the desire to retain simplicity in the procedure and by studies of Brockway et al. [7], who concluded that small static systems can produce results similar to larger more complex (flow-through) systems. In the baseline experiments, Darby Creek water was dosed with 10 mg/L DCP. The ETPU solution was mixed by magnetic stirrers and positioned under a mixed light source (combination of incandescent, "grow light" fluorescent, and black light) in an environmental chamber.

Ultraviolet (UV) radiation was essentially that of a clear summer day, whereas the visible radiation was approximately one-fifth that of a clear, midsummer day.

Reaeration studies of our ETPU system using Darby Creek water established that our baseline or control ETPU tank had an absorption coefficient (k) of 0.20/h (regression equation, $y = -0.008 - 0.2x$; $r = -0.999$). This value translates into an f value (exchange coefficient) of 2.5 cm/h. Examination of the literature for measured "real world" values indicated our ETPU reaeration rate fell into a realistic environmental range [8]. The basic environmental conditions for the ETPU were: 22°C, 95% humidity, simulated natural lighting, a 16-h light/8-h dark photoperiod with gentle surface airflow. Both DCP concentration (GLC) and toxicity (Microtox) were measured daily.

Our initial fate investigations had shown that the DCP disappeared by a pseudo first order process. Also, it was found that the toxicity disappearance as measured with Microtox (Mt) paralleled the degradation of the toxicant. Thus, with DCP as the toxicant, a log plot of normalized "toxicity" as measured by Microtox, $[Tox]_{Mt}$, versus time was examined and found to be essentially linear ($r = 0.94 \pm 0.08$, $n = 14$). Thus a convenient method to compare experiments in the development and calibration studies was to determine the half-life ($t_{\frac{1}{2}}$) and rate constant for the disappearance of toxicity, and then to compare these values from run to run to determine effects.

After initial evaluation of the ETPU operation, a series of experiments was conducted to determine how variation of environmental parameters in the laboratory might affect toxicity persistence/degradation in the ETPU. The environmental parameters manipulated were (1) biodegradation, (2) surface/volume ratio, (3) turbulence, (4) surface airflow, (5) biosurface (surface area available for microbial colonization), (6) temperature, (7) humidity, (8) photoperiod, (9) light intensity, and (10) reaeration rate.

Following the environmental parameter manipulation studies, several industrial and municipal wastewater effluents and receiving water samples were collected and evaluated in the ETPU. Toxicity of each effluent was screened using Microtox to determine the effluent/receiving water dilutions most appropriate for the toxicity persistence/degradation assay. Two replicated dilutions (for example, 75 and 50% effluent in receiving water) and one replicated control (100% receiving water) treatment were prepared in the ETPUs (total volume of 10 L each). Each assay was conducted until toxicity degraded completely, or for at least five days. Water quality parameters (temperature, pH, dissolved oxygen, and conductivity) were measured at the start, during, and at the end of each assay. Evaporated water was replaced daily with refrigerated receiving water to maintain the ETPU volume of 10 L.

Field Validation of ETPU with Chlorine

The first field validation of the toxicity persistence protocol was conducted at the EPA Duluth's Monticello field research station. The purpose of this validation was to determine, under semicontrolled "field conditions," how well the ETPU-predicted toxicity persistence/degradation profile would compare with environmental toxicity persistence/degradation profiles using a simulated effluent (NaOCl solution) and natural receiving water. The experimental streams at Monticello were utilized to access semicontrolled field conditions, and one was dosed at its head with 500 µg/L of chlorine (as sodium hypochlorite solution). Toxicity and total residual chlorine (TRC) were measured on site periodically (from 17 to 137 min postdosing) during daylight and night conditions, downstream from the chlorine addition. Upstream (undosed) Mississippi River water was collected, placed on ice in sealed, dark containers, and shipped to Battelle's Environmental Research Laboratory. Immediately upon arrival at the laboratory, 10 L of Mississippi River water were placed in duplicate ETPUs, dosed with chlorine as at Monticello, and incubated under conditions previously described. At times representative of the Monticello field sampling times, the ETPUs were sampled, and both toxicity and TRC were measured in the laboratory and compared with the results from the field.

Results

Individual Fate Processes and Toxicity Screens

The results of the individual fate studies are summarized in Table 2. The results compare favorably with literature data. In addition, toxicity generally followed the concentration of DCP due to individual fate path degradations or compartmentalizations.

TABLE 2—*Summary of the results for individual fates for DCP comparing experimental data with literature values.*

Fate Process	Literature Value [Reference]	Experimental Value
Sorption partition coefficient	~12 [*9*]	Very little adsorption
Hydrolysis	NES[a] [*10*]	None observed
Biodegradation (half-life)	~4 days [*11*]	~9 to 10 days
Volatilization (half-life)	~6 days [*12*]	~4 days
Photolysis (half-life)[b]	Negligible, unless sensitized [*13*]	~24 h

[a] Not environmentally significant.
[b] Reference indicates that conflicting literature information has been reported for the photolysis of DCP.

The results of the biodegradation fate tests with DCP indicate that DCP biologically degraded in Darby Creek water after a lag period of approximately seven days (Table 3). In septic flasks, the toxicity decreased over twelve days from 1.6 to 1.0 (baseline), and from 2.1 to 1.1 in the flasks containing 5 and 10 mg/L DCP, respectively. The decrease in toxicity coincided with the decline in DCP concentration (Fig. 1). In the aseptic flasks, however, neither the toxicity nor the DCP concentrations changed substantially through twelve days of incubation.

In both septic and aseptic flasks DCP concentrations in filtered Darby Creek samples from settled and nonsettled treatments were nearly identical after twelve days of incubation. Therefore, sorption was not a major fate path for DCP loss, and the values presented in Table 3 represent the means of settled and nonsettled treatments. The results for aseptic flasks also indicate that hydrolysis was not a major fate path for DCP because of its persistence in water under sterile conditions.

Volatilization was an important fate path for DCP in our ETPU containing Darby Creek water. The concentration of DCP dropped from 10.1 to 4.5 mg/L in open ETPUs over three days, but loss in covered ETPUs over this time period was only 0.2 mg/L. Thus the half-life of DCP in Darby Creek water due to volatility in our ETPU was approximately three days. Toxicity was not measured during the volatility experiments.

The results also indicate that photolysis is an important fate path for DCP; approximately 70%

TABLE 3—*Biodegradation of DCP (measured by GLC) compared to toxicity (measured by Microtox).*

	Septic Flasks				Aseptic Flasks			
	5 mg/L[a]		10 mg/L		5 mg/L		10 mg/L	
Day	NMU[b]	DCP[c]	NMU	DCP	NMU	DCP	NMU	DCP
0	1.6	4.0	2.1	9.2	1.4	4.4	1.8	8.8
1	1.7	. . .[d]	2.2	. . .	1.2	. . .	1.5	. . .
2	1.8	5.1	1.8	10.2	1.5	4.6	1.9	11.1
4	1.4	. . .	1.9	. . .	1.7	. . .	2.2	. . .
7	1.4	4.7	1.8	9.1	1.3	4.98	1.7	10.8
12	1.0	<0.05	1.1	<0.05	1.3	4.3	1.6	5.9

[a] Treatment concentration of DCP on Day 0.
[b] Normalized Microtox unit, that is, the recriprocal of the mean light output of the dosed flasks compared to 0 mg/L flasks, so that no toxicity = 1.0.
[c] Concentration of DCP (mg/L) measured analytically.
[d] Not measured.

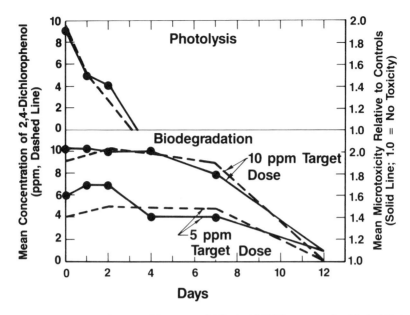

FIG. 1—*Comparison of toxicity (Microtox; solid line) with DCP concentration (dashed line).*

of the DCP was photolyzed within 24 h in Darby Creek water (Fig. 1). This degradation may have been influenced by the presence of sensitizing agents such as humic compounds in Darby Creek water. Figure 1 also shows that toxicity tracked the measured concentrations of DCP.

Environmental Toxicity Persistence Unit Development and Calibration

The predicted half-life for DCP in our ETPU system (all fate process in operation) based upon DCP degradation data collected from the individual fate screens was estimated to be 1.3 days for a nonacclimated system and 0.6 days for a biologically acclimated system. The measured half-life values were 1.1 days and 0.8 days for nonacclimated and acclimated ETPUs, respectively, demonstrating a good rate agreement between predicted and measured combined fate paths. In addition, the disappearance of toxicity paralleled the disappearance of DCP.

The results of the calibration studies are summarized in Table 4. These experiments not only allowed us to determine how changing environmental parameters would affect degradation of DCP, but also provided data on the reproducibility of DCP degradation in the ETPUs. Five baseline experiments produced a mean $t_{\frac{1}{2}}$ of 1.2 days with a standard deviation of ± 0.2 days ($r = -0.94 \pm 0.04$) with toxicity degradation measured over a four-day period in each experiment. The mean time-zero toxicity value found for all 14 of the experiments in the calibration studies was 2.4 ± 0.2 normalized Microtox Unit (NMU), which indicates good reproducibility with the ETPU system.

Two suitable industrial discharges with their respective receiving waters were evaluated to determine their applicability for studies in the ETPU system. One effluent's toxicity was due to the presence of inorganic metallic compounds, while the other effluent's toxicity was primarily caused by organic constituents. The toxicity degradation/persistence profiles for those effluents in the ETPU system are shown in Figs. 2 and 3. The inorganic effluent profile shows persistence of toxicity while the organic effluent studied shows nonpersistent characteristics.

TABLE 4—*Summary of ETPU calibration results.*

ETPU Study[b]	Manipulation	d(Tox)Mt/dt[a] Regression Coefficient, r	$t^{1/2}$, day	k, day^{-1}
Baseline	None[c]	−0.94	1.2	0.58
Surface/Volume Ratio	None (baseline, this series)	−0.98	1.2	0.58
	Small surface, large depth	−0.71	3.3	0.21
	Baseline surface, shallow depth	−0.98	1.5	0.46
Turbulence	None (baseline this series)	−0.87	1.1	0.63
	Static	−0.89	1.1	0.63
	Turbulent	−0.95	0.68	1.02
Airflow over Surface	None (baseline this series)	−0.97	1.5	0.46
	Low	−1.00	1.4	0.50
	High	−0.99	1.2	0.58
Biosurface[d]	None (baseline this series)	−0.96	1.2	0.58
	Sterile surface	−0.93	1.3	0.53
	Acclimated surface	−0.98	0.95	0.73

[a] Change in toxicity with time as measured by Microtox; treated as a pseudo first-order reaction.

[b] The ETPU consisted of a 20-L all glass tank. Baseline experiments utilized 10 L of receiving water containing 10-mg/L 2,4-DCP. Mixing was achieved by use of a magnetic stirrer, and the ETPU was positioned under the light source in an environmental chamber. The basic environmental conditions were: 22°C, 95% humidity, simulated natural lighting with a 16-h light/8-h dark photoperiod.

[c] These data represent an average of five ETPU runs.

[d] Glass beads; acclimated surface represents beads that have colonies of cultured microorganisms acclimated to biodegrade 2,4-DCP on their surface.

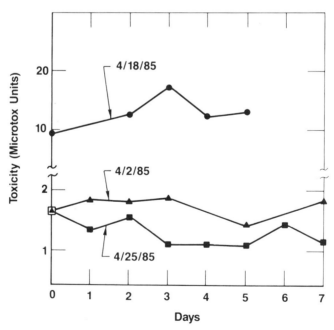

FIG. 2—*Persistence of toxicity in an industrial effluent which had primarily inorganic toxicants. Diluted 1:1 with upstream receiving water. Toxicity expessed as Microtox units normalized to receiving water. Dates shown are dates of sampling.*

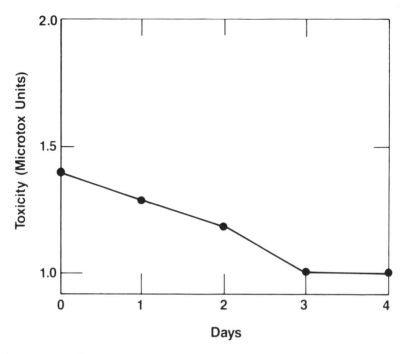

FIG. 3—*Persistence of toxicity in an industrial effluent which had primarily organic toxicants. Diluted 1:1 with upstream receiving water. Toxicity expressed as Microtox units normalized to receiving water.*

Field Validation: Monticello

The results of the field validation at Monticello are presented in Tables 5 and 6. Upstream of the riffle (from 17 to 60 min postdosing), the toxicity degraded similarly in the field and ETPU simulation. However, as Tables 5 and 6 illustrate, between 60 and 62 min downstream from dosing in the field, both toxicity and TRC decreased rapidly. Similar decreases in toxicity and TRC were not measured in the ETPUs. At Monticello, the period between 60 and 62 min postdosing represents a riffle zone with increased turbulence and much higher surface-to-volume ratio when compared with the rest of the experimental stream and the laboratory ETPUs. In the riffle zone, the turbulence and higher surface-to-volume ratio may have been responsible for the rapid loss of chlorine, possibly due to increased volatilization. Since a similar turbulent zone was not incorporated in the ETPUs, we did not observe similar rapid reductions in TRC and toxicity between 60 to 62 min postdosing.

The data from Tables 5 and 6 were used to calculate first-order half-lives for the rate of degradation of toxicity and TRC (Table 7). Half-lives were calculated for the entire stream segment studied (137 min) by regressing TRC or toxicity versus time. Riffle-corrected half-lives were also calculated for the field trials (values in parentheses in Table 7). Regression coefficients for the riffle-corrected half-lives ranged from -0.88 to -0.98. These results show good agreement between field and ETPU results when the correction is made for the riffle zone and illustrates the need to develop a riffle simulation technique in the ETPU.

Finally, the toxicity and TRC results from Tables 5 and 6 were compared using linear regression in order to determine how well toxicity correlated with the disappearance of TRC. Good correlations were found in all cases (Table 7), suggesting that the Microtox system is valuable for monitoring the degradation of chlorine in the environment.

TABLE 5—*Monticello field validation: day run.*

Time Post Dosing, min	Microtox Toxicity, NMU		TRC, μg/L	
	Field	ETPU	Field[a]	ETPU[b]
17	2.5	2.2	550	450
26	2.4	2.0	600	...
48	2.1	2.0	360	320
60	2.2	...	370	320
62	1.6	1.8	270	305
92	1.5	...	240	255
137	...	2.1[c]	...	205

[a] Concentration at dosing, 3.2 mg/L.
[b] Concentration at dosing, 2.3 mg/L.
[c] Suspect value. A new batch of Microtox reagent was used for this assay, which may have accounted for the high value.

Discussion

The agreement between the degradation rates for DCP found in the literature and those measured in our laboratory experiments was reasonably good for the fate paths examined with the exception of photolysis (Table 2). One literature source indicated that photolysis was not an important fate path for DCP, but in our experiment a half-life of ~ 24 hours was measured. The literature did indicate, however, that if sensitizers (such as fulvic and humic acids) were present in solution with DCP, DCP could degrade photolytically. Since Darby Creek drains a mixed agricultural and wooded watershed, we believe that sensitizing agents are present in Darby Creek water (as a result of vegetation decomposition) and that the presence of those sensitizing agents accounts for the difference between our observed photolytic DCP degradation rate and literature values.

When all fate pathways for DCP were examined concurrently, we found good correlations between the predicted degradation rate (based upon our individual fate experiments) and the observed degradation rate in our ETPUs for both biologically acclimated and nonacclimated systems. This correlation was encouraging, because it indicated that the ETPU had good potential for duplicating environmental toxicity persistence/degradation in a receiving water. We were further encouraged by the fact that the rate of toxicity degradation (as measured by Microtox) paralleled DCP degradation as measured by GLC. This good correlation allowed us to continue

TABLE 6—*Monticello field validation: night run.*

Time Post Dosing, min	Microtox Toxicity, NMU		TRC, μg/L	
	Field	ETPU	Field[a]	ETPU[b]
17	2.4	2.5	540	685
26	...	...	...	...
48	...	2.0	...	450
60	2.1	...	430	...
62	1.5	...	350	...
92	1.3	2.1	290	395
137	...	1.7	...	355

[a] Concentration at dosing, 3.2 mg/L.
[b] Concentration at dosing, 3.1 mg/L.

TABLE 7—*Half-lives and correlation values for the disappearance of toxicity and TRC for the Monticello field validation.*

Parameter	Day Run		Night Run	
	Field[b]	ETPU	Field[b]	ETPU
Half-life, h[a,b]				
Toxicity	1.1(2.4)	1.8	1.0(2.3)	2.2
TRC	1.2(1.6)	1.9	1.7(2.4)	2.2
Correlation Coefficient, r[c]	0.92	0.91	0.98	0.93

[a] Based on pseudo first-order reaction.
[b] Value in brackets is an estimated half-life in the field for pool conditions only, that is, an adjusted value using only samples taken from the pool section of the stream. This value was obtained by averaging the $t\frac{1}{2}$ from points prior to the riffle zone and the $t\frac{1}{2}$ found by adjusting the pool points after the riffle zone. The adjustment was made by adding the loss in toxicity or TRC found in the riffle zone to all points after the riffle zone and removing the time period of the riffle zone.
[c] Linear regression correlation of toxicity values as measured by Microtox against TRC values for samples on which both measurements were made.

to pursue the EPTU concept with complex effluents, where the toxicant(s) frequently cannot be identified and therefore cannot be easily measured analytically.

The results of the ETPU persistence/degradation study with the industrial effluent containing organic toxicants were predictable (Fig. 3). Many toxic industrial organic compounds degrade in the environment given sufficient time, and as Fig. 3 illustrates, the toxicity of this effluent (as measured by Microtox) degraded in three days. In contrast, the toxicity of the industrial effluent containing primarily inorganic metallic toxicants persisted for the duration of our experiments (Fig. 2). This result was also not surprising given that many industrial inorganic contaminants (particularly heavy metals) persist in the environment. Thus, we believe that the ETPU realistically represented the anticipated "fate" of toxicity for these two industrial effluents.

Chlorine (sodium hypochlorite solution) was selected for the field validation study at Monticello for several reasons. First, the Microtox assay is sensitive to chlorine, secondly, chlorine readily degrades in the aquatic environment via several well-established fate paths, and thirdly, chlorine is one of the chemicals most commonly discharged into surface waters in the United States. In the field measurements, chlorine degraded (as measured by toxicity or TRC) at a constant rate in the pool environments at Monticello, both upstream and downstream of the riffle zone. However, in the short (2 min flow time) riffle zone chlorine concentrations and toxicity degraded much more quickly than in the immediately upstream or downstream pool sections of the artificial stream (Tables 5 and 6). The difference in chlorine degradation rate was apparent both in the experiment conducted during daylight hours and in the night experiment, suggesting that increased photolysis was not the mechanism responsible for the much more rapid chlorine degradation rate in the riffle zone. Since individual chemical fate measurements were not made in the field, we can only speculate that increased volatilization in the riffle zone must be primarily responsible for the high rate of chlorine degradation in the riffle section of the artificial stream.

The rate of chlorine degradation in the Monticello pool environments (upstream and downstream of the riffle) was very seimilar to the chlorine degradation rate measured in our laboratory ETPUs (Tables 5 and 6). This good correlation was anticipated, given that the pool conditions at Monticello and our ETPUs are similar in that they are both nearly static, relatively deep environments. On the other hand, our ETPUs did not effectively duplicate the chlorine degradation rate that we observed in Monticello's riffle environment. Our inability to duplicate the riffle effect was understandable, given that the ETPUs more closely simulate a static pond environment. However, we believe that by either agitating or aerating the ETPUs in the future we may be able to come much closer to simulating a natural riffle environment.

Overall, for the artificial stream segment which included both pool and riffle environments, the half-lives for chlorine degradation in the field and in the laboratory were within a factor of 2 for TRC and toxicity both during the day and at night (Table 7). We felt that this was a good correlation, given that site-specific adjustments were not made to compensate for the riffle. When the results were adjusted, however, to remove the riffle measurements (by comparing only the field pool degradation rates with the ETPUs) the correlation between field and laboratory chlorine degradation rates was very good (Table 7). Consequently, we believe that the ETPU in its present configuration can be used to simulate environmental toxicity degradation for receiving waters which are primarily pool environments as opposed to high-velocity, high-gradient receiving streams. Our ETPUs will require further refinements to accomodate those receiving waters which contain riffles or high-gradient, high-velocity reaches. We are presently conducting experiments in our laboratory to accomplish this objective as well as addressing the need for incorporating substrate into the ETPU. In addition, we are evaluating the importance of dilution as a natural process which may account for "toxicity reduction" in receiving waters flowing downstream from the effluent source.

Acknowledgements

We gratefully acknowledge funding of this work by the U.S. Environmental Protection Agency-Environmental Research Laboratory (EPA-ERL), Duluth, Minnesota, under contract no. 811703-01-0, and the encouragement and advice of the project officers, Donald Mount and Nelson Thomas. Bruce Vigon provided important initial program design and key technical advice throughout the early stages of the project. We would also like to acknowledge John Steichen for technical assistance, Robin Faulk and Rhonda Pullen for assistance with the manuscript, and the staff at the EPA-ERL field station in Monticello, Minnesota, including Jack Arthur, Steve Hedtke, Kathleen Allen, and Chuck Kroll.

References

[1] "Development of Water Quality-Based Permit Limitations for Toxic Pollutants: National Policy," *Federal Register*, Vol. 49, No. 48, 1984, pp. 9016–9019.
[2] "Technical Support Document for Water Quality-Based Toxics Control," Office of Water, U.S. Environmental Protection Agency, Washington, DC, 1985.
[3] Lloyd, R. in *Environmental Hazard Assessment of Effluents*, H. L. Bergman, R. A. Kimerle, and A. W. Maki, Eds., Pergamon Press, New York, 1986, pp. 42–53.
[4] Ladd, E. C., Chen, D. H., Grant, J., Hernandez, J. W., Jr., Huckabee, J. W., Lloyd, R., Macek, K. J., and Tebo, L. B., Jr. in *Environmental Hazard Assessment of Effluents*, H. L. Bergman, R. A. Kimerle, and A. W. Maki, Eds., Pergamon Press, New York, 1986, pp. 54–58.
[5] Mount, D. I. in *Environmental Hazard Assessment of Effluents*, H. L. Bergman, R. A. Kimerle, and A. W. Maki, Eds., Pergamon Press, New York, 1986, pp. 61–65.
[6] Beckman Instruments, Inc., "Operating Instructions Microtox® Analyzer Model 2055 Interim Manual 100697," Microbics Operations, Carlsbad, CA, 1979.
[7] Brockway, D. L., Smith, P. D., and Stancil, F. E., "Fate and Effects of Atrazine in Small Aquatic Microcosms," *Bulletin Environmental Contamination and Toxicology*, Vol. 32, 1984, pp. 345–353.
[8] Klein, L., *River Pollution II. Causes and Effects*, Chapter 6, "Biochemical and Physico-chemical Aspects of River Pollution," Butterworth and Co., Ltd., London, 1962, pp. 232–238.
[9] Schellenberg, K., Levenberger, C., and Schwarzenbach, R. P., "Sorption of Chlorinated Phenols by Natural Sediments and Aquifer Materials," *Environmental Science and Technology*, Vol. 18, 1984, pp. 652–657.
[10] "Aquatic Fate Process Data for Organic Priority Pollutants," final report, U.S. EPA-OWRS, U.S. Environmental Protection Agency, Washington, DC.

[11] Blades-Fillmore, L. W., Clement, W. H., and Faust, S. D., "An Investigation into Hazardous Phenolic Compounds from Petroleum Sources and Urban Runoff," publication of the New Jersey Water Resources Research Institute, Rutgers University, New Brunswick, NJ, 1980.
[12] Mackay, D. and Leinnen, P. J., "Rate of Evaporation of Low-Solubility Contaminants from Water Bodies to Atmosphere," *Environmental Science and Technology,* Vol. 9, 1975, pp. 1178–1180.
[13] Plimmer, J. R. and Klingebiel, U. I., "Riboflavin Photosensitized Oxidation of 2,4-Dichlorophenol: Assessment of Possible Chlorinated Dioxin Formation," *Science,* Vol. 174, 1971, pp. 407–408.

William J. Rue,[1] James A. Fava,[2] and Donald R. Grothe[3]

A Review of Inter- and Intralaboratory Effluent Toxicity Test Method Variability

REFERENCE: Rue, W. J., Fava, J. A., and Grothe, D. R., "**A Review of Inter- and Intralaboratory Effluent Toxicity Test Method Variability,**" *Aquatic Toxicology and Hazard Assessment: 10th Volume, ASTM STP 971*, W. J. Adams, G. A. Chapman, and W. G. Landis, Eds., American Society for Testing and Materials, Philadelphia, 1988, pp. 190–203.

ABSTRACT: Due to the increased emphasis on the use of aquatic toxicity tests to evaluate the quality of complex effluents, there is a need to objectively evaluate the precision of effluent toxicity test methods. Based upon an extensive search of the published and unpublished literature (available through April 1985), the intralaboratory and interlaboratory precision of effluent toxicity test methods was evaluated.

Most of the inter- and intralaboratory studies obtained address the acute toxicity of effluents to three standard test organisms: *Daphnia* spp. (water fleas); *Salmo gairdneri* (rainbow trout); and the marine bacterium *Photobacterium phosphoreum* (Microtox). Only limited data exist for *Pimephales promelas* (fathead minnows); *Mysidopsis bahia* (opposum shrimp); and *Cyprinodon variegatus* (sheepshead minnows). Based on LC_{50} or EC_{50} values for 141 effluents for which interlaboratory data were available, 81.6% had coefficients of variation ≤40%, and 74.5% had coefficients of variation ≤30%. For 46 effluents for which intralaboratory data were available, 89.2% had coefficients of variation ≤40%, and 89.2% also had coefficients of variation ≤30%. To put this variability in perspective, coefficients of variation for these replicated toxicity tests with effluents are compared to published precision estimates for analytical chemistry methods.

KEYWORDS: effluent toxicity testing, variability, precision, interlaboratory, intralaboratory, acute toxicity

Acute toxicity tests are intended to provide numerical estimates of the effect (EC_{50}) or lethality (LC_{50}) of a chemical, or mixture of chemicals, to the test organisms during short-term exposure. If the same toxicant is tested by many laboratories, each of which conducts replicate tests, it is unlikely that all the EC_{50} or LC_{50} values for the test material will be the same. How much the EC_{50} or LC_{50} values vary between laboratories (interlaboratory) and within each laboratory (intralaboratory) can be affected by numerous factors (e.g., test species, exposure concentrations, test temperatures).

Most analytical methods include information on the level of "precision" or "accuracy" that can be expected from utilizing the methodology. As defined by Bainbridge [1] and used in this study: *precision* is the closeness of agreement between randomly selected individual measurements or test results; and *accuracy* is the closeness of agreement between an observed value and an accepted reference value. While the accuracy of effluent toxicity test measurements cannot be

---

[1] Senior scientist, Environmental Toxicology, EA Engineering, Science, and Technology, Inc., Sparks, MD 21152.

[2] Section manager, Environmental Technology and Assessment, Battelle Columbus Laboratories, Columbus, OH 43201.

[3] Senior environment specialist, Monsanto Co., St. Louis, MO 63167.

determined because there is no single "reference value" against which to compare test results, effluent toxicity test methods do allow for estimates of precision. This study assembled both published and unpublished effluent toxicity test data in order to evaluate the inter- and intralaboratory precision of commonly utilized aquatic acute toxicity test methods.

Methods

Effluent variability data were obtained through extensive searches of the published literature and telephone interviews with more than 25 federal, industrial, academic, and consulting professionals who are active in the field of biomonitoring. This information was then critically reviewed to determine its appropriateness to this study. Only data from replicate testing of identical effluent samples were acceptable. The assembled data base was then summarized to quantify both inter- and intralaboratory variability.

Based on LC_{50} (or EC_{50}) values, the precision of the test methods are presented as interlaboratory and intralaboratory variabilities. They are summarized using the following statistics: arithmetic mean, range of reported values, standard deviation, maximum-minimum ratio, and coefficient of variation. Note that for interlaboratory results, the range represents the highest and lowest of the arithmetic means reported by the laboratories. The coefficient of variation is the standard deviation expressed as a percentage of the mean $[(SD \div X) \times 100]$.

While the coefficient of variation is a useful tool for ascertaining the variability of test data, in order to avoid potentially misleading conclusions about a dataset, it may be useful to use both max-min ratios and coefficients of variation. For example, coefficients of variation may be reasonably high (for example, 79%) simply because the sample mean is low. The min-max ratio helps identify this situation.

Results and Discussion

Following extensive reviews of the assembled data base, 13 studies were identified that evaluated the variability of acute toxicity test methods using effluents. Test design features of these studies are briefly summarized in Table 1.

Table 2 summarizes the interlaboratory test variability for each study. Note that the mean, standard deviation, range, max-min ratio, and coefficient of variation are calculated by first determining the mean values for each laboratory and then calculating subsequent statistics on those laboratory means. In most cases this approach had little effect on the summary statistics. Where large differences did occur, they are discussed under the comments column of Table 2.

Intralaboratory variability calculations are presented in Table 3. These statistical summaries are based on the data values presented by each of the authors in each of the studies evaluated. However, to summarize and interpret the results of these studies in a consistent manner, some of the summary data had to be recalculated based upon information presented in each study.

The ability of a test method to provide reproducible data within and between laboratories should be known before a method is utilized for regulatory purposes. While more work is needed to better understand the precision of effluent toxicity test methods, the assembled data base suggests that the coefficients of variation for several effluent toxicity test methods are comparable to accepted analytical methodologies. Presented below are analyses of: (1) the distribution of the coefficients of variation for the inter- and intralaboratory variability data; (2) the coefficients of variation for acute toxicity tests with different tests organisms; and (3) a comparison of coefficients of variation for effluent toxicity test methods with accepted analytical chemistry procedures.

The results of acute effluent toxicity tests are generally reported as the LC_{50} (or EC_{50}) value. It is important to point out, however, that these values are averaging statistics which are significantly affected by the concentration series used in each test. Assume, for example, that the dose-response

TABLE 1—*Effluent toxicity study design features.*

Description	Number of Labs	Species Tested	Test Type	Protocol	Effluent Source
Grothe and Kimerle [9]	9	*Daphnia magna*	Acute, static	Peltier [16]	Raw process waste stream
Buikema [10]	3	*Daphnia magna*	Acute, static	Peltier [16] and EPA [17] draft	Refinery effluents
Dorn [11]	2	*Daphnia pulex*	Acute, static renewal	EPA—Region VI	Manufacturing
		Lepomis macrochirus	Acute, static		
		Mysidopsis bahia	Acute, static renewal		
		Cyprinodon variegatus	Acute, static		
EA [12]	2	*Pimephales promelas*	Acute, static	EA [12]	Chemical
Peltier [13]	3	*Pimephales promelas*	Acute, static	Peltier [13]	Metal plater
EA [6]	1	*Ceriodaphnia* sp.	Acute, static renewal	Mount and Norberg [18]	Sewage treatment plant
Grothe [14]	1	*Daphnia magna*	Acute, static	Peltier [16]	Manufacturing
	2	*Daphnia pulex*	Acute, static	Peltier [16]	Manufacturing
	3	*Daphnia magna*	Acute, static	Peltier [16]	Manufacturing
	1	*Pimephales promelas*	Acute, static	Peltier [16]	Manufacturing
	3	*Daphnia magna*	Acute, static	Peltier [16]	Manufacturing
	1	*Daphnia magna*	Acute, static	Peltier [16]	Manufacturing
Strosher [15]	2 to 3	*Photobacterium phosphoreum*	Acute, static	Beckman Instruments, Inc. [19]	Waste drilling fluids
	2 to 3	*Photobacterium phosphoreum*	Acute, static	Beckman Instruments, Inc. [19]	Waste drilling fluids
	2 to 3	*Salmo gairdneri*	Acute, static	Standard two-fish bioassay	Waste drilling fluids
	2 to 3	*Salmo gairdneri*	Acute, static	Standard ten-fish bioassay	Waste drilling fluids

curves for Effluents A and B were known with absolute certainty (Fig. 1) and yielded LC_{50} values of 40% effluent and 90% effluent, respectively. However, if tests with Effluents A and B were set up using control and exposure concentrations of 1, 3, 10, 30 and 100% effluent, the binomial calculation procedure would yield LC_{50} values of 54.8% effluent for both tests. As a result, effluent toxicity test precision estimates that are based on LC_{50} (or EC_{50}) values may be different from what the results would be based on knowledge of the true dose-response curves or the percent mortality at a single exposure concentration.

Distribution of Coefficients of Variation

Based on the studies summarized in Table 2, the results from 141 effluent toxicity tests can be evaluated with regard to interlaboratory variability. Using the data for all 141 effluent tests for which interlaboratory data were available, 81.6% (115/141) yielded coefficients of variations ≤40%, and 74.5 % had coefficients of variation ≤30% (Table 4). If the data are deleted where coefficients of variation are 0.0% simply because all replicate LC_{50} values are ≥100% effluent, then 78.5% (95/121) of the tests yielded coefficients of variation ≤40%, and 70.2% had coefficients of variation ≤30% (Table 4).

A similar analysis of the coefficients of variation for all of the intralaboratory studies presented in Table 3 indicates that 89.2% of the laboratory studies yielded coefficients of variation ≤40% and also ≤30%, with 78.3% ≤20% (Table 4). Again, if the data are filtered to remove those values where the coefficient of variation is 0.0% simply because all replicate LC_{50} values are ≥100%, 87.5% of the studies yielded coefficients of variation ≤40% and also ≤30%, and 75.0% of the values had coefficients of variation ≤20% (Table 4).

Coefficients of Variation for Acute Toxicity Tests with Different Test Organisms

To evaluate the precision of effluent toxicity test methods using different test organisms, the data presented in Table 4 were segregated by test species. This permits an examination of the distribution of coefficients of variation for the more common test species. The distribution of coefficients of variation by species were only considered when the sample size was greater than ten.

Interlaboratory Precision. Data in Tables 2 and 4 were sufficient to permit a comparison of the distribution of coefficients of variation for three species: *Salmo gairdneri, Daphnia* spp., and *Photobacterium phosphoreum* (Microtox). As summarized in Table 5, the distribution of the coefficients of variation for acute tests with these three taxa was very similar. Interlaboratory tests using rainbow trout, *Daphnia* spp., and *P. phosphoreum* yielded coefficients of variation ≤40% for 80.3, 78.6, and 81.4% of the tests, respectively. Similarly, coefficients of variation ≤30% occurred for 72.4, 78.6, and 72.1% of the tests for rainbow trout, *Daphnia* spp., and *P. phosphoreum,* respectively (Table 5).

Intralaboratory Precision. Data summarized in Tables 3 and 4 were sufficient to permit closer analyses of only one test organisim, *Daphnia* spp. No other taxa had more than ten intralaboratory coefficients of variation. Generally, the distribution of coefficients of variation for acute tests with *Daphnia* spp. (Table 6) was similar to the distribution of coefficients of variation for all taxa summarized in Table 4. Almost 95% of the tests using *Daphnia* spp. yielded coefficients of variation ≤30%. In comparison, Table 4 indicates that approximately 90% of tests using several species yielded intralaboratory coefficients of variation ≤30%. Based on the assembled data, it is noteworthy that 73.7% of the *Daphnia* spp. intralaboratory effluent tests yielded coefficients of variation ≤10% and that none of these tests had coefficients of variation above 50% (Table 6).

TABLE 2—Interlaboratory variability for effluent toxicity test studies.

	Species	No.	LC_{50} or EC_{50} Mean, %	SD, %	Range, %	Max ÷ Min	Coefficient of Variation, %	Comments
			GROTHE AND KIMERLE [9]					
48-h data	D. magna	9	5.27	1.50	4.2 to 9.0	2.12	28.46	1,2
			BUIKEMA [10]					
24-h screening (endpoint is percent survival, not LC_{50} value)								3
Effluent A	D. magna	3	76.67	8.32	67.5 to 83.75	1.24	10.85	
Effluent B	D. magna	3	87.92	14.91	71.25 to 100	1.40	16.96	
Effluent C	D. magna	3	98.75	1.25	97.5 to 100	1.03	1.26	
Effluent D	D. magna	3	88.75	6.61	83.75 to 96.25	1.15	7.45	
Effluent E	D. magna	3	80.42	13.83	67.5 to 95	1.41	17.19	
48-h EC_{50}s (definitives)								4
Effluent A	D. magna	3	18.47	15.04	4.6 to 34.45	7.49	81.43	2,5
Effluent D	D. magna	2	98.00	0.0	98.0 to 98.0	1.00	0.0	
Effluent E	D. magna	2	24.65	27.08	5.5 to 43.8	7.96	109.86	2,5
			DORN [11]					
Site 1, Week 1	M. bahia	2	3.68	1.79	2.41 to 4.94	2.05	48.61	6,7
Week 2	M. bahia	2	10.77	2.88	8.73 to 12.8	1.47	26.72	7,8
Week 3	M. bahia	2	7.23	1.31	6.3 to 8.15	1.29	18.09	6,7
Site 2, Week 1	D. pulex	2	>100	0	>100	1.0	0	
Week 2	D. pulex	2	>100	0	>100	1.0	0	
Week 3	D. pulex	2	>100	0	>100	1.0	0	
			PELTIER [13]					
	P. promelas	3	21.83	2.57	19 to 24	1.26	11.75	

EA Engineering, Science, and Technology, Inc. [12]

Sample 1	P. promelas	2	18.3	4.10	15.4 to 21.2	1.38	22.41
Sample 2	P. promelas	2	17.45	0.35	17.2 to 17.7	1.03	2.01
Sample 3	P. promelas	2	16.95	3.61	14.4 to 19.5	1.35	21.28
Sample 4	P. promelas	2	19.75	5.16	16.1 to 23.4	1.45	26.14

Grothe [14]

D. pulex	2	53.36	3.73	50.72 to 56.0	1.10	7.00
D. magna	3	51.46	14.90	42.33 to 68.65	1.62	28.95

Strosher [15]

Microtox 15-min test 9
 Results of 48 separate replicated tests were reported (2 to 6 labs/test).
Two fish acute test
 Results of 38 separate replicated tests were reported (2 to 3 labs/test).
Ten fish acute test
 Results of 38 separate replicated tests were reported (2 to 3 labs/test).

COMMENTS:

1—Mean, standard deviation, and CV recalculated from Table 3 of the author's paper. Author also presents 24-h test results.
2—Analysis of variance indicates that there are statistically significant differences between the results reported by the laboratories. Results are significant at the $P < 0.01$ confidence level.
3—Calculated on percent survival data presented in Table 7 of Buikema [10].
4—Calculations based on Table 9 of Buikema [10].
5—If individual data are used instead of laboratory means, max-min ratios for effluents A and E are 8.51 and 12.5, respectively.
6—"Less than" values are calculated as one half of the reported number.
7—If individual data are used instead of laboratory means, the max-min ratio and coefficient of variation for Weeks 1, 2, and 3 mysid data are 77.82 and 78.46%, 1.65 and 22.36%, and 2.78 and 31.82%, respectively.
8—Analysis of variance indicates that there are statistically significant differences between the results reported by the laboratories. Results are significant at the $P < 0.01$ confidence level.
9—Strosher [15] reported both 5-min and 15-min Microtox results for each sample. Only the 15-min results are presented here.

TABLE 3—*Intralaboratory variability for effluent toxicity test studies.*

	Species	No.	Mean, %	SD, %	Range, %	Max ÷ Min	Coefficient of Variation, %	Comments
48-hr data				GROTHE AND KIMERLE [9]				
Lab I-1	*D. magna*	2	4.55	0.49	4.2 to 4.9	1.17	10.77	
Lab I-2	*D. magna*	3	6.33	0.40	6.1 to 6.8	1.11	6.38	
Lab I-3	*D. magna*	2	5.30	2.55	3.5 to 7.1	2.03	48.03	
Lab G-1	*D. magna*	3	4.30	0.17	4.1 to 4.4	1.07	4.03	
Lab G-2	*D. magna*	2	4.5	0.0	4.5 to 4.5	1.00	0.0	
Lab G-3	*D. magna*	1	...	...	4.2	...	...	
Lab C-1	*D. magna*	2	4.8	0.14	4.7 to 4.9	1.04	2.95	1
Lab C-2	*D. magna*	2	4.65	1.34	3.7 to 5.6	1.51	28.82	
Lab C-3	*D. magna*	3	8.9	0.26	8.6 to 9.1	1.06	2.97	
				BUIKEMA [10]				
24-hr Screening Tests (endpoint is percent survival, not LC_{50} value)								
Refinery A—Lab 1	*D. magna*	2	78.75	8.84	78.5 to 85	1.08	11.22	2
Lab 2	*D. magna*	2	67.50	17.68	55 to 80	1.45	26.19	2,3
Lab 3	*D. magna*	2	83.75	1.77	82.5 to 85	1.03	0.21	2
Refinery B—Lab 1	*D. magna*	2	71.25	12.37	62.5 to 80	1.28	17.37	2
Lab 2	*D. magna*	2	100.0	0	100 to 100	1.00	0	2
Lab 3	*D. magna*	2	92.5	0	92.5 to 92.5	1.00	0	2
Refinery C—Lab 1	*D. magna*	2	100.0	0	100 to 100	1.00	0	2
Lab 2	*D. magna*	2	97.5	3.54	95 to 100	1.05	3.63	2
Lab 3	*D. magna*	2	98.75	1.77	97.5 to 100	1.03	1.79	2
Refinery D—Lab 1	*D. magna*	2	86.25	1.77	85 to 87.5	1.03	2.05	2
Lab 2	*D. magna*	2	83.75	1.77	82.5 to 85	1.03	2.11	2
Lab 3	*D. magna*	2	96.25	5.30	92.5 to 100	1.08	5.51	2
Refinery E—Lab 1	*D. magna*	2	67.5	3.54	65 to 70	1.08	5.24	2
Lab 2	*D. magna*	2	95.0	7.07	90 to 100	1.11	7.44	2
Lab 3	*D. magna*	2	78.75	19.45	65 to 92.5	1.42	24.69	2,3
				BUIKEMA [10]				
Definitive tests yielding 48-h EC_{50} values								
Refinery A—Lab 1	*D. magna*	2	16.35	0.49	16 to 16.7	1.04	3.03	4
Lab 2	*D. magna*	2	34.45	3.04	32.3 to 36.6	1.13	8.83	4
						1.14	9.22	4

Refinery B—Lab 1	D. magna	2	83.90	9.05	77.5 to 90.3	1.17	10.79	4
Refinery E—Lab 1	D. magna	2	43.8	1.70	42.6 to 45.0	1.06	3.87	4
Lab 3	D. magna	2	5.50	2.69	3.6 to 7.4	2.06	48.85	4
				DORN [11]				
Site 1								
Lab 1, Week 1	M. bahia	3	4.94	3.14	3.125 to 8.56	2.74	63.52	5
Week 2	M. bahia	3	8.73	0.56	8.18 to 9.29	1.14	6.36	
Week 3	M. bahia	3	6.30	2.77	3.125 to 8.21	2.63	43.89	
Lab 2, Week 1	M. bahia	2	2.41	3.25	0.11 to 4.7	42.73	134.67	5
Week 2	M. bahia	2	12.80	0.99	12.1 to 13.5	1.12	7.73	
Week 3	M. bahia	2	8.15	0.78	7.6 to 8.7	1.14	9.54	
Site 2								
Lab 1, Week 1	D. pulex	3	100	0	>100 to >100	1.00	0	
Week 2	D. pulex	3	100	0	>100 to >100	1.00	0	
Week 3	D. pulex	3	100	0	>100 to >100	1.00	0	
Lab 2, Week 1	D. pulex	2	100	0	>100 to >100	1.00	0	
Week 2	D. pulex	2	100	0	>100 to >100	1.00	0	
Week 3	D. pulex	2	100	0	>100 to >100	1.00	0	
			EA ENGINEERING, SCIENCE, AND TECHNOLOGY, INC. [6]					
48-h LC$_{50}$ (lethality)	Ceriodaphnia	7	26.14	7.45	15.0 to 35.0	2.33	28.49	
				GROTHE [14]				
	D. magna	3	12.02	0.91	10.97 to 12.57	1.15	7.57	
	P. promelas	4	5.99	0.79	5.27 to 6.89	1.31	13.18	
	D. magna	3	74.83	0.0	74.83 to 74.83	1.0	0.0	
	D. magna	3	12.82	2.80	9.61 to 14.78	1.54	21.88	6

COMMENTS:

1—Laboratory 1-3 used two different daphnid cultures in their tests, one raised on *Selanastrum*, the other on trout chow/yeast.
2—Results are for the 24-h screening test using 100% effluent. All values are in percent survival and are based on Table 7 in Buikema [10].
3—Results for duplicate samples were reported to be statistically different ($P < 0.05$).
4—Results are for the 48-h definitive tests which yielded EC$_{50}$ values. All values are based on Table 9 in Buikema [10].
5—Less than values (for example, <6.25) are computed as one half the reported value.
6—This experiment was designed to evaluate the effect of *D. magna* age on acute toxicity. The tests utilized Daphnia <24, ≤48, and ≤96 h old and indicated that the 96-h-old specimens were the most sensitive.

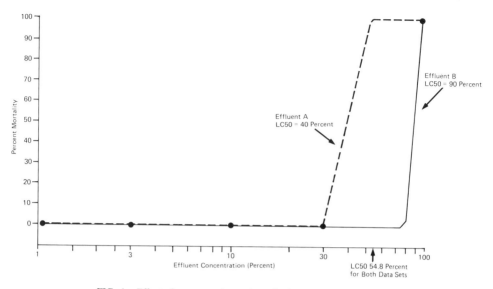

FIG. 1—*Effect of concentration series selection on LC_{50} calculation.*

Comparisons With Other Analytical Methods

To put this variability in perspective, the coefficients of variation from replicated effluent toxicity test studies were compared with coefficients of variation for accepted analytical chemistry methods. By using coefficients of variation as a relative measure of method variability, the precision estimates for effluent toxicity tests (Tables 4 through 6) and analytical chemistry methods (Table 7) can be compared. Note, however, that the two factors are not entirely equivalent: the distribution of precision estimates for replicated tests with effluents is compared to the distribution of precision estimates for chemical methods.

To address chemical method variability, inter- and intralaboratory precision data were calculated for each of the 126 chemicals in U.S. EPA's priority pollutant list [2] for which necessary data were available. These substances were chosen to demonstrate the range of variability that exists for the quantification of routine analytical parameters. Precision data for inorganics were obtained from EPA's "Method for Chemical Analysis of Water and Wastes" [3]. Inter- and intralaboratory precision data for the toxic organics were obtained from EPA's "Guidelines Establishing Test Procedures for the Analyses of Pollutants Under the Clean Water Act" [4]. For the toxic organics, EPA [4] presents single analyst precision (intralaboratory) and overall precision (interlaboratory) estimates in the form of linear equations which yield standard deviations (SD). Because variability estimates often differ with chemical concentration, coefficients of variation for these chemistry methods were calculated based on concentrations determined using two different approaches:

1. *Approach 1*—EPA's acute water quality criterion for the protection of freshwater aquatic life [2,5], or, if no criterion existed, a concentration equal to five times the Agency's reported method detection level [4].

2. *Approach 2*—concentrations equal to 50 times the reported method detection level [3,4].

Using benzene as an example, single analyst (intralaboratory) precision is calculated using the following linear equation

$$SD = 0.09 \text{ (concentration)} + 0.59$$

TABLE 4—*Summary of coefficient of variation data for interlaboratory and intralaboratory effluent toxicity test results.*

C.V., %	Interlaboratory Coefficients of Variation			
	Number of Tests (Using All Data)	Cumulative Percent	Number of Tests (Filtered)[a]	Cumulative Percent
0.0	23	16.3	3	2.5
0.1 to 10	15	27.0	15	14.9
10.1 to 20	40	55.3	40	47.9
20.1 to 30	27	74.5	27	70.2
30.1 to 40	10	81.6	10	78.5
40.1 to 50	8	87.2	8	85.1
50.1 to 60	2	88.6	2	86.8
60.1 to 70	7	93.6	7	92.6
70.1 to 80	2	95.0	2	94.2
80.1 to 90	3	97.2	3	96.7
90.1 to 100	1	97.9	1	97.5
<100	3	100.0	3	100.0
	141		121	

C.V., %	Intralaboratory Coefficients of Variation (from Table 3)			
	Number of Tests (Using All Data)	Cumulative Percent	Number of Tests (Filtered)[a]	Cumulative Percent
0.0	11	23.9	5	12.5
0.1 to 10	20	67.4	20	62.5
10.1 to 20	5	78.3	5	75.0
20.1 to 30	5	89.2	5	87.5
30.1 to 40	0	89.2	0	87.5
40.1 to 50	3	95.7	3	95.0
50.1 to 60	0	95.7	0	95.0
60.1 to 70	1	97.9	1	97.5
70.1 to 80	0	97.9	0	97.5
80.1 to 90	0	97.9	0	97.5
90.1 to 100	0	97.9	0	97.5
>100	1	100.0	1	100.0
	46		40	

NOTE: C.V. = coefficient of variation.
[a] Excludes data where coefficients of variation are 0.0% because all replicate LC_{50} values are ≥100%.

and overall (interlaboratory) precision is calculated as

$$SD = 0.21 \text{ (concentration)} + 0.56$$

Using Approach 1, coefficients of variation (defined here as (SD/concentration) × 100) for benzene were calculated using concentrations equal to U.S. EPA's acute water quality criteria for freshwater (5300 µg/L) [2]. Based on these equations, intra- and interlaboratory coefficients of variation for benzene are 9.0 and 21.0%, respectively. Using Approach 2, a concentration equal to 50 times the method detection level (MDL = 0.2 µg/L) yields intra- and interlaboratory coefficients of variation of 14.9 and 26.6%, respectively.

Coefficient of variation data for these U.S. EPA chemical methods are summarized in Table 7. The number of chemicals evaluated in each case is ≤126 because precision data were either not available for certain chemicals (for example, acrolein, acrylonitrile), or the standard deviation regression equations yielded erroneous values (for example, ≤zero). Based on the data presented

TABLE 5—*Summary of interlaboratory coefficients of variation for three common test species.*

	Cumulative Percent		
C.V., %	S. gairdneri	Daphnia spp.[a]	P. phosphoreum
0.0	25.0	28.6	0.0
0.1 to 10.0	32.9	50.0	11.6
10.1 to 20.0	57.9	71.4	51.2
20.1 to 30.0	72.4	78.6	72.1
30.1 to 40.0	80.3	78.6	81.4
40.1 to 50.0	86.8	85.7	86.0
50.1 to 60.0	89.5	85.7	86.0
60.1 to 70.0	96.0	85.7	90.7
70.1 to 80.0	96.4	85.7	93.0
80.1 to 90.0	96.4	92.8	97.7
90.1 to 100.0	98.7	92.8	97.7
>100	100.0	100.0	100.0
	n = 76	n = 14	n = 43

Note: n = number
[a] Includes data for both *D. magna* and *D. pulex*.

in Table 7, 76 to 83% of the interlaboratory test data yielded coefficients of variation ≤50%. For intralaboratory measurements, coefficients of variation ≤50% were achieved for between 90 to 94% of the priority pollutant chemical methods. Comparisons of the distributions of coefficients of variation for the effluent toxicity data reviewed in this study and the analytical methods for EPA's priority pollutants indicate they are generally within the same ranges (Table 8).

Conclusions

For the test species reviewed in this study, quantitative inter- and intralaboratory variability comparisons between the results from standardized effluent acute toxicity test methods and priority

TABLE 6—*Summary of intralaboratory coefficients of variation data for* Daphnia *spp.*

C.V., %	Number of Tests (Using All Data)	Cumulative Percent	Number of Tests (Filtered)[a]	Cumulative Percent
0.0	11	28.9	3	10.0
0.1 to 10	17	73.7	17	66.7
10.1 to 20	4	84.2	4	80.0
20.1 to 30	4	94.7	4	93.3
30.1 to 40	0	94.7	0	93.3
40.1 to 50	2	100.0	2	100.0
50.1 to 60	0	100.0	0	100.0
60.1 to 70	0	100.0	0	100.0
70.1 to 80	0	100.0	0	100.0
80.1 to 90	0	100.0	0	100.0
90.1 to 100	0	100.0	0	100.0
>100	0	100.0	0	100.0
	38		30	

[a] Excludes data where coefficients of variation are 0.0% because all replicate LC_{50} values are ≥ 100%.

TABLE 7—*Summary of coefficient of variation data for priority pollutant chemistry methods.*

APPROACH 1—CALCULATIONS BASED ON EPA'S FRESHWATER ACUTE CRITERION OR FIVE TIMES THE EPA'S REPORTED MINIMUM DETECTION LEVEL FOR THE METHOD

	Interlaboratory		Intralaboratory	
C.V., %	Number of Chemicals	Cumulative Percent	Number of Chemicals	Percent Cumulative
0.0	0	0	0	0
0.1 to 10	0	0	16	14.4
10.1 to 20	20	19.2	37	47.7
20.1 to 30	19	37.5	27	72.0
30.1 to 40	22	58.7	15	85.5
40.1 to 50	18	76.0	5	90.0
50.1 to 60	9	84.7	2	91.8
60.1 to 70	5	89.5	0	91.8
70.1 to 80	1	90.4	1	92.7
80.1 to 90	1	91.3	2	94.5
90.1 to 100	2	93.2	1	95.4
>100	7	100.0	5	99.9
	104		111	

APPROACH 2—CALCULATIONS BASED ON FIFTY TIMES THE EPA'S REPORTED MINIMUM DETECTION LEVEL FOR THE METHOD

	Interlaboratory		Intralaboratory	
C.V., %	Number of Chemicals	Cumulative Percent	Number of Chemicals	Percent Cumulative
0.0	0	0	0	0
0.1 to 10	0	0	11	10.2
10.1 to 20	9	9.7	34	41.7
20.1 to 30	28	39.8	35	74.1
30.1 to 40	20	61.3	20	92.6
40.1 to 50	20	82.8	1	93.5
50.1 to 60	5	88.2	2	95.4
60.1 to 70	6	94.6	0	95.4
70.1 to 80	2	96.8	1	96.3
80.1 to 90	1	97.8	1	97.2
90.1 to 100	0	97.8	1	98.1
>100	2	100.0	108	100.0
	93		108	

pollutant chemical methods indicate that they are generally within the same range. It is important to recognize, however, that the levels of precision presented above may not be representative of all effluent acute or chronic toxicity test methods and complex chemical mixtures. Although there is some data on the precision of chronic toxicity testing with effluents [6–8], the data base is small.

Finally, it should be emphasized that this evaluation of effluent toxicity test precision is based on the results of only 13 studies, most of which utilize only two or three replicates per sample. Further work is needed to better understand: (1) the levels of precision that can be expected from effluent toxicity test methods, particularly those commonly used for regulatory purposes (for

TABLE 8—*Summary of coefficients of variation data for effluent toxicity testing and analytical chemistry methods.*

INTERLABORATORY COEFFICIENTS OF VARIATION

C.V., %	ETT Methods[a]	Chemical[b] Methods	Chemical[c] Methods
		Cumulative Percent	
0.0	2.5	0	0
0.1 to 10	14.9	0	0
10.1 to 20	47.9	19.2	9.7
20.1 to 30	70.2	37.5	39.8
30.1 to 40	78.5	58.7	61.3
40.1 to 50	85.1	76.0	82.8
50.1 to 60	86.8	84.7	88.2
60.1 to 70	92.6	89.5	94.6
70.1 to 80	94.2	90.4	96.8
80.1 to 90	96.7	91.3	97.8
90.1 to 100	97.5	93.2	97.8
>100	100.0	100.0	100.0

INTRALABORATORY COEFFICIENTS OF VARIATION

C.V., %	ETT Methods[a]	Chemical[b] Methods	Chemical[c] Methods
		Cumulative Percent	
0.1	12.5	0	0
0.1 to 10	62.5	14.4	10.2
10.1 to 20	75.0	47.7	41.7
20.1 to 30	87.5	72.0	74.1
30.1 to 40	87.5	85.5	92.6
40.1 to 50	95.0	90.0	93.5
50.1 to 60	95.0	91.8	95.4
60.1 to 70	97.5	91.8	95.4
70.1 to 80	97.5	92.7	96.3
80.1 to 90	97.5	94.5	97.2
90.1 to 100	97.5	95.4	98.1
>100	100.0	99.9	100.0

[a] Effluent toxicity test variability using filtered data set from Table 4.
[b] Priority pollutant test methods variability based on Approach 1—concentrations equal to freshwater acute criteria or five times the method detection limit (from Table 7).
[c] Priority pollutant test methods variability based on Approach 2—concentrations equal to 50 times the method detection limit (from Table 7).

example, *Daphnia, Ceriodaphnia*, fathead minnows, algae); and (2) how test method precision can be incorporated into effluent safety assessments.

Acknowledgments

The authors gratefully acknowledge the Chemical Manufacturers Association (Washington, D.C.) for their support of this study.

References

[1] Bainbridge, T. R., "The Committee on Standards: Precision and Bias," *ASTM Standardization News*, American Society for Testing and Materials, Philadelphia, January 1985, pp. 44–46.

[2] U.S. Environmental Protection Agency, "Water Quality Criteria Documents," *Federal Register*, Vol. 45, 28 Nov. 1980, pp. 79318–79341.

[3] "Methods for Chemical Analyses of Water and Wastes," EPA-600/4-79-020, U.S. Environmental Protection Agency, Environmental Monitoring and Support Laboratory, Cincinnati, OH, 1979.

[4] U.S. Environmental Protection Agency, "Guidelines Establishing Test Procedures for the Analyses of Pollutants Under the Clean Water Act" *Federal Register*, Vol. 49, 26 Oct. 1984, pp. 43234–43442.

[5] U.S. Environmental Protection Agency, "Water Quality Criteria; Availability of Documents," *Federal Register*, Vol. 50, 29 July 1985, pp. 30784-30796.

[6] EA Engineering, Science, and Technology, Inc., "*Ceriodaphnia* Reproductive Potential Tests on Ambient Stations and Selected Effluents, Naugatuck River, CT," prepared for U.S. EPA's Monitoring and Data Support Division, Washington, DC., 1984.

[7] Horning, W. B. and Weber, C. I, "Methods for Estimating the Chronic Toxicity of Effluents and Receiving Waters to Freshwater Organisms," EPA-600/4-85-014, Draft dated May 1985, U.S. Environmental Protection Agency, Washington, DC.

[8] DeGraeve, G. M., Cooney, J. D., Pollock, T. L., Reichenbach, N. G., and Dean, J. H., "Round Robin Study to Determine the Reproducibility of the 7-day Fathead Minnow Larval Survival and Growth Test," prepared for the American Petroleum Institute (API) and the Electric Research Institute, published by API, Washington, DC.

[9] Grothe, D. R. and Kimerle, R. A., "Inter- and Intralaboratory Variability in *Daphnia magna* Effluent Toxicity Test Results;" *Environmental Toxicology and Chemistry*, Vol. 4, 1985, pp. 189–192.

[10] Buikema, A. L., "Variation in Static Acute Toxicity Test Results with *Daphnia magna* Exposed to Refinery Effluents and Reference Toxicants," *Oil and Petrochemical Pollution*, Vol. 1, No. 3, 1983, pp. 189–198.

[11] Dorn, P. B., "Biological Assessments and Their Applications in NPDES Permits," presented at the International Society of Petroleum Industry Biologists, October 1984, Houston, TX, unpublished manuscript.

[12] "*Pimephales promelas* Acute Toxicity Test, EA Engineering, Science, and Technology, Inc., Sparks, MD, confidential client, 1984.

[13] Peltier, W., "Round Robin Effluent Test," U.S. EPA, Athens, GA, unpublished data.

[14] Grothe, D. R., six unpublished studies, Monsanto Co., St. Louis, MO, May 1985.

[15] Strosher, M., "A Comparison of Biological Testing Methods in Association with Chemical Analyses to Evaluate Toxicity of Waste Drilling Fluids in Alberta," Canadium Petroleum Association, Calgary, Alberta, 1983.

[16] Peltier, W., "Methods for Measuring the Acute Toxicity of Effluents to Aquatic Organisms," EPA-600/4-78-012, U.S. EPA, Environmental Monitoring and Support Laboratory, Cincinnati, OH, 1978.

[17] "Effluent Toxicity Screening Test Using *Daphnia* and Mysid Shrimp," U.S. Environmental Protection Agency, Washington, DC, 1980 draft.

[18] Mount, D. I. and Norberg, T. J., "A Seven-Day Life-Cycle Cladoceran Toxicity Test," *Environmental Toxicology and Chemistry*, Vol. 3, 1984, pp. 425–434.

[19] "Toxicity Testing of Hazardous Waste," Microtox Application Notes, No. M105, Beckman Instruments, Inc., Fullerton, CA, 1 Sept. 1982.

Avital Gasith,[1] Krzysztof M. Jop,[2] Kenneth L. Dickson,[3] Thomas F. Parkerton,[3] and Stan A. Kaczmarek[4]

Protocol for the Identification of Toxic Fractions in Industrial Wastewater Effluents

REFERENCE: Gasith, A., Jop, K. M., Dickson, K. L., Parkerton, T. F., and Kaczmarek, S. A., **"Protocol for the Identification of Toxic Fractions in Industrial Wastewater Effluents,"** *Aquatic Toxicology and Hazard Assessment: 10th Volume, ASTM STP 971,* W. J. Adams, G. A. Chapman, and W. G. Landis, Eds., American Society for Testing and Materials, Philadelphia, 1988, pp. 204–215.

ABSTRACT: A toxic fraction identification protocol (TFIP) was developed to identify the fractions(s) of a complex wastewater effluent contributing to toxicity. The TFIP is based on sequential physical/chemical fractionations of the effluent with associated toxicity testing using the aquatic invertebrate test organism *Daphnia*. Fractionation treatments used in the TFIP include filtration, aeration, activated carbon, and cation and anion exchange treatments. Before and after each treatment, acute toxicity tests are conducted to determine the efficacy of the treatments in removing toxicity. Selected chemical analyses are used along with toxicity test results to identify the toxic fraction(s) in the wastewater. The TFIP can determine whether or not the toxicity is associated with filterable solids, volatile and/or biodegradable organics, carbon adsorbable compounds, cationic or anionic inorganic constituents, and charged organic compounds. The TFIP was developed using six industrial effluents possessing a diversity of chemical constituents. It was validated using a nontoxic industrial effluent spiked with hexavalent chromium and naphthenic acids.

KEYWORDS: wastewater effluent toxicity, *Daphnia pulex,* toxic fraction identification, toxicity reduction evaluation, naphthenic acids, chromium

The U.S. Environmental Protection Agency's recently released "Policy for the Development of Water Quality–Based Permit Limitations for Toxic Pollutants" integrates traditional chemical analyses and whole effluent toxicity methods for assessing and controlling the discharge of toxic chemicals into U.S. waterways [1]. States are now being faced with the task of identifying discharges where effluent toxicity should be characterized. Where appropriate, states may require toxicity-based permit limits, incorporating effluent toxicity as a compliance parameter in wastewater discharge permits. Guidance for the appropriate application as well as the advantages and disadvantages of both specific chemical and whole effluent toxicity approaches have been provided [2].

Toxicity tests by their nature provide useful information about an important property of an effluent. However, they do not identify the causative agent(s) responsible for the toxicity. Thus, an industrial or municipal wastewater discharger who determines via toxicity testing that a particular

---

[1] Professor, Institute for Nature Conservation Research, Tel Aviv University, Ramat Aviv, Israel 69978.
[2] Research scientist, Battelle Ocean Sciences, 397 Washington St., Duxbury, MA 02332.
[3] Professor and research associate, respectively, Department of Biological Sciences and Institute of Applied Sciences, North Texas State University, Denton, TX 76203.
[4] Environmental engineer, SAK Environmental Technologies, Morris Plains, N. J.

wastewater discharge has an unacceptable level of toxicity frequently has little idea of how to initiate a program to eliminate the observed toxicity. This difficulty is due to the potentially large number of compounds and variable concentration levels present in some wastewaters. In such cases the regulatory authority may require the discharger to conduct what EPA refers to as a toxicity reduction evaluation (TRE). The goal of TRE is to determine what control options could be implemented so that toxicity or chemcial concentration requirements could be met. As part of the TRE, EPA has identified effluent fractionation followed by chemical and/or toxicological testing to identify causes of toxicity.

With this in mind, a project was conducted to develop a protocol which can be used by a wastewater discharger to determine the chemical fractions(s) causing toxicity. The protocol could also be used to evaluate the feasibility of various wastewater treatment approaches to remove toxic constitutents. The protocol utilizes simple effluent fractionation treatments such as filtration, aeration, carbon adsorption, and ion exchange in conjunction with invertebrate acute toxicity tests to determine the chemical fraction(s) contributing to the observed toxicity. Several other studies have indicated the potential utility oft this approach [3–6].

The identification of a toxic effluent fraction is based on the principal of sequential removal of a chemical fraction coupled with a toxicity test of the fractionated effluent. The procedure is followed until the effluent is nontoxic. Routine effluent water quality analyses are conducted at each fractionation step to assist in interpretation of the results. Water quality analyses may also be used to establish indicator parameters as surrogates for other toxicity-related parameters that are more difficult or costly to monitor.

To ensure the practicality of the protocol as a routine procedure, the fractionation was based on simple and widely used laboratory techniques including filtration, aeration, powdered activated carbon (PAC), and ion exchange. These treatments allow selective removal of nonfilterable residues, volatile and readily degraded compounds, activated carbon adsorbable compounds, and charged compounds, respectively. Ultimate identification of the toxic constituent(s) would usually require detailed and sophisticated analytical methods.

The toxic fraction identification protocol (TFIP) was developed while testing final and partially treated refinery wastewaters and chemical plant effluents. It was validated by testing nontoxic effluents spiked with toxic constituents. The wastewater samples from various petroleum industry sources were fractionated, tested for toxicity, and analyzed at the Aquatic Toxicology Laboratory at North Texas State University (NTSU). Ion analyses were performed by Exxon Research and Engineering Co., Baytown, Texas. Organic analyses by gas chromatography mass spectroscopy (GCMS) were performed by Radian Corp.

Materials and Methods

Experimental Design

Approximately 75 L of the freshly collected effluents were shipped in glass containers via overnight express. Toxicity testing and chemical fractionation of the wastes began immediately after arrival.

A flow diagram outlining the TFIP is presented in Fig. 1. The fractionation sequence requires filtration and aeration as the first two treatments; activated carbon, cation exchange, and anion exchange treatments then are applied in parallel following aeration. The reason to apply filtration and aeration early in the sequence is to reduce the concentration of consititutents that can complicate other results. For example, filtration not only reduces the levels of potentially toxic suspended solids, but also reduces biomass concentrations. Aeration not only strips volatile organics, but also equilibrates carbon dioxide concentrations (and thus pH) with the lab atmosphere.

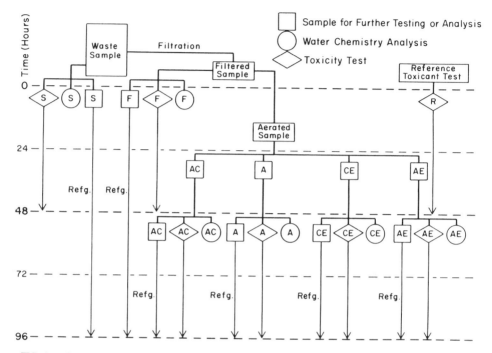

FIG. 1—*Flow diagram of toxic fraction identification protocol: S = raw wastewater; F = filtered sample; R = reference sample; A = aeration and filtration; CE = cation exchange; AE = anion exchange; and AC = activated carbon.*

The time required for cascading through the protocol is determined by the number of fractions tested, by the duration of each toxicity test, and by laboratory capability of performing simultaneous toxicity tests. An optional 24-h screening toxicity test using undiluted effluent may be performed when immediate information on sample toxicity is required. Ultimate evaluation of wastewater toxicity should be based, however, on a definitive test. The TFIP as outlined in Fig. 1 takes four days to complete, assuming a capability exists, to conduct four simultaneous *D. pulex* acute toxicity tests per day. If more toxicity tests can be performed, the time required to complete the TFIP may be shortened.

The first toxicity test was performed on raw effluent (S-fraction). The remaining effluent sample was vacuum filtered (F-fraction) through a prewashed 0.3-μm glass fiber filter (Schleicher and Schuell No. 30). Twenty litres of the filtered effluent sample were then aerated (A-fraction) for 24 h at a rate of 155 mL/min using stainless steel diffusers. A filtered and aerated 7.5-L effluent sample aliquot was treated with 0.5-g/L powdered activated carbon (AC-fraction) (Sigma Chemical Co.). Other aliquots of the filtered and aerated effluent were eluted separately through glass columns containing either strong-acid cation exchange resin (Amberlite IR-1200, Rohm and Haas) resulting in the CE-fraction or a strong-base anion exchange resin (Amberlite IRA-402, Rohm and Haas) resulting in the AE-fraction. Following cation or anion exchange treatments, the pH for these samples was adjusted to 7 ± 1 with NaOH or H_2SO_4. Attempts were made to return the pH as close to the aerated sample pH as possible. To ensure the presence of major minerals in the eluates from the ion exchange columns, $CaSO_4$ (60 mg/L), $NaHCO_3$ (96 mg/L), and KCl (4 mg/L) were added prior to toxicity testing. An aliquot of each effluent fraction was refrigerated (4°C) for additional testing and chemical analyses if warranted. Concomitant with the effluent testing, chromium reference toxicity tests (R) were conducted for each effluent tested to insure adquate quality control of the toxicity data generated. Stock hexavalent chromium solutions were prepared from reagent grade potassium chromate (Fisher Scientific Co.).

During the protocol validation experiments, a nontoxic effluent maxtrix was spiked with potassium chromate. The initial total chromium concentrations were verified by atomic absorption spectrophotometry. Hexavalent chromium was also determined by diphenylcarbozide (DPC) chelation and selective extration into methylisobutylketone (MIBK) [7]. In the second set of validation experiments, a nontoxic effluent matrix was spiked with naphthenic acids (Eastman Kodak Co.). Concentrations of naphthenic acids were measured with a (Waters ALC-201) high performance liquid chromatograph (HPLC). An aliquot of the effluent was spiked with 1000 mg/L of naphthenic acids and aerated in the dark at room temperature for three days. The aeration facilitated the dispersal and solubilization of the oily naphthenic acids complex in the effluent matrix. The naphthenic acid–spiked effluent sample used to validate the TFIP was obtained by siphoning the lower two thirds of the spiked sample, leaving behind the undissolved oily residue.

Toxicity Tests

Static acute toxicity tests (48-h EC_{50}) were conducted at selected points in the TFIP with less than 48-h-old *Daphnia pulex* Richard in accordance with standard procedures [8]. Chemical fractionation and toxicity testing were conducted in the North Texas State University—Aquatic Toxicology Laboratory. Test conditions included a temperature of 20 ± 2°C and a 16:8h light and dark photoperiod. Each test consisted of five waste dilutions and a control in which moderately hard reconstituted water was used as diluent [7]. Five replicates, each containing ten animals, were used for each dilution and the control. Daphnids were counted after 48-h exposure by pouring each replicate through a 0.365-mm mesh Nitex screen. Organisms caught on the screen were quickly placed in a petri dish containing dilution water and examined for immobilization or death using a dissecting microscope. The *D. pulex* culture, which provided test organisms, was fed vitamin-enriched *Selenastrum capricornutum* at densities of 2×10^6 cells day^{-1}.

Chemical Analyses

Prior to initiation of the toxicity tests, alkalinity, pH, hardness, conductivity, salinity, dissolved oxygen, ammonia, total organic carbon, total suspended solids, total dissolved solids, free cyanide, oil and grease, biochemical oxygen demand, and chemical oxygen demand were measured in each of the tested fractions. Additional metals analyses (Ag, Al, As, Ba, Be, Ca, Ce, Co, Cr, Cu, Fe, K, Li, Mg, Mn, No, Na, Ni, Pb, Pt, Sb, Se, Si, Sn, Sr, Ti, V, Zn) were conducted using inductively coupled plasma emission spectoscopy (ICPES) on each of the tested effluent fractions. GCMS characterization of organics followed EPA Methods 624 and 625.

Calculations of EC_{50} values (percent effluent concentration immobilizing 50% of the test organisms) and 95% confidence limits were performed using a program developed by Stephan [9]. Data were also analyzed using the statistical analysis system (SAS) [10].

Results and Discussion

The purpose of this study was to develop a simple protocol to identify toxic fraction(s) in an industrial effluent. The results of toxicity tests conducted on eight effluent samples are summarized in Table 1. The first two effluents tested (EF 1 and EF 2) were highly treated wastewater samples. They were not toxic to *D. pulex*, as reflected by the 48-h EC_{50} values, which were greater than 100% effluent. Since mortality did not exceed 20% in either sample, no further testing was initiated.

TABLE 1—*Responses of* Daphnia pulex *to different effluent samples and fractions.*

Sample Label	Treatment	48-h EC_{50} as Percent Effluent	Percent Immobilization/ mortality in 100% Effluent Sample after 48 h
EF 1	S	>100	13
EF 2	S	>100	8
EF 3	S	80	69
	F	94	56
	A	>100	43
	AC	65	77
	EM	>100	2
EF 4	S	71	97
	F	60	100
	A	>100	45
	AC	>100	32
	CE	>100	28
	AE	>100	46
EF 5	S	23	100
	F	38	100
	A	58	98
	AC	>100	25
	CE	76	65
	AE	75	94
Ef 6	S	3	100
	F	>100	40
	A	>100	17
	AC	>100	12
EF 7	S	67	98
	F	65	100
	A	64	100
	AC	71.0	100
	CE	93	57
	AE	>100	2
EF 8	S	57	95
	F	84	69
	A	76	85
	AC	>100	4
	CE	91	59
	AE	>100	0

NOTE: S = raw effluent; F = filtration; AC = activated carbon; A = aeration and filtration; CE = cation exchange; AE = anion exchange; EM = combined anion and cation exchange.

The toxicity test conducted on effluent sample EF 3, a biologically treated effluent, showed higher than expected toxicity (Table 1). Filtration, 24-h aeration, and activated carbon treatment did not reduce toxicity. However, removal of anions and cations by a combined ion exchange treatment eliminated toxicity. Increase in effluent toxicity following the PAC treatment was associated with incomplete separation of PAC and effluent. Remaining carbon particles were actively filtered by animals and contributed to the higher mortality of *Daphnia* in this treatment. Ion analyses of sample EF 3 revealed 3 mg/L of vanadium present in all fractions except the fraction obtained after the ion exchange treatment (Table 2). The results of these experiments showed that the toxic fraction was composed of charged constituents, most of which were represented by vanadium. Shu-Rong [*11*] studied the effects of short-term exposure to vanadium on *Daphnia magna*. The reported 48-h EC_{50} value for vanadium is 2.3 mg/L, which corresponds with the acute value for *D. pulex* in the EF 3 matrix.

Sample EF 4 was a biologically treated wastewater discharged from a covered biological treatment unit. It was toxic to *D. pulex* as reflected by a 97% immobilization/mortality in the undiluted effluent and by a 48-h EC_{50} value of 71% of effluent dilution (Table 1). Removal of nonfilterable solids did not reduce effluent toxicity (48-h EC_{50} value was 60%). However, subsequent aeration reduced *D. pulex* mortality in the undiluted sample by 50%. These results suggested that the effluent toxicity was partially associated with volatile compounds. That is, toxic volatiles that are not readily biodegradable could have existed in the water because chances for stripping them from the covered reactor were low. However, GCMS analysis failed to identify any volatile organics. Another possibility for the decreased toxicity via aeration included removal of toxic surfactants in a foam layer, but foam presence was not researched when these tests were run. The rapid degradation of nonvolatile organic constituents was not a possibility because this effluent was biologically treated. Interestingly, aeration raised pH and thus increased unionized ammonia concentrations to a potentially toxic level. Thus, the decrease in toxicity caused by aeration may have been greater if ammonia were not present.

After filtration and aeration, activated carbon and cation exchange treatments both slightly further reduced toxicity in sample EF 4 (Table 1). Cation exchange reduced ammonia concentrations, whereas activated carbon reduced organic concentrations as measured by total organic carbon and chemical oxygen demand tests. The specific organic contribution to toxicity was not identified. However, the fact that anion exchange did not reduce toxicity probably removes organic acids and anionic surfactants from consideration. It should also be noted that metals were not present at acutely toxic concentrations in sample EF 4 (Table 2).

Wastewater effluent EF 5 was collected prior to biological treatment from a dissolved air flotation unit. It was toxic, with a *D. pulex* 48-h EC_{50} value of 23% dilution (Table 1). Removal of nonfilterable solids slightly reduced the effluent toxicity as was evident by the lower *D. pulex* mortality in the 25 and 12.5% effluent dilutions (Fig. 2) and the higher EC_{50} values (38%). This effect was associated with a reduction in total suspended solids from 30 to 2.5 mg/L. The 24-h aeration treatment further reduced toxicity in sample EF 5, resulting in a 48-h EC_{50} value of 58% of waste dilution. This latter reduction in toxicity was attributed to a decrease in unionized ammonia from 1.2 mg/L to below 1.0 mg/L, probably caused by stripping (Table 2). Interestingly, a drop in volatile aromatics from 1.00 mg/L to less than 0.01 mg/L during aeration probably contributed insignificantly to this reduced toxicity, because the original contaminant levels were probably too low to be acutely toxic.

Subsequent removal of anions and cations by ion exchange treatments from sample EF 5 both further reduced toxicity as reflected by 48-h EC_{50} values of 75 and 76%, respectively. In both cases, the ion exchange treatments reduced the organic fraction in the samples. Total organic carbon (TOC) and chemical oxygen demand (COD) in the raw wastewater were 82 and 345 mg/L, respectively. After anion and cation exchange treatments, total organic carbon concentrations were reduced to 25 and 31.5 mg/L, and chemical oxygen demand concentration levels were lowered to 171 and 226 mg/L, respectively (Table 2). A substantial reduction in toxicity was also observed after activated carbon treatment (Table 1). Since activated carbon completely removed the organic carbon components in the effluent, toxicity of sample EF 5 was judged to be predominantly associated with high concentrations of nonvolatile organics. In fact, GCMS identified 165 mg/L of varied acid and base/neutral toxic organic constituents. This illustrates a case in which TOC or COD might serve as a useful indicator parameter for assessing and monitoring effluent toxicity.

The toxicity test conducted on effluent sample EF 6 showed that the effluent was highly toxic

TABLE 2—Results of water quality analyses of effluent samples evaluated using the TFIP.

Sample	pH	Conductivity, μS	Alkalinity, mg/L as CaCO$_3$	Hardness, mg/L as CaCO$_3$	NH$_3$ mg/L as N	TSS, mg/L	TOC, mg/L	COD, mg/L	Oil and Grease, mg/L	TCr, mg/L	V, mg/L
					EF No. 1						
S	7.4	4500	157	125	0.02	20	51	...	...	0.0	0.019
F	7.7	4450	115	117	...	...	...	...	...	0.0	0.024
					EF No. 2						
S	7.4	312	109	166	0.15	44	38	...	...	0.04	0.475
F	...	...	...	...	...	...	...	...	...	0.02	0.440
					EF No. 3						
S	7.1	1700	175	38	0.001	22	11.0	48	...	0.00	2.98
F	7.4	1610	175	36	0.005	...	7.0	36	...	...	...
A	8.4	1710	162	40	0.061	...	8.5	38	...	...	...
AC	8.3	1690	160	30	0.058	...	1.0	9.5	...	0.00	3.03
EM	8.8	350	...	122	...	...	0.0	...	...	0.00	0.00
					EF No. 4						
S	6.9	3135	517	22	0.026	28	90	100	...	...	...
F	7.5	3085	512	36	0.062	6	90	83	...	0.00	0.678
A	8.5	3200	522	38	0.590	...	73	78	...	...	...
AC	8.5	3180	512	32	0.669	...	59	25.5	...	...	...
CE	...	2900	460	74	...	...	...	...	...	0.008	0.658
AE	10.0	2750	790	84	...	...	...	...	...	0.009	0.160
					EF No. 5						
S	7.8	2800	225	148	1.184	30	82	345	...	...	...

	pH									
F	7.8	2800	222	142	1.100	3	44	293	...	...	...
A	7.7	2770	235	138	0.879	8	50	264	...	...	0.009
AC	8.0	2780	198	136	...	3	17	82	...	0.000	0.000
CE	7.6	2740	102	84	0.004	4	32	226	...	0.006	0.000
AE	7.8	3800	...	...	...	...	25	171	...	0.000	0.000

EF No. 6

S	6.6	3830	65	352	0.009	52	40	392	...	0.107	0.353
F	6.6	4100	63	358	0.018	13	29	250	20.3	0.063	0.389
A	6.4	4220	35	336	0.012	32	35	171	0.9	...	...
AC	6.6	4140	25	334	0.014	12	10	34	...	...	...
CE	6.6	4100	45	108	...	...	...	...	...	0.044	0.358
AE	6.8	4000	50	92	...	...	...	...	...	0.006	0.022

EF No. 7

S	7.7	4000	88	138	0.700	13	7	87	...	1.150	0.313
F	7.8	4000	93	136	0.800	2	7	63	...	...	...
A	8.0	3900	90	138	0.700	2	6	69	...	1.150	0.281
AC	8.0	4020	85	64	1.000	1	2	10	...	1.160	0.280
CE	7.2	3700	60	114	0.026	2	5	55	...	1.060	0.047
AE	7.3	3700	355	242	0.200	2	5	33	...	0.006	0.087

EF No. 8

S	7.7	2830	73	82	0.000	7	28	236	...	...	...
F	7.8	2810	80	94	0.000	3	20	226	...	...	...
A	7.8	2810	28	94	0.000	1	20	201	...	...	...
AC	7.7	2810	20	80	0.000	0	7	34	...	...	...
CE	8.1	2750	60	58	0.000	1	21	126	...	...	...
AE	8.1	2750	73	20	0.000	0	12	63	...	...	...

NOTE: S = raw effluent; F = filtration; AC = activated carbon; A = aeration and filtration; CE = cation exchange; AE = anion exchange; EM = combined anion and cation exchange; TSS = total suspended solids; TOC = total organic carbon; COD = chemical oxygen demand; and TCr = total chromium.

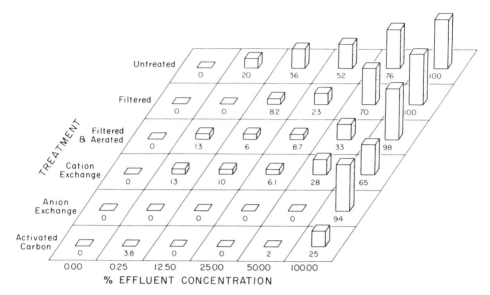

FIG. 2—*Toxicity of Sample EF 5 at various dilutions and after different fractionization treatments (numbers in the blocks indicate percent* Daphnia pulex *mortality).*

to *D. pulex*. This sample, like EF 5, was collected prior to biological treatment, in this case from an American Petroleum Institute (API) separator effluent. The 48-h EC_{50} value was 3%. Filtration significantly reduced toxicity, indicating that filterable residue was a potential source of toxicity. In fact, observation of a surface film in the raw effluent sample implicated insoluble oil and grease as possible sources of toxicity. Indeed, oil and grease were present in a raw wastewater at 20.3 mg/L (Table 2). Filtration effectively removed oil and grease to a low level (0.9 mg/L). Subsequent aeration and activated carbon treatments slightly reduced the toxicity further (Table 1). These results suggested that the major toxic fraction(s) of effluent sample EF 6 were insoluble hydrocarbons. Moderately degradable and/or volatile organics contributed only slightly to toxicity.

In conjunction with effluent testing, acute toxicity tests with the chromium as a reference toxicant were conducted with *D. pulex*. The 48-h EC_{50} values for hexavalent chromium ranged from 0.16 to 0.12 mg Cr^{6+}/L and fell within EC_{50} values for hexavalent chromium obtained in previous quality assurance/quality control tests [*12*]. In eight Cr^{6+} reference tests with *D. pulex,* the mean 48-h EC_{50} was 0.14 mg Cr^{6+}/L, with a coefficient of variation of 11%. The coefficient of variation in reference tests over a six-month period was less than previously reported for hexavalent chromium with *D. pulex* [*12*].

Nontoxic effluents spiked with hexavalent chromium (EF 7) and naphthenic acids (EF 8), respectively, were used to validate the TFIP. Since organic matter in effluents has been shown to reduce hexavalent chromium to the much less toxic trivalent form [*13*], the nontoxic, biologically treated effluent matrix was spiked with potassium chromate to obtain a final concentration of 1.0 mg/L total chromium. The hexavalent chromium concentration was approximately 0.2 mg/L. As expected, a significant fraction of the hexavalent chromium was reduced. The 48-h EC_{50} value of the spiked sample was 67% of dilution, which corresponds to 0.78 mg/L total or approximately 0.16 mg/L hexavalent chromium. Neither filtration, aeration, nor activated carbon treatment reduced effluent toxicity. Water quality analyses showed reductions in total suspended solids from 13.2 to 2.4 mg/L following filtration and total organic carbon reductions from 6.8 to 2.2 mg/L following activated carbon treatment (Table 2). Cation exchange slightly reduced toxicity. However, this result could be associated with removal of ammonia, which was probably created after chromate was added. Toxicity was most reduced following anion exchange treatment (Table 1). Follow-up analyses of the cations and anions in the sample revealed 1.16 to 1.06 mg/L of total chromium in all fractions except for the anion exchange sample (Table 2). Since hexavalent chromium occurs as an anion in aqueous solution [13] and would thus be removed during anion exchange treatment, the TFIP successfully showed that toxicity was associated with the spiked chromium.

In the second validation experiment a nontoxic effluent matrix was spiked with naphthenic acids. The spiked effluent had a concentration of 556 mg/L of naphthenic acids, 236-mg/L chemical oxygen demand, and 28-mg/L total organic carbon compared to 0.0-mg/L naphthenic acids, 69-mg/L COD, and 15-mg/L TOC before spiking (Table 2). The reason for apparently low TOC and COD recoveries is not known. The 48-h EC_{50} value for the nontoxic, biologically treated effluent spiked with naphthenic acid (EF 8) was 57% of effluent dilution (Table 1). A slight reduction in wastewater toxicity followed the filtration and aeration treatments and was associated with a 15% reduction in chemical oxygen demand and a 30% reduction in total organic carbon (Tables 1 and 2). However, toxicity was most dramatically mitigated by activated carbon and anion exchange treatments (Table 1), which markedly reduced COD concentrations by 85 to 67%, respectively (Table 2). Cation exchange treatments reduced the COD concentration by only 22% and subsequent toxicity by only 10%. These results indicate that negatively charged organics (for example, naphthenic acids) may be responsible for observed toxicity. HPLC analysis confirmed that the concentration of naphthenic acid complexes dropped from 556 mg/L in the initial spike to 125 mg/L after activated carbon and 90 mg/L after anion exchange, but remained at 516 mg/L following cation exchange treatment.

Conclusions

The TFIP was developed to assist dischargers of treated wastewater in identification of effluent constituents contributing to toxicity. Application of the TFIP to toxic effluents and nontoxic effluents matrices spiked with known toxicants demonstrated its utility in directing users to the generic identity of toxicants. When combined with analytical determinations guided by the results of the TFIP, users can identify the causes of effluent toxicity and can use this information as appropriate for corrective action.

In order to avoid confusion when using the TFIP, the following should be noted. First, the protocol provided the least confusing results when only one toxic chemical or related chemicals existed. Therefore, it should be applied mostly to final treated effluents. In addition, methods of selectively removing ammonia from samples should be investigated to reduce concentrations of this oft-interfering toxicant. Second, changes in water appearance that occur during aeration should be recorded to determine if foaming or precipitation, as examples, are responsible for toxicity changes. Finally, changes in pH during the ion exchange treatments may change water characteristics in unexpected ways. For example, anion exchange may increase pH and thus cause some cations to precipitate from solution. Changes in ion exchange treatments are not recommended, since one of the major purposes of the tests is to direct a user to a series of chemical tests. Rather, this is pointed out only to indicate that unexpected results should be anticipated.

Identification of specific toxicants within a toxic fraction is often possible through various chemical screening techniques and with knowledge of the sample sources. Fractions containing inorganic toxicants can be rapidly screened by ICPES according to EPA Method 200.7 [14]. Fractions containing organic toxicants can be screened using GC/MS for purgeables by EPA Method 624 [14] and for nonpurgeables by EPA Method 625 [15]. Finally, organic and inorganic ions can be screened by various ion chromatographic, liquid chromatographic, and direct aqueous injection gas chromatographic techniques. Application of these analytical procedures to the appropriate fraction identified as toxic, using the TFIP, will often pinpoint a single pollutant or a small number of related pollutants as the candidate constituents responsible for the observed toxicity. Additional confirmation is possible by spiking the effluent after removal of the toxic fraction(s) with the suspect toxicant(s) and observing the extent to which the original toxicity is mimicked. Once the toxic fraction has been identified, dischargers can then focus efforts on treatability studies for determining the most cost-effective control strategy for solving effluent toxicity problems. Likewise, regulatory agencies can require that these parameters or related indicator parameters be monitored to demonstrate that problems have been rectified.

The cost of conducting the TFIP is dependent on whether or not toxicity tests can be conducted in-house or have to be contracted and the cost of chemical analyses. The TFIP requires at least seven 48-h acute toxicity tests with *Daphnia*. Sample preparation (that is filtration, aeration, carbon sorption, cation and anion exchange treatments) require only a few hours of laboratory time. It is estimated that implementation of the TFIP on an effluent sample would cost approximately $3500, not including chemical analyses. Cost of chemical analyses will vary with the effluent and the identification of the toxic fraction(s) per the TFIP. Analytical costs can vary considerably, from <$100 for simple ion analyses to several thousand dollars for various gas chromatography scans, depending on the type of analytical procedures selected. Thus it is difficult to estimate the cost of chemical analyses which are needed to complement the TFIP. However, the application of the TFIP identifies where chemical analytical efforts should be focused and thus should result in considerable savings in analytical costs.

Acknowledgments

This research was supported by the Exxon Research and Engineering Co. We thank E. Price, D. Chang, and K. Inyoung for their laboratory assistance throughout the study and B. Venables for his critical review of the manuscript.

References

[1] U.S. Environmental Protection Agency, "Policy for the Development of Water Quality-Based Permit Limitations for Toxic Pollutants," *Federal Register*, Vol. 49, No. 9016, 9 Mar. 1984.
[2] U.S. Environmental Protection Agency, "Technical Support Document for Water Quality-Based Toxics Control," EPA Office of Water, Washington, DC, September 1985.
[3] Walsh, G. E. and Garnas, R. L., "Determination of Bioavailability of Chemical Fractions of Liquid Wastes Using Freshwater and Saltwater Algae & Crustaceans," Environmental Science and Technology, Vol. 17, 1983, pp. 180–182.
[4] Fava, J. A., Gift, J. J., Maciorowski, A. F., McCulloch, W. L., and Reisinger II, H. J., "Comparative Toxicity of Whole and Liquid Phase Sewage Sludges to Marine Organisms," *Aquatic Toxicology and Hazard Assessment: Seventh Symposium, ASTM STP 854*, R. D. Cardwell, R. Purdy, and R. C. Bahner, Eds., American Society for Testing and Materials, Philadelphia, 1985, pp. 229–252.
[5] Reece, C. H. and Burks, S. L., "Isolation and Chemical Characterization of Petroleum Refinery Wastewater Fractions Acutely Lethal to *Daphnia magna*," *Aquatic Toxicology and Hazard Assessment: Seventh Symposium, ASTM STP 854*, R. D. Cardwell, R. Purdy and R. C. Bahner, Eds., American Society for Testing and Materials, Philadelphia, 1985, pp. 319–334.
[6] Lentzen, D. E., Wagoner, D. E., Estes, E. D., and Gutkencht, W. F. "IERL-RTP Procedures Manual, Level 1: Environmental Assessment," 2nd ed., EPA-600/7-78-201, Environmental Protection Agency, Washington, DC, October 1978.
[7] "Standard Methods for the Examination of Water and Wastewater," 15th ed., American Public Health Association, Washington, DC, 1980.
[8] Peltier, W. and Weber, C. I., "Methods for Measuring the Acute Toxicity of Effluents to Freshwater and Marine Organisms," Environmental Monitoring and Support Laboratory, U.S. Environmental Protection Agency, Cincinnati, OH, 1985, pp. 1–216.
[9] Stephan, C. E. in *Aquatic Toxicology and Hazard Evaluation (First Conference), ASTM STP 634*, American Society for Testing and Materials, Philadelphia, 1979, pp. 65–84.

[10] "SAS User's Guide: Statistics, SAS Institute Statistical Analysis System," Statistical Analysis System Institute, Raleigh, 1982.
[11] Shu-Rong, D., *Environmental Quality*, Vol. 1, 1980, pp. 1–20.
[12] Jop, K. M., Rodgers, Jr., J. H., Dorn, P. B., and Dickson, K. L., "Use of Hexavalent Chromium as a Reference Toxicant in Aquatic Toxicity Tests," in *Aquatic Toxicology and Environmental Fate: 9th Volume*, American Society for Testing and Materials, Philadelphia, 1986, pp. 390–403.
[13] Schmidt, R. L., "Thermodynamic Properties and Environmental Chemistry of Chromium," Battelle Pacific Northwest Laboratory, Richland, WA, 1984.
[14] U.S. Environmental Protection Agency, *Federal Register*, 1984, Vol. 49, No. 209, Appendix C, 26 Oct. 1984.
[15] "Methods for Organic Chemical Analysis of Municipal and Industrial Wastewater," EPA-600/4-82-057, U.S. Environmental Protection Agency, Washington, DC, July 1982.

Environmental Modeling and Exposure Assessment

Frederick E. Brinckman,[1] *Gregory J. Olson,*[1] *William R. Blair,*[1] *and Edwin J. Parks*[1]

Implications of Molecular Speciation and Topology of Environmental Metals: Uptake Mechanisms and Toxicity of Organotins

REFERENCE: Brinckman, F. E., Olson, G. J., Blair, W. R., and Parks, E. J., "**Implications of Molecular Speciation and Topology of Environmental Metals: Uptake Mechanisms and Toxicity of Organotins,**" *Aquatic Toxicology and Hazard Assessment: 10th Volume, ASTM STP 971,* W. J. Adams, G. A. Chapman, and W. G. Landis, Eds., American Society for Testing and Materials, Philadelphia, 1988, pp. 219–232.

ABSTRACT: This paper compares predictions of environmental fate and effect parameters derived from quantitative structure-activity relationships (QSAR) using estimates of molecular total surface area (TSA) with experimental data. Organotins are used as an example. In addition, a simple linear free-energy relationship with TSA is demonstrated to be applicable to organotin aqueous solubility, chromatographic retentivity, octanol-water partition coefficients and bacterial uptake, and aquatic toxicity. New measurement methodology providing nondestructive optical imaging *in vivo* of tin employing a fluorescent, tin-specific ligand (3-hydroxyflavone) is used to evaluate a likely mechanism of uptake for triorganotins on cells. Finally, the laboratory results are extended to a preliminary appraisal of environmental persistence of tributyltin, which involves both uptake and degradation processes.

KEYWORDS: topology, molecular predictors, organotins, bioaccumulation, chromatography, organometallic compounds, fluorescence detection, metal-specific imaging, persistence

Many metals and metalloids are widely distributed in the environment as organic derivatives bearing *sigma* carbon-element bonds. These organometallic molecules may originate from industrial processes or wastes [1,2] and from natural biogeochemical cycles [3,4]. Organometallic species of metals and metalloids, as compared to their inorganic forms, display distinctive chemical and physical properties that exert a profound influence on their bioactivity, mobility, and persistence. Therefore, it is essential to quantify organometal species in the environment and not simply total metal concentration [5]. Such critical metal speciation is possible because longevity or durability of such metal- and metalloidal-carbon architectures controls their conformational, stereochemical, and related physicochemical properties, which also dictate their environmental fate or impacts. This same bonding longevity can also yield greater insights on each element's behavior because appropriate molecular speciation techniques are therefore possible.

This manuscript examines some unique molecular measurement requirements associated with the study of environmental organometals, employing the example of tributyltin compounds which are the subject of much current industrial and environmental interest worldwide. Though we will be focusing on tin, these results yield obvious implications for detection of other metal and metalloid species and their related environmental modeling and exposure assessments. This present

---

[1] Ceramics Chemistry and Bioprocesses Group, National Bureau of Standards, Gaithersburg, MD 20899.

work describes results merging organotin chromatographic separation related to molecular speciation, with the intrinsic and essential physical-chemical information derived therefrom, to provide quantitative structure-activity relationships (QSAR). These relationships are interpreted in terms of the mode of organotin uptake and degradation by living (bacterial) cells and are reviewed in the context of recent organotin toxicity studies described elsewhere [6,7].

Methods

Chemicals and Materials

Organometals were obtained from commercial suppliers and examined as appropriate for purity and isomeric composition by proton nuclear magnetic resonance (NMR). Solvents employed were of "spectroquality" grade or reagent, each lot being checked for possible metal interference by appropriate atomic absorption (AA) analysis before use. Stock solutions of organotins were prepared and maintained in nitric acid–washed [8] glass bottles with Teflon-lined closures. Wetted surfaces exposed to biological materials consisted of glass, Teflon, or polycarbonate, all sterilized prior to use. Inorganic salts, 3-hydroxyflavone (flavonol), and organic nutrients were obtained from commercial suppliers.

Liquid Chromatographic Measurement and Separations

Liquid chromatographs (HPLC) coupled on-line with tandem ultraviolet (UV) and graphite furnace atomic absorption (GFAA) detectors have been described previously [9,10] in connection with separation and quantitation of either charged organometallic cations and anions [10,11], or for separation and quantitation of neutral, lipophilic organometals [11,12]. Determination of individual capacity factors [13], k'—which is an equilibrated retention coefficient, for each of the organotin compounds discussed in the present work was performed under standardized HPLC-GFAA conditions: mobile phase—methanol/water = 93:7 (volume %); flow rate = 0.5 mL/min; column = C_{18} reverse bonded phase on 10-μm silica (3 by 250 mm). Reported log k'-values were derived from means of two or three duplicate HPLC runs. Separation and quantitative speciation by HPLC-GFAA of butyltin moieties taken up by bacterial strains have been described elsewhere [14].

Topological Calculations

The SAREA program [15], suitably modified for use with organometallic molecular systems on our computer, was employed to calculate total molecular surface areas (TSA) in square angstroms for complete molecular arrays in several anticipated isomeric or conformational geometries as well as for functional group fragments. Details of our procedures are described elsewhere [6,7]. Normal input data were conventional bond distances, bond angles, and Van der Waals' radii available in handbooks. Both holistic molecular calculations [16] and assembly of molecular TSA values from addition of "mean functional group and atom" TSA values [17] were employed. Typical examples of an ORTEP printout and stick figures for the organotins examined in this study with corresponding TSA values are illustrated in Fig. 1.

Fluorescence Detection of Tin Bioaccumulation

Bacterial cultures were grown in Nelson broth [18] to late-logarithmic stage and harvested by centrifugation at 6000 × g for 10 min. Cells were then resuspended in 5 mM PIPES buffer [piperazine-N,N'-bis(2-ethane sulfonic acid), pH 7.0, Ref 19] and recentrifuged. Cells were again

TSA = 334 Å²   1 Å   TSA = 343 Å²

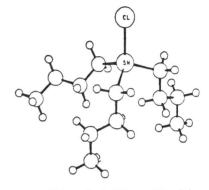

Tri-*n*-butyltin chloride hydrate   Tri-*n*-butyltin chloride

A

endo-n-Butyltriphenylstannane   *exo-n*-Butyltriphenylstannane

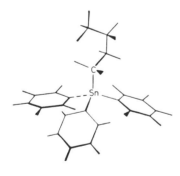 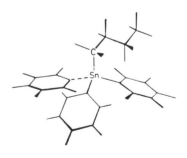

TSA = 388.3 Å²   TSA = 397.0 Å²

B

FIG. 1—*An ORTEP representation of a pentacoordinate tributyltin hydrate compared with its tetracoordinate neutral form along with respective total surface areas* (A). *Neutral butyltriphenylstannane in its closed* (endo) *and extended* (exo) *conformations* (B) *are compared in stick presentation for clarity showing small differences in calculated TSA values.*

TABLE 1—*Treatment protocol for cell suspension[a] of* Pseudomonas *(Strain 244) in fluorescence experiments.*

	Tube 1	Tube 2	Tube 3	Tube 4	Tube 5	Tube 6
1. Add:	flavonol[b]	$SnCl_4$[c]	$SnCl_4$	flavonol + $SnCl_4$	MeOH[d]	nothing
2. Wait 30 min, centrifuge cells, wash 2 times in buffer, resuspend in 0.4 mL buffer.						
3. Add:	nothing	flavonol	nothing	nothing	nothing	flavonol
4. Wait 5 min, centrifuge cells, wash 3 times in buffer, examine cell pellet under 365-nm illumination.						
5. Fluorescence color of cells:	green	blue	none	blue	none	green

[a] 0.4 mL of cells in 5-mM PIPES buffer in 1.5-mL plastic centrifuge tubes, cell density about 10^{10}/mL.
[b] 20 µL of 2 × 10^{-4} M in methanol.
[c] 15-mg Sn/L, final concentration.
[d] 20 µL.

resuspended in PIPES buffer and were then exposed to combinations of tin species (15 mg/L, final concentration) and 3-hydroxyflavone (2 × 10^{-4} M, in methanol), according to the protocol summarized in Table 1.

In aqueous solution, uncomplexed flavonol fluoresces in the green region of the spectrum (λ_{em} = 530 nm) when excited at 365 nm, but forms soluble blue fluorescent complexes (λ_{em} = 460 nm) with tin (Sn)(IV) compounds through a charge transfer bidentate chelation [20]. Following treatments, the cells were visually examined under a hand-held ultraviolet lamp (365 nm) or under a Zeiss Universal microscope fitted for epifluorescence microscopy. The microscope was equipped with a monochromator (400 to 700 nm) and photometer system for quantitative microspectrofluorimetry. An excitation wavelength of 365 nm was obtained using a 100-W mercury lamp and a 365-nm exciter filter. Some experiments employed porous glass beads with covalently bound surface mercapto (—SH) groups

$$\text{glass(Si)—O—Si—CH}_2\text{—CH}_2\text{—CH}_2\text{—SH} \rightleftharpoons$$
$$\text{glass(Si)—O—Si—CH}_2\text{—CH}_2\text{—CH}_2\text{—S}^- + \text{H}^+ \quad (2)$$

These were obtained from W. Haller, National Bureau of Standards [21]. The beads were treated with stannic chloride ($SnCl_4$), washed repeatedly in deionized water, then treated with flavonol.

Biodegradation of Tributyltin

Chesapeake Bay surface water samples (depth 1.0 m) were distributed into 0.25 to 1.2-L sterile glass bottles and spiked with a chromatographically purified aqueous tributyltin solution [22] to give a final concentration (as tin) of 1 to 100 µg/L, depending on the experiment. Samples were incubated aerobically at *in situ* field seasonal temperatures (5° to 28°C) in the dark or under a bank of 40-W incandescent lamps. Samples were periodically removed for butyltins analysis using a conventional column gas chromatograph (GC) directly coupled with a flame photometric detector (FPD) operating under modified reducing flame conditions promoting transformation of exit organotin analytes to gaseous SnH as detected flame species [8]. Using a direct methylene chloride (CH_2Cl_2) extraction technique combined with simultaneous sodium borohydride ($NaBH_4$) reduction [23], samples could be assayed for all butyltin species in the range of 2 to 10 ng/L (as tin).

Results and Discussion

Correlations Between Molecular Shapes of Organotins and Experimental Data

A well-behaved liquid chromatographic system provides an effective means for reliable measurement of many equilibrium parameters and their thermodynamic functions [12,13,24]. Many authors have demonstrated the utility of HPLC for estimation of solubility [25], octanol-water partition coefficients [26,27], and bioconcentration [28]. Essential to understanding the relationships with HPLC retentivity are the specific structural features of sparingly soluble, lipophilic organic molecules [29]. However, not until our recent toxicity studies on a wide range of commercial organotin compounds, relative to survival of crab larvae [6,7], had direct computational approaches to quantitating xenobiotic molecular conformation and topologies to biological effects of metals been shown.

A highly correlated predictor of bulk aqueous and lipid phase solubilities, based upon the effective "exclusion volume" or molecular surface area [30], is clearly applicable for assessing the relative toxicities of a homologous series of organotins towards crab larvae (Table 2). Not

TABLE 2—*QSARs for predicting physicochemical and toxicity properties of organotin from their total surface areas.*

Property, Organotins	QSAR, Measured Units	Slope	Constant	N	r^2	Source Data Reference (See Below)
Aqueous solubility: R_4M (R = alkyl; M = C,Si,Ge,Sn)	$-\log S$ molal	0.0224	0.442	9	0.992	1,2
Partition coefficient octanol/water: $R_nR^1{}_{4-n}Sn$	$\log P$	0.0263	−1.98	12	0.991	1,3
Liquid chromatographic retention: $R_nR^1{}_{4-n}Sn$, HPLC	$\ln k^1$	0.0117	−1.94	12	0.995	1
Gas chromatographic retention: $R_nR^1{}_{4-n}Sn$, (R = alkyl, vinyl)	RI, min	0.0525	3.64	7	0.983	1,4
Microbial uptake: R_3SnCl (R = Et,Pr,Bu)	% uptake, g	0.459	−69.4	3	0.944	5
Algal reproduction (growth): $R_3SnX \cdot H_2O$ (X = Cl,Br,CO$_3$)	$\ln$ IC50 Sn, mg/L	−0.0365	7.69	5	0.957	6
Toxicity (crab larvae survival): $R_3SnX \cdot H_2O$ (X = alkyl,Ph,c-Hx)	$\ln$ LD50, nM	−0.0146	9.18	8	0.938	7

[1] This work.
[2] C. L DeLigny and N. G. Van der Veen, *Recueil des Travaux Chimiques des Pays-Bas,* Vol. 90, 1971.
[3] C. Hansch and A. Leo, "Substituent Constants for Correlation Analysis in Chemistry and Biology," Wiley, New York, 1979.
[4] M. V. Budahegyi, *Journal of Chromatography,* Vol. 271, 1983, p. 214.
[5] J. Yamada, *Agricultural and Biological Chemistry,* Vol. 48, 1981, p. 997.
[6] P. T. S. Wong et al., *Canadian Journal of Fisheries Aquatic Science,* Vol. 39, 1982, p. 483.
[7] R. B. Laughlin et al., *Chemosphere,* Vol. 13, No. 575, 1984.

apparent, however, and certainly not predicted by such a simplistic physical topological approach is the precise modality of cellular uptake or transport and toxic effect. Such investigations require solution and cellular chemical speciation of organotins described below.

For preliminary assessments of aquatic toxicity with both organic and organometallic xenobiotic molecules, which share properties of low aqueous solubility and structural longevity, we find that combinations of TSA predictors correlated with HPLC-generated log k' data, the latter representing thermodynamic information [10,12,25,30], provide a broad basis for evaluating key environmental parameters (PROPERTY) prerequisite to toxicity modeling. Employing HPLC retention (k') data and conventional quantitative structure-activity relationships (QSAR) derived from independent equilibrium and thermodynamic measurements [31], we demonstrated [12] by regression analyses a very simple linear free-energy relationship, viz.,

$$\ln (\text{PROPERTY}) = m (\text{QSAR}) + \text{constant} \tag{1}$$

where m = slope.

Table 2 illustrates this basic relationship yielding correlations of unusually high degree ($r^2 > 0.94$) for neutral or charged organotin molecular shapes expressed as TSA with: key physicochemical properties of aqueous solubility; octanol-water partition coefficients; and chromatographic retention for liquid and gas separations. Lastly, TSA is correlated by Eq. 1 with a range of diagnostically important biological effects including cellular uptake, growth inhibition, and aquatic toxicity by a class of polar, charged triorganotins (R_3SnX) based on the limited data so far available in the literature.

Unfortunately, comparable biological effects data to correlate QSAR or TSA with other homologous series of organometals are not available. Nonetheless, use of TSA to estimate biological effects of aquatic organometals or organometalloids, analogous to the present case of organotins, seems warranted on the basis of our previous appraisal of QSAR to estimate organoarsenical chromatographic properties [12].

Modes of Organotin Bioaccumulation and Persistence

Based on physical/chemical principles, the topological predictors derived from TSA for organotin uptake and toxicity appear promising. However, they generally presume equilibrium conditions (for example, capacity factor) [22] and do not provide an understanding of molecular properties controlling mechanisms of biological uptake and degradation dependent on transient nonequilibrium events.

We previously discussed [6] molecular size as a rate-limiting factor for molecular diffusion into the cell lipid pool, and Mackay [32] has observed that short-lived xenobiotic molecules may occur in steady state distributions between lipid uptake, depuration, and degradation. Such conditions may render a free-energy analysis invalid. Therefore, we address the questions of organotin solution longevity and availability versus cellular uptake mechanisms.

The work of Yamada [33], which represents the only previous quantitative study on organotin uptake on bacterial membranes (spheroplasts), is encouraging in light of our reappraisal of their data summarized in Fig. 2. Here we employ Eq. 1 to fit the property of uptake (1 h) of R_3SnCl on *Escherichia coli* sphaeroplasts in a linear regression with TSA, Hansch's π parameter [31], and a Taft-Hammet σ^ϕ function [12]. The latter two variables represent, respectively, a solvophobic-lipophilic and an electronic charge density attribute of the R_3Sn^+ cation. For all the fitting functions, it is clear that the TSA description for the charged pentacoordinate organotin moiety or the σ^ϕ term are best, suggesting that neither charge transfer nor lipid solvolysis is a distinguishable mode of uptake.

Previously, we demonstrated that several tributyltin-resistant bacterial isolates accumulated both

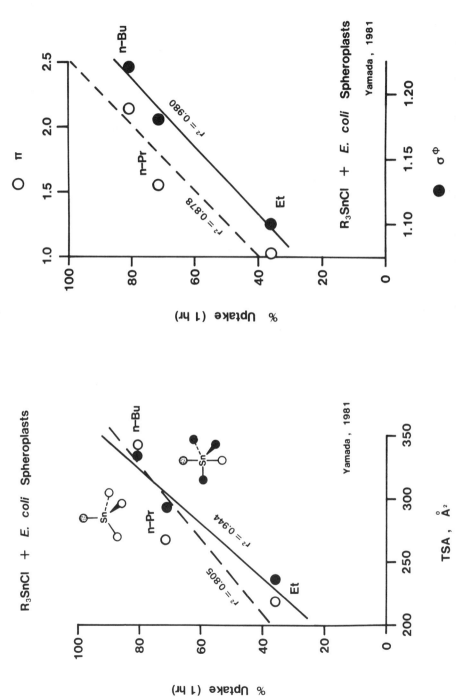

FIG. 2—*Correlation (Eq 1) of R_3SnCl uptake (PROPERTY) by Escherichia coli membranes (spheroplasts) with QSAR: TSA (left) or with Hansch's π parameter and a Taft-Hammett σ^ϕ function (right). The uptake data was taken from Yamada [31], and Eq 1 was used to fit the data in the linear regressions shown.*

tributyltin and inorganic tin species [14]. We have found that tin accumulation by such bacteria can be imaged and monitored *in vivo* directly by epifluorescence microscopy, by complexation with flavonol.

Cell pellets exhibited green fluorescence following flavonol treatment. However, if the cells were allowed to accumulate $SnCl_4$ prior to, or concurrently with, flavonol exposure, they gave blue fluorescence. In the absence of tin or flavonol, no cell fluorescence was observed. In order to be certain that the blue fluorescence associated with the tin and flavonol treatment cell pellets was not due to soluble tin carryover, three additional centrifugation and washing steps [between Steps 2 and 3, Table 1] were performed on the cells. Results showed no visible diminution in blue fluorescence intensity of cells.

Microspectrofluorograms of cells treated with tin and flavonol are shown in Fig. 3. These provide quantitative measurements of cell fluorescence and confirm the results of the macroscopic experiments. Glass beads doped with $SnCl_4$ groups fluoresced blue under 365-nm illumination, with an emission maximum at 460 nm, nearly identical to that of treated cells [Fig. 4]. As depicted, beads treated with tin only did not fluoresce.

Accumulation of tributyltin and monobutyltin on bacterial cells was also monitored using flavonol. However, fluorescence intensities (λ_{em} = 460 nm) of cells that were exposed to equimolar concentrations of these tin species were only 7 and 50%, respectively, of the intensity with $SnCl_4$. In part this difference is explained by anticipated [20] lower fluorescence intensities (quantum efficiencies) for the flavonol-tributyltin complex and flavonol-monobutyltin complex, which give only 5 and 20%, respectively, of the $SnCl_4$-flavonol complex fluorescence value in flow injection experiments. These values suggest that monobutyltin may be accumulated to a greater extent than tributyltin and $SnCl_4$ on cells. This has not yet been verified by atomic absorption analysis. The relatively low fluorescence of flavonol-treated *Pseudomonas* cells that accumulated tributyltin is also consistent with our previous findings using HPLC-GFAA that tributyltin is accumulated but not degraded by this organism [14]. Degradation products of tributyltin (di- and monobutyltin, inorganic tin) are more fluorescent than tributyltin at 460 nm [34]. Further studies in this direction are warranted for prospective *in situ* abiotic calibration substrates in metals bioaccumulation studies.

The relationship of topological predictors and their utility for modeling and assaying effects of environmental metals (for example, organotins) on key organisms holds promise on the basis of the foregoing laboratory studies. Carrying on with the example of organotins, we have sought to relate tributyltin persistence (longevity) and uptake modes in environmental waters.

Tributyltin Persistence in Chesapeake Bay Waters

Our previous research [14] with tributyltin-resistant bacterial isolates showed no detectable tributyltin biodegradation. Similarly, little degradation of tributyltin was detected in Chesapeake Bay waters spiked with tributyltin at a final concentration of 100 μg/L and incubated in the dark for 12 weeks at 20°C (Table 3). However, addition of yeast extract (0.05% final concentration) to the samples often resulted in substantial degradation of tributyltin to di- and monobutyltin species. Water samples spiked with yeast extract become somewhat turbid. Microscopic examination of these samples showed development of bacterial populations. Recoveries of butyltins in these extended experiments were somewhat low, perhaps due to: (1) adsorption of tin to microorganisms attached to the glass container walls; (2) difficulties in completely extracting butyltin species from cells; or (3) complete dealkylation of tributyltin to inorganic tin which is not detected by this analytical method.

Subsequently, we have been conducting experiments using more environmentally realistic levels of tributyltin and shorter term experiments to identify immediate degradation products. These

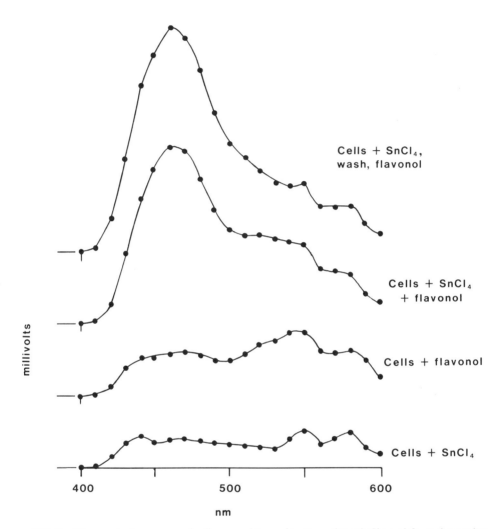

FIG. 3—*Microspectrofluorograms of cells exposed to combinations of tin (SnCl₄) and flavonol according to the protocol in Table 1. Cell fluorescence intensity was measured by drawing the cell suspensions into capillary tubing and recording the emission intensities at varying wavelengths. Excitation wavelength was 365 nm.*

recent experiments have shown that tributyltin at µg/L levels are quite stable in samples collected at Annapolis and Baltimore Harbor in winter and incubated at 4°C. However, samples collected in summer (22 to 28°C) showed much more rapid degradation of tributyltin to mono- and dibutyltin species, with half lives for tributyltin of 1 to 2 weeks. Incubation under incandescent light accelerated the degradation, suggesting the involvement of photosynthetic microorganisms.

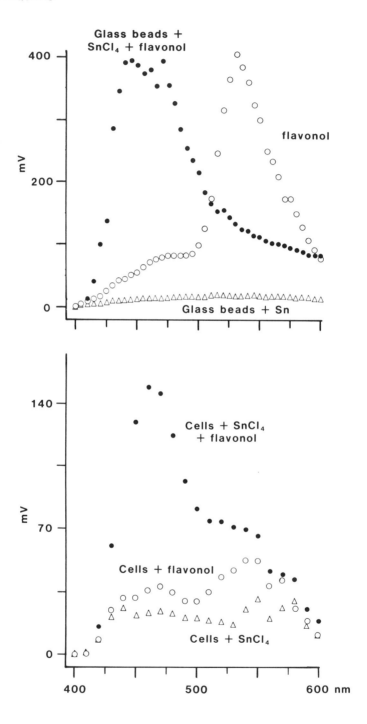

FIG. 4—*Microspectrofluorograms of cells and glass beads exposed to combinations of $SnCl_4$ and flavonol. The emission spectrum of flavonol alone is shown in the top figure. Excitation wavelength was 365 nm.*

TABLE 3—*Degradation of tributyltin in spiked (100 μg/L) Chesapeake Bay waters, % of Total Butyltins 3-Month Incubation at 20°C.*

	Bu_3Sn	Bu_2Sn	BuSn	% Recovered
C & D Canal	89	6	5	57
+ yeast ext.[a]	43	12	45	40
sterile	86	6	8	33
sterile + yeast ext.	95	3	2	81
Jones Falls	86	5	9	55
+ yeast ext.	...	...	...	...
sterile	90	6	4	55
sterile + yeast ext.	...	...	...	...
Colgate Creek	92	5	3	73
+ yeast ext.	15	6	79	67
sterile	94	4	2	67
sterile + yeast ext.	95	2	3	72
Sparrows Point	89	5	6	64
+ yeast ext.	50	14	36	39
sterile	91	6	3	56
sterile + yeast ext.	95	3	3	79
821[b]	92	3	5	67
+ yeast ext.	87	5	8	53
sterile	95	3	2	79
sterile + yeast ext.	96	2	2	87
Little Ck.	93	4	3	74
+ yeast ext.	45	10	45	26
sterile	95	3	2	79
sterile + yeast ext.	94	2	2	81
716[b]	86	6	8	67
+ yeast ext.	87	5	8	51
sterile	92	5	3	43
sterile + yeast ext.	96	2	2	88
656[b]	92	4	4	58
+ yeast ext.	86	6	8	34
sterile	94	5	1	45
sterile + yeast ext.	94	3	3	69

[a] Yeast extract—0.05% final concentration in all cases.
[b] Central Chesapeake Bay, midchannel locations, numbers are stations designed by the Chesapeake Biological Laboratory.

Summary and Conclusions

Our results with tin and flavonol show that fluorescent metal ligands have the potential to be used to monitor accumulation of metals by microorganisms. Apparently cell-bound tin is still available for interaction with flavonol, since cells that accumulated tin and were subsequently treated with flavonol fluoresced blue. This applies to nonliving particles also, since glass beads with surface —SH groups gave similar results.

These results provide insight into the mode of bacterial uptake of organotins as summarized in

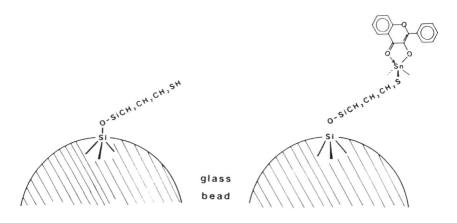

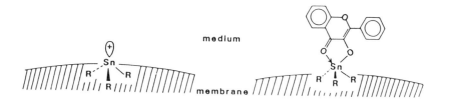

FIG. 5—*Summarized interaction of flavonol with inorganic (A) or organotin (B) compounds bound to nonbiological and biological surfaces by dative coordination (charge transfer) or salvophobic (lipid retention), respectively.*

Fig. 5, where we contrast the reaction for inorganic tin bound to an inorganic substrate with flavonol and a proposed mechanism for organotin binding on the living cell envelope. The latter case is consistent with the results derived from Fig. 2 showing that uptake may occur by either charge transfer or solvophobic processes, or by both. We have interpreted our results to follow the latter mode (Fig. 2B) since the R_3Sn^+ moiety is otherwise coordinately saturated. We know from Blunden and Smith [20] that the unoccupied tin orbital must remain available in order to accommodate the charge transfer bidentate fluorescent chelate while on the cell membrane.

Microorganisms in Chesapeake Bay water samples can degrade tributyltin to mono- and dibutyltin species, especially when supplied with nutrients. The rates of *in situ* degradation of tributyltin in Chesapeake Bay surface waters are not yet certain. However, preliminary experiments with winter and summer collected samples suggest, not unexpectedly, that degradation is exceedingly slow in winter but can be significant in summer, with tributyltin half-lives on the order of 1 to 2 weeks in collected samples.

Acknowledgments

The authors are grateful to the Office of Naval Research for sustained support of our studies on environmental transformations and fate of organotins, partly described in this paper. We thank Wolfgang Haller for his aid in preparing chemically bonded glass particles and interpreting data. Mention of certain commercial equipment is for purposes of adequately describing experimental procedures and does not imply endorsement or recommendation by the National Bureau of Standards, nor that the item is necessarily the best available for the purpose. Contributions from the National Bureau of Standards are not subject to copyright.

References

[1] *Environmental Speciation and Monitoring Needs for Trace Metal-Containing Substrates from Energy-Related Processes*, F. E. Brinckman and R. H. Fish, Eds. NBS Special Publication 618, National Bureau of Standards, Washington, DC, 1981.
[2] *Organometallic Compounds in the Environment*, Craig, P. J., Ed., Longman, London, 1986.
[3] *Organometals and Organometalloids: Occurrence and Fate in the Environment*, F. E. Brinckman and J. M. Bellama, Eds., American Chemical Society, Washington, DC, 1978.
[4] Thayer, J. S., *Organometallic Compounds and Living Organisms*, Academic Press, New York, 1984.
[5] *The Importance of Chemical Speciation in Environmental Processes*, M. Bernhard, F. E. Brinckman, and P. Sadler, Eds., Springer-Verlag, W. Berlin, Germany, 1986.
[6] Laughlin, R. B., Jr., French, W., Johannesen, R. B., Guard, H. E., and Brinckman, F. E., *Chemosphere*, Vol. 13, 1984, pp. 575–584.
[7] Laughlin, R. B., Jr., Johannesen, R. B., French, W., Guard, H., and Brinckman, F. E., *Environmental Toxicology and Chemistry*, Vol. 4, 1985, pp. 343–351.
[8] Jackson, J. A., Blair, W. R., Brinckman, F. E., and Iverson, W. P., *Environmental Science and Technology*, Vol. 16, 1982, pp. 110–119.
[9] Brinckman, F. E., Blair, W. R., Jewett, K. L., and Iverson, W. P., *Journal of Chromatographic Science*, Vol. 15, 1977, pp. 493–503.
[10] Jewett, K. L. and Brinckman, F. E., *Journal of Chromatographic Science*, Vol. 19, 1981, pp. 583–593.
[11] Fish, R. H., Brinckman, F. E. and Jewett, K. L., *Environmental Science and Technology*, Vol. 16, 1982, pp. 172–179.
[12] Weiss, C. S., Jewett, K. L., Brinckman, F. E., and Fish, R. H. in *Environmental Speciation and Monitoring Needs for Trace Metal-Containing Substrates from Energy-Related Processes*, Chapter 16, National Bureau of Standards, Washington, DC, pp. 197–216.
[13] Committee D-19, *ASTM Standards on Chromatography*, ASTM, Philadelphia, 1981, pp. 684–693 (see Table 1 for a specific definition of k).
[14] Blair, W. R., Olson, G. J., Brinckman, F. E., and Iverson, W. P., *Microbial Ecology*, Vol. 8, 1982, pp. 241–251.
[15] Pearlman, R. S., QCPE Bulletin 1, Program No. 413, Indiana University, Bloomington, IN, 1981.

[16] Eng, G., Brinckman, F. E., Johannesen, R. B., Tierney, L., and Bellama, J. M., *Journal of Chromatography*, Vol. 403, 1987, pp. 1–9.
[17] Craig, P. J. and Brinckman, F. E. in *Organometallic Compounds in the Environment*, Chapter 1, P. J. Smith, Ed., Longman, London, 1986, pp. 1–64.
[18] Nelson, J. D., Blair, W. R., Brinckman, F. E., Colwell, R. R., and Iverson, W. P., *Applied Microbiology*, Vol. 26, 1973, pp. 321–326.
[19] Good, N. E., Winget, D., Winter, W., Connolly, T. N., Izawa, S., and Singh, R. M. M., *Biochemistry*, Vol. 5, 1966, pp. 467–477.
[20] Blunden, S. J. and Smith, P. J., *Journal of Organometallic Chemistry*, Vol. 226, 1982, pp. 157–163.
[21] Haller, W. in *Solid Phase Biochemistry—Analytical and Chemical Aspects*, W. H. Scouten, Ed., Wiley, New York, 1983, p. 535.
[22] Blair, W. R., Olson, G. J., and Brinckman, F. E., NBSIR 86-3321, National Bureau of Standards, Gaithersburg, MD, 1986.
[23] Matthias, C. L., Olson, G. J., Brinckman, F. E., and Bellama, J. M., *Environmental Science and Technology*, Vol. 20, 1986, pp. 609–615.
[24] Tomlinson, E., *Journal of Chromatography*, Vol. 113, 1975, pp. 1–45.
[25] Hansch, C., Quinlan, J. E., and Lawrence, G. L., *Journal of Organic Chemistry*, Vol. 33, 1968, pp. 347–350.
[26] Veith, G. D., Austin, N. M., and Morris, R. T., *Water Research*, Vol. 13, 1979, pp. 43–47.
[27] Harnish, M., Mockal, H. J., and Schulze, G., *Journal of Chromatography*, Vol. 282, 1983, pp. 315–332.
[28] Neely, W. B., *Chemosphere*, Vol. 13, 1984, pp. 813–819.
[29] Wise, S. A., Bannett, W. J., Guenther, F. R., and May, W. E., *Journal of Chromatographic Science*, Vol. 19, 1981, pp. 457–465.
[30] Valiani, S. C., Yalkowsky, S. H., and Amidon, G. L., *The Journal of Physical Chemistry*, Vol. 80, 1976, pp. 829–835.
[31] Hansch, C. and Leo, A. J., in *Substituent Constants for Correlation Analysis in Chemistry and Biology*, Wiley, New York, 1979.
[32] Mackay, D., *Environmental Science and Technology*, Vol. 16, 1982, pp. 274–278.
[33] Yamada, J., *Agricultural and Biological Chemistry*, Vol. 45, 1981, pp. 997–999.
[34] Blair, W. R., Parks, E. J., Olson, G. J., Brinckman, F. E., Valeiras-Price, M.C., and Bellama, J. M., *Journal of Chromatography*, Vol. 410, 1987, pp. 383–394.

John F. McCarthy[1] *and Marsha C. Black*[2]

Partitioning Between Dissolved Organic Macromolecules and Suspended Particulates: Effects on Bioavailability and Transport of Hydrophobic Organic Chemicals in Aquatic Systems

REFERENCE: McCarthy, J. F. and Black, M. C., "**Partitioning Between Dissolved Organic Macromolecules and Suspended Particulates: Effects on Bioavailability and Transport of Hydrophobic Organic Chemicals in Aquatic Systems,**" *Aquatic Toxicology and Hazard Assessment: 10th Volume, ASTM STP 971,* W. J. Adams, G. A. Chapman, and W. G. Landis, Eds., American Society for Testing and Materials, Philadelphia, 1988, pp. 233–246.

ABSTRACT: Hydrophobic organic contaminants in aquatic systems bind to sediments or to suspended particles, and these associations affect the transport and bioavailability of contaminants and thus alter the exposure received by biota. We present evidence demonstrating the importance of another less obvious natural sorbent—dissolved organic macromolecules (DOM), such as dissolved humic materials—in altering the environmental fate of contaminants. Binding of a polycyclic aromatic hydrocarbon, benzo[a]pyrene, to DOM and to particles was measured for each sorbent and in a mixed system containing both sorbents. DOM competed with particles for binding of the dissolved contaminant and reduced the amount bound to particles. Binding to each sorbent was independent and noninteractive. Based on these results, equations were developed to indicate those environments and ecosystems for which failure to account for the role of DOM as a sorbent would result in 50 or 90% errors in steady state predictions about the exposure of aquatic organisms to contaminants.

KEYWORDS: sorption, polycyclic aromatic hydrocarbons, benzo[a]pyrene, partitioning, environmental transport, bioavailability, organic contaminants, dissolved organic matter, humic acid, particles, sediment

The exposure of biota to organic contaminants in aquatic environments is greatly influenced by the physicochemical behavior of the contaminant in aqueous systems. For hydrophobic compounds, interaction of the contaminant with natural or anthropogenic sorbents plays a major role in both the contaminant's distribution in the environment and its bioavailability. Sediments or suspended particles have been recognized as a sink for many hydrophobic contaminants, effectively reducing the amount of chemical dissolved in the water. This reduction in water-borne contaminant will reduce the exposure to pelagic organisms while increasing the exposure to organisms living or feeding in the benthos. Recently, there has been a growing recognition of the role of another naturally occurring sorbent—dissolved organic macromolecules (DOM), such as dissolved humic materials—that can affect the environmental fate and the bioavailability of hydrophobic contaminants

---

[1] Research staff member, Environmental Sciences Division, Oak Ridge National Laboratory, Oak Ridge, Tennessee 37831.
[2] Graduate Program in Ecology, The University of Tennessee, Knoxville, TN 37996.

[*1–9*]. DOM can bind contaminants, stabilize them in the water column, and enhance their advective transport from point sources, thus permitting further dilution by incoming water sources.

Binding of contaminants to particulate and dissolved sorbents has been demonstrated for a number of hydrophobic compounds, including herbicides, pesticides, polycyclic aromatic hydrocarbons (PAHs), and polychlorinated biphenyls (PCBs) [*3–6,8,10–18*]. In general, the affinity for binding is directly related to the hydrophobicity of the contaminant, measured as the compound's octanol-water partition coefficient or the inverse of its aqueous solubility [*3,8,11,12*], and the organic content of a particulate sorbent [*11,12*]. Chiou et al. [*10*] describe the binding of contaminants to sorbents as a solvophobic partitioning between the contaminant and the organic matter of the sorbent. Because these interactions involve primarily hydrophobic or van der Waals forces or hydrogen bonds rather than the formation of true covalent bonds, thermodynamics predict a reversible equilibrium between the sorbent and sorbate, with a linear relationship between the bound and freely dissolved concentrations of contaminant.

However, many researchers report nonlinear sorptive behavior such as hysteresis or adsorbent concentration effects for binding of organic chemicals to particles. Many of these apparent effects may be explained in terms of the presence of a "hidden" sorbent. The apparent decrease in particle-binding affinity at higher particle concentrations is explained by the release of microparticles or DOM from particles and subsequent binding of the chemical to this alternate sorbent [*19–23*]. Gschwend and Wu [*22*] demonstrated that removal of "nonsettling particles" or removal of sediment-associated DOM eliminated the apparent adsorbent concentration effects and resulted in a constant binding affinity (K_p) for selected PCBs over a wide range of particle concentrations. Similarly, Caron et al. [*23*] corrected for the apparent decrease in binding of dichlorodiphenyltrichloroethane (DDT) to particles by calculating the fraction of the nonparticulate fraction bound to DOM versus that freely dissolved, based on independent measurements of the partitioning of DDT between water and DOM in a two-phase system. These results suggest the importance of quantifying the binding of contaminants to DOM to determine the true affinity of particles to bind contaminants.

Quantitative determination of the influence of DOM on binding is complicated by the spatial and temporal variability in DOM and by the influence of environmental variables on binding to DOM. Variations in local water chemistry may also be important, because ionic strength and pH of the aqueous media affect binding affinity [*4,16–17*].

In this study, we measured the partitioning of a model hydrophobic compound, the PAH benzo[a]pyrene (BaP), between water, particles, and DOM within a mixed three-phase system. This multiple-phase system permits direct measurement of the binding affinities of the dissolved and particulate sorbents and indicates quantitatively any competitive interactions between the sorbents for binding of chemicals or for interactions between the sorbents themselves, such as binding of a DOM-contaminant complex to a particulate sorbent. Based on these interactions, we estimated the types of environmental situations where the presence of DOM could significantly alter the steady state partitioning of contaminants between the particle-bound phase and the dissolved phase. In these scenarios, failure to consider the role of DOM as a sorbent could produce serious errors in model predictions of the transport, fate, and bioavailability of a contaminant, and thus alter estimates of the exposure received by aquatic biota.

Experimental Procedures

Humic acid (Aldrich Chemical Co., Milwaukee, Wisconsin) was dissolved in deionized water, centrifuged for 30 min at 10 000 × g, and filtered through a precombusted glass-fiber filter (Type A-E, 0.3 μm, Gelman Sciences, Inc., Ann Arbor, Michigan) to remove particulate material. The organic carbon content of the solution of humic acid (DOM) was determined using a carbon analyzer (OI Corporation, College Station, Texas); DOM concentrations are reported as milligrams of carbon per liter (mg C/L). Yeast cells were cultured for 3 days at 30°C in sterile media containing 1% nutrient broth (Difco Laboratories, Detroit, Michigan) and 0.5% dextrose (Fisher Scientific, Fair Lawn, New Jersey) adjusted to pH 5.6. Yeast cells were prepared for each experiment by centrifuging the cultured cell suspension at 2000 rpm followed by washing with deionized water and recentrifugation at 2000 rpm to form a yeast pellet. The washed yeast cells were determined to have a uniform spherical shape approximately 5 μm in diameter by microscopic analysis.

Radio-labeled ^{14}C-benzo[a]pyrene (BaP) with a specific activity of 1.4×10^3 mega becquerels (MBq) (39 mCi)/mM was obtained from California Bionuclear (Sun Valley, California). The BaP solution was repurified by high-performance liquid chromatography using a C-18 column and redissolved in glass-distilled p-dioxane (MCB Reagents, Cincinnati, Ohio) before use in the experiments.

Dialysis tubing (Spectra/Por 6; molecular weight cutoff of 1000 daltons) was rinsed in deionized water to remove the sodium azide preservative. Equilibrium dialysis experiments were performed as illustrated in Fig. 1. Two lengths of dialysis tubing were filled with 8 mL of either a DOM solution or filtered (precombusted Gelman A-E) deionized water and secured with a clamp. The two dialysis bags were then suspended in a 1-L glass jar containing a solution of ^{14}C-BaP, DOM, and yeast in filtered deionized water (bulk water). Sodium azide was added to each jar (0.02%

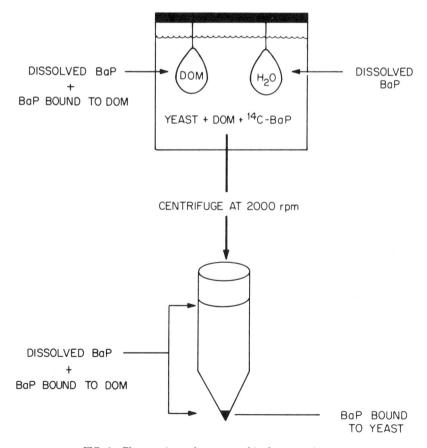

FIG. 1—*The experimental system used in these experiments.*

final concentration) to prevent bacterial growth. Jars were sealed with Teflon-lined caps, incubated at 23°C for 3 days, and gently mixed, using a magnetic stirrer to ensure equilibration of the solutions. All experiments were performed under gold fluorescent lights to prevent photodegradation of the BaP.

After the 3-day incubation period, samples from both dialysis bags and from the bulk water were dissolved in scintillant (ACS, Amersham Corp., Arlington Ht., Illinois) and analyzed for ^{14}C radioactivity using a Packard CD 460 scintillation counter. In addition, a sample from the bulk water was centrifuged to separate the yeast cells from the supernatant. The yeast cells were resuspended in deionized water, and a portion of the yeast suspension and of the particle-free supernatant were also analyzed for ^{14}C radioactivity.

The concentration of freely dissolved BaP (expressed as pmol BaP/mL) was determined from the radioactivity measured in the dialysis tubing containing deionized water. The concentration of BaP bound to DOM was determined from the radioactivity both in the dialysis bag containing DOM and in the supernatant of the centrifuged bulk water, each corrected for the freely dissolved fraction. These measurements are expressed as picomoles per gram of carbon (pmol BaP/g C). BaP bound to yeast was determined from the radioactivity in the yeast pellet and is expressed as picomoles per gram of yeast (pmol BaP/g yeast). Details and validation of the equilibrium dialysis method have been published [9].

Statistical analyses were performed using SAS (Statistical Analysis System, Cary, North Carolina).

Results and Discussion

Binding of BaP to DOM

The isotherm describing the binding of BaP to DOM is illustrated in Fig. 2. The concentration of contaminant bound to DOM can be calculated from measurements of the radioactivity in the dialysis bag containing DOM and in the centrifuged bulk water containing freely dissolved BaP as well as BaP bound to DOM (Fig. 1). Differences in isotherms derived from measurements in these compartments of the experimental system suggest that interactions within the dialysis bag or between DOM and the particles in the bulk water can alter the binding of the BaP with the DOM. There were, however, no differences ($p < 0.05$) between the isotherms derived from the two measurements. Therefore, the isotherm in Fig. 2 represents data pooled from both measurements. The association coefficient, K_p, for DOM-bound BaP was calculated as $1.62 \, (\pm 0.04) \times 10^6$ from the least squares estimate of the slope of the association curve (Table 1).

In a separate experiment, the K_p describing the binding of BaP to DOM remained constant in the presence of 0.5 to 50 mg/L of yeast particles. This K_p was not significantly different ($P < 0.05$) from the K_p measured in a two-phase system with no yeast particles present. These data suggest that the binding affinity of BaP for DOM is unaffected by the presence of particulate material over a wide range of particulate concentrations.

Binding of BaP to Particulate Matter

Yeast cells were used as particulate matter in these experiments because of their uniform size, shape, surface characteristics, and ease of culture. The isotherms describing the binding of BaP to yeast cells in the presence of DOM (5 mg C/L) are shown in Fig. 3. The solid curve represents the partitioning of BaP between a particle-bound and freely dissolved phase, based on measurements of the radioactivity in the centrifuged pellet from the bulk water compared to those in the dialysis bag containing deionized water. The K_p was $7.2 \, (\pm 0.4) \times 10^4$, which is identical ($p > 0.05$) to that measured when no DOM was added.

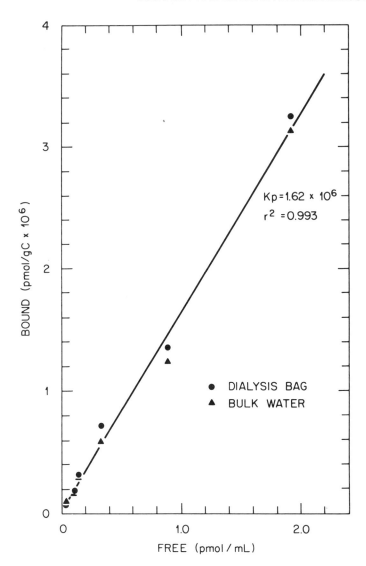

FIG. 2—*Isotherm for the binding of BaP to DOM in the presence of 5 mg particles/L. The DOM concentration was 5 mg C/L. The circles are data derived from measurements of the radioactivity present in the dialysis bag containing DOM (see Fig. 1), and the triangles are from measurements of the radioactivity in the centrifuged bulk water.*

The dashed curve in Fig. 3 represents the partitioning of BaP between a particulate and nonparticulate phase and is based on measurements of radioactivity in the yeast pellet compared to those in the supernatant of the bulk water after centrifugation (Fig. 1). This measurement of the "apparent" K_p (K_{app}, Fig. 3) reflects the procedure normally employed to determine an association coefficient, but ignores the role of DOM as a competing, but dissolved, natural sorbent. Including DOM-bound BaP in the "apparently dissolved" phase, although resulting in a linear relationship, underestimates the true binding affinity of BaP for yeast by almost an order of

TABLE 1—*Estimates of K1 and K2, based on nonlinear fitting of the data in Fig. 4 to Eq 3 and estimates of K1, K2, and K3, based on fitting of data in Fig. 4 to Eq 5. The numbers in parentheses are the 95% asymptotic confidence intervals for the estimates and the standard error of the mean for the measured values.*

Parameter	Estimate (Eq 3)	Estimate (Eq 5)	Measured (Fig. 2 or 3)
$K1$	6.9×10^4 (4.9, 8.8)	7.0×10^4 (5.1, 8.9)	7.2×10^4 (± 0.6)
$K2$	3.0×10^6 (1.3, 4.7)	3.6×10^6 (1.3, 5.9)	1.6×10^6 (± 0.04)
$K3$	...	0.1×10^4 ($-0.9, 0.3$)	...

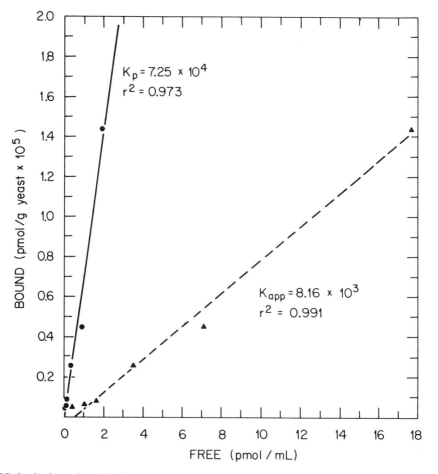

FIG. 3—*Isotherms for the binding of BaP to yeast particles in the presence of DOM (5 mg C/L). The circles and solid line represent the isotherm based on the concentration of freely dissolved BaP, derived from measurements of radioactivity in the dialysis bag containing water (see Fig. 1). The triangles and dashed line represent the isotherm based on the concentration of nonparticulate BaP in the supernatant of the centrifuged bulk water.*

magnitude in the presence of 5 mg C/L of DOM [K_{app} = 0.82 (±0.04 × 10⁴]. Only when the actual freely dissolved fraction can be separated from the DOM-bound fraction can a K_p be correctly determined for contaminants bound to particulates.

Interactions Between Particulate and Dissolved Sorbents

The presence of DOM lowered the apparent K_p for binding of BaP to the yeast (K_{app}, Fig. 3), probably due to competition by the DOM for the dissolved BaP. If the contaminant binds independently and noninteractively between its dissolved form and that bound to either sorbent, then the multiple-sorbent system can be analyzed in terms of individual interactions with each sorbent

$$K_p \text{ for BaP-yeast} = K1 = \frac{C_p}{C_d \times P} \quad (1)$$

$$K_p \text{ for BaP-DOM} = K2 = \frac{C_{dom}}{C_d \times \text{DOM}} \quad (2)$$

$$K_{app} = \frac{C_p}{(C_d + C_{dom}) \times P} = \frac{K1}{1 + (K2 \times \text{DOM})} \quad (3)$$

where
- P = mass of particles/mL,
- C_p = pmol of contaminant bound to particles/mL,
- C_d = pmol of freely dissolved contaminant/mL,
- C_{dom} = pmol of contaminant bound to DOM/mL, and
- DOM = concentration of DOM (g C/mL). $K1$ is the isotherm illustrated in Fig. 2, and $K2$ is the isotherm illustrated by the solid line in Fig. 3.

The apparent binding of BaP to particles (K_{app}, Eq 3) was measured in the presence of different concentrations of DOM, and the results are illustrated in Fig. 4. The data in Fig. 4 were fit to Eq 3 using a nonlinear least squares fitting procedure to estimate values of $K1$ and $K2$ (Table 1), along with the independent measurements of the association coefficients from Figs. 2 and 3. The measured K_ps fall within the 95% asymptotic confidence intervals for the estimate.

Hassett and Anderson [20] suggested the possibility that there is an interaction between the contaminant-DOM complex and the particle

$$K3 = \frac{C_{p,dom}}{C_{dom} \times P} \quad (4)$$

$$K_{app} = \frac{K1 + (K2 \times K3 \times \text{DOM})}{(K2 \times \text{DOM}) + 1} \quad (5)$$

where $C_{p,dom}$ = pmol of DOM-bound contaminant binding to particles/g of particles. The data in Fig. 4 were fit to Eq 5 and the values of $K1$, $K2$, and $K3$ were estimated (Table 1). $K3$ was estimated to be very low compared to $K1$ and $K2$, with a 95% asymptotic confidence interval that included zero. Comparison of the residual sums of squares (incremental F-test) for the two parameter and three parameter models (Eqs 3 and 5, respectively) indicate that the extra parameter did not significantly improve the fit of the data to the model ($p < 0.05$). In this system the

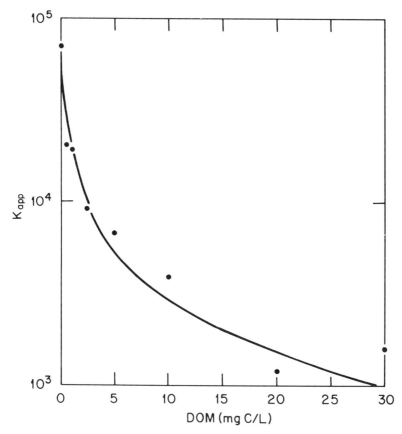

FIG. 4—*The decrease in the apparent K_p (K_{app}) for the binding of BaP to particles in the presence of DOM. The circles are the data derived from the ratio of BaP bound to particles compared to BaP in the nonparticulate supernatant of the centrifuged bulk water. The curve represents model prediction based on fitting the data to Eq 3.*

interaction between the DOM and the particles exerts little influence on contaminant partitioning. This result is also consistent with the observation that the DOM concentration in the supernatant of the centrifuged bulk water before the addition of yeast did not differ significantly from that remaining at the end of the incubation period ($p < 0.05$), suggesting that there was no measurable binding of DOM to the particles.

The results of our study suggest that DOM can compete independently with particulate sorbents for binding of hydrophobic organic contaminants. This result is also consistent with that of Gschwend and Wu [22], who analyzed "complex" sorptive behavior such as hysteresis and adsorbent concentration effects in terms of a three-phase system with dissolved, particulate, and "nonsettling microparticle" or macromolecule components.

Predictive fate and transport models and simpler equilibrium models used to assess the environmental impact of contamination generally account for the partitioning of hydrophobic contaminants between a particulate and nonparticulate phase. In some situations, failure to include a third phase (a dissolved macromolecular or microcolloidal sorbent such as DOM) can lead to serious errors in estimates of the form and distribution of contaminants. Errors could include:

1. Overestimation of the amount of contaminant bound to particles.
2. Overestimation of contaminant removal from the water column, resulting from sedimentation of particles.
3. Underestimation of the extent of downstream advection of contaminants stabilized in the water column by DOM.

Stabilization of the chemicals in the water column could result in dilution of the contaminant concentration during downstream advection in lotic systems, or by diffusive mixing and gradual water turnover in lentic systems. However, binding to DOM is fully reversible [9], and rates of both binding and dissociation are very rapid compared to interactions with particles [9,18]. Because of these rapid kinetics, dissolved and DOM-bound contaminant would be continuously in equilibrium within environmentally relevant time frames. As dilution occurs, the contaminant would dissociate from the DOM, leading perhaps to a chronic low level of widely distributed contamination. Later in this paper, an approach will be developed that will suggest the range of natural conditions under which these errors could be significant.

Effect of DOM on Bioavailability

The physicochemical partitioning of contaminants also affects their bioavailability. Contaminants bound to particles are far less available for uptake by biota than are dissolved contaminants. The rate coefficient for uptake of particle-bound PAHs is only 10% of that for uptake of dissolved chemical [1]. By reducing the amount of contaminant bound to particles, the presence of DOM would increase the concentration of contaminant in the nonparticulate phase. However, that increase in the nonparticulate concentration does not increase the uptake and accumulation, because contaminant bound to DOM appears to be unavailable for uptake by aquatic invertebrates or fish [3,7,8]. The presence of DOM results in an apparent decrease in the bioconcentration factor ($BCF = C_a/C_w$, where C_a = pmol of contaminant/g of animal, and C_w = pmol of nonparticulate contaminant/mL of water = $C_d + C_{dom}$).

The decrease in the apparent BCF in the presence of DOM can be calculated for compounds over a range of hydrophobicities based on the following relationships:

1. Regression equations have been reported which relate the BCF to physicochemical properties of the contaminant, such as the octanol-water partition coefficient (K_{ow}) or water solubility [24–26]. Equation 6 is based on data from 84 bioaccumulation experiments using compounds that ranged in log K_{ow} from 1 to 8.

$$\log BCF = 0.76 \times \log K_{ow} - 0.23 \quad [24] \quad (6)$$

2. The fraction of the total contaminant which is bound to DOM (f_{dom}), and which is thereby unavailable for uptake, can be calculated if the K_p for binding of the contaminant to DOM ($K2$, Eq 2) and the concentration of DOM are known.

$$f_{dom} = \frac{C_{dom}}{C_w} = \frac{C_{dom}}{C_d + C_{dom}} = \frac{K2 \times \text{DOM}}{1 + (K2 \times \text{DOM})} \quad (7)$$

3. To a first approximation, it is possible to estimate the magnitude of $K2$ based on the K_{ow} of the contaminant. For suspended solids or for sediment, the K_p for the binding of a solute to particles is related to the K_{ow} of the solute and the fractional organic content of the particle (f_{oc}) [11,13,15]. The carbon-referenced partition coefficient, K_{oc} accounts for the differences in the f_{oc} of different sorbents and permits the binding affinity to be predicted from a physicochemical

property of the contaminant. Equation 8 is a regression equation based on data from sorption experiments using 5 PAHs. If the concentration of DOM is expressed on a carbon-referenced basis (mg C/L), then $f_{oc} = 1.0$, and the $K2$ for DOM in Eq. 7 is equal to, and can be predicted from, the K_{oc}

$$\log (K_{oc}) = \log \left(\frac{K_p}{f_{oc}}\right) = \log K_{ow} - 0.35 \quad [13] \quad (8)$$

4. In an analysis based on multicompartment models, McCarthy et al. [9] showed that the observed BCF measured in the presence of DOM can be calculated from the "potential" BCF (no DOM, Eq 6). If the contaminant bound to DOM is unavailable for biological uptake [9]

$$\text{observed } BCF = (1 - f_{dom}) \times \text{potential } BCF \quad [9] \quad (9)$$

By substituting Eq 7 and solving Eq 9 over a range of concentrations of DOM for compounds with different K_{oc}s, the decrease in observed BCF due to the binding of the contaminant to DOM can be calculated. Figure 5 illustrates the fractional decrease in BCF due to DOM for compounds with K_{oc}s of 10^5 to $10^{6.5}$. Clearly, the presence of DOM can exert a large influence on contaminant accumulation, and the magnitude of the effect is directly related to the hydrophobicity of the contaminant.

Environmental Relevance of DOM

The apparent decrease in both the observed BCF (Fig. 5) and the K_{app} (Fig. 4) results from a failure to explicitly distinguish between the amount of the total nonparticulate contaminant (C_w) that is freely dissolved in water (C_d) and that which is bound to DOM (C_{dom}). While separation of particulate and nonparticulate contaminant using either filtration or centrifugation is a simple and routine procedure, measurement of the amount of contaminant bound to DOM is more difficult, time consuming, and costly. Several procedures have been used for these separations, including equilibrium dialysis [4,9], diafiltration [6,15], and head-space analysis [5,18]. It would be useful, therefore, to predict those environmental conditions in which DOM would significantly affect either the amount of contaminant binding to particles or the amount of freely dissolved (and readily bioavailable) contaminant.

Since the association of hydrophobic contaminants with particulate and dissolved sorbents appears to be independent and noninteractive (Eq 3, Figs. 2–4), the fraction of the total contaminant associated with each phase of the system can be calculated from the K_p for each sorbent

$$f_d = \frac{1}{1 + (K1 \times P) + (K2 \times \text{DOM})} \quad (10)$$

$$f_p = \frac{K1 \times P}{1 + (K1 \times P) + (K2 \times \text{DOM})} \quad (11)$$

$$f_{dom} = \frac{K2 \times \text{DOM}}{1 + (K1 \times P) + (K2 \times \text{DOM})} \quad (12)$$

where f_d, f_p, and f_{dom} are the fractions of the total contaminant associated with the dissolved, particulate, and DOM phases, respectively. These equations can be used to predict those situations

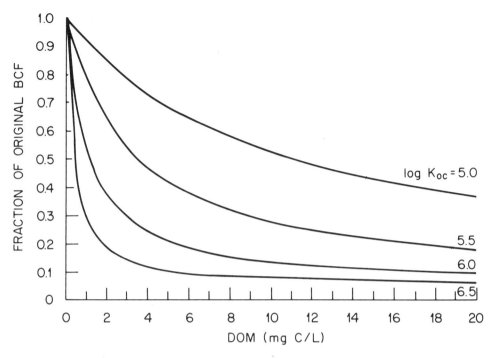

FIG. 5—*The fractional decrease in the apparent BCF due to the presence of DOM for contaminants having a K_{oc} ranging from 10^5 to $10^{6.5}$. The "original" BCF is that derived from the regression equation for BCF in the absence of DOM (Eq 6). The fractional decrease was calculated as the ratio of the "observed" BCF" in the presence of DOM (C_a/C_w, from Eq 9) compared to the "original" BCF.*

in which failure to include the effect of DOM (that is, by setting the term $K2 \times \text{DOM}$ in Eqs 10 and 11 equal to zero) will alter the calculated estimates of f_d and f_p. Errors in estimating f_d may result in overestimates of bioaccumulation, whereas errors in estimating f_p may significantly alter predictions on the transport and fate of contaminants. The percent error due to failure to include the term $K2 \times \text{DOM}$ can be calculated

$$\% \text{ Error} = 1 - \frac{1 + (K1 \times P)}{1 + (K1 \times P) + (K2 \times \text{DOM})} \times 100\% \tag{13}$$

This equation can be expressed in terms of the concentration of carbon in the particulate and dissolved phase and the carbon-referenced association coefficient, K_{oc}

$$\% \text{ Error} = 1 - \frac{1 + (K_{oc} \times POC)}{1 + (K_{oc} \times POC) + (K_{oc} \times \text{DOM})} \times 100\% \tag{14}$$

where the particulate organic carbon concentration, $POC = f_{oc} \times P$. The shaded areas of Fig. 6 illustrate the ranges of *POC* and DOM concentrations for which failure to consider the effect of DOM would lead to a 50% (Fig. 6A) or 90% (Fig. 6B) error in predictions of the fractions of contaminants in the dissolved and particulate phases for contaminants having K_{oc}s of 10^7 to 10^3. When *POC* levels are very high and DOM levels are low, binding to *POC* dominates and DOM

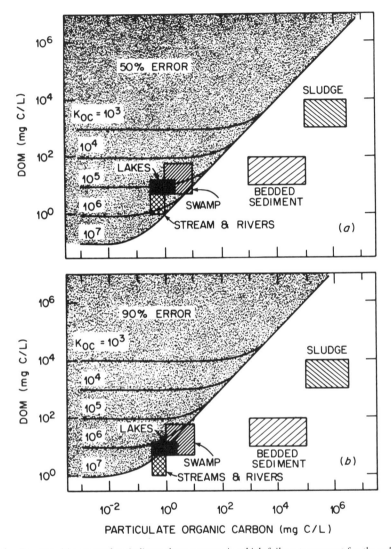

FIG. 6—*Equation 14 was used to indicate those systems in which failure to account for the role of DOM as a sorbent would result in significant errors in predictions about exposure of aquatic organisms. The shaded areas indicate the range for which interactions between DOM and particular organic carbon (POC) would lead to a (A) 50% or (B) 90% error in steady state estimates of the fraction of contaminant freely dissolved or bound to particles if the role of DOM as a competing sorbent were ignored. The 50 and 90% error regions are shown for contaminants with K_{oc}s from 10^3 to 10^7; DOM-POC combinations above or to the left of the line indicate the error region for each K_{oc}. The range of DOM-POC for different types of environmental systems is indicated to illustrate those situations in which the role of DOM is significant.*

has little effect (lower right-hand corner of figures). As *POC* concentrations decrease and DOM levels increase, DOM becomes increasingly important for accurate predictions.

Different types of ecosystems have relatively characteristic ranges of *POC* and DOM. Several systems are mapped onto Fig. 6. Although a municipal waste treatment sludge is characterized by very high levels of DOM (10^3 to 10^4 mg C/L), organic solids so dominate the binding of contaminants in this system that failure to consider DOM would not result in a large error. Likewise, the organic coating on particles in a bedded sediment have a sufficient capacity to bind contaminants that DOM would appear to have little influence on partitioning, even assuming that the carbon content of the interstitial waters reached levels of 100 mg C/L. Lakes, swamps and bogs, and rivers and streams fall into the range in which, for compounds with K_{oc}s above $10^{4.5}$, failure to consider DOM can result in a 50% error in predicting partitioning equilibria (Fig. 6A). In some lakes and swamps, estimates can be in error by an order of magnitude for contaminants with a K_{oc} above $10^{5.5}$ (Fig. 6B).

Our analysis assumes that binding of hydrophobic contaminants to dissolved or particulate sorbents is controlled by solvophobic associations with sorbent carbon, and that dissolved carbon has about the same binding affinity as does the particulate carbon. For DOM in particular, this assumption needs to be qualified. Low molecular weight organic acids and simple carbohydrates can contribute to the overall concentration of dissolved organic carbon, but probably have little effect on binding of contaminants. DOM from different natural waters vary in their binding affinities [3,4,6,15,27]. If the components or properties of DOM that are active in binding of contaminants were known, the DOM concentration in Eq 14 should be expressed in terms of the carbon concentration of that "active DOM." In this sense, the analysis in Fig. 6 estimates conservatively the influence of DOM, and may overestimate its influence if based on total dissolved carbon levels in the water. As the binding affinity of total DOM decreases, the shaded "error" zones of Fig. 6 shift upward.

Another assumption in this analysis is that the environment is at steady state with respect to distribution of contaminant among biotic and abiotic components of the system. Because of the long periods of time required for some compartments to achieve equilibrium (up to months for binding to particles or for bioaccumulation of very hydrophobic chemicals [14]), steady state may be an unusual condition in the real world. Binding to DOM, however, appears to be quite rapid, and is likely to approach equilibrium in less than a day [8,18]. One of the major roles of DOM may be to alter the kinetics of contaminant distribution during the approach to equilibrium. The influence of DOM on the environmental transport and fate of chemicals may be greatest in systems far from equilibrium. These effects may be seen when perturbations occur in time frames which are short compared to the characteristic time for contaminant exchange with particles or biota. Under these conditions, a larger fraction of the total contaminant will be bound to DOM than would be predicted from Eq 12, and the "error" zone indicated in Fig. 6 would increase by shifting downward and to the right.

Nevertheless, it is clear that there are realistic environmental pollution scenarios in which the influence of DOM is an important determinant of contaminant transport and bioavailability. The types of compounds for which the presence of DOM is important are those which raise the greatest concerns in terms of their environmental persistence and their potential danger to human health. PAHs, including very potent carcinogens such as BaP, as well as PCBs, dioxins, dibenzofurans, and many hydrophobic pesticides, are examples of contaminants whose transport and bioavailability can be greatly altered by the presence of DOM. Models for predicting exposure and transport as well as experiments examining the environmental behavior of these compounds need, therefore, to account for the potential interactions with DOM.

Summary and Conclusions

Binding of hydrophobic contaminants to dissolved and particulate sorbents appears to be independent and can be predicted from the association coefficients between the freely dissolved contaminant and each individual sorbent. Significant errors in estimating the environmental partitioning and bioavailability of very hydrophobic compounds can result if the role of DOM as a competitive sorbent is ignored. The influence of DOM is greatest when suspended particle concentrations are low, such as in lakes and streams, or when DOM levels are high, such as in swamps and bogs. In most cases only the very hydrophobic contaminants ($K_{oc} > 10^5$) will be significantly affected by binding to DOM, but these compounds constitute the greatest concerns in terms of human health and environmental persistence.

Acknowledgments

We wish to thank S. M. Bartell and G. R. Southworth for technical review of this manuscript. We also wish to thank them, in addition to J. E. Breck and A. J. Stewart, for valuable discussions and suggestions. This research was sponsored by the Ecological Research Division of the Office of Health and Environmental Research, U.S. Department of Energy, under Contract No. DE-AC05-840R21400 with Martin Marietta Energy Systems, Inc. Publication No. 2903, Environmental Sciences Division, Oak Ridge National Laboratory.

References

[1] McCarthy, J. F., *Archives of Environmental Contamination and Toxicology*, Vol. 12, 1983, pp. 559–568.
[2] Leversee, G. J., Landrum, P. F., Giesy, J. P., and Fannin, T., *Canadian Journal of Fisheries and Aquatic Science*, Vol. 40, Supplement 2, 1983, pp. 63–69.
[3] Landrum, P. F., Reinhold, M. D., Nihart, S. R., and Eadie, B. J., *Environmental Toxicology and Chemistry*, Vol. 4, 1985, pp. 459–467.
[4] Carter, C. W. and Suffet, I. H., *Environmental Science and Technology*, Vol. 6, 1982, pp. 735–740.
[5] Diachenko, G. W., "Sorptive interactions of selected volatile halocarbons with humic acids from different environments," Ph.D. Thesis, University of Maryland, College Park, Md., 1981.
[6] Means, J. C. and Wijayaratne, R. D., *Science*, Vol. 215, 1982, pp. 968–970.
[7] McCarthy, J. F. and Jimenez, B. D., *Environmental Toxicology and Chemistry*, Vol. 4, 1985, pp. 511–521.
[8] McCarthy, J. F. and Jimenez, B. D., *Environmental Science and Technology*, Vol. 19, 1985, pp. 1072–1076.
[9] McCarthy, J. F., Jimenez, B. D., and Barbee, T., *Aquatic Toxicology*, Vol. 7, 1985, pp. 15–24.
[10] Chiou, C. T., Peters, L. J., and Freed, V. H., *Science*, Vol. 206, 1979, pp. 831–832.
[11] Karickhoff, S. W., Brown, D. S., and Scott, T. A., *Water Research*, Vol. 13, 1979, pp. 241–248.
[12] Means, J. C., Wood, S. G., Hassett, J. J., and Banwart, W. L., *Environmental Science and Technology*, Vol. 14, 1980, pp. 1524–1528.
[13] Karickhoff, S. W., *Chemosphere*, Vol. 10, 1981, pp. 833–846.
[14] Karickhoff, S. W., *Journal of Hydraulic Engineering*, Vol. 110, 1984, pp. 707–735.
[15] Means, J. C. and Wijayaratne, R. D., *Marine Environmental Research*, Vol. 11, 1984, pp. 77–89.
[16] Wijayaratne, R. D. and Means, J. C., *Environmental Science and Technology*, Vol. 18, 1984, pp. 121–123.
[17] Whitehouse, B., *Estuarine, Coastal, and Shelf Science*, Vol. 20, 1985, pp. 393–402.
[18] Hassett, J. P. and Milicic, E., *Environmental Science and Technology*, Vol. 19, 1985, pp. 638–643.
[19] O'Connell, D. J. and Connolly, J. D., *Water Research*, Vol. 14, 1980, pp. 1517–1523.
[20] Hassett, J. P. and Anderson, M. A., *Water Research*, Vol. 16, 1982, pp. 681–686.
[21] Voice, T. C., Rice, C. P., and Weber, W. J., *Environmental Science and Technology*, Vol. 17, 1983, pp. 513–518.
[22] Gschwend, P. M. and Wu, S. C., *Environmental Science and Technology*, Vol. 19, 1985, pp. 90–96.
[23] Caron, G., Suffet, I. H., and Belton, T., *Chemosphere*, Vol. 14, 1985, pp. 993–1000.
[24] Veith, G. D., Macek, K. J., Petrocelli, S. R., and Carroll, J., in *Aquatic Toxicology, (third symposium)*, ASTM STP 707, American Society for Testing and Materials, Philadelphia, 1980, pp. 116–129.
[25] Chiou, C. T., Freed, V. H., Schmedding, D. W., and Kohnert, R. L., *Environmental Science and Technology*, Vol. 11, 1977, pp. 475–478.
[26] Neely, W. B., Branson, D. R., and Blau, G. E., *Environmental Science and Technology*, Vol. 8, 1974, pp. 1113–1115.
[27] Carter, C. W. and Suffet, I. H., in *Fate of Chemicals in the Environment*, R. L. Swann and A. Eschenroeder, Eds., ACS Symposium Series No. 225, American Chemical Society, Washington, DC, 1983, pp. 215–229.

B. W. Vigon,[1] G. B. Wickramanayake,[1] J. D. Cooney,[1] G. S. Durell,[1] A. J. Pollack,[1] T. Shook,[2] and M. Frauenthal[2]

Techniques for Environmental Modeling of the Fate and Effects of Complex Chemical Mixtures: A Case Study

REFERENCE: Vigon, B. W., Wickramanayake, G. B., Cooney, J. D., Durell, G. S., Pollack, A. J., Shook, T., and Frauenthal, M., "**Techniques for Environmental Modeling of the Fate and Effects of Complex Chemical Mixtures: A Case Study,**" *Aquatic Toxicology and Hazard Assessment: 10th Volume, ASTM STP 971*, W. J. Adams, G. A. Chapman, and W. G. Landis, Eds., American Society for Testing and Materials, Philadelphia, 1988, pp. 247–260.

ABSTRACT: A mathematical model along with physical models were used to determine the fate of complex chemical mixtures released to the multimedia environment. Transport and transformation of phosphorus compounds generated during munition testing were studied by employing the Environmental Protection Agency's (EPA) TOX-SCREEN model. The species modeled were phosphine and a representative linear condensed polyphosphate (LCP). The model predicted that a significant fraction of phosphine was volatilized. The fate of LCP is mainly governed by the sorption capacity of soils. The data gaps identified during the preliminary modeling phase included estimates of polyphosphate sorption capacity onto soil.

Laboratory experiments indicated that the adsorption of total phosphorus onto sand was well represented by Freundlich isotherms. Batch and column studies indicated that sand enhances the oxidation of burn products to orthophosphate in the aqueous environment. The leachate generated from the combustion residue was highly toxic to *Daphnia magna*. Passage of the leachate through a sand/gravel column removed the toxic compounds and/or transformed them to less toxic products.

A combination of mathematical and physical modeling and toxicity monitoring proved highly effective in evaluating and mitigating potential environmental impacts of phosphorus obscurant smokes.

KEYWORDS: elemental phosphorus, red phosphorus, obscurant smokes, adsorption, environmental fate, mathematical modeling, aquatic toxicity, *Daphnia magna*

The behavior of complex mixtures of chemicals released to the environment can be difficult to assess using only individual techniques, such as toxicological tests or mathematical modeling, in isolation. Physical models, such as microcosms, can be used to study the fate of complex chemical mixtures in the environment, but this approach can be time consuming and costly, especially when the evaluation must be applicable to a wide range of environmental conditions. Consequently, mathematical models based on deterministic and/or stochastic approaches are widely employed to predict the fate and transport of chemicals in single or multimedia environments. For complex mixtures, however, the current models fall short of the capability necessary to model multiple concurrent and sequential reactions. Therefore, laboratory scale physical models can be used either to generate input data for mathematical modeling of selected fate reactions or to provide an

---

[1] Environmental Sciences Department, Battelle Columbus Division, Columbus, OH 432-2693.
[2] Engineering and Technology Directorate, Pine Bluff Arsenal, Pine Bluff, AR 71611.

integrated perspective on the chemical fate of the mixture. The purposes of this study were: (1) to apply a combination of physical and mathematical models to determine the fate and transport of a complex chemical mixture of phosphorus compounds released to the environment and (2) to evaluate a physical treatment system for mitigating the environmental impacts using a combination of modeling and toxicological techniques.

Red phosphorus/butyl rubber (RP/BR) and white phosphorus/felt (WP/F), obscurant smokes are used by the military for screening purposes. Spanggord and Podoll [1] and Spanggord et al. [2] reported that approximately 50% of the starting quantity of RP/BR was consumed in the burns to generate phosphorus trioxide (P_2O_3) and phosphorus pentoxide (P_2O_5), which comprise the smoke as well as various other linear or polymeric phosphates, phosphine, and residual elemental phosphorus (P_4) homologs (Fig. 1). Brazell et al. [3] ignited RP/BR under different environmental conditions and reported that the smokes were composed of monomeric and polymeric phosphoric acids varying from 1 to 13 or more phosphate units. Burn residue, which is about 25% of the initial charge, also contained a significant amount of condensed linear polyphosphates. These compounds, when deposited on the ground, undergo biodegradation and/or hydrolysis to yield orthophosphate as the final product. The unconsumed elemental phosphorus and its partially

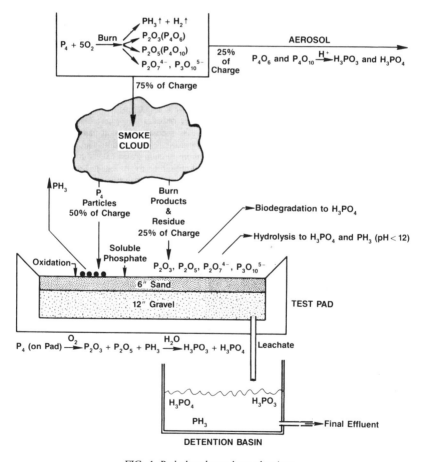

FIG. 1–*Red phosphorus burn chemistry.*

oxidized products deposit on the soil and react with atmospheric oxygen to produce water soluble phosphates. These oxidation reactions, especially for elemental phosphorus, depend on relative humidity and may proceed over a long period of time.

A certain percentage of RP/BR munitions require testing under simulated field deployment conditions to evaluate their performance. In order to minimize the ground and surface water contamination that is likely to occur from the frequent deposition of burn residue, a test facility has been designed. This facility consists of a bed of sand, underlain by gravel and a drainage system, with the leachate collected and sent to a holding pond for testing and/or treatment prior to discharge to the environment. Smoke grenades will be tested over the bed with residual RP/BR and by-products settling onto the bed over time. Rain falling on the bed will solubilize some of the materials and carry them through the bed. Assessment of the fate and transport of this complex mixture of chemical compounds in the laboratory under different environmental conditions and different test bed configurations would be very tedious. Thus, the U.S. Environmental Protection Agency's (EPA) TOX-SCREEN computer model, a multimedia screening level program for assessing the potential fate of chemicals released to the environment, was used in this study to follow the fate of selected burn by-products. Based on the previous literature review, the species selected were a gaseous, water-soluble by-product, phosphine, and a representative linear condensed polyphosphate (LCP), pyrophosphate.

Because of the wide distribution of species resulting from RP/BR combustion, it is quite difficult to relate specific chemical species to the consequent aquatic toxicity. The burn products, when dissolved in water, have been found to be highly toxic to the freshwater macroinvertebrate *Daphnia magna* [4]. Thus, it appeared desirable to conduct a set of bench-scale experiments to study changes in toxicity and chemical characteristics of leachate passed through a sand and gravel bed simulating the test facility. Based on the TOX-SCREEN results, a set of adsorption experiments was conducted to determine the effect of the sand on chemical adsorption and transformation of the RP/BR burn mixture. These data were then combined with operational loading estimates to design the column experiments.

Materials and Methods

TOX-SCREEN is a screening level multimedia model that can be used to assess the potential fate of toxic chemicals released to air, surface water, or soil. Consequently, this model was chosen to study the fate and transport of toxic phosphorus compounds in air, soil, and water. Because of the large number of phosphorus compounds generated during the phosphorus burn, variations in by-product composition, and interaction between the different species, it is difficult to model the fate of the overall toxicity of burn products. Therefore, in the preliminary steps reported in the present study, the fate of some representative chemicals was examined. It is important to note that the model has a limited capacity to incorporate the effects of other phosphorus compounds on the fate of the chemical compound being modeled. Consequently, some batch and column studies were conducted to determine parameters, such as sand adsorption capacity and transformation rates of the relevent phosphorus compounds. These physical models were also used to determine the change in the burn solution toxicity when passed through the sand layers.

Modeling Studies

A considerable amount of input data is required for the TOX-SCREEN model itself. Much of the chemistry data pertaining to RP/BR combustion products were obtained from a series of reports authored by Spanggord and Podoll [1] and Spanggord et al. [2]. Some of the remaining input data were obtained from the TOX-SCREEN user's guide [5,6] and SESOIL manual [7]. The rest of the required input data was obtained from open or unpublished literature.

Mass loading of pyrophosphate and phosphine to the bed surface was determined from the data generated by Spanggord et al. [2]. Since the proportions of the chemical species that resulted from a burn have not been quantified, it was conservatively assumed that all the reduced and partially oxidized products were in the form of phosphine and pyrophophate, respectively. The site specific hydrologic inputs were used for the analysis. Annual precipitation and evaporation were set at 129 and 27.31 cm (50.79 and 10.75 in.), respectively.

Spanggord et al. [2] reported that the partition coefficients for phosphorus compounds generated from RP/BR burns ranged from 1 to 160 mL/g. Thus, a high value of 160 mL/g and a low value of 1 mL/g were used for modeling phosphine behavior. An adsorption coefficient of 120 mL/g, which was derived by Gerritse et al. [8] for orthophosphate, was used for pyrophosphate. Gas phase diffusivity for phosphine was assumed to be 8×10^{-2} cm²/s.

The potential effects of alternative bed materials and depth on the removal or transformation of phosphine and pyrophosphate were studied by the changing permeability and thickness of the sand and gravel layers. For pyrophosphate, the low intrinsic permeabilities for clean sand and gravel were 1×10^{-7} and 1×10^{-4} cm², respectively [9]. The high values were fixed one order of magnitude greater than the low values. Transport of phosphine was examined under slightly different engineering conditions where the intrinsic permeabilities ranged from 1×10^{-8} cm² (low) to 1×10^{-7} (high). Each of the above conditions was tested at two different sand depths, 15 and 30 cm (6 and 12 in.), representing the range of design conditions. The associated gravel depths varied from 30 to 15 cm so that the total depth of sand and gravel was a constant 45 cm (18 in.).

Adsorption Experiments

The adsorption capacity of sand was determined by conducting a series of laboratory batch experiments similar to the procedures given in the EPA's "Chemical Fate Test Guidelines" [10]. A typical river sand sample was washed several times with deionized Barnstead water, air dried, at 40°C, and stored at 4°C until use. For equilibration experiments, 0.27 g of RP/BR was burned in an apparatus designed to allow humidity, oxygen, and mass balance control. The solid residue was then dissolved in 1 L of deionized Barnstead water. The undissolved fraction was sedimented by centrifugation, and the supernatant was collected. This solution was mixed with water to generate 50, 20, 10, 5, 2, 1 and 0.5 weight % dilutions for RP/BR adsorption experiments. A parallel set of experiments also was conducted to study the adsorption of polyphosphoric acid (PPA) onto sand. Approximately 90 mg of PPA was dissolved in 1 L of Barnstead water and diluted subsequently to generate 50, 25, 5, 2.5, 1.25 and 0.5% solutions. The RP/BR solution was analyzed for orthophosphate and total phosphorus, whereas the PPA solution was analyzed only for total phosphorus. The phosphates were analyzed by the Automated Ascorbic Acid Method (Methods 424C2 and 424G) given in "Standard Methods" [11].

In the equilibration experiments, 100 mL volumes of each RP/BR and PPA dilution and a water (control) were added to 250 mL Erlenmeyer flasks (reactors) that contained 10 g of cleaned sand. The reactors were placed on a tabletop shaker (at $20 \pm 1°C$) and swirled at 1450 rpm to keep the contents in a well-mixed condition. After 40 h, the liquids in the reactors were separated from suspended matter and analyzed for total phosphorus. In addition, the RP/BR solutions were analyzed for orthophosphate.

Column Studies

The column study was conducted to investigate the change in both the chemistry and toxicity of an RP/BR burn solution as it passed through sand and gravel layers simulating field operations. The columns consisted of glass tubes 10 cm in diameter and 60 cm long, sealed on the bottom

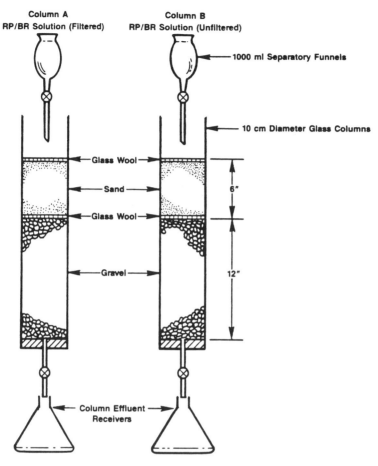

FIG. 2—*Laboratory setup for column study.*

with an inert plug that was fitted with a glass take-off tube. The columns were packed with a 15-cm layer of clean river sand and an underlying 30-cm layer of washed gravel (Fig. 2).

The columns were initially eluted with reconstituted water. Samples were analyzed for total phosphorus and orthophosphate to determine baseline levels associated with the column materials. Static acute toxicity tests with *Daphnia magna* were also run on the effluents to verify that the column materials were not introducing any toxicity at the start of the study.

During the experiment, approximately 0.39 g of RP/BR was burned under controlled environmental conditions. When combustion was complete, the burn residue was dissolved in 3 L of reconstituted water. This water had its pH adjusted to 4.5 to simulate the pH reported at the location of the future test bed.

A 900-mL aliquot of the burn solution was passed through a layer of glass wool to remove the suspended particles, and the filtered liquid was applied to the column over a 2-h period. Another 900-mL aliquot was administered to the second column without filtration over the same time period. The remaining raw influent was saved for toxicity testing and chemical analysis.

The column loadings were based on an annual rainfall of 129 cm (50.8 in.) and an annual projected RP/BR residue loading of 946 kg (2535 lb). The daily column loading of 900 mL was equivalent to an average monthly loading to the test bed.

The raw influent and two column effluents were then comparatively tested for static acute toxicity and levels of total phosphorus and orthophosphates to determine the effects of column exposure on burn solution toxicity and chemistry.

Aquatic Toxicity Studies

The aquatic toxicity studies were conducted to see whether the sand/gravel columns are effective in toxicity reduction. These toxicity studies consisted of a series of static acute tests with *Daphnia magna* following the EPA-recommended procedure [*12*]. During the tests, young *Daphnia* (less than 24 h old) were continuously exposed for 48 h to at least five dilutions of the test solution and a reconstituted water control. In these tests, *Daphnia* were exposed in groups of five animals in 100-mL beakers containing 80 mL of test solutions or control water. Four replicates, each containing 5 animals, were used for each test concentration and the control. The beakers were placed in environmental chambers where the following conditions were maintained: temperature, 20 ± 2°C; photoperiod, 16-h light and 8-h dark; light intensity, 323 to 1,076 lx). The specific conductivity of the test samples was measured at the beginning of the test, whereas temperature, dissolved oxygen, and pH were measured at both the beginning and the end of the test. The number of live and dead (or immobilized animals) were counted at 24 and 48 h. The viability data were used to estimate the median lethal concentration (LC_{50}), which is the concentration of test solution that kills or immobilizes 50% of the exposed *Daphnia* after 48 h [*12*].

Results and Discussion

Modeling Studies

The TOX-SCREEN model estimates pollutant concentrations in the various media based on chemical property data, climatic and soil data, simple water body descriptions, and specific chemical release scenarios. It also generates an annual summary of mass loading and distribution of the pollutant in air, soil, and water.

Because TOX-SCREEN was developed as a screening tool for evaluating chemical fate and effects, it has been simplified to minimize input data requirements and is designed to be highly conservative. Another factor that contributes to the conservative results is that conservatively estimated input values were used in the absence of real data. For example, from the limited literature reviewed, data are not available on the biodegradability of phosphine. Thus, the rate of biodegradation of phosphine was assumed as zero. These conservative assumptions lead to overestimates of the leachate concentrations of the chemical species of interest. Conclusions drawn (based on the values reported as results from a single analysis) are to be interpreted as order-of-magnitude data. A sensitivity analysis can also be performed using the TOX-SCREEN model to compare the effect of different input variables or the effect of variation in the magnitude of one engineering or environmental parameter on the mass distribution/concentration in air, soil, and water.

Phosphine.—The effects of sand depth, permeability, and partition coefficient on the phosphine concentrations in the leachate are presented in Fig. 3. Depending on the scenario, the leachate concentration of phosphine ranged from less than 0.01 to nearly 10 ppb. An important parameter controlling leachate quality is the partition coefficient. Increasing the partition coefficient from 1 to 160 caused the leachate concentration to decrease by about 2 orders of magnitude. Changing

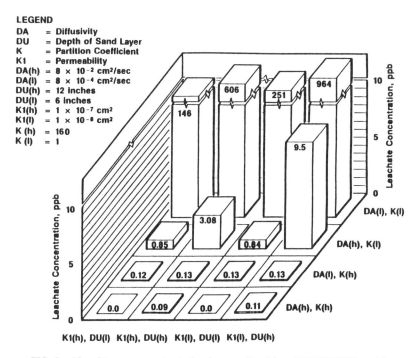

FIG. 3—*Phosphine concentration in leachate predicted from TOX-SCREEN model.*

permeability did not have a significant effect on the leachate concentration, especially when the sand depth was 15 cm. Increasing sand depth accompanied by decreasing gravel depth resulted in increased leachate concentrations. When the degradation and hydrolysis rates are low, the leachate concentration of the chemical increases with increasing sand depth. This effect is significant for compounds, such as phosphine, with high gas phase diffusivity and can be attributed to the following two factors. First, the increase in sand depth results in increased evaporation that can be attributed to the extended retention of moisture in the sand bed. This increased evaporation of water in turn increases the pollutant concentration in the leachate. Second, when the sand depth is high, the depth of the gravel layer becomes smaller since the total depth of both layers is kept constant in the facility design. The decrease in gravel depth results in reduced volatilization of the dissolved compound, which causes a consequent increase in leachate concentration.

Pyrophosphate.—The fate of pyrophosphate deposited on the bed was analyzed for two different sand/gravel permeabilities ($1 \times 10^{-7}/1 \times 10^{-4}$ cm^2 and $1 \times 10^{-6}/1 \times 10^{-6}/1 \times 10^{-3}$ cm^2) and sand/gravel depths (15/30 cm and 30/15 cm). The leachate concentrations resulting from a uniformly distributed annual loading of 870 kg of pyrophosphate is given in Fig. 4. The model predicts lower leachate concentrations for 15 cm sand/30 cm gravel than for 30 cm sand/15 cm gravel. The lowest effluent concentration of pyrophosphate is expected when lower permeability sand and gravel are present.

The annual data summary from TOX-SCREEN indicated that the amount of pyrophosphate adsorbed by the sand and gravel bed was significantly higher than the amount volatilized. However, it is important to note that the model assumes a linear relationship between the increase in concentration of pollutant in soil moisture and the amount adsorbed, but does not consider an upper-bound adsorption value (saturation capacity). A second limitation in the current version of

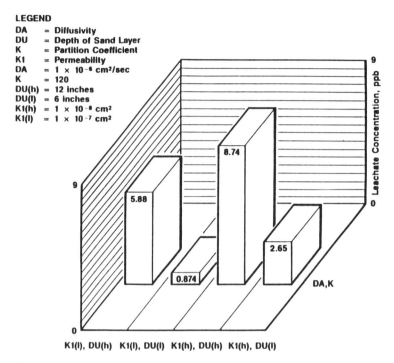

FIG. 4—*Pyrophosphate concentration in leachate predicted from TOX-SCREEN model.*

the TOX-SCREEN model is that it does not have the capability to account for competitive adsorption between condensed polyphosphate species in the leachate and orthophosphate resulting from previous burns. Thus, supplemental laboratory experiments were designed to address these deficiencies.

Phosphorus Adsorption to Sand

In the RP/BR equilibration experiments, the solutions were analyzed for orthophosphate and total phosphate at the beginning and end of the tests. The change in orthophosphate concentration during the 42-h equilibration period is presented in Fig. 5. In the absence of sand, the orthophosphate concentration increased by 13 to 35% of the original concentration as a function of initial concentration. In the presence of sand, however, the transformation of other phosphorus compounds into orthophosphate increased significantly (50 to 220%). While the experimental data are inadequate to explain the mechanism or chemistry involved in the increased oxidation of phosphorus compounds when sand was present, the observations are consistent with a surface hydrolysis model involving hydroxyl groups on the siliceous sand surface.

The total P adsorption isotherms were developed for RP/BR solution and PPA. The log-log plots of specific adsorption (mass sorbed per unit mass of adsorbant) against equilibrium concentration of adsorbate in the solution for RP/BR and PPA are given in Figs. 6 and 7, respectively. The sorption constants were calculated using the Freundlich equation:

$$x/m = C_s = KC_e^{1/n} \qquad (1)$$

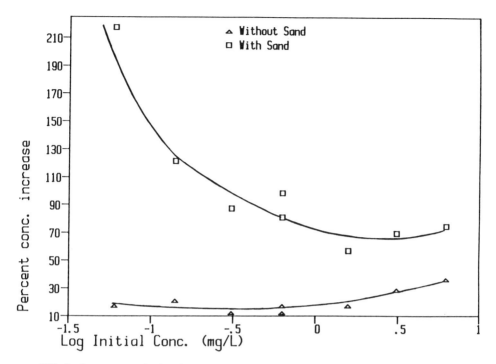

FIG. 5—*Increase in orthophosphate concentration after 42-h equilibration in batch experiments.*

where

- C_e = equilibrium concentration of the chemical in the solution phase, μg/mL,
- C_s = equilibrium concentration of the chemical in solid phase, μg/g,
- K = Freundlich adsorption coefficient,
- m = mass of adsorbant, gram,
- $1/n$ = exponent, where n is a constant, and
- x = mass in micrograms of chemical adsorbed by m grams of adsorbant.

The constants for the Freudlich isotherms are given in Table 1. Results indicate that adsorption of total phosphorus from RP/BR solution and PPA solution onto sand is not significantly different.

Column Study

During this study, the test columns were loaded with RP/BR solutions five days per week. The test was continued for 38 days. The cumulative loading for this period was approximately the typical loading that could be expected at the test facility for three years.

Total P concentrations in the effluent gradually increased during the first five days and then remained fairly steady for the rest of the test period (Fig. 8). The concentration of orthophosphate in the leachate, however, increased within three days (Fig. 9). After that, the orthophosphate concentration was higher in the column effluent than in the influent. These results, which are comparable to the observations in the batch equilibration experiments, confirm that sand and gravel enhance the oxidation of phosphorus compounds to orthophosphate.

The acute toxicity test data are given in Table 2. During the 38-day study period, while considerable toxicity was found in the influent, no toxicity was observed in the effluent. Further

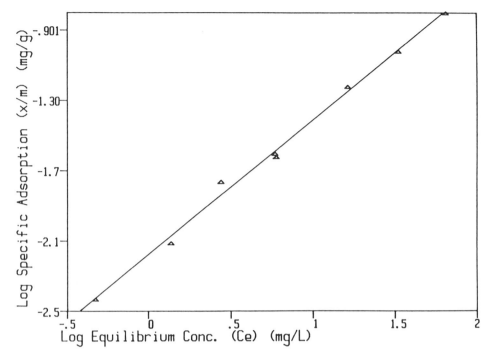

FIG. 6—*Total phosphorus adsorption isotherm for RP/BR burn solution.*

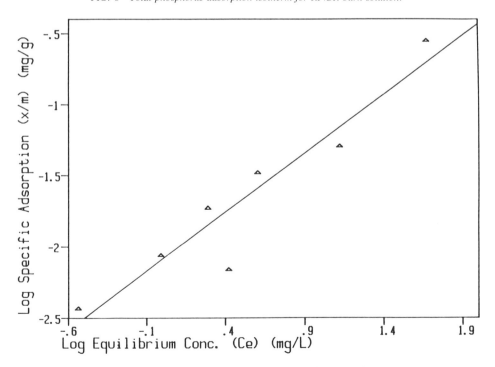

FIG. 7—*Total phosphorus adsorption isotherm for PPA solution.*

TABLE 1—*Freundlich adsorption constants for total phosphorus.*

	K	n
RP/BR burn solution	9.4	1.026
PAA	10	1.215

experiments are planned to determine whether the toxicity reduction resulted from a chemical process or a physicochemical process such as sedimentation, volatilization, or adsorption. Work is also underway to chemically regenerate the test bed.

Toxicity Mitigation

Previous experimentation has demonstrated that training/testing exercises using white phosphorus/felt screening munitions release potentially toxic materials to the environments [13]. Since this material is disseminated over an area that is uncontrolled with regard to ground and surface water contamination, efforts to mitigate the toxicity by accelerated conversion of the uncombusted residue are underway. These efforts are focused on addition of oxidants to the soil to enhance the conversion effect of the soil on phosphorus residues. Preliminary results indicate that oxidant-dosed residues exhibit enhanced conversion rates relative to undosed water controls.

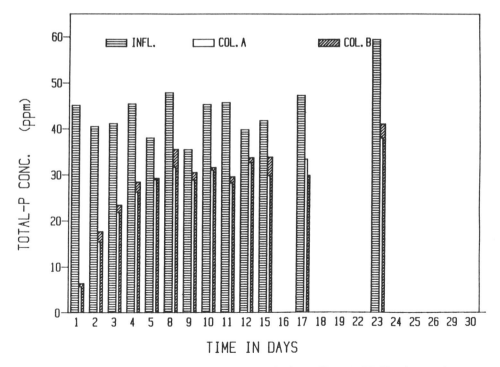

FIG. 8—*Total phosphorus levels in the influent and column effluents in RP/BR column study*

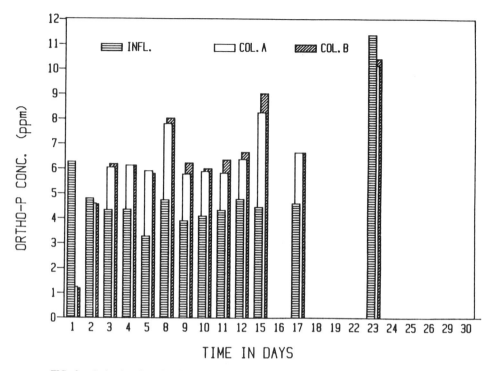

FIG. 9—*Orthophosphate levels in the influent and column effluents in RP/BR column study.*

Conclusions

A combination of mathematical and physical modeling and toxicity monitoring proved highly effective in evaluating and mitigating potential environmental impacts of phosphorus obscurant smoke testing as well as the efficacy of alternative engineering designs. For complex mixtures, available chemical models are inadequate, although judicious choice of modeled components of the mixture can contribute to the delineation of laboratory and/or field experiments.

Laboratory experiments allowed the effect of specific fate processes to be understood as well as simulating the behavior of the integrated system over short-to-moderate time frames. These types of chemical experiments should be coupled with toxicity measurements to gain maximum benefits. The chemistry investigations themselves do not allow a complete prediction of impact due to the difficulty of relating the fate of individual components of the effluent mixture to toxic effects. The toxicity measurements, on the other hand, are easier to interpret with respect to causation and organism sensitivity if chemical data are available.

The synergism of this approach is clearly demonstrated by the finding of no toxicity in the column effluent. While both the batch and column chemistry tests suggested substantial effects due to the presence of the solid phase, the related dramatic toxicity reduction could not have been predicted.

Acknowledgments

This work was supported under an Environmental Engineering Services Agreement with Pine Bluff Arsenal, U.S. Department of the Army under Contract No. DACA 56-84-0035 and a contract

TABLE 2—*Toxicity data for influent and column effluents in RP/BR column study.*

Day	Influent, LC_{50} (% Leachate Dilution)	Effluent[a] Column A (Filtered)	Effluent[a] Column B (Unfiltered)
0	...	100% survived	100% survived
3	2.6	100% survival at 50% dilution[b]	...
9	15.8	100% survived	95% survived
12	3.0	100% survived	100% survived
17	11.0	100% survived	100% survived
	>32[c]		
23	6.1	100% survived	90% survived
	24.3[c]		

[a]Survival is at 100% effluent concentration unless otherwise noted.
[b]Concentration range was not completely bracketed.
[c]Influent samples were filtered through 0.3-μm pore size glass fiber filters.

with the Chemical Research and Development Center, U.S. Department of the Army under Contract DAAG29-81-D-0100-1850. The authors wish to thank Janet Dean, John Steichen, Bill Palmer, Michele Moore, Ed Moore, and Kelly O'Brien for laboratory assistance during this investigation. We also wish to acknowledge the computer modeling support of Ann Langham and technical discussions with William Clement.

References

[1] Spanggord, R. J. and Podoll, R. T., "Environmental Fate of White Phosphorus/Felt and Red Phosphorus/Butyl Rubber Military Screening Smokes," reports from SRI International to U.S. Army Medical Bioengineering Research Development Laboratory, Frederick, MD, Contract No. DAMD 17-82-C-2320, 1983.

[2] Spanggord, R. J., Podoll, R. T., Rewick, R., Backovshy, J. Wilson, R., and Chou, T., "Environmental Fate of White Phosphorus/Felt and Red Phosphorus/Butyl Rubber Military Screening Smokes," reports from SRI International to U.S. Army Medical Bioengineering Research Development Laboratory, Frederick, MD, Contract No. DAMD 17-82-C-2320, 1984.

[3] Brazell, R. S., Holmberg, R. W., and Moneyhun, J. H., "Application of High-Performance Liquid Chromatography-Flow Injection Analysis for the Determination of Polyphosphoric Acids in Phosphorus Smokes," *Journal of Chromatography*, Vol. 290, 1984, pp. 163–172.

[4] Vigon, B. W., Wickramanayake, G. B., Clement, W. H., Langham, A., Cooney, J. D., Pollack, A. J., Goss, L. B., Shook, T., and Frauenthal, M., "Environmental Protection and Elemental Phosphorus Fate Behavior," *Proceedings of the Chemical Defense Research Conference*, CRDC-SP-86007, Aberdeen Proving Grounds, MD, Nov. 1985.

[5] McDowell-Boyer, L. M. and Hetrick, D. M., "A Multimedia Screening-Level Model for Assessing the Potential Fate of Chemicals Released to the Environment," ORNL/TM-8334, Oak Ridge National Laboratory, Oak Ridge, TN, 1982.

[6] Hetrick, D. M. and McDowell-Boyer, L. M., "User's Manual for TOX-SCREEN: A Multimedia Screening Level Program for Assessing the Potential Fate of Chemicals Released to the Environment," EPA-560/5-83-024, Office of Toxic Substances, U.S. Environmental Protection Agency, Washington, DC, 1984.

[7] Bonazountas, M. and Wagner, J., SESOIL: "A Seasonal Soil Compartment Model," EPA/OTS Contract No. 68-01-6271, Arthur D. Little, Inc., Cambridge, MA, 1982.

[8] Gerritse, R. G., Vriesema, R., Dalenberg, J. W., and DeRoss, H. P., "Effect of Sewage Sludge on Trace Element Mobility in Soils," *Journal of Environmental Quality*, Vol. 11, No. 3, 1982, pp. 359–364.

[9] Chow, V. T., *Handbook of Applied Hydrology*, McGraw Hill, Inc., New York, 1966.

[10] "Chemical Fate Test Guidelines," EPA-560/6-82-003, Office of Pesticides and Toxic Substances, US Environmental Protection Agency, Washington, DC, 1982.

[11] "Standard Methods for the Examination of Water and Wastewater," APHA-AWWA-WPCF, 16th Ed., American Public Health Assn., Washington, DC, 1985.
[12] Peltier, W. H. and Weber, C. I., "Methods for Measuring the Acute Toxicity of Effluents to Freshwater and Marine Organisms," 3rd Ed., EPA/600/4-85/013, Environmental Protection Agency, Washington, DC, 1985.
[13] Tolle, D. A., Arthur, M. F., Chesson, J., Duke, K. M., Jackson, D. R., Kogan, V., Kuhlman, M. R., and Margeson, D. P., "Ecological Effects Evaluation of Two Phosphorus Smokes Using Terrestrial Microcosms," draft final report from Battelle Columbus Division to U.S. Army Medical Research and Development Command, Maryland, Contract No. DAMD 17-84-C-4001, 1984.

Steven M. Bartell,[1] *Robert H. Gardner,*[1] *and Robert V. O'Neill*[2]

An Integrated Fates and Effects Model for Estimation of Risk in Aquatic Systems

REFERENCE: Bartell, S. M., Gardner, R. H., and O'Neill, R. V., **"An Integrated Fates and Effects Model for Estimation of Risk in Aquatic Systems,"** *Aquatic Toxicology and Hazard Assessment: 10th Volume, ASTM STP 971,* American Society for Testing and Materials, Philadelphia, 1988, pp. 261–274.

ABSTRACT: An integrated fate and effects model was developed to mathematically simulate the toxic effects of naphthalene on the growth of interacting populations in an aquatic system. Daily effects on biomass production were calculated for each model population as a function of its dynamic body burden of naphthalene. Separate simulations using constant environmental loading rates of 0.0001, 0.001, 0.01, and 0.10 g m^{-2} d^{-1} demonstrated changes in production that could not be extrapolated directly from naphthalene toxicity measured for related laboratory populations of aquatic organisms. Estimated risks of 50% reduction in fish production ranged from 0.08 to 0.65 for naphthalene loading rates of 0.000025 to 0.10 g m^{-2} d^{-1}. Examination of naphthalene flux through the model system indicated that the combined effects of photolysis and volatilization were more important than biological processes in determining the fate of naphthalene for the 0.10 loading rate. This relative importance of physicochemical versus biologial processes reversed for the 0.0001 loading rate. At intermediate loading rates, the relative importance of these processes varied seasonally. The integrated model demonstrated potential contributions of population-specific rates of naphthalene uptake and depuration and naphthalene toxicity to estimation of possible ecological risks posed by naphthalene in aquatic systems.

KEYWORDS: fates and effects model, ecological risk, naphthalene, toxic effects, biomass production, dose-response functions

Ecological risk may be defined as the likelihood of occurrence of a prespecified ecological response to toxic chemical exposure. The response might range from the local extinction of one or more populations to changes in total system productivity or patterns of energy flow or material cycling [1]. Estimation of ecological risk requires accurate characterization of chemical fate and subsequent effects of accumulated toxicant on populations interacting within highly connected ecological systems. The large number of potentially toxic chemicals and variety of ecosystem types prohibit an exclusively experimental approach to estimation of ecological risk [2]. Simulation models, because of their comparative economy in implementation, should play an increasing role in ecological risk analysis [3].

Estimation of ecological risk has remained difficult in part because the modeling of chemical fates has proceeded somewhat independent of modeling chemical effects [4]. An implicit hypothesis was that toxic effects on resident populations produced minimal feedback on physical-chemical processes that determined future distribution, exposure, and bioaccumulation of the toxicant.

---

[1] Research staff members, Environmental Sciences Division, Oak Ridge National Laboratory, Oak Ridge, TN 38731.
[2] Senior scientist, Environmental Sciences Division, Oak Ridge National Laboratory, Oak Ridge, TN 38731.

Recent modeling efforts have focused on the integration of the ecologial fates and effects of toxicants. For example, Mancini [5] derived a model that simulated toxic effects on biota in relation to kinetics of toxicant uptake and metabolism for time-varying concentrations of dissolved toxicant. Models have also been constructed to estimate the frequency and duration of exposures of organisms to toxicant concentrations in excess of lethal concentrations measured for laboratory populations [6]. O'Neill et al. [7] derived a method for extrapolating toxic effects in laboratory populations to estimates of risk for natural populations, assuming steady-state exposure concentrations.

With few exceptions (for example, Ref 7), previous efforts have focused on direct lethal effects of toxicants on organisms. Less attention has been directed towards developing the capability to forecast the effects of long-term, chronic inputs of toxicants to structurally complex, dynamic ecological systems. Patterns of connectedness among system components in the form of energy flow or material cycling might influence the effects of chemical perturbations [1,3,8]. A simple example of potential feedback between effects and fate of toxicants in an ecosystem would be the complete annihilation of an intermediate trophic level (for example, zooplankton) followed by an observation that little or no toxicant was measureable in planktivorous fish.

Another characteristic of effects models (for example, Refs 6,7) is that toxicity is commonly extrapolated from concentrations of the dissolved toxicant. This approach is not unexpected given the nature of current acute toxicity assays where mortality is quantified in relation to the initial concentration of dissolved toxicant. Realistically, toxicity may be more strongly correlated with chemical actually accumulated at some metabolic site than with the dissolved toxicant concentration. However, the necessary explicit representations of differential accumulation, metabolism, and depuration of toxicants by aquatic organisms are not routinely included in effects models.

Our research focused on the development of an integrated fates and effects model that emphasized interrelations among physical, chemical, and biological processes that determine toxicant fate and effects in a hypothetical aquatic system. The resulting model is innovative in that sublethal toxic effects on modeled populations are predicted in relation to temporally changing body burdens of toxicant. Equally innovative, the body burden dynamics are simulated in relation to temporal changes in the growth rates of the populations, variations in toxicant loading rate, and changes in the rates of principal fate-determining processes in a seasonal light and temperature environment. A secondary objective was to determine from this new model whether biological activity could, at least in theory, outweigh the influence of physiocochemical processes in determining the fate of organic toxicants in aquatic systems.

Description of the Model

The integrated fates and effects model (IFEM) represents a synthesis of the Fates of Aromatics Model (FOAM) [9,10] and the Standard Water Column Model (SWACOM) [7]. Because FOAM was designed specifically for polycyclic aromatic hydrocarbons (PAHs), we were initially constrained to these chemicals in the development of the IFEM. We chose naphthalene as a model PAH because of its comparatively greater water solubility and toxicity relative to other PAHs.

Model Structure

IFEM describes the accumulation and effects of PAHs on growing populations of primary producers and consumers in a freshwater aquatic system (Fig. 1). The primary producers consist of phytoplankton, periphyton, and rooted macrophytes. The consumer populations incude zooplankton, benthic insects, other larger benthic invertebrates (for example, clams, crayfish), pelagic omnivorous fish, and a benthic detritivorous fish. IFEM also simulates the sorption of PAH by dissolved organic matter, suspended and settled particulates, detritus, and sediments. Modeled

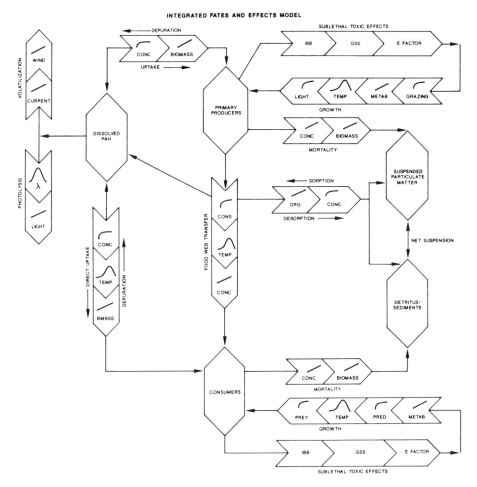

FIG. 1—*Schematic representation (after Park et al. [19]) of the combined fates and effects model. Hexagons identify aggregated state variables, for example, dissolved PAH concentration (g PAH/L), biomass of primary producers (g dry wt/m²), PAH concentration in primary producers (g PAH/g dry wt). Broad arrows identify model processes and pathways. Subdivisions of these arrows indicate important model factors and point out linear and nonlinear relations in the model. For sublethal toxic effects, IBB = 48-h integrated body burden, GSS = the general stress syndrome, and the E factor is the effects factor calculated from the internal dose-response functions (for example, Table 3).*

chemical fate processes include volatilization, photolysis, and sorption-desorption, in addition to the uptake and depuration of the PAH by the model populations. Structure-activity regression models were used to estimate these process rates for PAHs in relation to molecular weight, melting point, octanol:water partition coefficient, and the light absorption spectrum for individual PAHs [11]. IFEM permits time-varying loading rates of dissolved PAH. Initial masses and toxicant concentrations of the populations, sediments, detritus, and particulate matter can be specified. Fate parameters for naphthalene are listed in Table 1. Model equations and parameters are detailed in the Appendix.

TABLE 1—*Parameter values for fate of naphthalene in the combined fates and effects model (IFEM).*

Parameter	Description	Value[a]
S	Log_{10} water solubility	-3.89
H	Henry coefficient	-1.66
k_G	Gas phase transfer, cm/d	853.0
k_L	Liquid phase transfer, cm/d	34.70
P	Photolytic yield coefficient	0.022
R_s	Sorption rate, 1/d	0.268
R_d	Desorption rate, 1/d	0.86
U_1–U_3	Uptake by plants, mg PAH g^{-1} d^{-1}	0.00069
U_4	Uptake by zooplankton, mg PAH g^{-1} d^{-1}	0.011
U_5	Uptake by benthic insects, mg PAH g^{-1} d^{-1}	20.0
U_6	Uptake by invertebrates, mg PAH g^{-1} d^{-1}	0.0039
U_7	Uptake by bacteria, mg PAH g^{-1} d^{-1}	0.00069
U_8–U_9	Uptake by fishes, mg PAH g^{-1} d^{-1}	0.10
D_1–D_3	Depuration rate for plants, 1/d	0.99
D_4,D_7	Depuration by zooplankton, bacteria, 1/d	1.00
D_5,D_6	Depuration by insects, invertebrates, 1/d	0.99
D_8,D_9	Depuration by fishes, 1/d	0.213

[a] Calculated from structure-activity regressions for PAHs [11].

Dose-Response Functions in IFEM

Population growth in IFEM is determined by the integration of physiological process equations adapted from FOAM [9]. Biomass of primary producers and consumers changes in relation to daily variation in incident light, water temperature, and predator-prey interactions. Producers grow as a function of available light, water temperature, and losses to excretion, mortality, and grazing. Consumers grow as a function of food availability and losses to respiration, egestion, excretion, mortality, and predation.

The physiological process formulation of the growth equations facilitated the expression of toxic effects as increases or decreases in rates of the processes that determine growth, analogous to the approach used in SWACOM [7]. Unlike SWACOM, where effects were modeled in relation to a constant concentration of dissolved chemical, toxic effects in IFEM were calculated in relation to changing body burdens of toxicant for each model population. Body burdens changed in relation to the time-varying growth rates of the populations, the concentration of dissolved PAH, and the population- and chemical-specific kinetics for PAH uptake and depuration. PAH availability changed in relation to varied loading rates and the fate processes that determine the concentration of dissolved PAH.

To derive the dose-response functions for naphthalene, growth of each population in IFEM was simulated in isolation under nonlimiting conditions of light, temperature, or food. Bioassays were subsequently simulated using ten exposure concentrations and acute toxicity data reported for naphthalene (Table 2). Toxic effects were represented as population-specific effects factors, E_i, that adjusted the rate of population i growth to simulate the effects measured in the bioassays, for example, a 50% decrease in biomass over a 48-h exposure period. An effects value of 0.1, for example, would decrease the rate of photosynthesis by 10% or increase respiration rate by 10%. Corresponding population parameters in IFEM would be redefined as 0.90 or 1.10 of their nominal values. Growth rates were adjusted according to a partially documented and otherwise assumed stress syndrome for aquatic plants and animals. For primary producers, rates of photosynthesis were decreased [12], and rates of mortality and susceptibility to grazing were increased. For

TABLE 2—*Acute toxicity[a] data used to derive dose-response functions for naphthalene for populations in IFEM.*

Assay Species	IFEM Population	Toxicity Data[b]
Selenastrum capricornutum	algae, periphyton, macrophytes, bacteria	33.0[c]
Daphnia magna	zooplankton, benthic insects, benthic invertebrates	8.6[c]
Pimephales promelas	detritivorous fish	6.6[d]
Salmo gairdneri	omnivorous fish	2.3[d]

[a] U.S. EPA 440/5-80-059 (1980). Ambient water quality criteria for naphthalene. Office of Water Regulations and Standards, Criteria and Standards Division, Washington, DC.
[b] mg/L.
[c] 48-h LC_{50}.
[d] 96-h LC_{50}.

consumers, feeding rates were decreased, whereas rates of respiration [*13*], mortality, and susceptibility to predation were increased. The net accumulation of naphthalene by each population was calculated in each bioassay simulation. The body burden was assumed a correlate of the concentration that produced toxic effects at specific metabolic sites of action, aggregated as population biomass in the model [*14*]. The results of the bioassay simulations produced values of the 48-h body burdens of naphthalene and corresponding values of E for each population (Fig. 2). Dose-response functions were constructed by regressing E values on body burden for the set of bioassay simulations (Table 3).

The dose-response functions for naphthalene were incorporated in IFEM. At the beginning of each simulated day, the 48-h integrated body burden was calculated for each population and the growth processes for that day were adjusted by the corresponding E value. The 48-h integrated

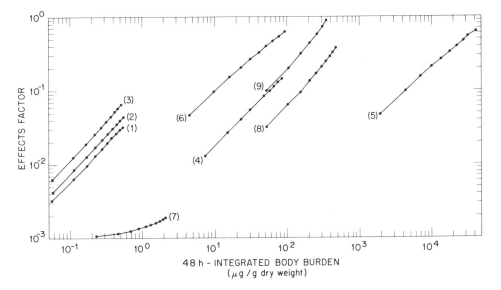

FIG. 2—*Dose-response functions for IFEM populations exposed to naphthalene. Dose is expressed as 48-h integrated body burden, ug/g dry weight. Effects factors quantify the fractional change in growth parameters of IFEM populations. Parenthetical numbers identify IFEM populations in order presented in Table 3 [for example, (1) is phytoplankton, (9) is detritivorous fish].*

TABLE 3—*Dose-response functions for populations in the combined fates and effects model (IFEM). Intercept (a) and slope (b) are for* $log_{10}(E) = a + b\ log_{10}(body\ burden)$.

IFEM Population	a	b
Phytoplankton	4.907	1.021
Periphyton	5.075	1.028
Macrophytes	5.486	1.062
Zooplankton	4.136	0.980
Benthic insects	0.989	0.845
Larger invertebrates	4.085	0.848
Bacteria	3.881	1.046
Omnivorous fish	3.249	1.111
Detritivorous fish	4.104	1.204

body burden reflects the history of exposure and the uptake and depuration kinetics specific to each CFEM population. Through calculation of effects in this way, the importance of a body burden at time t was weighted by the body burden selected arbitrarily at the time $t - 1$. Thus, the magnitude of the toxic effect lagged an increasing body burden. Similarly, a population could continue to exhibit a toxic response under conditions of rapid depuration or transformation of a previously high toxicant burden.

Parameter Uncertainty and Risk Estimation

Estimation of risk requires consideration of several sources of uncertainty including the toxicity of the chemical, incomplete knowledge of system function, biological variability and measurement error in toxicity testing, and extrapolation of toxic effects from laboratory populations to field situations. These uncertainties were represented in IFEM by assigning statistical distributions to the model parameters that determined fates and effects. The implications of parameter variation on model results were then quantified through repeated simulations using values chosen independently from these distributions. Risk was calculated as the frequency of model solutions that exceeded some prespecified endpoint, for example, a doubling of annual phytoplankton production or a 50% decrease in fish production [7].

Results

Several model experiments were performed with IFEM to evaluate the efficacy of an integrated fates and effects model for naphthalene.

Deterministic Effects on Fish Production

The model was used to examine the effects of four different, but constant, naphthalene loading rates (0.0001, 0.001, 0.01, and 0.10 g m^{-2} d^{-1}) on ecological production for a hypothetical aquatic system. Annual biomass production for each IFEM population was calculated for comparison with production in the absence of naphthalene (Table 4).

Examination of model results revealed unexpected patterns of effects given the population sensitivities of the laboratory toxicity data (Table 2). Consistent with the corresponding dose-response functions (Fig. 2), bacterial and benthic insect production was largely unaffected. Bacteria were least sensitive to naphthalene based upon the bioassay simulations. Benthic insects required body burdens on the order of 1 mg/g dry weight to exhibit toxic effects. At the lowest loading rate, phytoplankton production increased by 0.4% relative to the control simulation. Periphyton

TABLE 4—*Annual biomass production (g dry wt m^{-2} $year^{-1}$) for the nine populations in IFEM in relation to constant loading rates for naphthalene (g m^{-2} d^{-1}).*

IFEM Population	Loading Rate				
	0	0.0001	0.001	0.01	0.10
Phytoplankton	2366	2376	2349	2129	1451
Periphyton	3695	3691	3633	3264	2647
Macrophytes	6882	6846	6505	5159	4075
Zooplankton	238	218	216	195	123
Benthic insects	46	48	48	48	47
Larger invertebrates	58	51	50	49	44
Bacteria	0.36	0.35	0.35	0.35	0.35
Omnivorous fish	378	190	188	175	127
Detritivorous fish	57	47	47	47	49

were largely insensitive to naphthalene at lower rates of loading, but decreased by 19% between the 0.01 and 0.10 rates. In contrast, the greatest toxic effect on macrophyte growth occurred at the lower 0.001 and 0.01 loading rates.

Direct extrapolation of the toxicity data (Table 2) suggested that benthic insects, crayfish, and zooplankton should have shown similar responses to naphthalene and that the omnivorous fish should have been the most severely affected population. For the 0.10 loading rate, zooplankton production decreased by 48% relative to the control simulation, but crayfish decreased by only 24% and benthic insects were largely unaffected. Omnivorous fish decreased by 66% relative to the control, while detritivorous fish decreased by only 14%. The severe effects of naphthalene on production by the omnivorous fish were consistent with the acute toxicity data. However, the disparate effects among populations of zooplankton, benthic insects, and crayfish demonstrated that expected effects could not always be directly extrapolated from laboratory tests. These unanticipated changes in production resulted from differential uptake and elimination rates and trophic interactions among these populations, as well as the differences in population growth rates and sensitivities to naphthalene.

Exposure to naphthalene altered the magnitude and pattern of biomass production by pelagic omnivorous fish in IFEM (Fig. 3). The 0.01 exposure caused an overall decrease in production, especially between Days 100 and 200, but the seasonal pattern of production was not altered. The biomass peak on Day 155 was 38% less than the peak for the zero exposure simulation. Toxic effects were less pronounced between Days 200 and 300, when the majority of the naphthalene input was lost through volatilization and photolytic degradation. The two lowest loading rates (not shown) generated decreases in biomass that were encompassed by the zero and 0.01 exposure results. Exposures from the 0.10 loading rate decimated seasonal production, delayed timing of the early summer peak by ~20 d, and moved the autumn biomass peak ahead by ~15 d. The direct toxic effects on the fish plus indirect reductions in prey availability reduced annual fish production by 75%.

Risks of Decreased Fish Production

To examine the implications of parameter uncertainties on forecasts of fish production, 100 Monte Carlo simulations were performed for each of the four naphthalene loading rates. The resultant distribution for net annual fish production at the 0.0001 loading rate suggested a slight toxic effect (Fig. 4); the distribution was skewed with reference to the zero exposure solution of 378 g dry wt/m^2. Nineteen percent of the simulations produced more fish biomass than the no exposure simulation, thus emphasizing the indirect effects of naphthalene that propagated throughout

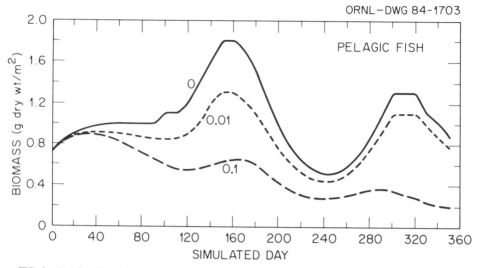

FIG. 3—*Simulated daily biomass values for IFEM omnivorous fish over annual production cycle in relation to loading rates of 0.0, 0.01, and 0.10 g naphthalene m^{-2} d^{-1}.*

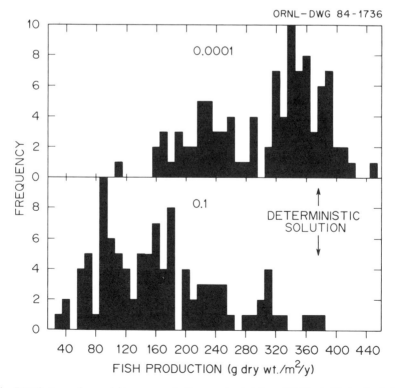

FIG. 4—*Distributions of annual biomass production constructed from 100 Monte Carlo simulations with IFEM for each of two naphthalene loading rates, 0.0001 and 0.1 g m^{-2} d^{-1}. The value of the deterministic solution was for zero naphthalene loading.*

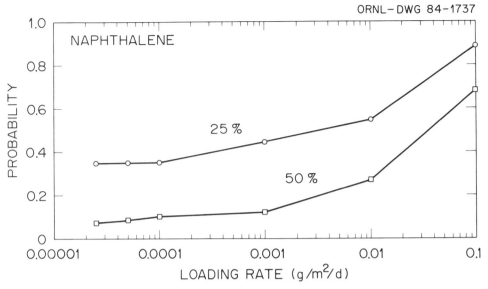

FIG. 5—*IFEM forecast of the risk of a 25 or 50% reduction in annual fish production in relation to naphthalene loading rates. Each point results from 100 Monte Carlo simulations.*

the food web. These indirect effects may assume the form of decreased competition of omnivorous fish for food or greater food availability as the result of the pattern of sensitivities of lower trophic level populations to naphthalene. The 0.10 loading rate produced severe toxic effects on fish biomass (Fig. 4). Only one simulation showed increased fish production. The median production value was 150 g dry wt/m², a 60% decrease from the zero exposure case.

From the frequency distributions of fish production (for example, Fig. 4), the probabilities or risks of observing at least a 25 or 50% reduction in annual fish biomass as a function of naphthalene loading rate were calculated (Fig. 5). Each point is the result of 100 IFEM simulations. Estimation of risk by this method has several desirable properties for ecological risk analysis [8]: (1) risk values calculated in this manner, like probabilities, can range from 0 to 1; (2) risk was an increasing function of loading rate; and (3) the risk of small (for example, 25%) reduction in fish biomass was consistently greater than that of a larger (for example, 50%) reduction, and this relationship was not simply multiplicative. The relatively toxic nature of naphthalene was evidenced by the risk value of 0.35 for observing a 25% fish reduction for loading rates as low as 0.025 mg m^{-2} d^{-1}. The model results suggested that risk did not extrapolate to zero for the no-exposure case. This result demonstrated the potential contribution of natural system variation to risk estimates. Natural variability was modeled in IFEM as parameter variability assigned independent of exposure to the temperature dependence of photosynthetic and feeding rates.

Interactions Between Naphthalene Fate and Effects

Values of daily naphthalene flux along individual pathways in the IFEM (for example, Fig. 1) were calculated to determine the relative importance of physicochemical and biological processing of naphthalene in the model aquatic system. Direct uptake, depuration, and food web transfers of naphthalene were summed daily in separate deterministic simulations using each of the four loading rates (Table 4). This percentage of biological flux was compared to the percentage of flux due to the sum of photolytic degradation and volatilization. Sorption of naphthalene was consistently

<1% of the total flux through the system and was therefore omitted. These percentages were compared using a two-dimensional phase space (Fig. 6a–d).

At the lowest rate of naphthalene input, biological processing dominated the overall flux of this PAH through the system (Fig. 6d). Photolytic degradation was important between Days 180 and 270. The toxic effects of naphthalene became increasingly apparent with higher loading rates. The combination of photolysis and volatilization accounted for a greater proportion of naphthalene flux (Fig. 6b,c,d), particularly during the later periods of each simulation. However, a resurgence of biological importance associated with the autumnal production peak was also evident (Fig. 6c,d). The greatest difference in the pattern of naphthalene processing occurred for the 0.10 g m^{-2} d^{-1} loading rate (Fig. 6a). The vernal biological production, characteristic of the lower loading rates (Fig. 6b,c,d), was never greater than 10% of total naphthalene flux. The toxic effects on annual production at this loading rate ranged between 3 and 41% of the zero exposure case (Table 4), with an average population reduction of 26%.

Discussion

The simulated patterns of change in biomass production for different naphthalene exposures importantly demonstrated that indirect toxic effects propagating throughout food webs can in theory influence estimates of ecological risk. These model results are consistent with previous

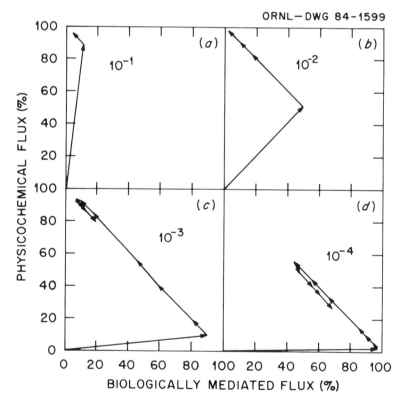

FIG. 6—*Percent of total naphthalene flux due to combined rates of volatilization and photolysis versus combined rates of biological processing plotted through time for IFEM deterministic solutions and four naphthalene loading rates (indicated on figure). Arrows indicate the time course of the 360-d simulations.*

observations that the combination of differential sensitivities of aquatic populations to toxicants and trophic interactions within ecological networks can confound predictions of toxic effects extrapolated directly from laboratory bioassays [15].

We must emphasize, however, that the indirect effects of naphthalene on production may be influenced by assumptions underlying the derivation of the model, particularly the dose-response functions. One assumption concerns the current formulation of the stress syndrome. The structure of the growth equations (Appendix) constrained the physiological resolution in simulating toxic effects to population biomass. As more data that describe specific modes of action become available, these equations may be refined. Similarly, calculation of accumulation and effects at the coarse scale of population biomass may be replaced by simulation of exposure of specific metabolic sites of activity within individual organisms. However, the current lack of information describing specific modes of action for toxic chemicals suggests the level of physiological detail represented in IFEM is an appropriate working hypothesis.

The log linear dose-response functions were derived by assuming that uptake and depuration of naphthalene were independent of toxic effects, that is, uptake and depuration parameters were not altered in relation to changing body burdens during the simulated bioassays. Alternative hypotheses include increasing or decreasing rates of uptake or depuration in relation to body burden. These alternatives could change the slopes of the dose-response functions for specific populations, but would not likely change the general functional form. The Monte Carlo simulations demonstrated the impact of variation in the slope and intercept of the functions on calculated fish production (Fig. 4) when these parameters were varied by as much as 20% of their mean values (Table 3).

Another important IFEM result was that biological uptake, depuration, and food-web transfers were frequently the dominant processes that determined the fate of naphthalene in this hypothetical system. Future research might experimentally identify the relative conditions of biological productivity, toxicant loading rates, and rates of physicochemical processes where control of toxicant fate shifts from physicochemical to biological control. The model might thus be effectively applied in the design of experiments for evaluating chemical fate [16] and effects [17,18]. Experimental results can be used to further evaluate and refine the model.

The IFEM represents an initial operational hypothesis concerning potential feedback between the fates and effects of toxicants in aquatic systems. The model permits exploratory analyses of the implications of population specific kinetics for the net accumulation of a toxicant and sensitivities to toxicants on the estimation of risk. Risk can be evaluated in the context of dynamic physicochemical processes that vary concentrations of dissolved toxicants. Variable exposure concentrations can be examined in relation to changing body burdens and subsequent toxic effects on population growth. Additional applications and evaluations of IFEM and other models (for example, Refs 6,14) may identify those circumstances where steady-state assumptions concerning population dynamics and chemical concentrations are sufficient or insufficient for accurate predictions of chemical effects. These capabilities have been specifically lacking in previous models of the fate and effects of PAHs in aquatic systems.

Acknowledgments

Research was sponsored jointly by the U.S. Environmental Protection Agency (EPA) under Interagency Agreement DW89930690-01-0 and the Ecological Research Division of the Office of Health and Environmental Research, U.S. Department of Energy, under Contract No. DE-AC05-840R21400 with Martin Marietta Energy Systems, Inc. D. L. DeAngelis and V. H. Dale provided helpful criticisms of an earlier draft of this manuscript. The research described in this document has not been subject to EPA review and therefore does not necessarily reflect the views of EPA and no official endorsement should be inferred. This publication is No. 2842, Environmental Sciences Division, Oak Ridge National Laboratory.

Appendix

The equations that determine rates of volatilization, sorption, and photolytic degradation of naphthalene in IFEM are the same as those defined for the FOAM [9]. Naphthalene parameters are listed in Table 1. The physiological process formulations for growth of the model populations are modified from SWACOM [8]. The change in biomass (B_i) of the primary producers was equated to

$$dB_i/dt = B_i(P_i - R_i - M_i - U_i - S_i - G_i), (i = 1,3) \quad (1)$$

where the subscript i designates the ith producer population and P, R, M, U, S, and G identify, respectively, the rates of photosynthesis, respiration, mortality, excretion, sinking (for phytoplankton only), and grazing or predation (Table A-1). Equations for photosynthesis, respiration, and consumption include nonlinear functions of light, water, temperature, and biomass of predator and prey [9]. Biomass units are grams dry weight per square metre. The growth processes are rates expressed as grams dry weight per square metre per day or 1/d. Dissolved naphthalene concentrations were modeled as grams per square metre. Concentrations in the populations were expressed as grams of naphthalene per gram dry weight.

Incorporation of toxic effects on growth of the primary producers was modeled through modification of Eq 1

$$dB_i/dt = E_i B_i (P_i - R_i - M_i - U_i - S_i - G_i) \quad (2)$$

where E_i was the effects factor for population i defined by the bioassay simulations (Fig. 2). The sign on E_i for each growth process was determined by the assumed stress syndrome described in the text.

Growth rates of the consumer populations were determined by

$$dB_i/dt = B_i (C_i - R_i - M_i - F_i - U_i - G_i), (i = 4,9) \quad (3)$$

TABLE A-1—*Values of parameters defining growth and net accumulation of napthalene by populations in the IFEM.*

Model Population	Parameter Value					
	R_i	U_i	P_i	C_i	Q_i	D_i
PRIMARY PRODUCERS						
Phytoplankton	0.40	0.03	1.90		0.69	0.99
Periphyton	0.30	0.03	1.90		0.69	0.99
Macrophytes	0.20	0.03	2.10		0.69	0.99
CONSUMERS						
Zooplankton	0.10	0.05		0.50	11.10	1.00
Benthic insects	0.10	0.05		0.30	1.10	0.99
Large invertebrates	0.10	0.05		0.48	3.90	0.99
Bacteria	0.42	0.0		0.62	0.69	1.00
Omnivorous fish	0.05	0.05		0.27	100.00	0.21
Detritivorous fish	0.05	0.05		0.13	100.00	0.21

where C_i and F_i designate consumption and egestion rates of the ith consumer population. The toxic effects of naphthalene were included through modifications analogous to Eq 3

$$dB_i/dt = E_i B_i (C_i - R_i - M_i - F_i - U_i - G_i) \qquad (4)$$

The accumulation and depuration of naphthalene (N_i) by the ith population was modeled as a second-order process

$$dN_i/dt = B_i (Q_i N_d/(K_{ni} + N_d) - D_i N_i) \qquad (5)$$

where Q_i defined a maximum specific rate of naphthalene uptake by producer i, N_d was the dissolved concentration of naphthalene, K_n was the naphthalene concentration where uptake equalled 0.5 Q_i, and D_i was the rate constant for depuration (Table A-1).

References

[1] Odum, E. P., "Trends Expected in Stressed Ecosystems," *BioScience*, Vol. 35, 1985, pp. 419–422.
[2] Maugh, T. H., "Chemicals: How Many Are There?," *Science*, Vol. 199, 1978, p. 162.
[3] Levin, S. A. and Kimball, K. D., Eds., "New Perspectives in Ecotoxicology," *Environmental Management*, Vol. 8, 1984, pp. 375–442.
[4] Hendrix, P. F., "Ecological Toxicology: Experimental Analysis of Toxic Substances in Ecosystems," *Environmental Toxicology and Chemistry*, Vol. 1, 1982, pp. 193–199.
[5] Mancini, J. L., "A Method for Calculating Effects, on Aquatic Organisms, of Time Varying Concentrations," *Water Research*, Vol. 17, 1983, pp. 1355–1362.
[6] Parkhurst, M. A., Onishi, Y., and Olsen, A. R., "A Risk Assessment of Toxicants to Aquatic Life Using Environmental Exposure Estimates and Laboratory Toxicity Data," in *Aquatic Toxicology and Hazard Assessment (Fourth Symposium), STP 737*, D. R. Bransen and K. L. Dickson, Eds., American Society for Testing and Materials, Philadelphia, 1981, pp. 59–71.
[7] O'Neill, R. V., Gardner, R. H., Barnthouse, L. W., Suter, G. W., Hildebrand, S. G., and Gehrs, C. W., "Ecosystem Risk Analysis: A New Methodology," *Environmental Toxicology and Chemistry*, Vol. 1, 1982, pp. 167–177.
[8] O'Neill, R. V. and Waide, J. B., "Ecosystem Theory and the Unexpected: Implications for Environmental Toxicology," in *Management of Toxic Substances in Our Ecosystems*, B. W. Cornaby, Ed., Ann Arbor Science, Ann Arbor, MI, 1981, pp. 43–73.
[9] Bartell, S. M., Landrum, P. F., Giesy, J. P., and Leversee, G. J., "Simulated Transport of Polycyclic Aromatic Hydrocarbons in Artificial Streams," in *Energy and Ecological Modeling*, W. J. Mitsch, R. W. Bosserman, and J. M. Klopatek, Eds., Elsevier, Amsterdam, 1981, pp. 133–144.
[10] Bartell, S. M., Gardner, R. H., O'Neill, R. V., and Giddings, J. M., Error Analysis of Predicted Fate of Anthracene in a Simulated Pond," *Environmental Toxicology and Chemistry*, Vol. 2, 1983, pp. 19–28.
[11] Bartell, S. M., "Forecasting Fate and Effects of Aromatic Hydrocarbons in Aquatic Systems," in *Synthetic Fossil Fuel Technologies; Results of Health and Environmental Studies*, K. E. Cowser, Ed., Butterworth, Boston, 1984, pp. 523–540.
[12] Soto, C., Hellebust, J. A., and Hutchinson, T. C., "Effect of Naphthalene and Aqueous Crude Oil Extracts on the Green Flagellate *Chlamydomonas anqulosa*. II. Photosynthesis and the Uptake and Release of Naphthalene," *Canadian Journal of Botany*, Vol. 53, 1975, pp. 118–126.
[13] Darville, R. G. and Wilhelm, J. L., "The Effect of Naphthalene on Oxygen Consumption and Hemoglobin Concentration in *Chironomus attenuatus* and on Oxygen Consumption and Life Cycle of *Tanytarsus dissimilis, Environmental Toxicology and Chemistry*," Vol. 3, 1984, pp. 135–141.
[14] Connolly, J. P., "Predicting Single Species Toxicity Tests in Natural Water Systems," *Environmental Toxicology and Chemistry*, Vol. 4, 1985, pp. 573–582.
[15] O'Neill, R. V., Bartell, S. M., and Gardner, R. H., "Patterns of Toxicological Effects in Ecosystems: a Modeling Study," *Environmental Toxicology and Chemistry*, Vol. 2, 1983, pp. 451–461.
[16] Rogers, J. H., Dickson, K. L., Saleh, F. Y., and Staples, C. A., "Use of Microcosms to Study Transport, Transformation and Fate of Organics in Aquatic Systems," *Environmental Toxicology and Chemistry*, Vol. 2, 1983, pp. 155–167.

[17] Franco, P. J., Giddings, J. M., Herbes, S. E., Hook, L. A., Newbold, J. D., Roy, W. K., Southworth, G. R., and Stewart, A. J., "Effects of Chronic Exposure to Coal-Derived Oil on Freshwater Ecosystems: I. Ponds," *Environmental Toxicology and Chemistry*, Vol. 3, 1984, pp. 447–463.

[18] Giddings, J. M., Franco, P. J., Cushman, R. M., Hook, L. A., Southworth, G. R., and Stewart, A. J., "Effects of Chronic Exposure to Coal-Derived Oil on Freshwater Ecosystems: II. Experimental Ponds," *Environmental Toxicology and Chemistry*, Vol. 3, 1984, pp. 465–488.

[19] Park, R. A., et al., "Modeling Transport and Behavior of Pesticides and Other Toxic Organic Materials in Aquatic Environments," Report No. 7, Center for Ecological Modeling, Rensselaer Polytechnic Institute, Troy, New York, 1980.

Alan R. Johnson[1]

Evaluating Ecosystem Response to Toxicant Stress: A State Space Approach

REFERENCE: Johnson, A. R., "**Evaluating Ecosystem Response to Toxicant Stress: A State Space Approach,**" *Aquatic Toxicology and Hazard Assessment: 10th Volume, ASTM STP 971*, W. J. Adams, G. A. Chapman, and W. G. Landis, Eds., American Society for Testing and Materials, Philadelphia, 1988, pp. 275–285.

ABSTRACT: Ecosystems can be regarded as complex biogeochemical systems maintained in a state of thermodynamic nonequilibrium by the flow of materials and energy. The state of such a system at any given time is generally assumed to be characterized by a finite set of measurable quantities. If these variables are taken to be the components of a vector, the instantaneous state of an ecosystem can be represented by a single point in an abstract multidimensional space. As an ecosystem undergoes changes in state, changes in the position of the corresponding vector will result, tracing out a state trajectory over time.

Within a state space representation, the response of an ecosystem to a perturbing influence, such as a toxicant, can be viewed as a displacement of the state vector away from its unperturbed trajectory. Such an approach was used to analyze data from a study of the response of experimental ponds and microcosms to chronic additions of a coal-derived synthetic oil. Ecosystem-level response surfaces and dose-response curves were derived based upon the average separation (distance of displacement) of exposed ecosystems relative to controls. The results exhibited patterns analogous to those observed in classical toxicology based on organismal response and could be used to define acceptable exposure conditions. The state space approach described here provides a coherent and objective framework for summarizing a large multivariate data set, and it should be of general use in providing both qualitative and quantitative descriptions of the behavior of perturbed ecosystems.

KEYWORDS: state space analysis, ecotoxicology, microcosms, ponds, synthetic oil

Scientific endeavor, at least since Newton's introduction of calculus, has been characterized by a heavy reliance on mathematical representations of nature. Most scientists recognize mathematical models to be extreme simplifications or idealizations of the real-world phenomena they are intended to represent, but this does not deny their usefulness. Historically, scientific progress often seems to depend upon selecting a particular model, despite its simplifying assumptions, and pushing it to its limits.

A commonly used geometric representation of dynamical systems is that of the state space trajectory [1,2]. If it is assumed that the state of a system at any instant, t, can be represented mathematically by a set of state variables, $[x_1(t), x_2(t), \ldots x_n(t)]$, then these variables can also be regarded as the components of a vector in an n-dimensional state space. The state space trajectory of a system is simply the set of points defined by the locations of the state vector as t is allowed to vary.

Ecological systems are frequently modeled as dynamical systems. As such, state space representations are applicable to ecosystem analysis. State space representations are widely used in the theoretical literature and are increasingly being used in the presentation of experimental

---

[1] Environmental Sciences Division, Oak Ridge National Laboratory, Oak Ridge, TN 37831.

results as well [3–8]. In the state space framework, the response of an ecosystem to an imposed perturbation, such as a toxicant, is reflected as a change in the state space trajectory. The objective of ecotoxicology, within this geometric metaphor, is the qualitative and quantitative description of the changes in state space trajectories occasioned by the exposure of ecosystems to toxicants. This paper describes one approach based upon the concept of vector displacement and applies the method to the analysis of experimental data on the effects of a coal-derived synthetic oil on aquatic microcosms and ponds.

Materials and Methods

Theoretical Background and Statistical Methodology

Let us assume that the unperturbed dynamics of an ecosystem can be modeled as a set of ordinary differential equations of the form

$$\dot{x}(t) = f(x,t) + z(t) \tag{1}$$

where
- x = the state vector,
- $\dot{x}$ = its time derivative,
- $f(x,t)$ = a set of linear or nonlinear functions describing the internal dynamics of the ecosystem, and
- $z(t)$ = normal exogenous inputs to the system.

Given a set of initial conditions, x_0, it is possible, at least numerically, to solve this system of differential equations. The solution corresponds to a single state space trajectory: $x(t) = F(x_0,t)$.

Consider the effect of a toxicant on the behavior of an ecosystem as modeled by Eq 1. Toxicant exposure may be modeled as an abnormal input to the system, requiring a modification of the input vector $z(t)$. Alternatively, the effects of the toxicant on the internal dynamics of the system may be modeled by replacing the function $f(x,t)$ with a new function, $g(x,t)$. In either case, the new system of differential equations may be solved to yield a new state space trajectory for the perturbed ecosystem. Henceforth $x_c(t)$ is used to denote the location of a state vector along an unperturbed (control) trajectory, and $x_p(t)$ is used to denote the location of a state vector along the trajectory of a toxicant-perturbed system. Changes in the state space trajectories induced by the toxicant can be described by calculating a displacement vector, $u(t)$, defined as the vector difference between the perturbed and control state vectors: $u(t) = x_p(t) - x_c(t)$ (Fig. 1). The word "displacement" as used here refers to a vector quantity possessing both magnitude and direction. The word "separation" is used to denote the associated scalar quantity, defined as the distance between perturbed and control state vectors or, equivalently, as the magnitude of the displacement vector.

The trajectory of the displacement vector, $u(t)$, from the time of toxicant introduction until the time the ecosystem recovers (assuming that such recovery occurs), characterizes the response of an ecosystem to a toxicant. It may be desirable to summarize the detailed information present in the displacement trajectory by some set of integrated measures that typify ecosystem response over a specified time interval. The following integrated measures are proposed which, when taken together, effectively summarize much of the information contained in the displacement trajectory.

1. *Maximum Displacement or Separation*—The maximum displacement vector is simply the largest displacement vector, $u(t)$, encountered over the specified time interval. The magnitude of this vector is defined as the maximum separation.

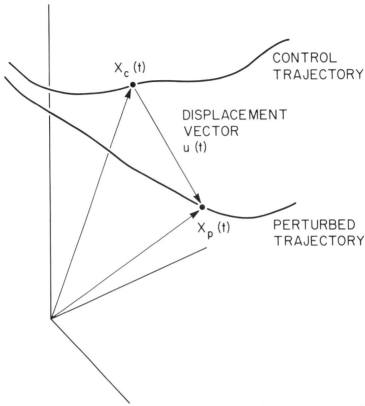

FIG. 1—*State space trajectories showing displacement vector as the difference between control and perturbed state vectors.* $x_c(t)$ = *control state vector;* $x_p(t)$ = *perturbed state vector;* $u(t)$ = *displacement vector.*

2. *Mean Separation*—Over a time interval beginning at t_1 and ending at t_2, the mean separation is defined as

$$\frac{1}{t_2 - t_1} \int_{t_1}^{t_2} |u(t)| \, dt$$

where $u(t)$ represents the magnitude of the displacement vector $u(t)$.

3. *Mean Displacement*—The mean displacement vector is defined as

$$\frac{1}{t_2 - t_1} \int_{t_1}^{t_2} u(t) \, dt$$

Note that in general the magnitude of the mean displacement vector will not equal the mean separation because a partial cancellation occurs as the vector $u(t)$ changes direction.

Comparisons among these measures may reflect aspects of the behavior of the displacement vector over time. For example, if the response of the system remains fairly constant over a given time interval, the mean displacement will approach the maximum displacement. Similarly, a comparison of the magnitude of the mean displacement with the mean separation measures the degree to which $u(t)$ wanders over time, since these quantities will be equal only if the direction of $u(t)$ is constant. Changes in direction will decrease the value of the mean displacement magnitude. This might be expected, for example, in systems which show a biphasic response to a toxicant, such as an initial decline in primary productivity due to direct phytotoxic effects, followed by indirect effects at higher trophic levels. The details of such a response can be reconstructed from an analysis of the displacement trajectory.

In experimental situations, the true state space trajectories and displacement vectors are not known but must be estimated from discrete samples in the presence of natural variability, measurement error, and other uncertainties. At any sampling time, t, the displacement vector $u(t)$ must be estimated based on observations of m replicate control ecosystems and n replicate perturbed ecosystems. The situation can be visualized as two clusters of m and n points distributed in state space which, in the absence of systematic bias, will tend to be centered around $x_c(t)$ and $x_p(t)$, respectively. In this paper, $u(t)$ is estimated as the average of the vectors calculated from all pairwise comparisons between clusters. Similarly, the summary measures were estimated by discrete summations corresponding to the integral formulae given above. The variability among the pairwise estimates gives an indication of the degree of uncertainty in the estimate of $u(t)$. More rigorous confidence intervals for $u(t)$ could be derived based on distributional assumptions or by means of resampling methods, such as the bootstrap procedure [9].

The coordinate system within which the state space trajectories are embedded requires consideration. In particular, two aspects of the coordinate system have important implications for practical applications of the state space approach: (1) the scaling of the coordinate axes, and (2) orthogonality of the axes. Both factors influence the proper interpretation of distance measures calculated in the space. Such considerations may necessitate a transformation of the data prior to analysis. Although no formal derivation of such transformations is presented here, a heuristic description of the underlying motivations is provided.

It has already been noted that the natural variability of ecosystems is likely to lead to stochastic fluctuations about their mean trajectory. It seems natural to allow the magnitude of these fluctuations to set the scale used to evaluate the deviations caused by toxicant exposure. This can be accomplished by standardizing the variances of the state variables about the mean trajectory. However, if the state variables are correlated (which corresponds geometrically to a nonorthogonal set of coordinate axes), our commonsense notion of distance (that is, Euclidean distance) will not provide an accurate measure of the severity of a perturbation. For example, the simultaneous increase of two positively correlated state variables is clearly not as significant as an identical response in two state variables that were originally negatively correlated. A common remedy in multivariate statistical analysis is to calculate Mahalanobis distances which are corrected for the effects of intercorrelation [10].

Both scaling and orthogonality considerations can be accounted for by performing a Mahalanobis transformation on the data. For each state vector $x(t)$ defined by the original variables, the transformed vector, $y(t) = \Sigma^{-1/2} x(t)$, is calculated, where $\Sigma^{-1/2}$ is the symmetric square root (Cholesky decomposition) of the inverse of the variance-covariance matrix for the controls. The transformed control vectors will be statistically independent and standardized to a unit variance in all dimensions. Perturbed state vectors will be scaled relative to the variability exhibited by the controls, and any remaining correlations can be attributed to the effects of the toxicant. It can also be readily demonstrated that Euclidean distances between vectors in the transformed space are equal to the corresponding Mahalanobis distances in the original space.

Experimental Materials and Methods

The state space approach outlined above was applied in the analysis of data on the effects of a coal-derived synthetic oil on aquatic microcosms and ponds. The experimental methods used in this study have been described in detail elsewhere [11,12], so the relevant aspects are only summarized here.

The toxicant used was an unrefined, coal-derived middle distillate from an H-Coal process, identified in the Oak Ridge National Laboratory repository as ACD No. 887. By weight, 12.4% of the oil consisted of water-soluble compounds. Approximately 95% of this water-soluble fraction was composed of phenolic compounds, particularly cresols, dimethylphenols, and other alkylphenols.

Aquatic microcosms were assembled in 72-L glass aquaria, using materials collected from a shallow, 0.04-ha pond. Each microcosm was filled to a depth of about 10 cm with sediment (sieved to remove large debris). Each microcosm also received 55 L of pond water and 100 g (drained wet weight) of the submerged aquatic macrophyte *Elodea canadensis*. A variety of other organisms, including algae, zooplankton, snails, and benthic invertebrates, was also introduced in the assembly process. The microcosms were maintained in a growth chamber under combined fluorescent and incandescent illumination (12 h light:12 h dark), with photosynthetically active radiation ranging from 160 to 215 μEi m^{-2} s^{-1} at the water surface. Air temperature was regulated at 21°C during the light period and at 15°C during the dark period.

In late April 1982, outdoor experimental ponds were assembled in 1-m-deep excavated depressions with sloping sides (5 by 5-m perimeter, 3.5 by 3.5-m bottom) lined with sheets of 0.036-in. reinforced potable-grade Hypolon (DuPont). Fine-grained sediment from a fish pond was placed on the bottom of each experimental pond to a depth of 15 cm. Water from the fish pond was pumped into the ponds to a depth of 80 to 90 cm (about 15 m^3 per pond). One week later 8 L of *Elodea canadensis* from a natural pond was added to each pond. The resulting assemblage of algae, macrophytes, zooplankton, and other invertebrates was typical of a shallow littoral ecosystem and generally similar to that established in the microcosms. On June 8, 35 immature and 4 adult mosquito fish (*Gambusia affinis*) were added to each pond. Fish were not included in the microcosms. Synthetic oil additions began 13 July 1982.

Both microcosms and ponds were subjected to chronic oil contamination over a 56-day exposure period. Duplicate microcosms were randomly assigned to controls or to one of seven treatment levels (M1–M7). Within each treatment level, one microcosm was dosed weekly and one was dosed daily, although the total amount of oil added per week was the same. Ponds were randomly assigned to controls or to one of five treatment levels (P1–P5), with two replicates at each level both dosed daily. Oiling rates in the ponds ranged from 1 to 16 mL m^{-2} d^{-1}, resulting in measured total phenol concentrations ranging from approximately 0.05 to 8 mg/L, averaged over the 56-day exposure period. Oiling rates in the microcosms encompassed a range from 0.07 to 18 mL m^{-2} d^{-1}, resulting in 56-day average total phenol concentrations of 0.01 to 10 mg/L.

The response of a wide variety of physical, chemical, and biological variables was monitored throughout the experiments. For this analysis, a subset of these variables was selected based upon two criteria: (1) only those variables which were routinely measured at weekly intervals during the dosing period were included: and (2) variables with values frequently missing or below detection limits were excluded. The few missing values in this subset were replaced by values interpolated from preceding and subsequent observations. The response variables analyzed from the pond experiment were pH, dissolved oxygen, conductivity, alkalinity, ammonium concentration, chlorophyll *a* in phytoplankton and in periphyton, and total abundance values of cladocerans, copepods, and rotifers. In the microcosm experiment, the same variables were analyzed except for alkalinity and periphyton chlorophyll, which were excluded.

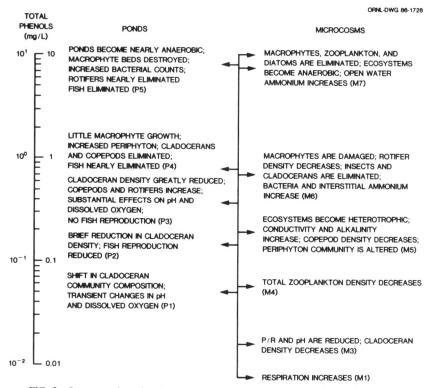

FIG. 2—*Summary of pond and microcosm responses to chronic synthetic oil exposure.*

Results and Discussion

Responses of both ponds and microcosms to the synthetic oil are summarized in Fig. 2; more detailed accounts of the response of individual state variables have been published by the original investigators [11,12]. This paper focuses on the results of an analysis of the data according to the state space approach.

The separation between control and perturbed trajectories provides a quantitative measure of overall ecosystem response. A plot of separation as a function of time and of oil input rate yields a response surface characterizing perturbed behavior for the pond and microcosm ecosystems (Fig. 3). The units on the separation axis are the standardized distance units which result from the Mahalanobis transformation of the data. In a univariate case, these would in fact be standard deviation units, where the standardization is calculated on the pooled sample of observations on the controls. For multivariate data, the interpretation of the numerical values is more subtle, but the calculated separation is an inverse measure of the probability that such a displacement would occur due to chance alone.

For this analysis, daily- and weekly-dosed microcosms at each treatment level were treated as replicates, as justified by previous statistical analysis [11]. Additionally, the transformations of both pond and microcosm data before state space trajectory analysis were based on sample variance-covariance matrices for data pooled over the entire exposure period. This procedure is valid as long as the covariance structure of the state variables was reasonably constant over the 56-day interval. Such assumptions are necessitated by the low degree of replication provided by this

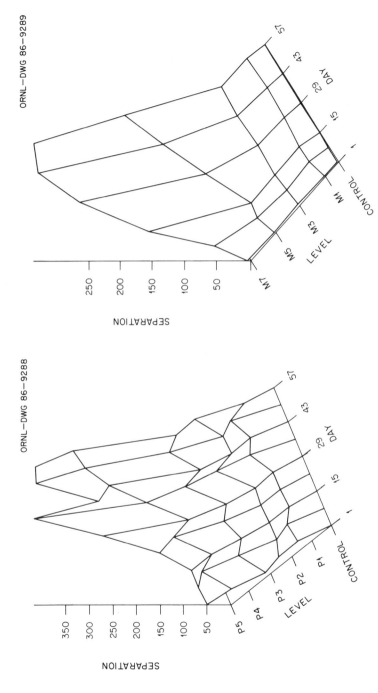

FIG. 3—*Response of (a) ponds and (b) microcosms to toxicant stress as measured by state space separation. Level = treatment level (a geometric progression of oil input rates). Day = day within the 56-day treatment period.*

experimental design but could be relaxed in other situations. All calculations were performed using the Statistical Analysis System (SAS), largely within the MATRIX procedure [*13*].

The response surfaces (Fig. 3) exhibit patterns similar to those seen in classical toxicology based on experiments at the organismal level. Specifically, (1) there is a generally monotonic increase in response with increasing dose or time of exposure; (2) exposure conditions exist below which response is negligible; and (3) there is a suggestion of a response plateau at higher doses or with prolonged exposure. A bivariate probit model was found to give a good fit to the pond data ($R^2 = 0.93$), providing further support of the use of classical toxicological methods in ecosystem analysis. Yet, analogies between ecosystem behavior and the behavior of individual organisms must be employed cautiously, as ecosystems and organisms respond differently to severe perturbations. The state assumed by even a heavily contaminated ecosystem is a dynamic state in which the possibility of future recovery or other change remains and, therefore, is not strictly analogous to organismal death.

The response surfaces do provide a basis for estimating acceptable exposure conditions (that is, conditions under which ecosystem response is below a specified threshold). Determining the value of this threshold involves a subjective judgment regarding the extent of ecosystem perturbation deemed acceptable and is therefore not entirely a scientific question; but, once determined, the exposure conditions which will maintain response below that threshold may be interpolated from the response surface. The use of a response surface based on state space trajectory separation is advantageous because it integrates many aspects of ecosystem behavior and is therefore likely to be more representative than a single state variable.

Dose-response curves were calculated over the 56-day exposure period for each of the three summary variables defined previously (Fig. 4). The time-averaged dose-response curves (Fig. 4) are also analogous to those found in organismal toxicology. A comparison of maximum separation with mean separation indicates that the magnitude of the displacement vector changed substantially over the exposure period, as can be seen in the response surfaces. A comparison of mean separation and mean displacement magnitude, on the other hand, indicates that the direction of the displacement vectors was relatively uniform, especially at the higher doses. It is possible that changes in direction, due to delayed responses and differential recovery rates, would have been observed if the analysis had extended beyond the dosing period.

The relationship of the calculated state space separation to the original variables was assessed using Pearson's correlation coefficient. In the ponds, separation was most highly correlated with conductivity, alkalinity, and cladoceran abundance ($r = 0.84$, 0.82, and -0.74, respectively). At the other extreme, phytoplankton chlorophyll *a*, rotifer abundance, and ammonium concentration showed the least correlation with separation ($|r| < 0.2$). All other variables exhibited moderate correlations ($0.5 < |r| < 0.7$). For the microcosms, correlations were generally higher than for the ponds, with all variables except for phytoplankton chlorophyll *a* having $|r|$ values > 0.7. The highest correlations were found for conductivity ($r = 0.99$), pH ($r = -0.93$), ammonium concentration ($r = 0.88$), and cladoceran abundance ($r = -0.84$). For most variables, the relationship to separation appeared similar in both ponds and microcosms, with the notable exceptions of ammonium concentration and rotifer abundance. Because r is a measure of the linear association between variables, it may be misleading if the true relationship is significantly nonlinear. Such may be the case for rotifer abundance, which increased at moderate doses but decreased at higher doses.

It is worth reviewing the strengths of state space analysis and suggesting certain possibilities for further development. State space analysis offers an objective methodology for quantitative and qualitative investigation of the behavior of ecosystems under toxicant stress. It effectively summarizes large multivariate data sets in a manner which can be given a simple geometric interpretation. The results are directly applicable to the investigation of the principles of ecosystem dynamics, particularly for resolving the components of relative stability.

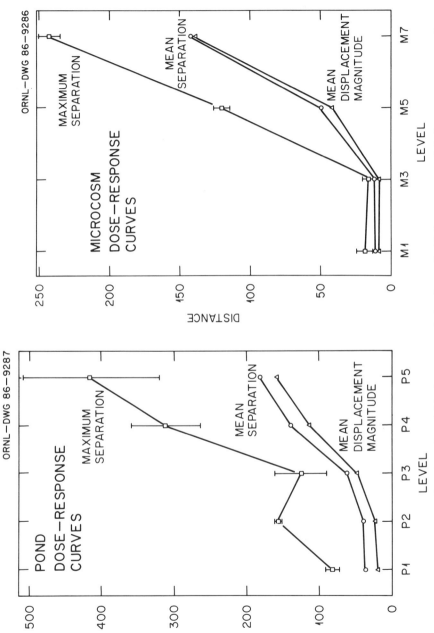

FIG. 4—*Dose-response curves for (a) ponds and (b) microcosms, calculated over the 56-day exposure period. Squares represent maximum separation, circles mean separation, and triangles mean displacement magnitude. Bars indicate the range of maximum separation values from all pairwise comparisons; ranges for the other variables are comparable.*

Following Webster et al. [14], we can define *resistance* as the ability of an ecosystem to resist change in the face of perturbation and *resilience* as the rate of recovery following perturbation. Quantifying resistance and resilience in terms of state space displacements provides a parsimonious description of inherently multidimensional phenomena.

Transformation of the data prior to the calculation of state space displacements is an important aspect of the analysis. In this paper a Mahalanobis transformation was used for two reasons: (1) the transformed data are then uncorrelated, so that a displacement of a given magnitude is associated with a unique probability of occurrence independent of direction; and (2) the transformed data are standardized relative to the variance of the controls. If displacement of the ecosystem along any axis of the state space is equally undesirable, this is indeed the appropriate standardization. In some cases, however, it may be appropriate to tolerate greater relative displacements in some state variables than in others. This situation can easily be incorporated in the analysis by weighting the original state variables relative to their importance.

A final important application of state space analysis is its potential usefulness in identifying diagnostic variables which can be used as indicators of ecosystem response. Following Patten [15], the task of ecosystem management can be thought of as insuring that ecosystem trajectories are restricted to acceptable regions of the state space. Diagnostic variables may then be defined as any subset of state variables such that maintaining the subset within defined bounds is both necessary and sufficient to insure that the trajectory remains in an acceptable region of the higher-dimensional state space. The problem is one of reducing the dimensionality of the system without sacrificing information. Identifying those state variables which correlate highly with the calculated separation is therefore a first step in identifying diagnostic variables. In most cases, ecosystem behavior is unlikely to be reducible to any single diagnostic variable. Stepwise multiple regression could then be used to select a set of diagnostic variables. Identification of diagnostic variables has important implications for both ecosystem management and ecological research, as concentrated monitoring of such variables should maximize the information gained for a given expenditure of effort.

Acknowledgments

I am indebted to all those who worked on the pond and microcosm experiments for providing the data used in this analysis, and especially to J. M. Giddings, with whom I had many productive discussions during the development of this method. S. M. Bartell also provided many useful suggestions as the work progressed. The manuscript has profited from reviews by S. M. Bartell, A. J. Stewart, and three anonymous reviewers. Research was sponsored by the Office of Health and Environmental Research, U. S. Department of Energy, under Contract No. DE-AC05-84OR21400 with Martin Marietta Energy Systems, Inc. This paper is Publication No. 2983, Environmental Sciences Division, Oak Ridge National Laboratory.

References

[1] Zadeh, L. A. and Desoer, C. A., *Linear System Theory; The State Space Approach*, McGraw-Hill, New York, 1963.

[2] Timothy, L. K. and Bora, B. E., *State Space Analysis: An Introduction*, McGraw-Hill, New York, 1968.

[3] Schindler, J. E., Waide, J. B., Waldron, M. C., Hains, J. J., Schreiner, S. R., Freedman, M. L., Benz, S. L., Pettigrew, D. R., Schissel, L. A., and Clark, P. J., in *Microcosms in Ecological Research*, J. P. Geisey, Jr., Ed., Symposium Series 52, U.S. Department of Energy, Washington, DC, 1980, pp. 192–203.

[4] Waide, J. B., Schindler, J. E., Waldron, M. C., Hains, J. J., Schreiner, S. R., Freedman, M. L., Benz, S. L., Pettigrew, D. R., Schissel, L. A., and Clark, P. J., in *Microcosms in Ecological Research*, J. P. Geisey, Jr., Ed., Symposium Series 52, Department of Energy, Washington, DC, 1980, pp. 204–223.

[5] Allen, T. H. F., Bartell, S. M., and Koonce, J. F., *Ecology,* Vol. 55, 1977, pp. 1076–1084.
[6] Gates, M. A., Zimmerman, P. A., Sprules, W. G., and Knoechel, R., *Canadian Journal of Fisheries and Aquatic Sciences*, Vol. 40, 1983, pp. 1752–1760.
[7] Marmorek, D. R. in *Early Biotic Responses to Advancing Lake Acidification*, G. R. Hendrey, Ed., Butterworth Publishing, Boston, 1984, pp. 23–41.
[8] Woltering, D. M. in *Aquatic Toxicology and Hazard Assessment: Sixth Symposium, ASTM STP 802*, American Society for Testing and Materials, Philadelphia, 1983, pp. 153–170.
[9] Efron, B. and Tibshirani, R., *Statistical Science,* Vol. 1, 1986, pp. 54–77.
[10] Mardia, K. V. in *Multivariate Analysis IV,* P. R. Krishnaiah, Ed., North-Holland Publishing, New York, 1977, pp. 495–511.
[11] Franco, P. J., Giddings, J. M., Herbes, S. E., Hook, L. A., Newbold, J. D., Roy, W. K., Southworth, G. R., and Stewart, A. J., *Environmental Toxicology and Chemistry*, Vol. 3, 1984, pp. 447–463.
[12] Giddings, J. M., Franco, P. J., Cushman, R. M., Hook, L. A., Southworth, G. R., and Stewart, A. J., *Environmental Toxicology and Chemistry*, Vol. 3, 1984, pp. 465–488.
[13] *The MATRIX Procedure: Language and Applications*, Technical Report P-135, SAS Institute, Inc., Cary, NC, 1985.
[14] Webster, J. R., Waide, J. B., and Patten, B. C. in *Mineral Cycling in Southeastern Ecosystems*, F. G. Howell, J. B. Gentry and M. H. Smith, Eds., ERDA Symposium Series, CONF-740513, Department of Energy, Washington, DC, 1975, pp. 1–27.
[15] Patten, B. C., *Ecological Modelling*, Vol. 23, 1984, pp. 313–340.

Short-Term Indicators
of Chronic Toxicity

Donald J. Versteeg,[1] *Robert L. Graney,*[2] *and John P. Giesy*[2]

Field Utilization of Clinical Measures for the Assessment of Xenobiotic Stress in Aquatic Organisms

REFERENCE: Versteeg, D. J., Graney, R. L., and Giesy, J. P., **"Field Utilization of Clinical Measures for the Assessment of Xenobiotic Stress in Aquatic Organisms,"** *Aquatic Toxicology and Hazard Assessment: 10th Volume, ASTM STP 971,* W. J. Adams, G. A. Chapman, and W. G. Landis, Eds., American Society for Testing and Materials, Philadelphia, 1988, pp. 289–306.

ABSTRACT: Histological, biochemical, and physiological measures of xenobiotic effects on aquatic organisms have been utilized extensively in laboratory exposures to document toxic effects. In spite of the ability of these measures of stress to integrate the effects of multiple stressors, and their utility to instantaneously assess the "health" of a population, to date few studies have used these methods in situ to document adverse effects of environmental stressors. This is not due to the lack of information on appropriate clinical methods. Sufficient laboratory research has developed clinical measures to the extent that they will be useful in field situations. A portion of the lack of field use of these methods is a lack of understanding of the utility and knowledge in the flexibility of these diagnostic tools. We have prepared a review of the clinical methods and present a rational scheme for the selection and use of these techniques. Examples of the use of these techniques are presented in the form of two case studies. Each case reviews the literature and recommends specific clinical measures which could be used to quantify the population level effects of the stressors involved in the pollution episode. The case studies involve assessment of the effects on aquatic organisms of pollution episodes involving acid rain and heavy metals.

KEYWORDS: fish, invertebrates, biochemistry, histology, acid, heavy metals, metallothionein, stress, enzymes, osmoregulation, glycogen, RNA, DNA

During the past 20 years, toxicological effects on the aquatic environment have been assessed from the suborganismal to the ecosystem level of organization [1,2]. However, research primarily at the population, community, and ecosystem levels are being used to monitor environmental effects, conduct hazard assessments, and make regulatory decisions. Research at these levels of organization is limited because ecologically important effects have already occurred. Research from the cellular to the organismal level has not been utilized extensively to address environmental questions. This is unfortunate since information not available from other methods can be obtained in a timely and cost-effective manner. Unfortunately, current methods and validation of methods are not sufficiently advanced to allow extensive use of these methods in hazard assessment and regulatory toxicology. Clinical methods can, however, be used more extensively to detect environmental effects.

Clinical measures or indicators of stress, developed for use with aquatic organisms, allow rapid

---

[1] Research scientist, Environmental Safety Department, The Procter and Gamble Co., Ivorydale Technical Center, Cincinnati, OH 45217.

[2] Graduate student and professor, respectively, Pesticide Research Center, Michigan State University, E. Lansing, MI 48824.

determination of the health of an organism. Although the population, community, and ecosystem are the important levels at which to monitor toxic effects, toxic effects are manifested at the organism level by impaired biological function. Clinical measures of stress are tools to monitor biological function. They are quantifiable biochemical, functional, behavioral, histological, or physiological measures which relate in a dose or time-dependent manner the degree of dysfunction the stressor has produced.

This paper concentrates on methods useful in the field to assess the health of an organism or population. We discuss the rationale for using clinical measures of stress in the field, review problems associated with the field use of clinical measures of effects, review the response of aquatic organisms to stressors, discuss experimental considerations in the design of field exposures, and present two case histories describing situations where clinical measures of stress can be utilized to understand the impact of contaminants on aquatic organisms. This paper will not discuss functional or behavioral tests as these tests are difficult to use to assess effects in situ. These tests include predator-prey, seawater challenge, swimming speed/stamina, and toxicant avoidance [3].

Rationale for Use of Clinical Methods in the Field

Clinical methods have several advantages over studies at other levels of organization which makes them uniquely suited to address current environmental issues. Stress begins at the molecular/cellular level and extends to tissues and organs [4]. Important ecologically relevant adverse effects of xenobiotics on the organism are the culmination of effects on biochemical and physiological processes and occur before changes in population or ecosystem level parameters. Measurement of effects at the biochemical, histological, and physiological levels of organization can be quantified antecedent to population or ecosystem level effects and allows remedial action to be initiated at the earliest date. The relative ease and speed of some clinical tests and the fact that general indicators of xenobiotic stress integrate all the stressors to which an organism is exposed make these procedures useful in determining synergism or antagonism of xenobiotics and the influence of accessory environmental factors on toxicity. The advantages of clinical measures suggest that well-conceived and executed research be funded to address the problems currently restricting use of clinical measures in the field.

Problems with Using Clinical Measures of Stress

Numerous clinical measures of stress have been developed and utilized successfully to quantify effects of toxicants on aquatic organisms in the laboratory; however, these techniques have not been used to address important environmental (that is, field) problems. The field use of biochemical, histological, and physiological measures of xenobiotic effects has four main problems:

1. The inability to relate a response in a clinical measure with important, ecologically relevant effects [5].
2. Variability in the measurement of the response at this level of organization.
3. Research attention has been focused on developing new methods, not on improving and validating existing methods so that they can be used in field situations.
4. Inability to discriminate the acclimation response from the toxic effect of a stressor.

To apply clinical measures of toxic effects in the field, the most critical need is for a correlation between the clinical measure and an important effect on the population, community, or ecosystem. This will be possible when toxicological tests correlate effects on clinical measures with effects on growth, reproduction, and survival. These tests can be conducted in the laboratory by extending exposures to measure one or all of these effects. Application of this information will allow prediction of ecosystem level effects by monitoring biochemical and physiological indicators.

Variability in the measurement of responses to environmental contaminants is a problem throughout aquatic toxicology. This variability has three sources: organismal, environmental (including the effects of the toxicant), and methodological. Some researchers believe that biological variability, a combination of variability due to the organism and the environment, is the greatest source of variability with clinical measures and that this variability masks treatment effects. This assumption is not always valid. A thorough assessment of variability will enable environmental and methods sources of variability to be reduced. Appropriate statistical design can further decrease the remaining variability. In fact, some clinical measures may be unacceptable due to high variability. Variability can be successfully addressed with proper selection of the clinical measures, research on experimental protocols and factors affecting variability, and appropriate statistical methods [6].

The third problem with the use of clinical measures in the field is the emphasis on developing new methods prior to field validation of existing methods. The utility of many clinical measures in a field situation is unknown due to a lack of field testing. Field research is more difficult than laboratory research; however, environmental relevance should be of greater concern to the scientist. We must apply our procedures to field situations and either validate or reject each technique in this arena. If a clinical method is validated in the field or fails field validation, the results must be communicated to other scientists.

The last problem with using clinical measures in the field is the difficulty in discriminating the acclimation or homeostatic response from the toxic effect. Xenobiotic-induced alterations at the suborganismal level of organization are not biologically relevant if the changes are fully compensated for by the organism and do not result in reduced reproductive performance [7]. Therefore, it is critical to differentiate between an acclimation response and a toxic response (Fig. 1). Certain specific alterations in an animal's physiology occur after stimulus by natural environmental stressors, exposure to nontoxic "doses" (exposure duration × concentration) of toxicants, and exposure to toxic doses. The first two responses do not result in serious effects on the individual or population and must be differentiated from the third. Biochemical and physiological changes caused by environmental cues and low doses of toxicants have survival value to the organism and are considered acclimation. The difficulty in discriminating acclimation response from the toxic response may have several causes (Fig. 1). Exposure to a low level of a stressor can cause an increase in a biochemical parameter without a change in the population level response due to the homeostatic capacity or acclimation response of the organism. As the stressor is increased so that toxicity occurs, the biochemical parameter may increase, as the organism attempts further acclimation, or decrease, representing exhaustion of the acclimation response. If the biochemical response increases as toxicity occurs, it may be difficult to statistically discriminate between a response in the acclimation range and a slightly increased response in the toxic range. It may, however, be possible to distinguish between the "normal" state and the acclimated state. If the biochemical parameter decreases as toxicity occurs, there will be levels of the biochemical parameter within the acclimation and toxic ranges which are equal. Resolution of this dilemma will involve appropriate selection of the clinical measure.

Response of an Aquatic Organism to Stress

Exposure of an aquatic organism to a xenobiotic can cause several different effects. The compound may directly and specifically affect an enzyme or tissue or nonspecifically elicit the general stress response. If the locus of effect is important for growth, reproduction, or survival, or if the effect reduces energy or nutrient flow for these processes, then a toxic effect may be observed. In the case of a direct-acting compound, clinical effects might include increase in a

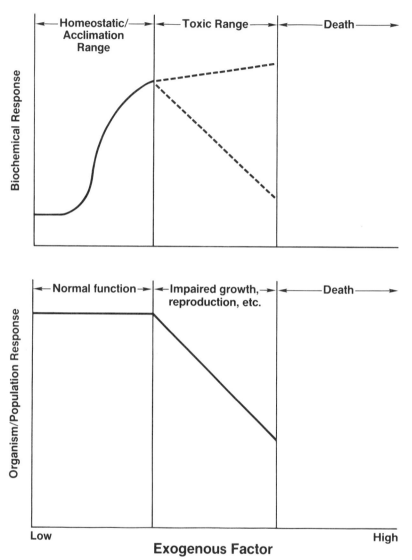

FIG. 1—*A comparison of the biochemical and organismal response of an organism to a stressor, demonstrating the difficulty in discriminating between a homeostatic response of a biochemical parameter and a "toxic" response.*

macromolecule or substrate [8,9], inhibition of enzyme activity [10], or a hyperplastic tissue response [11]. Even though the effect occurs directly on some critical enzyme or tissue, there can be a large acclimation response by the organism.

In addition to the direct effects of the compound, the compound may elicit the stress response (Fig. 2). Currently, little is known about the stress response in invertebrates. In higher vertebrates, the response is mediated through the hypothalamus-pituitary-interrenal axis and the hypothalamus-chromaffin tissue axis [12–14]. Stressors, in general, elicit a cascade of homeostatic alterations in

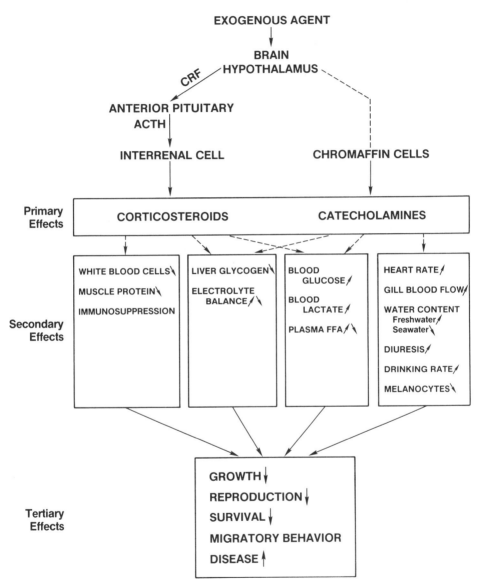

FIG. 2—*The physiological, biochemical, and population level responses of organisms to stress, CRF corticotropic releasing hormone, ACTH adrenocorticotropic hormone, and FFA free fatty acid (modified from Mazeuad et al. [14]).*

critical body systems. These responses usually improve the organism's capacity to survive during stress in the short term. If the stressor is persistent, adverse effects on the organism can occur due to stress-induced changes in the biochemistry and physiology of the organism. This cascade of compensatory alterations has been divided into three levels: primary, secondary, and tertiary

(Fig. 2) [3,14]. Stress, whether due to toxicants or environmental/physical factors, elicits these responses.

Attempts have been made to measure primary, secondary, and tertiary responses as indicators of toxicant effects. Measurement of primary responses has not been applicable to the field situation as primary responses are part of the acute response to most "stressors," including capture and handling stressors. Secondary level measures are less affected by sampling and capture. For example, hematological changes due to cortisol exposure do not occur until days after blood cortisol levels have been increased [15,16]. However, with some measures of the secondary stress response, it is difficult to discriminate acclimation from toxic effects. For these reasons, extrapolation of effects on the primary and secondary responses from the organism to the population is currently difficult, unless the effect noted will influence reproduction or growth. Liver glycogen, lipid, RNA/DNA ratio, adenylate energy charge (AEC), and protein production rates are potentially useful measures as effects on these measures indicate potential future effects on reproduction and growth.

Clinical Measures

Clinical measures indicative of a nonspecific response to stress include any measure which is altered by exposure to a variety of stressors. These nonspecific clinical measures, glycogen, RNA/DNA, radio-labelled amino acid or nucleotide incorporation, and AEC, give direct information as to the growth rate or potential of an organism. Nonspecific clinical measures of stress can be used to integrate the simultaneous impacts of multiple toxicants or environmental factors on the organism since all types of stressors can affect these endpoints. These measures cannot be used to identify the toxicant causing an effect. However, in many environmental situations, multiple stressors are present which cause effects in an interactive manner.

Specific indicators of stress are of two types, organ specific and toxicant specific. Organ-specific measures include organ function tests, organ-specific enzymes, histopathology, and isoenzymes. Organ function tests include p-amino hippurate uptake by the kidney [2] and bromosulfothalein clearance by the liver [17]. To date, these tests have not been used extensively in aquatic toxicology; however, they are potentially useful for understanding toxicant-induced alterations in organ function.

Histopathology allows the detection of multiple effects, including cell proliferation, cell death, and infectious and cancerous responses. This approach has been used extensively in the laboratory as well as the field [11,18,19].

Certain enzymes are located at increased concentrations in a few organs. These enzymes appear in blood when those organs are damaged and are indicative of the presence and extent of damage. Lactate dehydrogenase, the transaminases, creatinine phosphokinase, and alkaline phosphatase are examples of enzymes used as organ-specfic indicators of toxic effects. Organ-specific isozymes such as lactate dehydrogenase and creatinine phosphokinase can be used to detect organ-related effects in fish. Relatively few studies have used these methods to identify the specific organ damaged by a toxicant [20,21].

Toxicant-specific measures involve quantification of specific enzyme activities or biomolecules, which when found in a tissue indicate exposure and possibly effects due to one or a related group of compounds. Examples include cytochrome P_{450}-monooxygenase [22], metallothionein and metal binding proteins [8,23], acetylcholinesterase [9], and amino levulinic acid dehydratase (ALAD) [24]. Toxicant-specific indicators have been correlated with exposure of aquatic organisms to environmental pollutants in the laboratory and the field. To date, however, alterations in these clinical measures have shown poor correlation with toxic effects. Therefore, we do not currently recommend the use of these measures to quantify effects.

Methods

Methods for clinical measures in aquatic organisms were largely derived from mammalian medicine. Few methods have been optimized for use on aquatic organisms. This has made comparison of results generated with different methods difficult. Further, use of the mammalian interpretation of a test result has led to inaccurate assessments of the health of an organism [25]. The use of clinical measures in aquatic toxicology presents several unique problems which include variability and selection of the test organism.

Results obtained with samples taken from the field are usually more variable than those taken from the laboratory [26]. To increase the effective use of clinical measures in the field, we must understand the affects of accessory environmental factors as well as sampling and handling to a greater extent [19,21,27].

Selection of the organism is critical in developing a monitoring program based on clinical measures. Factors to be considered include: (1) exposure of caged versus sampling of indigenous organisms; (2) knowledge of organisms' physiology and biochemistry; (3) size; (4) ease of sampling, which includes availability of sufficient numbers and aged organisms; (5) trophic level; (6) sensitivity; and (7) importance in the ecosystem. Both fish and invertebrates can be used successfully, each with advantages and disadvantages. Fish have received the greatest amount of physiological and biochemical research; however, invertebrates may be more important in many ecosystems and are easier to conduct cage studies with. Cage or transplant studies give a high degree of control to the exposure and allow direct interpretation of the effect of a treatment, but cannot take into account the effects of long-term adaptation or acclimation to current environmental conditions. In sampling indigenous biota, care must be taken to select appropriate control stations. Even then, adaptation or acclimation-associated alterations in a given clinical measure may cause statistically but not biologically significant changes in a parameter. This process may not occur in all parameters in a monitoring program; measurement of a variety of parameters coupled with a large background data base should enable the identification of a stressed population.

Criteria for Selecting the Clinical Measure for Use in the Field

Choice of the appropriate biochemical parameters for monitoring toxic effects in aquatic organisms requires consideration of a variety of factors. For a clinical measure to be useful in the field, we believe it should fulfill six criteria.

1. The clinical measure must respond to the toxic component of interest.

2. The clinical measure should be relatively easy to measure to allow quantification from a number of individuals so that a population estimate can be made. Parameters which are expensive, time consuming, and require sophisticated expertise to measure will not be useful in quantifying effects on field populations.

3. The measure should respond in a dose or time-dependent manner to the toxicant so that the magnitude of the toxic effect can be determined. This will aid in the extrapolation to ecologically relevant effects.

4. The clinical measure should be sensitive. There is no need to develop measures of acutely lethal concentrations of a compound.

5. The variability due to accessory factors should be understood and within acceptable limits.

6. The measure must have a biological significance. This is a major reason why we advocate the use of general measures of toxicity like liver glycogen or radiolabelled amino acid incorporation.

Some of these criteria cannot be fulfilled, even for a clinical measure with a relatively large data base. However, being unable to fulfill these criteria does not eliminate that measure from consideration, but rather identifies areas of weakness which need to be considered during study planning, implementation, and data interpretation.

Acidification

Acidification of freshwater ecosystems via acid precipitation has become a serious environmental problem in recent years [28,29]. Throughout northeastern United States and Canada, thousands of lakes are suspected of being affected by acid precipitation. Industrial gaseous emissions containing sulfuric acid (H_2SO_4) and nitric acid (HNO_3) are considered to be the primary sources of low pH precipitation [30].

The primary toxic component of acidified freshwater habitats will vary both temporally and spatially. Under most circumstances, increased acidity caused primarily by increased H^+ ion input and effects on aluminum speciation and concentration are the main causes of toxicity. The susceptibility of a specific ecosystem to acidification depends upon the buffering capacity of the waters. Waters with high alkalinity will require a greater H^+ ion input before a significant decrease in pH is observed.

Increased concentrations of heavy metals can be an important toxic component of lake acidification. Although aluminum has received the greatest attention, other metals found to be increased in acidified lakes include cadmium, copper, lead, manganese, nickel, and zinc [29,31]. The effect of pH on metal mobilization is exacerbated by the increased percent of metal in the toxic, ionic form. The relative contribution of metals to the total toxicity of acidified systems can be important and needs to be considered when choosing a clinical measure for monitoring.

Toxic Effects

Selection of indicators most appropriate for monitoring acid stress requires a thorough understanding of these effects. A brief review of selected effects is provided to demonstrate the use of certain parameters as stress indicators. Specific effects to be discussed include: (1) ionoregulation; (2) bone mineralization; (3) respiration; (4) acid-base regulation; (5) histopathology; (6) enzymes and substrates; and (7) energy metabolism.

Freshwater organisms are hyperosmotic regulators maintaining a higher osmolyte concentration in their tissues relative to the ambient water. Maintenance of constant internal ion (sodium, chloride, calcium, magnesium, etc.) concentrations are essential, requiring active regulation of water influx and ion efflux. Acid exposure inhibits ionoregulation through inhibition of ion uptake and increases in ion efflux [6,32–36]. Alterations in Ca^{2+} metabolism have also been observed following chronic acid exposure [37]. Calcium is an important component of bone, and alterations in Ca^{2+} regulation may be related to the high frequency of skeletal deformities often associated with acid stress [37]. Bone demineralization has been proposed as a buffering mechanism used by organisms to counteract acidosis [29], although Lockhart and Lutz [38] found no evidence of bone demineralization in acid-stressed white suckers. Further, water Ca^{2+} concentrations affect blood Ca^{2+} levels, decreasing the potential utility of this indicator of acid effects in fish [37].

Utilization of blood ion levels to discriminate acid from metal effects on fish will be difficult due to the interference of metals with ionoregulation. Calcium metabolism in rainbow trout was altered by cadmium [39], and decreased plasma osmolytes have been reported in fish exposed to copper, aluminum, and zinc [40,41].

Acid exposure reduces gas exchange at the gill through direct gill damage or increased mucus secretions [42–46]. Direct gill damage can be assessed histologically and mucus production can be quantified biochemically by measuring sialic acid, a major component of gill epithelial mucus [47]. Increased mucus production has also been observed in fish exposed to hypercapnic stress and heavy metals [48,49]. Decreased blood pH (acidosis), along with the increased blood carbon dioxide concentrations, can reduce the blood oxygen transport capacity via the Bohr and Root effects. These effects cannot be directly measured, but they do induce hematological changes including increases in hematocrit, hemoglobin, and red blood cell (RBC) count in acid-stressed fish [50].

Documentation of histopathological lesions caused by acid exposure is limited. Daye and Garside [44,51] observed lesions in the gills of salmonids exposed to acid stress including hypertrophy of the mucus cells and necrosis and sloughing of the epithelial layer. Similar damage to gill epithelia was observed in trout exposed to acute acid stress [52]. Many of the histological changes caused by acid exposure are similar to those which occur during metal exposure [53]. Although histological alterations of gill tissue are known to occur in acid and metal-exposed organisms, it can be difficult to incorporate this parameter into a general monitoring program. Due to gill lamellar recruitment, significant damage can occur to the gills before the overall health of the organism is impaired. Histopathological assessments of pollutant stress can be expensive and time consuming. In addition, the qualitative nature of this parameter makes extrapolation of histological effects to higher levels of organization difficult [54]. Despite these limitations, measurement of structural alterations can provide useful information on the organs affected and the mode of toxic action.

The activity of a variety of enzymes has also been shown to decrease with decreasing pH exposure. Membrane-bound gill Na^+/K^+-adenosine triphosphatase (ATPase) activity is reduced in salmonids exposed to acid water and aluminum-containing acid water [55,56]. Other enzymes influenced by acid stress include carbonic anhydrase [56], chorionidase [57], serum aspartate aminotransferase [6], and fructose 1,6-bisphosphatase [58]. Many of these enzymes are also influenced by metal exposure [21,59]. Thus, discrimination of metal from acid effects based on enzymes will be very difficult.

A large number of other physiological/biochemical alterations have been observed in aquatic organisms exposed to acid stress. Changes in energy metabolism include increases in plasma glucose concentrations [6,60,61], alterations in tissue glycogen levels [60–62], and decreases in tissue AEC [61,63]. The majority of these changes reflected the stress-induced mobilization of energy reserves. Hormonal responses have also been measured in acid-stressed fish [6,50,64] but, in general, have not been useful indicators of chronic stress in aquatic organisms.

Clinical Indicators of Acid Stress

Changes in the osmolyte (Na^+ and Cl^-) concentration of blood and/or tissue can be indicative of acid stress and has potential as an in situ clinical measure. Since both metals and pH alter the concentrations of Na^+ and/or Cl^-, this parameter cannot be used to differentiate between these two components. Ion concentrations are easily measured, and considerable confidence in the numbers can be attained. Both acute lethality and long-term, mild acid stress have been related in a dose-dependent manner with altered ionoregulation [32,65,66].

Lee et al. [60] discussed changes in osmolyte concentrations with respect to zones of tolerance, resistance, regulation, and reproductive success. Based on these zones, reproductive stress lies in the tolerance zone where partial osmolyte regulation is possible. Thus, changes in osmolyte levels would not be dependably predictive of reproductive effects. In addition, physiological adaptation can occur during mild acid stress such that osmolyte concentrations eventually return to normal or new steady state concentrations are attained [66]. Thus, osmoregulation may not be a sensitive measure of reproductive impairment in acid-exposed fish.

Calcium regulation can also be affected during acid stress and should be considered as a potential clinical measure. Serum concentrations in female fish may represent a sensitive sublethal indicator of acid stress [29,37]. We recommend the measurement of serum Ca^{2+} concentration as a clinical measure. In addition, recent evidence indicates that the mechanism of aluminum toxicity may involve interference with the intracellular Ca^{2+} regulating protein calmodulin [67]. The increased aluminum concentrations often associated with acidification may interfere with Ca^{2+} metabolism and be responsible for a portion of the observed alterations.

Since the gills are the primary site of toxicity in acid-stressed organisms, parameters which assess gill damage have potential as clinical indicators of acid stress. Mucus secretion is a protective reaction in response to a variety of stressors, including acid, metals, and hypercapnia [28]. The pH threshold for increased mucus production appears to be pH 5.6 [43]. This would seem to represent greater sensitivity than changes in osmolyte concentrations and is close to the zone of reproductive failure identified by Lee et al. [60]. Measurement of gill sialic acid content as an indicator of mucus secretion has potential as a clinical measure of acid exposure in fish.

The decreased oxygen availability associated with acid stress can cause compensatory hematological changes which maintain the oxygen carrying capacity of the blood. Hematocrit (Ht) is a general indicator responding to a variety of toxicants [15]. It is easily measured and, in combination with ionoregulatory and mucus secretion data, can give a good indication of whether gill damage caused hypoxic stress. Presently, specific data on sensitivity to chronic exposure to low pH are not available as the majority of the work has been conducted with fish exposed to acutely toxic acid levels. However, since rather mild acid stress can upset acid-base balance, Ht may be altered by sublethal acid exposure. The interpretive problems associated with acclimation are compounded as changes in Ht may represent a physiological adaptation to acid stress. Seasonal changes have been noted in Ht [68]. Since handling stress can cause hemodilution and possibly affect Ht, this must be considered during sampling. Overall, there are problems with using Ht as a clinical measure; however, in combination with other parameters, it may be a useful indicator of acid stress.

Measurement of parameters involved with acid-base regulation, such as blood pH, and the concentration of blood gases, such as carbon dioxide (CO_2) and oxygen (O_2) are not recommended. These parameters may represent the most sensitive indicators of acid exposure; however, they are also the most difficult to measure and have the greatest degree of variability due to sampling.

The activity of gill Na^+/K^+-ATPase decreases during acid exposure and may represent a useful clinical measure. This enzyme is easily measured and is inhibited by both acid stress and heavy metal exposure. The relative sensitivity of the enzyme to acid stress is unknown; however, it has been correlated with decreased blood ion composition [61]. The ecological significance of changes in ATPase activity is difficult to assess due to the large reserve capacity of ATPase. This being the case, measurement of changes in enzyme activity may not provide information any more useful than osmolyte concentrations.

In addition to the above "specific" indicators of acidification, we feel that measurement of a "nonspecific" indicator of the relative health status of the organism would enhance data interpretation. Many nonspecific indicators have been proposed. Choice of the appropriate indicator will depend upon the advantages and disadvantages of each and the question addressed. In the present situation, a primary criteria is that it be indicative of long-term, sublethal stress. Parameters indicative of changes in energy metabolism have been used extensively and have been shown to respond to acid stress. For example, changes in glucose and glycogen concentration during acid stress have been related to alterations in carbohydrate metabolism. Plasma glucose levels might be of limited use in a field situation due to high variability; however, tissue glycogen levels may reflect long-term stress caused by increased glycogen mobilization or decreased synthesis [9]. It is recommended that tissue glycogen levels, or a similar parameter, be included as a clinical measure.

In summary, any monitoring program using biochemical endpoints as indicators of environmental stress will require the measurement of a variety of parameters. There is no single parameter responsive to all the potential stressors in an acidified ecosystem. We recommend measurement of serum Na^+, Cl^-, and Ca^{2+} concentrations, gill sialic acid levels as an indicator of mucus secretion, and hematocrit. Many of these clinical measures have been shown to vary seasonally

[68]. However, information on seasonal effects can be addressed with appropriate experimental and statistical designs. A general indicator of stress, such as tissue glycogen levels, should also be measured. The success of these measurements as an indicator of acid stress will depend on the existence of appropriate reference stations and a strong background data base.

Heavy Metals

Trace metals reach the environment through a variety of routes: acid leachates of soil, smelting, electroplating, power generation, mining, consumer product, and pigment industries [69–71]. In the aquatic environment, metals exist in a variety of forms, including ionic, hydrated, sorbed, as salts, and complexed with a variety of naturally occurring organic acids [71]. Generally, it is considered the ionic form of the metal which is toxic [72]; however, it now appears that water calcium concentration affects metal uptake and toxicity [73–76]. A number of factors reduce the amount of ionic metals in solution, including pH, suspended particles, other complexing materials, and the redox potential of the water [71,76].

Toxic Effects

Although heavy metal toxicity has been extensively studied in aquatic organisms, the specific biochemical, histological, or physiological lesion which causes chronic health affects has not been identified for any of the metals. Histopathological assessments of fish chronically exposed to metals have failed to demonstrate this technique as a sensitive indicator of metal-induced toxic effects [11,77]. Acute metal toxicity appears to be due to gill damage and hypoxia. Chronic metal exposure causes a number of effects, including alterations in enzyme activities, changes in energy substrates, osmoregulatory changes, and induced metal-binding protein levels. These effects can be and have been used to quantify the effects of metals in the laboratory, but very few attempts have been made to utilize these methods in the field. With the lack of information of the specific lesions involved in heavy metal toxicity, it has been difficult to correlate effects on clinical measures with ecologically relevant effects on the population.

Alterations in internal ion concentrations due to metal exposure occur through a number of physiological mechanisms, making blood ion levels sensitive indicators of metal effects. Osmoregulatory alterations can result from hormonally mediated response to metals [14], gill damage, and effects on the ion transporting enzyme Na^+/K^+-ATPase. Acute exposure of fish in freshwater to heavy metals usually results in decreased plasma ion levels [56,78]. In addition, long-term exposure of freshwater and marine fish causes altered plasma ion composition [59,79–81]. These studies support the use of blood ion levels for the quantification of toxic effects of metals. However, during long-term exposures of fish to metals, osmoregulation is not always impaired. Initial effects may be transient and ion concentrations can return to normal values [82–84]. Therefore, under field conditions where the source term and the hormonal impact on ionoregulation are unknown, concentrations of blood ions will not be effective measures of metal-induced toxicity. Extrapolation of effects on ionoregulation to effects on growth and reproduction will be difficult as studies correlating alterations in ion levels with measures of chronic toxicity have not been conducted. Finally, many factors affect osmoregulation, decreasing the utility of osmoregulatory status as an indicator of metal effects. Little research has been conducted on the effects of metals on invertebrate hemolymph ion concentrations.

Mg^{++} activated, Na^+/K^+-adenosine triphosphatase (Na^+/K^+-ATPase) is a membrane bound protein which transports Na^+ and K^+ ions across the cell using the energy gained from the conversion of ATP to ADP [85]. This enzyme is primarily responsible for maintaining the transmembrane Na^+ and K^+ gradients. Effects on enzyme activity may occur due to a reduction in ATP availability, or a direct toxic effect on the enzyme. Exposure of Na^+/K^+-ATPase to a variety of metals in vitro inhibits this enzyme [86,87]. In vivo, acute exposures of aquatic organisms to metals have resulted in variable effects on gill Na^+/K^+-ATPase. Acute exposure of the rainbow trout to aluminum decreased Na^+/K^+-ATPase activity [56], while exposure to mercury or methylmercury did not alter gill Na^+/K^+-ATPase [78]. Cadmium exposure of the American lobster [86] and zinc exposure of trout [84] and salmon [88] have all resulted in increased ATPase activity in gill tissue. Due to this variability in response and the inability to extrapolate effects on ATPase to the population, we do not currently recommend Na^+/K^+-ATPase as a clinical measure to quantify metal effects in the field.

Serum enzymes indicative of specific organ damage, LDH and the transaminases, have not been shown to be altered by sublethal exposure of fish to metals [11,89]. Serum enzyme activities which have been used successfully as indicators of metal toxicity include several enzymes of lysosomal origin. Acid phosphatase (ACP) and N-acetyl-B,D-glucosaminidase (NAG) are increased in the serum of fish exposed to metals [11]. Bouck [4] has found plasma concentrations of the lysosomal enzyme leucine amino napthylamidase (LAN) to be useful for quantifying tissue damage in fish. Lysosomes contain increased concentrations of metals and may be active in the degradation of metal binding proteins [90,91]. Metals have also been shown to increase lysosome numbers and reduce lysosomal membrane stability, possibly leading to enzyme leakage [11,21,92,93]. In general, increased serum lysosomal enzymes appear to be indicative of metal toxicity and may be due to increased enzyme production, decreased lysosomal stability, or tissue damage. At present, much additional research is needed to understand the mechanism of lysosomal enzyme appearance in serum and to correlate effects on blood enzyme levels with reproduction or growth.

ALAD (delta amino levulinic acid dehydratase) is a cytosolic enzyme found in many tissues and is active in the synthesis of hemoglobin, catalyzing the formation of porphobilinogen, a precursor of heme [94]. Lead exposure causes a dose-dependent decrease in erythrocyte ALAD activity in fish through direct inhibition of the enzyme [95]. With long-term exposure, compensatory mechanisms do not increase ALAD activity [96,97]. Blood lead concentrations are directly correlated with ALAD inhibition [98,99]. Because toxic lead concentrations based on ALAD inhibition are similar to laboratory-derived MATC values, Hodson et al. [98] advocated this method to relate exposure and toxicity. However, ALAD is not the rate-limiting enzyme; therefore, ALAD inhibition does not greatly affect heme synthesis. Thus, toxic effects of lead on fish have not been causally related to ALAD inhibition, and it is speculated that compensatory mechanisms in fish will counteract whatever effect lead has on hemoglobin production [95]. Field studies have demonstrated ALAD inhibition in fish from lead-polluted lakes. However, there were no other signs of toxicity, indicating that ALAD is a good monitor of lead exposure but not of lead toxicity [100]. Thus, we currently recommended that ALAD not be used to quantify lead *effects* in the field.

The quantification of ribonucleic acid (RNA) and deoxyribonucleic acid (DNA) have been used to assess growth rates in aquatic organisms [101–103]. RNA content is directly related to protein synthesis, while DNA content per cell is fairly constant. Normalization of the RNA content to DNA reduces the influence of cell number and size and is believed to reflect growth rate. Thus, the RNA:DNA ratio is an integrative measure of stress in that it is sensitive to toxicants which directly reduce growth and toxicants which reduce the energy expenditure on growth. MATCs determined with early life-stage tests and nucleotide measurements made after 96-h exposure to a variety of toxicants agreed well. Kearns and Atchinson [104] correlated RNA/DNA ratios and growth rates in yellow perch with cadmium and zinc pollution in a lake. Whole fish cadmium concentrations were significantly negatively correlated with RNA/DNA ratios. Feeding has a large effect on nucleotide measurement in fish [105,106]. Thus, the RNA:DNA ratio must be carefully controlled to avoid misleading results, but it may be useful in determining effects of metals on growth rates.

Glycogen is a branched chain polymer of glucose residues and represents the readily mobilizable storage form of glucose for most organisms. During toxicity, the increased energy demand associated with stress and the hormonal-induced hyperglycemia results in depletion of glycogen stores [107,108]. Glycogen content of fish is affected by both acute and chronic in vivo exposure to metals. Chronic exposure to a single metal or a mixture of metals generally causes a decrease in liver, brain, and skeletal muscle glycogen concentrations [109–112]. However, in a number of exposures of fish to metals, tissue glycogen levels either remained constant [113] or increased [114]. These results can be explained by an affect of the metal on insulin production, alterations in endocrine control over carbohydrate metabolism [115], or the lack of a toxic metal exposure in these studies. We recommend the use of glycogen in liver or hepatopancreas as a monitor of chronic toxicity due to metals. This clinical measure will be readily correlated with important effects on the population, is easily quantified, and integrates the effects of other environmental stressors. Effects due to sex, season, and nutrition will have to be controlled. It is important to remember that a decrease in body glycogen does not necessarily indicate a decrease in whole-body energy reserves since lipid and cholesterol concentrations must be considered [115]. Research into whole-body calorimetry may have the greatest potential of success as an indicator of toxicity [116].

Metal Binding Proteins

Metal binding proteins (MBP) are cytosolic proteins which are associated with metals and function in modifying the availability of metals. Metallothionein, chelatins, and probably numerous other proteins within the cell are specific MBP [117]. Currently, terminology on these proteins is not uniform and it is difficult to distinguish among the various types of MBPs in the literature. In fact, some research has questioned the formation of MBPs during metal exposure through the water column [118–120].

MBPs are cytosolic proteins which are induced by, and bind, the metals copper, cadmium, silver, mercury, and zinc [121]. MBPs function in the uptake, subcellular compartmentalization and transport of the essential metals, copper and zinc [122]. MBPs are active in the sequestration and detoxification of the toxic metals, cadmium and mercury, and high concentrations of the essential metals, copper and zinc [123–125]. Finally, there is evidence that some MBPs may have another function not directly involved in metal regulation. Metallothionein (Mt) is an efficient scavenger of hydroxyl radicals, and it may be this vital function which is conserved in nature [126].

Metal Binding Proteins as Indicators of Toxicity

MBPs have been proposed as indicators of metal exposure and toxicity for the following reasons [127]:

1. MBP production occurs in a wide variety of organisms after exposure to increased concentrations of cadmium, copper, mercury, and zinc. This makes MBPs a specific indicator of exposure to these metals.
2. MBP production occurs rapidly following exposure.
3. MBPs are induced by environmentally relevant concentrations of metals.

For any biochemical response to be accepted as an index of sublethal stress, it must lead to, or be associated with, a decrease in animal performance [127]. The spillover hypothesis proposes a mechanism for metal-induced toxicity. It states that MBPs will bind and sequester metals until the binding capacity is filled. Then metals spill over onto other binding sites (for example, proteins), compromising the function of the proteins and causing toxicity [9,128–130]. Knowledge of the maximum amount of MBP which could be produced and the measurement of MBP levels could be used to estimate the reserve capacity of the cell and serve as an index of toxicity. Evidence for this hypothesis has been mixed.

Fish and zooplankton were exposed to mercury until MBP levels reached a maximum. Exposure to increased concentrations of mercury caused the appearance of mercury in the high-molecular-weight (HMW), enzyme-containing, and low-molecular-weight (lMW) subcellular fractions [129]. The concentration of mercury appearing in the HMW fraction of the cell was related to mercury-induced pathology in fish. In other studies metal exposure has been observed to induce a maximum level of MBP synthesis. With continued exposure, metals are observed to "spillover" into the HMW pool [8,131]. In neither of these last two studies was metal binding in the HMW fraction associated with toxic effects. This suggests that spillover can occur without toxicity.

Other research does not support the spillover hypothesis. Viarengo et al. [132] sampled mussels from a control and metal-polluted environment. Tissue metal concentrations indicated increased exposure to copper and mercury in the contaminated area. Biochemical measures indicated toxicity in the mussels collected from the polluted area, while cytosolic metal distribution in the digestive gland demonstrated the presence of copper only in the MBP fraction. In a study on the crab (*Rhithropanopeus harrisii*) exposed to copper, reduction in growth was correlated with metal associated with the MBP and IMW fractions and not with the HMW fraction [133,134]. In coho salmon (*Oncorhynchus kisutch*) exposed to 0.25 and 0.5 times the LC_{50} concentration of copper for 14 weeks, copper concentrations were increased in the MBP fraction and eventually in the HMW fraction. However, the appearance of copper in the HMW fraction was not associated with increased copper sensitivity. In fact, fish preexposed to copper could withstand exposure to greatly increased copper concentrations [135]. Finally, McCarter and Roch [136] demonstrated a positive correlation between MBP concentrations and the LC_{50} value in salmon preexposed to copper. Thus, MBPs can be used as an index of resistance but not toxicity. The correlation between MBPs and acclimation to metals is much stronger than a correlation with toxicity and is currently well accepted [127].

Clearly, the presence of MBPs cannot be related with metal toxicity. Thus, we do not recommend the use of MBPs as an indicator of metal-induced toxicity. Future research should investigate the relationship between the rate of MBP formation and the detoxification of metals.

In summary, there are relatively few clinical measures which can be used to quantify the toxic effects of metal exposure in the field. We recommend that whole-body metal concentrations, metal-binding proteins, or ALAD, for lead exposure, be used to quantify *exposure* to heavy metals. Currently, there are no toxicant specific measures of metal *toxicity*. In addition, since none of the metals appear to specifically affect a single organ, organ specific indicators of toxicity will not be useful. We recommend the use of general indicators of toxicity to monitor the overall effects of metals on the health of a population of aquatic organisms. General measures such as liver or hepatopancreas glycogen levels, adenylate energy charge, or RNA/DNA ratios appear to be most useful and will be most readily extrapolated from the organism to the population level. Use of serum enzymes as indicators of metal-induced toxicity may have some utility in the future. To date, sufficient research to link effects on serum enzyme levels with population level effects in feral populations has not been conducted.

Summary

We have briefly reviewed the rationale for use, advantages and disadvantages, and the state of the art in the utilization of clinical measures for assessing the health of aquatic organisms. We have also discussed two situations which consider the use of clinical methods in field studies. The potential for utilization of these methods are great and not limited to the study of toxic mechanisms or mode of action. The problems with the field use of clinical measures can be overcome through proper clinical measure selection and additional research on applying these methods in field situations.

In conclusion, we believe clinical measures of toxic effects will be important, effective tools for assessing the environmental impacts of toxic substances in the field; however, at present we agree with Bouck [4] in recognizing that relatively few tests for stress have progressed beyond the investigative stages, and thus few tests lend themselves to field application. Only with focused, purposeful research will it be possible to develop assays for use in the field. No research effort on clinical measures of stress should be considered concluded until the technique is validated or rejected in the field. The potential benefits of clinical methods are great and warrant additional attention.

References

[1] Larsson, A., Haux, C., Sjobeck, M., *Ecotoxicology and Environmental Safety,* Vol. 9, 1985, pp. 250–282.
[2] Miller, D. S., *The Journal of Pharmacology and Experimental Therapeutics,* Vol. 219, 1981, pp. 428–434.
[3] Wedemeyer, G. A., McLeay, D. J., and Goodyear, C. P., in *Contaminant Effects on Fisheries,* V. W. Cairns, P. V. Hodson, and J. O. Nriagu, Eds., Wiley, New York, 1984, Chap. 12, pp. 163–195.
[4] Bouck, G. R., in *Contaminant Effects on Fisheries,* V. W. Cairns, P. V. Hodson, and J. O. Nriagu, Eds., Wiley, New York, 1984, Chap. 6, pp. 61–71.
[5] Mehrle, P. M. and Mayer F. L., *Environmental Health Perspectives,* Vol. 34, 1980, pp. 139–143.
[6] Adams, S. M., Burtis, C. A. and Beauchamp, J. J., *Comparative Biochemistry and Physiology,* Vol. 82C, 1985, pp. 301–310.
[7] Livingstone, D. R., *Marine Pollution Bulletin,* Vol. 13, 1982, pp. 261–263.
[8] Nolan, C. V. and Duke, E. J., *Aquatic Toxicology,* Vol. 4, 1983, pp. 153–164.
[9] Lowe-Jinde, L. and Niimi, A. J., *Archives of Environmental Contamination and Toxicology,* Vol. 13, 1984, pp. 759–764.
[10] Coppage, D. L., Matthews, E., Cook, G. H., and Knight, J., *Pesticide Biochemistry and Physiology,* Vol. 5, 1975, pp. 536–542.
[11] Versteeg, D. J. and Giesy, J. P., *Ecotoxicology and Environmental Safety,* Vol. 11, 1986, pp. 31–43.
[12] Turner, C. D. and Bagnara, J. T., *General Endocrinology,* W. B. Saunders Co., Philadelphia, PA, 1976.
[13] Donaldson, E. M., in *Stress in Fish,* A. D. Pickering, Ed., Academic Press, London, 1981, pp. 11–47.
[14] Mazeaud, M. M., Mazeaud, F., and Donaldson, E. M., *Transactions of the American Fisheries Society,* Vol. 106, 1977, pp. 201–212.
[15] Casillas, E. and Smith, L. S., *Journal of Fish Biology,* Vol. 10, 1977, pp. 481–491.
[16] Johansson-Sjobeck, M., Dave, G., Larsson, A., Lewander, K., and Lidman, U., *Comparative Biochemistry and Physiology,* Vol. 60A, 1978, pp. 165–168.
[17] Gingerich, W. H. and Weber, L. J., "Assessment of Clinical Laboratory Procedures to Evaluate Liver Intoxication in Fish," EPA Report EPA-600/3-79-088, Environmental Protection Agency, Duluth, MN, Aug. 1979.
[18] Hinton, D. E. and Couch, J. A., "Pathobiological Measures of Marine Pollution Effects,'' EPA Report EPA-600/D-85/123, Environmental Protection Agency, Gulf Breeze, FL, June 1985.
[19] Mitz, S. V. and Giesy, J. P., *Ecotoxicology and Environmental Safety,* Vol. 10, 1985, pp. 22–39.
[20] Lockhart, W. L., Wagemann, R., Clayton, J. W., Graham, B., and Murray, D., *Environmental Physiology and Biochemistry,* Vol. 5, 1975, pp. 361–369.
[21] Versteeg, D. J., "Lysosomal Membrane Stability, Histopathology, and Serum Enzyme Activities as Sublethal Bioindicators of Xenobiotic Exposure in the Bluegill Sunfish (*Lepomis macrochirus* Rafinesque)," Dissertation, Michigan State University, East Lansing, MI, 1985.
[22] Gooch, J. W. and Matsumara, F., *Toxicology and Applied Pharmacology,* Vol. 68, 1983, pp. 380–391.
[23] Dixon, D. G. and Sprague, J. B., *Aquatic Toxicology,* Vol. 1, 1981, pp. 69–81.
[24] Berglind, R., Dave, G., and Sjobeck, M., *Ecotoxicology and Environmental Safety,* Vol. 9, 1985, pp. 216–229.
[25] Mehrle, P. M. and Mayer, F. L., in *Aquatic Toxicology,* G. M. Rand and S. R. Petrocelli, Eds., Hemisphere Publishing Co., New York, 1985, Chap. 10, pp. 264–282.

[26] Lockhart, W. L. and Metner, D. A., in *Contaminant Effects on Fisheries,* V. W. Cairns, P. V. Hodson, and J. O. Nriagu, Eds., Wiley, New York, 1984, Chap. 7, pp. 73–85.
[27] Barnhardt, R. A., *Transactions of the American Fisheries Society,* Vol. 3, 1969, pp. 411–448.
[28] Fromm, P. O., *Environmental Biology of Fishes,* Vol. 5, 1980, pp. 79–83.
[29] Spry, D. J., Wood, C. M., and Hodson, P. V., "Canadian Technical Report of Fisheries and Aquatic Sciences," No. 999, April 1981.
[30] Galloway, J. N., Likens, G. E., and Egerton, E. S., *Science,* Vol. 194, 1976, pp. 722–724.
[31] Campbell, P. G. C. and Stokes, P. M., *Canadian Journal of Fisheries and Aquatic Sciences,* Vol. 42, 1985, pp. 2034–49.
[32] Packer, R. K. and Dunson, W. A., *Journal of Experimental Zoology,* Vol. 174, 1970, pp. 5–72.
[33] Neville, C. M., *Canadian Journal of Fisheries and Aquatic Sciences,* Vol. 42, 1985, pp. 2004–2019.
[34] Leivestad, H. and Muniz, I. P., *Nature,* Vol. 259, 1976, pp. 391–392.
[35] Booth, J. H., Jansz, G. F., and Holeton, G. F., *Canadian Journal of Zoology,* Vol. 60, 1982, pp. 1123–1130.
[36] McDonald, D. G., *Canadian Journal of Zoology,* Vol. 61, 1983, pp. 691–703.
[37] Beamish, R. J., Lockhart, W. L., VanLoon, J. C., and Harvey, H. H., *Ambio,* Vol. 4, 1975, pp. 98–102.
[38] Lockhart, W. L. and Lutz, A., *Water, Air and Soil Pollution,* Vol. 7, 1977, pp. 327–332.
[39] Roch, M. and Maly, E. S., *Journal of the Fisheries Research Board of Canada,* Vol. 36, 1979, pp. 1297–1303.
[40] Lewis, S. D. and Lewis, W. M., *Transactions of the American Fisheries Society,* Vol. 100, 1971, pp. 639–643.
[41] Muniz, I. P. and Leivestad, H., "Proceeding of the International Conference on the Ecological Impact of Acid Precipitation," 1980, pp. 320–321.
[42] Ultsch, G. R., *Copeia,* Vol. 1978, 1978, pp. 272–279.
[43] Packer, R. K., *Journal of Experimental Biology,* Vol. 79, 1979, pp. 127–134.
[44] Daye, P. G. and Garside, E. T., *Canadian Journal of Zoology,* Vol. 54, 1976, pp. 2140–2155.
[45] Ultsch, G. R. and Gros, G., *Comparative Biochemistry and Physiology,* Vol. 62A, 1979, pp. 685–689.
[46] Plonka, A. C. and Neff, W. H., *Proceedings of the Pennsylvania Academy of Science,* Vol. 43, 1969, pp. 33–55.
[47] Arillo, A., Margiocco, C., and Melodia, F., *Journal of Fish Biology,* Vol. 15, 1979, pp. 405–410.
[48] Janssen, R. G. and Randall, D. J., *Respiratory Physiology,* Vol. 25, 1975, pp. 235–45.
[49] Schofield, C. L. and Trojnar, J. R. in *Polluted Rain,* T. Y. Torebara, M. W. Miller, and P. E. Morrow, Eds., 1980, pp. 341–366.
[50] Vaala, S. S. and Mitchell, R. B., *Proceedings of the Pennsylvania Academy of Science,* Vol. 44, 1970, pp. 41–44.
[51] Daye, P. G. and Garside, E. T., *Canadian Journal of Zoology,* Vol. 58, 1980, pp. 27–43.
[52] Jagoe, C. H. and Haines, T. A., *Transactions of the American Fisheries Society,* Vol. 112, 1983, pp. 689–695.
[53] Vander Putte, I., Laurier, M. B. H. M., and Van Eijk, G. J. M., *Aquatic Toxicology,* Vol. 2, 1982, pp. 99–112.
[54] Malins, D. C., *Canadian Journal of Fisheries and Aquatic Sciences,* Vol. 39, 1982, pp. 877–889.
[55] Saunders, R. L., Henderson, E. B., Harmon, P. R., Johnston, C. E., and Davidson, K., in *Workshop on Acid Rain,* R. H. Peterson and H. H. V. Hord, Eds., 1983, p. 49.
[56] Staurnes, M., Sigholt, T., and Reite, O. B., *Experientia,* Vol. 40, 1984, pp. 226–227.
[57] Haya, K. and Waiwood, B. A., *Bulletin of Environmental Contamination and Toxicology,* Vol. 27, 1981, pp. 7–12.
[58] Arillo, A., Maniscalco, N., Margiocco, C., Melodia, F., and Mensi, P., *Comparative Biochemistry and Physiology,* Vol. 63C, 1979, pp. 325–331.
[59] Christensen, G., Hunt, E., and Fiandt, J., *Toxicology and Applied Pharmacology,* Vol. 42, 1977, pp. 523–530.
[60] Lee, R. M., Gerking, S. D., and Jezierska, B., *Environmental Biology of Fishes,* Vol. 8, 1983, pp. 115–123.
[61] Haya, K., Waiwood, B. A., and Van Eckhaute, L., in *Workshop on Acid Rain,* R. H. Peterson and H. H. V. Hord, Eds., 1983, pp. 57–70.
[62] Murthy, V. K., Reddanna, P., Bhaskar, M., and Govindappa, S., *Canadian Journal of Zoology,* Vol. 59, 1981, pp. 1909–1915.
[63] MacFarlane, R. B., *Comparative Biochemistry and Physiology,* Vol. 68B, 1981, pp. 193–202.
[64] Mudge, J. E., Dively, J. L., Neff, W. H., and Anthony, A., *General and Comparative Endocrinology,* Vol. 31, 1977, pp. 208–215.

[65] Leivestad, H., Hendrey, G., Muniz, I. P., and Snekvik, E., in *Impact of Acid Precipitation on Forest and Freshwater Ecosystems in Norway*, F. H. Braekke, Ed., SNSF-project FR 6/76, 1976.
[66] McWilliams, P. G., *Journal of Experimental Biology*, Vol. 88, 1980, pp. 269–280.
[67] Haug, A., *CRC Critical Reviews in Plant Science*, Vol. 1, 1984, pp. 345–373.
[68] Dively, J. L., Mudge, J. E., Neff, W. H., and Anthony, A., *Comparative Biochemistry and Physiology*, Vol. 57A, 1977, pp. 347–351.
[69] Leland, H. V. and Kuwabara, J. S., in *Aquatic Toxicology*, G. M. Rand and S. R. Petrocelli, Eds., Hemisphere Publishing Co., Washington, 1984, Chap. 13, pp. 374–415.
[70] Aylett, B. J., in *The Chemistry, Biochemistry and Biology of Cadmium*, M. Webb, Ed., Elsevier/North Holland Biomedical Press, New York, Chap. 1, pp. 1–43.
[71] Connell, D. W. and Miller, G. J., *Chemistry and Ecotoxicology*, Wiley, New York, 1984, pp. 289–293.
[72] Anderson, R. W., in *Toxicity to Biota of Metal Forms in Natural Waters*, R. W. Andrew, P. V. Hodson, and D. E. Konasewich, Eds., International Joint Commission, Windsor, Ontario, 1975, Chap. 6, pp. 127–142.
[73] Wright, D. A., *Journal of Experimental Biology*, Vol. 67, 1977, pp. 163–173.
[74] Carroll, J. J., Ells, S. J., and Oliver, W. S., *Bulletin of Environmental Contamination and Toxicology*, Vol. 22, 1979, pp. 575–581.
[75] Wright, D. A. and Frain, J. W., *Archives of Environmental Contamination and Toxicology*, Vol. 10, 1981, pp. 321–328.
[76] Pagenkopf, G. K., *Environmental Science and Technology*, Vol. 17, 1983, pp. 342–347.
[77] Sippel, A. J. A., Geraci, J. R., and Hodson, P. V., *Water Research*, Vol. 17, 1983, pp. 1115–1118.
[78] Lock, R. A. C., Cruijsen, P. M. J. M., and van Overbeeke, A. P., *Comparative Biochemistry and Physiology*, Vol. 68C, 1981, pp. 151–159.
[79] Giles, G. A., *Canadian Journal of Fisheries and Aquatic Sciences*, Vol. 41, 1984, pp. 1678–1685.
[80] Sugatt, R. H., *Archives of Environmental Contamination and Toxicology*, Vol. 9, 1980, pp. 41–52.
[81] Larsson, A., Bengtsson, B.-E., and Haux, C., *Aquatic Toxicology*, Vol. 1, 1981, pp. 19–35.
[82] McKim, J. M., Christensen, G. M., and Hunt, E. P., *Journal of the Fisheries Research Board of Canada*, Vol. 27, 1970, pp. 1883–1889.
[83] Skidmore, J. F., *Journal of Experimental Biology*, Vol. 52, 1970, pp. 481–494.
[84] Watson, T. A. and Beamish, F. W. H., *Comparative Biochemistry and Physiology*, Vol. 66C, 1980, pp. 77–82.
[85] Haya, K. and Waiwood, B. A., in *Aquatic Toxicology*, J. O. Nriagu, Eds., Wiley, New York, 1983, Chap. 10, pp. 307–334.
[86] Tucker, R. K., *Bulletin of Environmental Contamination and Toxicology*, Vol. 23, 1979, pp. 33–35.
[87] Watson, T. A. and Beamish, F. W. H., *Comparative Biochemistry and Physiology*, Vol. 66C, 1980, pp. 77–82.
[88] Lorz, H. W. and McPherson, B. P., *Journal of the Fisheries Research Board of Canada*, Vol. 33, 1976, pp. 2023–2030.
[89] Roberts, K. S., Cryer, A., Kay, J., Solbe, J. F. De L. G., Warfe, J. R., and Simpson, W. R., *Comparative Biochemistry and Physiology*, Vol. 62C, 1979, pp. 135–140.
[90] Fowler, B. A. and Nordberg, G. F., *Toxicology and Applied Pharmacology*, Vol. 46, 1978, pp. 609–623.
[91] Koenig, H., *Journal of Histochemistry and Cytochemistry*, Vol. 11, 1962, pp. 120–121.
[92] Leland, H. V., *Environmental Toxicology and Chemistry*, Vol. 2, 1983, pp. 353–368.
[93] Moore, M. N. and Stebbing, A. R. D., *Journal of the Marine Biological Association of the United Kingdom*, Vol. 56, 1976, pp. 995–1005.
[94] Hammond, P. B. and Belile, R. P., in *Cassarett and Doull's Toxicology: The Basic Science of Poisons*, J. Doull, C. D. Klassen, and M. O. Amdur, Eds., Macmillan, New York, 1980, pp. 409–467.
[95] Johansson-Sjobeck, M.-L. and Larsson, A., *Archives of Environmental Contamination and Toxicology*, Vol. 8, 1979, pp. 419–431.
[96] Jackim, E., *Journal of the Fisheries Research Board of Canada*, Vol. 30, 1973, pp. 560–562.
[97] Hodson, P. V., *Journal of the Fisheries Research Board of Canada*, Vol. 33, 1976, pp. 268–271.
[98] Hodson, P. V., Blunt, B. R., Spry, D. J., and Austen, K., *Journal of the Fisheries Research Board of Canada*, Vol. 34, 1977, pp. 501–508.
[99] Hodson, P. V., Hilton, J. W., Blunt, B. R., and Slinger, S. J., *Journal of the Fisheries Research Board of Canada*, Vol. 37, 1980, pp. 170–176.
[100] Haux, C., Larssin, A., Lithner, G., and Sjobeck, M.-L., *Environmental Toxicology and Chemistry*, Vol. 5, 1986, pp. 283–288.
[101] Haines, T. A., *Journal of the Fisheries Research Board of Canada*, Vol. 30, 1973, pp. 195–199.
[102] Buckley, L. J., *Journal of the Fisheries Research Board of Canada*, Vol. 36, 1979, pp. 1497–1502.

[103] Sower, S. A., Schreck, C. B., and Evenson, M., *Aquaculture,* Vol. 32, 1983, pp, 243–254.
[104] Kearns, P. K. and Atchison, G. J., *Environmental Biology of Fishes,* Vol. 4, 1979, pp. 383–387.
[105] Bulow, F. J., *Journal of the Fisheries Research Board of Canada,* Vol. 27, 1970, pp. 2343–2349.
[106] Bulow, F. J., *Iowa State Journal of Science,* Vol. 45, 1971, pp. 71–78.
[107] Nakano, T. and Tomlinson, N., *Journal of the Fisheries Research Board of Canada,* Vol. 24, 1967, pp. 1701–1715.
[108] Pickering, A. D., Pickering, T. G., and Christie, P., *Journal of Fish Biology,* Vol. 20, 1982, pp. 229–244.
[109] Dubale, M. S. and Shah, P., *Environmental Research,* Vol. 26, 1981, pp. 110–118.
[110] Arillo, A., Margiocco, C., Melodia, F., and Mensi, P., *Chemosphere,* Vol. 11, pp. 47–57.
[111] Gill, T. S. and Pant, J. C., *Toxicology Letters,* Vol. 18, 1983, pp. 195–200.
[112] Sastry, K. V. and Subhadra, K., *Environmental Research,* Vol. 36, 1985, pp. 32–45.
[113] Haux, C. and Larsson A., *Aquatic Toxicology,* Vol. 5, 1984, pp. 129–142.
[114] Sastry, K. V. and Sunita, K., *Toxicology Letters,* Vol. 16, 1983, pp. 9–15.
[115] Holland, D. L., in *Biochemical and Biophysical Perspectives in Marine Biology,* Vol. 4, D. C. Malins and J. R. Sargent, Eds., Academic Press, New York, 1978.
[116] Kruger, H. M., Saddler, J. B., Chapman, G. A., Tinsley, I. J., and Lowry, R. R., *American Zoology,* Vol. 8, 1968, pp. 119–129.
[117] Viagengo, A., Pertica, M., Mancinelli, G., Zanicchi, G., and Bouquegneau, J. M., *Molecular Physiology,* Vol. 5, 1984, pp. 41–52.
[118] Thomas, D. G., Solbe, J. F. De L. G., Kay, J., and Cryer, A., *Biochemical and Biophysical Research Communications,* Vol. 110, 1983, pp. 584–592.
[119] Thomas, D. G., Solbe, J. F. De L. G., Kay, J., and Cryer, A., *Comparative Biochemistry and Physiology,* Vol. 76C, 1983, pp. 241–246.
[120] Squibb, K. S., Chignell, C. F., and Fowler, B. A., *Marine Environmental Research,* Vol. 14, 1984, p. 453.
[121] Winge, D. R., Premakumar, R., Wiley, R. D., and Rajagopalan, K. V., *Archives of Biochemistry and Biophysics,* Vol. 170, 1975, pp. 253–266.
[122] Evans, G. W., in *Metallothionein,* H. R. Kagi and M. Nordberg, Eds., Birkhouser Verlag, Boston, 1979, pp. 321–329.
[123] Ley, H. L., Failla, M. L., and Cherry, D. S., *Comparative Biochemistry and Physiology,* Vol. 74B, 1983, pp. 507–513.
[124] Klaverkamp, J. F., Macdonald, W. A., Duncan, D. A., and Wageman, R., in *Contaminant Effects on Fisheries,* V. W. Cairns, P. V. Hodson, and J. O. Nriagu, Eds., Wiley, New York, 1984, Chap. 9, pp. 99–113.
[125] Nordberg, M., and Kojima, Y., in *Metallothionein,* H. R. Kagi and M. Nordberg, Eds., Birkhauser Verlag, Boston, 1979, pp. 41–117.
[126] Thornally, P. J. and Vasak, M., *Biochimica et Biophysica Acta,* Vol. 827, 1985, pp. 36–44.
[127] Livingstone, D. R., in *The Effects of Stress and Pollution on Marine Animals,* B. L. Bayne, D. A. Brown, K. Burns, D. R. Dixon, A. Ianovici, D. R. Livingstone, D. M. Lowe, M. N. Moore, A. R. D. Stebbing, and J. Widdows, Eds., Praeger, New York, 1985, Chap. 4, pp. 81–132.
[128] Winge, D., Krasno, J., and Colucci, A. V., in *Trace Element Metabolism in Animals,* Vol. 2, W. G. Hoekstra, J. W. Suttie, H. E. Ganther, and W. Mertz, Eds., University Park Press, Baltimore, 1974, pp. 500–502.
[129] Brown, D. A. and Parsons, T. R., *Journal of the Fisheries Research Board of Canada,* Vol. 35, 1978, pp. 880–884.
[130] Brown, D. A., Bay, S. M., Alfafara, J. F., Hershelman, G. P., and Rosenthal, K. D., *Aquatic Toxicology,* Vol. 5, 1984, pp. 93–108.
[131] Harrison, F. L. and Lam, J. R., "Concentrations of Copper-Binding Proteins in Livers of Bluegills from the Cooling Lake at the H. B. Robinson Nuclear Power Station," NUREG/CR-2822 UCLR-53041, U.S. Nuclear Regulatory Commission, Washington, DC, 1982.
[132] Viarengo, A., Pertica, M., Mancinelli, G., Palmero, S., Zanicchi, G., Orunesu, M., *Marine Environmental Research,* Vol. 6, 1982, pp. 235–243.
[133] Sanders, B. M., Jenkins, K. D., Sunda, W. G., and Costlow, J. D., *Science,* Vol. 222, No. 4619, 1973, pp. 53–54.
[134] Sanders, B. M. and Jenkins, K. D., *Biological Bulletin,* Vol. 167, 1984, pp. 704–712.
[135] Buckley, J. T., Roch, M., McCarter, J. A., Rendell, C. A., and Matheson, A. T., *Comparative Biochemistry and Physiology,* Vol. 72C, 1982, pp. 15–19.
[136] McCarter, J. A. and Roch, M., *Comparative Biochemistry and Physiology,* Vol. 74C, 1983, pp. 133–137.

William H. van der Schalie,[1] Tommy R. Shedd,[1] and Maurice G. Zeeman[2]

Ventilatory and Movement Responses of Bluegills Exposed to 1,3,5-Trinitrobenzene

REFERENCE: van der Schalie, W. H., Shedd, T. R., and Zeeman, M. G., "**Ventilatory and Movement Responses of Bluegills Exposed to 1,3,5-Trinitrobenzene,**" *Aquatic Toxicology and Hazard Assessment: 10th Volume, ASTM STP 971*, W. J. Adams, G. A. Chapman, and W. G. Landis, Eds., American Society for Testing and Materials, Philadelphia, 1988, pp. 307–315.

ABSTRACT: TNB (1,3,5-trinitrobenzene) is a by-product of the TNT (2,4,6-trinitrotoluene) manufacturing process and is also formed by the photolysis of TNT in natural waters. The effects of TNB on the ventilatory patterns and whole-body movement rates of bluegills (*Lepomis macrochirus*) were determined. Fish were exposed for six days to concentrations of TNB ranging from 6% (0.03 mg/L) to 108% (0.61 mg/L) of the 96-h LC_{50}. The lowest TNB concentrations causing significant changes in the parameters monitored were 0.13 mg/L (ventilatory depth) and 0.61 mg/L (cough rate and body movement). These responses occurred within the first 4 h of exposure. No effects on ventilatory rate were found. The large differences is sensitivity between the parameters monitored indicate that automated water or wastewater toxicity monitoring systems that utilize fish ventilatory or movement responses should use several end points to determine the presence or absence of toxicity. The most sensitive, short-term ventilatory changes measured in this test occurred at concentrations near the estimated chronic toxicity levels of TNB for fish. The range between the highest TNB concentration causing no responses and the lowest concentration causing short-term ventilatory responses was 0.06 to 0.13 mg/L. This is comparable to reported no effect/effect ranges for TNB in early life stage tests of 0.08 to 0.12 mg/L for fathead minnows and 0.08 to 0.17 mg/L for rainbow trout.

KEYWORDS: biological monitoring, fish, bluegill, *Lepomis macrochirus*, ventilation, cough, movement, 1,3,5-trinitrobenzene, chronic toxicity

As fish move their buccal and opercular cavities to ventilate their gills, they generate an electrical signal that can be monitored by remote electrodes placed in the water with the fish. From this ventilatory signal, several parameters can be determined, including the rate of ventilatory movements, the relative strength (or depth) of the movements, coughing (or gill purge) rates, and whole body movements. A review of the aquatic toxicology literature demonstrates that changes in the ventilatory patterns of fish can be rapid, sensitive indicators of the presence of toxic materials in water [1,2]. A major problem with interpreting these ventilatory studies is that, in virtually all cases, toxicant effects on but one of the four possible parameters were determined. Thus, the parameter selected may not have been the one most sensitive to the toxicant being tested. Changes in fish ventilatory patterns (primarily ventilatory rate) have also been used in automated biomonitoring systems for continuously determining the toxicity of wastewater and surface waters [2]. Once

---

[1] Research biologists, Health Effects Research Division, U.S. Army Biomedical Research and Development Laboratory, Fort Detrick, MD 21701-5010.

[2] Environmental toxicologist, U.S. Food and Drug Administration, Center for Veterinary Medicine, Rockville, MD 20857.

again, only single ventilatory parameters have been used, and it would seem that utilization of all four possible end points could greatly improve the sensitivity of the biomonitors.

A computerized data acquisition system capable of simultaneously monitoring ventilatory rate and depth, cough rate, and movement of the bluegill (*Lepomis macrochirus*) has been developed and described previously [*3*]. The objectives of the present study were: (1) to establish the accuracy of the computer monitoring technique with respect to visual analysis of the ventilatory signal; (2) to compare the lowest concentration of a toxicant tested causing short-term changes in bluegill ventilatory patterns with chronically toxic concentrations; (3) to determine which of the four ventilatory parameters was most sensitive to toxicant-induced changes; and (4) to establish how rapidly ventilatory patterns change after the start of toxicant exposure.

The toxicant chosen for this study was 1,3,5-trinitrobenzene (TNB). TNB has been found in wastewaters associated with the production of 2,4,6-trinitrotoluene (TNT) and is also formed as a result of environmental transformations of the latter compound [*4*]. The acute toxicity of TNB to bluegills and the estimated chronic toxicity (early life stage tests) of TNB to both rainbow trout (*Salmo gairdneri*) and fathead minnows (*Pimephales promelas*) was previously determined in well water from the same source as was used in the present study [*5*]. An acute flow-through toxicity test was conducted with TNB using the same batch of bluegills used in the present ventilatory test. In the ventilatory test, ventilatory signals of individual acclimated fish were monitored in clean water during a preexposure period to establish the normal ventilatory and movement patterns for each fish. The fish were then exposed to TNB concentrations, and changes in the ventilatory parameters relative to the controls were determined.

Materials and Methods

TNB Analysis

TNB was synthesized and purified to a final purity of 99.97% as measured by gas chromatography [*6*]. TNB stock solutions were prepared by dissolving TNB in distilled, deionized water. Test solution concentrations of TNB for the flow-through acute toxicity test were measured by high-pressure liquid chromatography (HPLC) (detection limit 0.10 mg/L), while TNB analyses for the ventilatory test were done by gas chromatography (GC) (detection limit 0.02 mg/L) [*6*]. Recovery rates for spiked samples were 92.4% ($\pm 4.03\%$) for the HPLC and 87.4% ($\pm 4.04\%$) for the GC.

Dilution Water Quality

The dilution water used for testing was taken from a well 62 m in depth. The water was filtered, sterilized with ultraviolet light, and temperature adjusted before being used in the toxicity tests. A water softener was used to reduce the water hardness by 40%, thus reducing precipitation of calcium carbonate on glassware and equipment. A summary of dilution water quality during testing is given in Table 1. For all testing, temperature was 22°C (± 2), dissolved oxygen concentration ranged from 88 to 98% of saturation, unionized ammonia was less than 10 µg/L, and total suspended solids were less than 2.0 mg/L. A comprehensive dilution water analysis is provided in Table 2.

Bluegills

Bluegills used in testing were obtained from Kurtz Fish Hatchery, Elverson, Pennsylvania. The fish were approximately one year old and 55 mm in length. They were acclimated to the test conditions for from one to three weeks prior to testing. During acclimation and testing, wide

TABLE 1—*Dilution water quality.*

Test Type	pH	Alkalinity, mg/L as CaCO$_3$	Hardness, mg/L as CaCO$_3$	Conductivity, µmhos/cm
Flow-through acute				
mean	...	241	173	622
range	8.2 to 8.3	227 to 250	170 to 174	602 to 644
n	4	4	4	4
Ventilatory				
mean	8.2	217	164	662
range	...	205 to 233	154 to 174	598 to 706
n	5	5	5	5

NOTE: n = number of observations.

TABLE 2—*Comprehensive dilution water analysis.*

Parameter	Concentration, mg/L	Parameter (Chlorinated Hydrocarbons)	Concentration, µg/L
Ammonia (as N)	<0.05	Aldrin	<0.05
Nitrate (as N)	0.16	p,p'-DDT	<0.05
Chloride	70.5	o,p-DDT	<0.05
Fluoride	0.19	DDD	<0.05
Sulfate	39.3	Dieldrin	<0.05
Aluminum	<0.002	Endrin	<0.05
Barium	0.089	Heptachlor	<0.05
Boron	<0.05	Heptachlor epoxide	0.05
Cadmium	<0.0005	Lindane	<0.05
Calcium	47.5	Chlordane	<0.20
Cobalt	<0.002	Alpha-HBC	<0.05
Copper	<0.002	Beta-BHC	<0.05
Iron	<0.1	Delta-BHC	<0.01
Lead	<0.002	Toxaphene	<1.01
Magnesium	14.7	Methoxychlor	<1.05
Manganese	<0.002	Polychlorinated biphenyls	<1.0
Mercury	<0.0005	2,4-D	<0.05
Potassium	1.25	.2,4,5-T	<0.05
Silicon	2.7	Silvex	<0.05
Silver	<0.0005	Diazinon	<0.20
Sodium	93.5	Malathion	<0.20
Zinc	<0.02	Parathion	<0.20
Cyanide	<0.002		
Arsenic	<0.002		
Molybdenum	<0.003		
Nickel	<0.002		
Phosphorus	<0.01		
Sulfide	<0.05		
Selenium	<0.002		

spectrum fluorescent bulbs with a color-rendering index of 91 were used. The diurnal photoperiod for the flow-through acute test was 16-h light and 8-h dark. Bluegills used for the ventilatory test were held under continuous light for at least two weeks prior to testing. Testing was done under continuous light to eliminate diurnal changes in bluegill ventilatory patterns. Prior to testing, all fish were fed Rangen's (No. 3) trout food plus frozen brine shrimp. Fish were not fed during tests. The acclimation temperature was 22°C (± 2).

Flow-Through Acute Toxicity Test

The bluegills were exposed to TNB in 9.5-L aquaria containing 7.6 L of test solution. Ten fish were randomized to each of two test aquaria at six treatment levels (<0.10, 0.14, 0.34, 0.69, 1.29, and 3.07 mg/L mean measured concentration). Fish were not fed beginning 48 h prior to the start of the test. About five tank volumes of test solution were delivered to each test tank per day by means of a proportional diluter. Dissolved oxygen concentrations, pH, and temperature were measured from at least one replicate of each treatment daily. TNB analyses were performed on samples from each tank initially and after 24 and 96 h. Test tanks were not aerated. The 96-h LC_{50} was calculated by the Trimmed Spearman-Karber method [7,8].

Ventilatory Toxicity Test

The test apparatus described by van der Schalie [3] was modified for these tests. Capute [9] demonstrated the advantages of a dorsal/ventral orientation of the ventilatory signal electrodes over an anterior/posterior electrode arrangement for monitoring fish ventilatory signals, so the fish chambers were rebuilt to accommodate submerged electrodes on the top and bottom of the test chamber. The electrodes from each chamber were connected to individual amplifier/filter boards [10] with a two-conductor shielded cable. The ventilatory signal from each amplifier was sent to a Cromemco model Z-80 microcomputer equipped with a multichannel multiplexer and analogue-to-digital converter. All connections were made with shielded cable to reduce background electrical interference. Individual chambers were constructed for each fish with 3-mm glass plates cemented with clear silicone sealer. A piece of black plastic was attached to the outside of each tank to prevent fish in one chamber from being disturbed by fish in the other chambers. Thirty-one test chambers were enclosed in a flat black plywood box with fluorescent light fixtures mounted above the chambers. Light intensity at the top of the test chambers ranged from about 500 to 1000 lx. Water flow (about 28 mL/min) was continuous through the test chambers. One extra test chamber was used for continuous temperature monitoring.

A proportional toxicant diluter provided test solutions to the ventilatory chambers. Six treatments were split five ways to individual test chambers. Only three fish were tested at the top concentration. There was about a 50% dilution between each of the five TNB concentrations and the next lower; the highest TNB concentration was set at the approximate 96-h LC_{50} for bluegills determined from the flow-through acute test. TNB samples were taken 3 h after exposure to TNB had begun and on days 2, 4, and 6 of the exposure period from one replicate of each treatment.

Ventilatory Data Collection and Analysis

The ventilatory parameters for each test fish were monitored continuously by the microcomputer and compiled every 15 min [3]. The 15-min data records were stored on a magnetic tape data logger and printed on a data terminal. Data stored on the data logger were transmitted to an AMDAHL 470-V7 computer for further analysis using the Statistical Analysis System (SAS) [11]. The test was divided into three periods: a three-day acclimation period (data not recorded); a four-day preexposure period (all fish received clean dilution water); and a six-day exposure period. Since ventilatory patterns vary widely between fish, the preexposure period was used to establish normal baseline values for each fish.

The four ventilatory parameters were calculated as follows from each 15-min data summary: ventilatory rate was equal to the number of ventilatory peaks per minute; cough rate was equal to the number of coughs per minute; ventilatory depth was calculated by dividing the sum of all peak heights by the total number of ventilatory peaks counted in each 15-min interval; percent movement was calculated by dividing the number of 15-s intervals of movement within the 15-min interval by 60 (the total number of 15-s records in 15 min). Ventilatory rate, ventilatory depth, and cough rate computations excluded any 15-s periods during which movement was detected.

Preliminary graphs of the ventilatory data showed sporadic high or low readings on most fish even during the preexposure period, so a moving average technique was used to smooth the data. The moving average period was selected on the basis of an initial review of the data and on cough response patterns described by Drummond and Carlson [12]. Comparing the overall mean values for the preexposure and exposure periods was unsuitable, since this masked significant toxicant-related signal changes with durations of perhaps 12 to 72 h. On the other hand, a 1-h moving average technique did not smooth the data sufficiently to remove short-term events unrelated to toxicant exposure. A 4-h moving average of the ventilatory parameters was found to be the best compromise for separating toxicant- and nontoxicant-related ventilatory events.

To calculate the 4-h moving average, 16 consecutive data points representing 4 h of each ventilatory signal were averaged. With each new 15-min data point, the first data point was dropped and the next data point was added. A new mean value was then generated. This process continued through the last data point of the preexposure period and started again with the first 16 data points of the exposure period.

Through use of the moving average data points, maximum and minimum values for each ventilatory parameter were established for the preexposure period and subtracted from the maximum and minimum values for the exposure period. The differences were averaged for all the fish at each treatment level. The SAS General Linear Models Procedure (GLM) was used [11]. Analysis of variance (ANOVA) was performed to determine treatment effects, and, where significant effects were indicated ($P < 0.05$), pairwise comparisons between each treatment level and the control were made with Student's t-tests with Bonferroni's correction for simultaneous comparisons. To determine which, if any, of the test concentrations elicited ventilation responses soon after the start of TNB exposure, preexposure data were compared to the first 4-h moving average values after the start of the exposure period using the procedures described above. Comparisons were made for only those TNB concentrations and ventilatory parameters that were found to be significantly different from the controls when data from the whole exposure period was used in the analysis.

Different procedures were used to evaluate the accuracy of the computer ventilatory monitoring techniques. Periodically during ventilatory testing a multichanneled strip chart recorder was used to simultaneously record 15 min of ventilatory signals from several fish monitored by the computer. Data output from the computer was then directly compared to visual counts of the ventilatory parameters. Fifty-two 15-min strip chart records were manually counted. The information gathered from each visual record included the number of ventilatory peaks, the number of coughs, and the number of 15-s blocks of irregular signal corresponding to fish body movement. The ventilatory peak height was not manually measured. Visual analysis of the ventilatory parameters from strip chart recordings introduces an element of subjectivity in signal interpretation. To reduce inconsistency, visual signal analysis followed methods described by Carlson [13]. The visual and computer data were compared by using the SAS GLM program [11]. Simple linear regression analysis was performed for each ventilatory parameter.

Results and Discussion

Ventilatory Monitoring System Accuracy

Visual and computer counts for ventilatory peaks were significantly correlated. Linear regression analysis of visual and computer ventilatory peak data gave an R^2 value of 0.97 and a slope of 0.94. The relationship demonstrates that ventilatory peak analysis by the computer was accurate. Although visual measurements of the ventilatory peak depth were not made, it follows that if the ventilatory peak counts were accurate, the average depth of the ventilatory signals was also accurate. Coughs were very difficult to count visually with consistency owing to signal variability. Others have also reported problems in counting coughs [14]. The visual-computer count relationship for coughs was not quite as good as for ventilatory peaks; the visual-computer regression line had an R^2 value of 0.77 and a slope of 1.18. Movement intervals were fairly infrequent. Most of the 15-min records had either zero or one 15-s movement interval for both the computer and visual count. The correlation for movement was the lowest of the three ventilatory parameters, with an R^2 value of 0.52 and a slope of 0.67.

TNB Toxicity Tests

The flow-through acute test that was conducted prior to the TNB ventilatory test resulted in a 96-h LC_{50} of 0.57 mg/L, with 95% fiducial limits of 0.50 to 0.65 mg/L. There was a diluter malfunction during the last 16 h of the test that caused a reduction in TNB concentrations, but we feel that it is unlikely that the malfunction substantially affected the LC_{50} value. The observed LC_{50} is similar to a previously reported bluegill 96-h static LC_{50} of 0.85 mg/L (95% fiducial limits 0.52 to 1.38) [5].

The highest TNB concentration in the ventilatory test was set at about the 96-h LC_{50} level (0.61 mg/L). During the ventilatory test, one of three fish died at this concentration after three days of the six-day exposure period. When effects of TNB on ventilatory patterns were determined, data generated by this fish after death (that is, zero values) were not used. A second fish, at the lowest TNB concentration (0.034 mg/L), died when it jumped out of the test chamber, so information from this fish was also excluded from the data analysis.

There were no statistically significant changes in ventilatory signal minima from the preexposure to exposure periods for any of the groups of TNB-exposed fish when compared to the changes in control fish minima (Table 3). While there is little evidence of trends with increasing TNB concentration, the mean differences for all fish, including the controls, were consistently negative for ventilatory rate, cough rate, and movement. Decreases in these parameters with time may reflect the effects of a lack of food or continuing acclimation of the fish to the test chamber.

Significant changes in ventilatory maxima were noted for all parameters except ventilatory rate (Table 4). Although there was a reversal from negative to positive in ventilatory rate maxima at the highest TNB concentration, the variability between fish was too great to establish a significant difference from the controls. Ventilatory rate was the least sensitive of the four parameters to TNB, but it is the one parameter most frequently used in automated field biomonitoring systems [2]. Ventilatory depth was the most sensitive parameter tested; significant responses were noted at concentrations as low as 0.128 mg/L, nearly five times lower than the observed responses in cough rate and percent movement. Although depth was most sensitive to TNB, cough rate has been found to be extremely sensitive to other compounds [12], so it is highly advisable to use all four ventilatory parameters when using ventilatory data either in toxicity screening or in automated toxicity biomonitoring systems.

The rapidity of ventilatory responses to TNB is important, especially when fish ventilatory

TABLE 3—*Differences between preexposure and exposure minima in the TNB ventilatory test.*

Parameter	Mean Measured TNB Concentration, mg/L	Mean Differences Between Minima[a,b]	Standard Deviation
Ventilatory rate	0.613	−6.58	4.14
	0.279	−14.56	8.48
	0.128	−11.13	4.05
	0.061	−7.83	5.55
	0.034	−8.93	5.89
	BDL[c]	−6.89	3.71
Average depth	0.613	−25.94	67.22
	0.279	15.93	28.50
	0.128	4.97	13.49
	0.061	−15.46	31.11
	0.034	2.11	18.89
	BDL	−12.04	8.58
Cough rate	0.613	−0.47	0.18
	0.279	−0.27	0.09
	0.128	−0.11	0.14
	0.061	−0.09	0.14
	0.034	−0.12	0.27
	BDL	−0.18	0.17
Percent movement	0.613	−0.45	0.12
	0.279	−0.58	0.52
	0.128	−0.04	0.75
	0.061	−0.21	0.13
	0.034	−0.54	0.53
	BDL	−0.44	0.52

[a] Mean value (exposure minima − preexposure minima) for all fish at each treatment level.
[b] $n = 5$ for all treatments except 0.613 mg/L ($n = 3$) and 0.034 mg/L ($n = 4$).
[c] BDL = Below detection limit (0.020 mg/L).

signals are utilized in toxicity biomonitors [2]. When the present test data were reanalyzed by comparing the preexposure data with only the first 4-h moving average data point after the start of TNB exposure, it was found that significant responses to TNB occurred for the same parameters and concentrations that showed significant differences from the controls over the entire six-day exposure period (Table 4). A different data analysis technique might demonstrate response times much less than 4 h, but any method would have to contend with the high variability in responses between fish. In this regard, use of more than five fish per treatment could improve the results, as recommended by Drummond and Carlson [12].

A comparison of the most sensitive ventilatory response to TNB with other toxicologic end points (Table 5) shows that the ventilatory effect-no effect concentration range for bluegills is quite similar to those determined in early life stage tests with two other species of fish [5]. Some investigators have suggested that toxicant-related changes in the ventilatory patterns of fish may be useful as short-term indications of chronic toxicity [3,12,15], but others have found a lack of ventilatory response to some materials at levels causing chronic toxic effects [16,17]. Those finding a lack of sensitivity have generally monitored only ventilatory rate, so it is possible that different results might have been found had all four types of information available from the ventilatory signal been utilized.

TABLE 4—*Differences between preexposure and exposure maxima in the TNB ventilatory test.*

Parameter	Mean Measured TNB Concentration, mg/L	Mean Differences Between Maxima[a,b]	Standard Deviation
Ventilatory rate	0.613	23.28	35.06
	0.279	−23.23	16.64
	0.128	−13.16	7.95
	0.061	−4.88	20.57
	0.034	−8.75	6.98
	BDL[c]	−7.51	15.89
Average depth	0.613	74.13[d]	50.36
	0.279	42.03[d]	16.98
	0.128	12.06[d]	22.70
	0.061	11.37	13.38
	0.034	8.07	25.11
	BDL	−8.30	19.59
Cough rate	0.613	4.71[d]	2.80
	0.279	0.16	0.37
	0.128	−1.41	1.59
	0.061	−0.10	0.24
	0.034	0.02	0.22
	BDL	0.29	0.69
Percent movement	0.613	14.89[d]	7.96
	0.279	1.10	5.72
	0.128	−3.58	2.05
	0.061	−1.01	1.09
	0.034	−0.05	2.58
	BDL	−0.54	2.00

[a] Mean value (exposure maxima − preexposure maxima) for all fish at each treatment level.
[b] $n = 5$ for all treatments except 0.613 mg/L ($n = 3$) and 0.034 mg/L ($n = 4$).
[c] BDL = Below detection limit (0.020 mg/L).
[d] Significantly different from control ($p < 0.05$).

TABLE 5—*TNB toxicity test comparisons.*

Test Type	Fish	Length of Exposure, Days	Test End Point	Results (95% Confidence Limits, μg/L)	Reference
Static acute	Bluegill	4	LC_{50}	0.85 (0.52 to 1.38)	5
Flow through acute	Bluegill	4	LC_{50}	0.57 (0.50 to 0.65)	This report
Early life stage	Rainbow trout	71	No effect—effect range[a]	0.08 to 0.17	5
Early life stage	Fathead minnow	32	No effect—effect range	0.08 to 0.12	5
Ventilatory	Bluegill	6	No effect—effect range	0.06 to 0.13	This report

[a] Range from the highest no-effect concentration to the lowest concentration tested causing a significant difference ($p < 0.05$) from the control.

References

[1] Cairns, J., Jr., and van der Schalie, W. H., *Water Research,* Vol. 14, 1980, pp. 1179–1196.
[2] van der Schalie, W. H. in *Aquatic Toxicology and Environmental Fate: Ninth Volume,* American Society for Testing and Materials, Philadelphia, pp. 107–121.
[3] van der Schalie, W. H. in *Aquatic Toxicology (Third Symposium), ASTM STP 707,* American Society for Testing and Materials, Philadelphia, 1980, pp. 233–242.
[4] Burlinson, N. E., "Fate of TNT in an Aquatic Environment," Report NSWC TR 79-445, Naval Surface Weapons Center, Silver Spring, MD, 1980.
[5] van der Schalie, W. H., "The Acute and Chronic Toxicity of 3,5-Dinitroaniline, 1,3-Dinitrobenzene, and 1,3,5-Trinitrobenzene to Freshwater Aquatic Organisms," Technical Report 8305, AD A138408, U.S. Army Medical Bioengineering Research and Development Laboratory, Fort Detrick, Frederick, MD, 1983.
[6] Rosencrance, A. B. and Brueggemann, E. E., "Liquid and Gas Chromatographic Determinations of 1,3,5-Trinitrobenzene in Water," Technical Report 8601, U.S. Army Medical Bioengineering Research and Development Laboratory, Fort Detrick, Frederick, MD, 1986.
[7] Hamilton, M. A., Russo, R. C., and Thurston, R. V., *Environmental Science and Technology,* Vol. 11, 1977, pp. 714–719.
[8] Published Correction to Reference 7: *Environmental Science Technology,* Vol. 12, 1978, p. 417.
[9] Capute, A. J., Jr, "Effect of Carbon Tetrachloride, Chloroform, and Tetrachloroethylene on the Respiratory Activity in Bluegill Sunfish (*Lepomis macrochirus*), M.S. thesis, University of Cincinnati, Cincinnati, OH, 1980.
[10] Gruber, D. et al., *Transactions of the American Fisheries Society,* Vol. 106, 1977, pp. 497–499.
[11] *SAS User's Guide: Statistics, Version 5 Edition,* SAS Institute, Inc., Cary, NC, 1985.
[12] Drummond, R. A. and Carlson, R. W., "Procedures for Measuring Cough (Gill Purge) Rates of Fish," EPA-600/3-77-133. National Technical Information Services, Springfield, VA, 1977.
[13] Carlson, R. W., *Environmental Pollutant Series A,* Vol. 29, 1982, pp. 35–56.
[14] Maki, A. W. in *Aquatic Toxicology (Second Symposium), ASTM STP 667,* American Society for Testing and Materials, Philadelphia, 1979, pp. 77–95.
[15] Cairns, M. A., Garton, R. R. and Tubb, R. A., *Transactions of the American Fisheries Society,* Vol. 111, 1982, pp. 70–77.
[16] Morgan, W. S. G. and Kuhn, P. C. in *Freshwater Biological Monitoring,* D. Pascoe and R. W. Edwards, Eds., Pergamon Press, Oxford, United Kingdom, 1984, pp. 65—73.
[17] Sloof, W., *Bulletin of Environmental Contamination Toxicology,* Vol. 23, 1979, pp. 517–523.

John A. Bantle[1] *and Douglas A. Dawson*[1]

Uninduced Rat Liver Microsomes as a Metabolic Activation System for the Frog Embryo Teratogenesis Assay—Xenopus (FETAX)

REFERENCE: Bantle, J. A. and Dawson, D. A., "**Uninduced Rat Liver Microsomes as a Metabolic Activation System for the Frog Embryo Teratogenesis Assay—*Xenopus* (FETAX),**" *Aquatic Toxicology and Hazard Assessment: 10th Volume, ASTM STP 971*, W. J. Adams, G. A. Chapman, and W. G. Landis, Eds., American Society for Testing and Materials, Philadelphia, 1988, pp. 316–326.

ABSTRACT: Current validation studies of FETAX (Frog Embryo Teratogenesis Assay—*Xenopus*) suggest that fewer than 15% of test compounds will prove to be false negatives or positives in this *in vitro* teratogenesis screening assay. We have developed a metabolic activation system employing uninduced rat liver microsomes to convert proteratogens, thus reducing the number of potential false negatives. Microsomes were prepared by homogenizing livers and centrifuging the homogenate first at 600 and then at 9000 $\times$ g avg. The supernatant from these centrifugations was then centrifuged twice at 120 000 $\times$ g avg. By not inducing the rat liver with Aroclor 1254, toxicity from Aroclor metabolites was avoided. *Xenopus* blastulae were exposed to different concentrations of cyclophosphamide together with the microsomes, generating system, and antibiotics for a period of 96 h. Activation reduced the 96-h LC_{50} by 4-fold from >11.0 to 2.8 mg/mL. The EC_{50} malformation was reduced 3.4-fold from 6.8 to 2 mg/mL. The severity of malformation was also increased by activation. Activation of cyclophosphamide also caused a decrease in growth. Uninduced rat liver microsomes can be used as an acceptable *in vitro* metabolic activation system for FETAX.

KEYWORDS: *Xenopus*, teratogenesis assay, metabolic activation system, cyclophosphamide, FETAX (Frog Embryo Teratogenesis Assay—*Xenopus*)

Approximately 60 000 to 80 000 chemicals are currently available in the marketplace with some 1700 new chemicals released each year. The safety of these chemicals must be firmly established. Because of the large numbers of chemicals that need to be tested and the even larger number of interactions possible when these chemicals are present in complex mixtures, *in vivo* assays employing mammals are not practical. The need for routine teratogenicity testing has led to the development of a number of *in vitro* teratogenesis assays that may prove useful in prioritizing compounds for further testing [1–4]. Recently, Dumont et al. [5] have developed and used the Frog Embryo Teratogenesis Assay—*Xenopus* (FETAX) and applied it to screening complex environmental mixtures. We propose to further develop FETAX and validate this assay using standard repository compounds from the National Toxicology Program (NTP) [6] as well as with other compounds whose developmental toxicity in mammals is known. Successful development and validation of this assay will make available to the scientific community a four-day screening test that will provide reliable data capable of predicting whether an unknown is a human teratogen.

---

[1] Associate professor and graduate research assistant, respectively, Department of Zoology, Oklahoma State University, Stillwater, Oklahoma 74078.

In its present form, FETAX meets most of the criteria set forth by Kimmel et al. [7] for the validation of *in vitro* teratogenesis assays. Endpoints such as mortality, malformation, growth, and development are easily quantifiable and capable of being analyzed statistically and of exhibiting a dose-response relationship with the establishment of narrow confidence limits. Since many of the stages of amphibian development are similar to mammalian development, the "developmental relevance" of FETAX is higher [7] than many of the other teratogenesis assays. At present FETAX has not been rigorously tested with the NTP repository compounds although work is in progress in several laboratories. Another difficulty with this assay is that it lacks a metabolic activation system to convert proteratogens to teratogens (T. D. Sabourin, Battelle Columbus Laboratories). Without such an activation system many samples would test as false negatives in FETAX but would cause abnormal development in humans because they would be converted to their teratogenic forms. We have developed an *in vitro* metabolic activation system for FETAX by using uninduced rat liver microsomes and an *in vitro* NADPH generating system. This approach has been used by Kitchin and Woods [8] in providing a metabolic activation system for cultured rat embryos. Employment of the metabolic activation system in the FETAX protocol should greatly increase the acceptance of FETAX as a screening assay.

Materials and Methods

Water Quality

The water for holding tanks of *Xenopus* adults and larvae was filtered through an activated carbon filter (Barnstead™) and aerated for 48 h prior to use. The water was routinely tested for pH, oxygen content, hardness, heavy metal content, and total organic carbon.

Animal Care and Breeding

Adult *Xenopus* were obtained from *Xenopus* I® (Ann Arbor, MI) and kept in glass aquaria for a minimum of 30 days until use. Adults were fed beef liver and lung obtained from a local packing house. The meat was supplemented with baby vitamins (Polyvisol®, Mead Johnson, Evansville, IN). Twelve h prior to mating, the female was injected with 1000 international units (IU) of human chorionic gonadotropin (Sigma®, St. Louis, MO) and the male received 400 I.U. into the dorsal lymph sac. Amplexus ensued within 4 to 6 h and egg deposition within 9 to 12 h postinjection.

After breeding, the adults and any fecal material were removed from the tank and the embryos collected in 55-mm plastic Petri dishes. The jelly coats of the embryos were removed by gentle swirling for 3 to 4 min in a 2% weight by volume cysteine solution made up in FETAX solution and the pH adjusted to 8.1 with NaOH. FETAX solution is composed of 625 mg NaCl, 96 mg NaHCO$_3$, 30 mg KCl, 15 mg CaCl$_2$, 60 mg CaSO$_4 \cdot$ 2H$_2$O), and 75 mg MgSO$_4$ per litre of deionized distilled water.

Normally developing blastulae were initially selected by sorting them from necrotic eggs and abnormally cleaving embryos. A second selection was then performed to ensure that only normal embryos were used in the test. For each toxicant concentration, two groups of 20 embryos each were placed into 55-mm plastic Petri dishes. All glassware used in this study had previously been washed in dilute HCl, rinsed, washed in dilute NaOH, and then rinsed in distilled water.

Preparation of Rat Liver Microsomes

Uninduced rat liver microsomes were prepared essentially by the method of Kitchin and Woods [8]. All procedures were carried out in the cold. One male Sprague-Dawley rat liver was washed and perfused with 50 mL of 0.15% KCl. The liver was homogenized and the homogenate centrifuged first at 600 × g avg and then at 9000 × g avg. The supernatant from these centrifugations was then centrifuged twice at 120 000 × g avg. The pelleted microsomes were resuspended in cold 0.05-M tris-HCl buffer and the protein content determined by the Bio-Rad protein assay [9]. The amount of P-450 complex in the microsomes was quantified spectrophotometrically as follows. Microsomes (0.1 mL) were diluted in 0.9 mL 0.05-M tris-HCl buffer (pH 7.5), and approximately 6 µg of sodium hydrosulfite was added. Two of these solutions were used to establish a zero on a Varian® scanning spectrophotometer. One microsomal sample was then gassed for 2 min with a continuous stream of carbon monoxide. The sample was then scanned for absorbance difference spectra between 390 to 550 nm. The observed absorbance maximum at 451 nm confirmed the successful purification of the cytochrome P-450 complex. After quantification, the microsomes were split into four aliquots and frozen in liquid nitrogen for use on each succeeding day of the FETAX experiment. Freezing of microsomes is an acceptable procedure [10]. We have also performed difference spectra on aliquots of microsomes that had been frozen for time periods up to one week and the spectra were the same, indicating that no degradation occurred.

Each test dish contained 20 embryos and 100 units/mL penicillin and 100 µg/mL streptomycin. We report here the lack of mortality, malformation, and growth inhibition observed with these and other antibiotics in FETAX. Penicillin was also listed as a nonteratogen by Smith et al. [6]. Each dish received 0.1 mg/mL of microsomal protein, 3.5 mM glucose 6-phosphate, 0.31 I.U./mL glucose 6-phosphate dehydrogenase, 0.007 mM NADPH and 0.1 mM NADP. Depending on the experiment performed between 2 to 11 mg/mL of cyclophosphamide were immediately added to the solution. For experiments requiring sublethal and subteratogenic levels of cyclophosphamide, 3 mg/mL was used. All concentrations listed above were final concentrations in 8 mL of FETAX solution. All solutions were changed every 24 h during the 96-h test using the same lot of materials.

Data Collection and Analysis

At 24, 48, 72 and 96 h, the number of dead embryos was recorded for each dish and then removed from the dishes. The number of live-malformed embryos and the stage of development according to Nieuwkoop and Faber [11] were determined. Malformed embryos which had died were not included in the number malformed. Death at 24 (Stage 27) and 48 h (Stage 35) was ascertained by the embryo's skin pigmentation, structural integrity, and irritability while at 72 (Stage 42) and 96 h (Stage 45), the lack of heartbeat in the transparent embryo was an unambiguous sign of death.

Surviving embryos were fixed in 0.5% formalin and the head-tail length of each embryo was measured using a Radio Shack® digitizer and the data input directly to a Radio Shack® model 16 microcomputer. The t-Test for grouped observations was used to test for a significant reduction in growth [12].

Results

The Use of Cyclophosphamide in the Development of the Assay

Cyclophosphamide is a compound that requires bioactivation for its teratogenic activity [6,13,14]. By using rat liver microsomes to convert cyclophosphamide to its more teratogenic and toxic metabolites, we can demonstrate that these microsomes can be used as part of the standard protocol for FETAX. Cyclophosphamide was also selected because an extensive literature exists on its metabolism as well as its mutagenic and teratogenic potential [15]. This literature base made it much easier to interpret results.

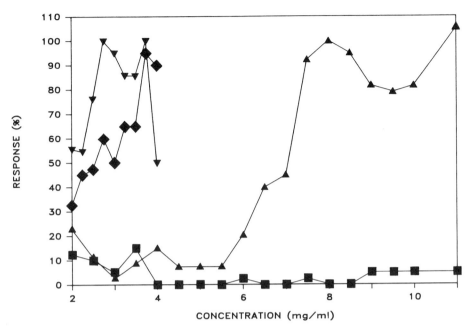

FIG. 1—*Unactivated and microsomal activated dose-response curves for mortality and malformation. Increasing concentrations of cyclophosphamide were used while all other system components were held to 1 X concentration. Microsomal protein was 0.12 mg/mL.* ■ *unactivated mortality,* ♦ *activated mortality,* ▲ *unactivated malformation,* ▼ *activated malformation.*

Activation of Cyclophosphamide Using Uninduced Rat Liver Microsomes

In a dose-response experiment the 96-h LC_{50} for unactivated cyclophosphamide was >11 (Fig. 1). In this experiment we were unable to obtain significant mortality at 11 mg/mL although previous FETAX experiments suggested a 96-h LC_{50} of 10 to 11 mg/mL. With the addition of microsomes and the NADPH generating system, the stimulation of embryotoxic activity was such that about 2.8 mg/mL was needed to kill 50% of the embryos after 96 h. Cyclophosphamide alone caused significant malformation in the concentration ranges tested with a 96-h EC_{50} (malformation) of approximately 6.8 mg/mL. Activation of cyclophosphamide with microsomes lowered this value to 2 mg/mL. Without bioactivation, the teratogenicity index [96-h LC_{50}/EC_{50} (malformation)] [5] was >1.7 but was reduced to about 1.4 after bioactivation. Controls for this experiment (no cyclophosphamide or metabolic activation system) had a 96-h malformation and mortality rate of 5 and 2.5%, respectively. In three other assays, we routinely activated cyclophosphamide with uninduced microsomes. The amount of activation varied in each case despite the same amount of added microsomal protein. Most of the variation was due to the amount of cytochrome P-450 present in the microsome preparation as higher P-450/protein ratios caused greater activation. We added the maximum tolerable amount of microsomal protein in order to see the highest possible activation of each microsomal preparation even though variation would be observed from experiment to experiment.

Activation of cyclophosphamide increased the severity of malformations observed in surviving embryos. The head-eye malformations and skeletal defects (as evidenced by the curved spinal cord) were typical of cyclophosphamide-induced malformations in mammals [14,15]. Additionally,

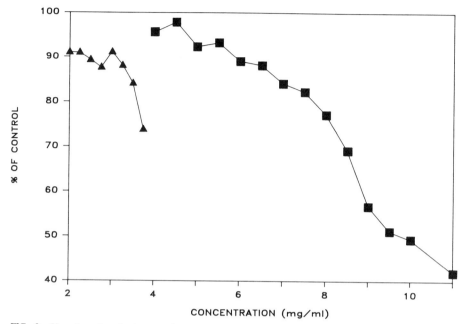

FIG. 2—*Unactivated and microsomal activated dose-response curves for growth after 96 h. Increasing concentrations of cyclophosphamide were used while all other system components were held to 1 X concentration. Microsomal protein was 0.12 mg/mL.* ■ *unactivated,* ▲ *activated.*

gut malformations were often observed. The involvement of most organ systems was readily apparent.

Figure 2 shows the effect of unactivated and activated cyclophosphamide on embryonic growth. Activation lowered the concentration at which growth effects were observed but did not change the slope of the curve. However, bioactivation led to the death of the embryos at about 3.8 mg/mL cyclophosphamide before their head-tail length was reduced greater than 30% compared to controls. This was in contrast to the unactivated group where 11 mg/mL cyclophosphamide inhibited growth by 58%.

Adjustment of System Components for Maximal Activation

System components. While it was clear that the uninduced rat liver microsomes converted cyclophosphamide into its embryotoxic and teratogenic forms, it was necessary to adjust all the components of the test solution so that the conversion rate was maximal. However, raising the components of the metabolic activation systems too high can cause mortality, malformation, and growth inhibition. This made it necessary to test all components of the metabolic activation system and adjust them to the highest level that did not cause malformation and death. A small amount of growth inhibition (≤7% of control) was considered tolerable. Microsomal protein, the NADPH generating system, and the antibiotics were all tested separately. The concentrations of system components used in the above dose-response experiment and listed in the "Materials and Methods" section proved to be optimal. When used in combination with one another but without cyclophosphamide, mortality and malformation rates were typically below 8% although 5% would have been preferred. Growth was significantly reduced by 7% compared to untreated control embryos as determined by the t-Test at $P = 0.05$. This was a fairly significant size reduction.

Microsomal protein. Microsomal protein can be used at concentrations between 0.1 and 0.12 mg/mL without raising the mortality and malformation rates to unacceptable levels (Fig. 3). In this particular experiment, the background mortality rate was very low at <4% but, as sometimes happens, the malformation rate for controls was high at 10%. Addition of microsomes without cyclophosphamide up to 0.12 mg/mL raised this malformation rate only 3% to an average of 13%. Addition of 3 mg/mL cyclophosphamide to the media indicated that increasing the amount of microsomal protein raised the mortality rate at protein concentrations above 0.08 to 1 mg/mL and the malformation rate above 0.06 mg/mL in a dose-response fashion. Figure 4 shows that increasing the concentration of microsomes to levels greater than 0.12 mg/mL begins to inhibit embryonic growth to unacceptable levels of >7%. Addition of 3 mg/mL cyclophosphamide results in growth being inhibited by >7% at concentrations of microsomal protein exceeding 0.035 mg/mL.

NADPH generator. Table 1 shows the effect of raising the amount of NADPH generator system on embryo mortality, malformation, and growth. The generator concentration (1 X) listed in "Materials and Methods" caused a 2.5% increase in mortality and 4.1% increase in malformation as well as a slight but significant reduction in growth. Higher concentrations (10 and 20 X) caused an unacceptable increase in the malformation rate.

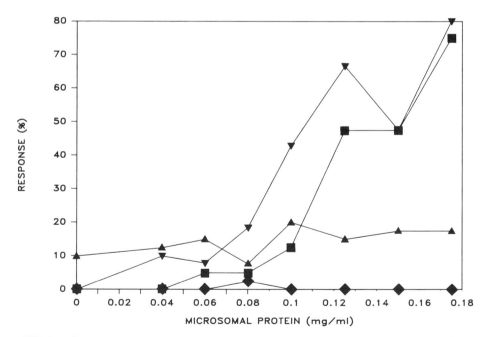

FIG. 3—*Effect of microsomal protein concentration on* Xenopus *embryo mortality and malformation in the presence or absence of cyclophosphamide. Increasing concentrations of uninduced rat liver microsomes were added and the effect on mortality and malformation recorded after 96 h.* ♦ *mortality; microsomes only;* ■ *mortality, microsomes, and 3 mg/mL cyclophosphamide;* ▲ *malformation, microsomes only;* ▼ *malformation, microsomes, and 3 mg/mL cyclophosphamide.*

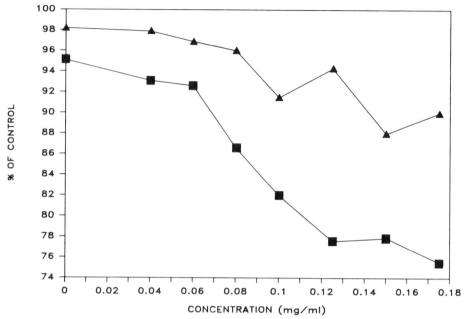

FIG. 4—*Effect of microsomal protein concentration on* Xenopus *embryo growth in the presence of cyclophosphamide. Increasing concentrations of uninduced rat liver microsomes were added and the effect on growth (head-tail length) recorded after 96 h.* ▲ *microsomes only,* ■ *microsomes and 3 mg/mL cyclophosphamide.*

Antibiotics. Each concentration of test substance also contained antibiotics in order to inhibit the growth of bacteria which would otherwise use the microsomes as a substrate and then destroy the embryos. We tested the effects of penicillin, streptomycin, gentamycin, and kanamycin on embryo mortality, malformation, and growth. At the 100 μg/mL (or 100 units/mL) concentration used in the experiments, the mortality and malformation rates were not above control values (data not shown). However, gentamycin caused significant mortality and malformation at concentrations of 200 μg/mL and above. For growth, the concentration of antibiotics used actually enhanced growth from what was observed for controls (Fig. 5). Figure 5 shows that for streptomycin and

TABLE 1—*Effect of the NADPH generator system on* Xenopus *96 h mortality, malformation, and growth.*[a]

Treatment	Mortality, %	Malformation, %	Length of Control, %
Control (with pen-strep)	2.5	3.8	100.0
0.1 X generator	2.5	5.1	98.7
1 X generator	5.0	7.9[b]	96.3[b]
10 X generator	5.0	18.4[b]	88.9[b]
20 X generator	5.0	86.8[b]	81.0[b]

[a] There were 40 embryos in each group. The X 1 generator concentration is that shown in "Procedure." Concentrations of each component were multiplied by the factor shown in the Table.

[b] Significantly reduced from the control length at $P = 0.05$ as determined by the tTest for grouped observations.

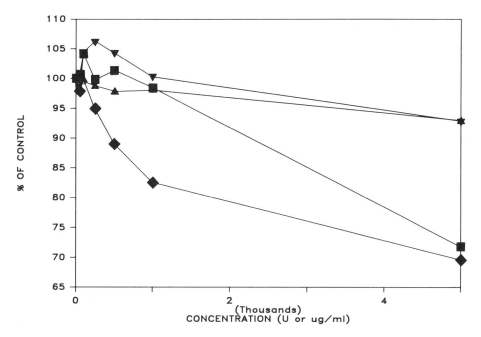

FIG. 5—*Effects of antibiotics on* Xenopus *growth. The effect of different concentrations of antibiotics on* Xenopus *embryo growth after 96 h of exposure. Concentration given is in μg/mL except for penicillin where the concentration was given in units/mL.* ■ *streptomycin,* ▼ *kanamycin,* ♦ *gentamycin,* ▲ *penicillin.*

kanamycin growth was actually increased by almost 5% over controls even at concentrations of 500 μg/mL. Gentamycin proved to be unacceptable compared to the other three antibiotics. For routine use in the assay we used penicillin and streptomycin at 100 units/mL and 100 μg/mL, respectively. When combined together at these concentrations, the two antibiotics had no more effect on any of the endpoints than when used alone.

Induction of Microsomal P-450 with Aroclor 1254

We performed a single experiment with Aroclor 1254–induced rat liver microsomes (Table 2). We used 0.1 mg/mL microsomal protein while the concentrations of the other metabolic activation system components remained the same. At 3 mg/mL cyclophosphamide, there was a large increase in mortality and malformation after only 48 h of exposure. This rate exceeded the 96-h results for uninduced rat liver microsomes at 3 mg/mL cyclophosphamide, demonstrating the high degree of activation obtained. By 96 h, all the embryos were dead in both the 3 and 3.5 mg/mL cyclophosphamide concentrations. Carbon monoxide pretreatment of microsomes greatly reduced mortality and malformation. This showed the involvement of cytochrome P-450, which was selectively inactivated by carbon monoxide. Aroclor metabolites remaining in the preparation after microsome preparation could not have caused the observed damage because carbon monoxide treatment allowed near normal development and growth.

TABLE 2—*Metabolic activation of cyclophosphamide by Aroclor 1254–induced rat liver microsomes as measured by FETAX.*[a]

Treatment	48 h		96 h	
	Mortality, %	Malformation, %	Mortality, %	Malformation, %
Control (with pen-strep)	1.25	8.0	5.0	9.21
3 mg/mL unactivated cyclophosphamide	6.25	6.25	6.25	10.67
3 mg/mL activated cyclophosphamide	72.5	95.45	100.0	...
3 mg/mL cyclophosphamide treated with CO-gassed microsomes[b]	2.5	19.2	2.5	12.82
3.5 mg/mL unactivated cyclophosphamide	3.75	6.25	6.25	9.33
3.5 mg/mL activated cyclophosphamide	86.25	100.0	100.0	...
3.5 mg/mL cyclophosphamide treated with CO-gassed microsomes[b]	11.25	39.4	26.25	23.73

[a] There were 80 embryos in each group. Rats were injected intraperitoneally with a single dose of 500 mg/kg (500 mg/mL in corn oil) of Aroclor 1254, five days prior to preparation of microsomes. The same procedure for microsome preparation was followed as described in "Procedure" except that 0.5% bovine serum albumin (Fraction V) was added to the homogenization and centrifugation buffers in order to bind possible Aroclor metabolites.

[b] P-450 was inactivated by treating the sample with a small amount of dithionite and then bubbling CO gas through the microsomal sample for 3 min.

Discussion

For *in vitro* teratogenesis screening assays to serve as suitable tests for prioritizing chemicals for further testing, they must duplicate mammalian metabolism as much as possible. Proteratogens that evaded detection in FETAX would be converted to their active teratogenic forms in humans and birth defects would result. This necessitated the development of a metabolic activation system for FETAX.

In our dose-reponse experiment using cyclophosphamide (Fig. 1), we were unable to obtain significant mortality for unactivated cyclophosphamide although early range finding experiments suggested a 96-h LC_{50} of 10 to 11 mg/mL. We attribute this failure mainly due to the wide range of concentrations used in the range-finding test. We also had to switch lots of cyclophosphamide and use a different pair of frogs for breeding. These changes may have contributed to our inability to obtain mortality in the range expected. For the purposes of this experiment, it was not necessary to obtain detailed dose-response data but only to show that cyclophosphamide was bioactivated. Figure 1 shows that this bioactivation clearly occurred.

We have successfully employed uninduced rat liver microsomes as an *in vitro* metabolic activation system for FETAX. The embryotoxic, teratogenic and growth-inhibiting potential of the proteratogen cyclophosphamide were enhanced by microsomal activation compared to cyclophosphamide alone. In using microsomes as a bioactivating system, the amount of added microsomal protein and NADPH must be kept low. High concentrations of NADPH by itself can cause abnormal rat embryo development [14]. Additionally, it is necessary to use antibiotics to control the growth of bacteria. The penicillin and streptomycin used exhibited little toxicity and teratogenicity and even allowed for greater embryonic growth.

The malformations observed in the *Xenopus* embryos were similar to those detected in mammals, and microsomal activation of cyclophosphamide increased the severity of malformations observed [*14,15*]. The teratogenicity index of 1.4 for activated cyclophosphamide indicated it was a compound that was more toxic to *Xenopus* embryos than teratogenic, although the severity of the malformations caused after activation would necessitate its classification as a teratogen [*7*].

Previous results in other laboratories have indicated that early *Xenopus* embryos do not contain measurable P-450 activity but that 96 h embryos have inducible aryl hydrocarbon hydroxylase and NADPH cytochrome C reductase activities (T. D. Sabourin, Battelle Columbus Labs). Attempts in our lab to use rat liver S9 supernatant (Litton Bionetics®) as a metabolic activation system proved unsuccessful because the S9 supernatant was extremely toxic to the embryos. This result suggested that by using uninduced rat liver microsomes, toxicity from Aroclor metabolites and excess liver proteins could be avoided. Another advantage of using uninduced microsomes is that induction with Aroclor 1254 may actually reduce the amount of P-450 enzyme present in the liver that bioactivates the test substance. This was apparently the case in thalidomide bioactivation [*16*]. However, Aroclor induction would allow the reduction of the microsomal protein to levels that did not inhibit embryonic growth. It would also permit a greater number of tests to be run with the same batch of microsomes.

We were successful in a single experiment in bioactivating cyclophosphamide using Aroclor 1254–induced microsomes. The 100 000 × g avg centrifugation helped reduce Aroclor metabolites by leaving them in the supernatant as the microsomes were pelleted. The addition of the 0.5% weight by volume bovine serum albumin may also have helped remove excess metabolites. Carbon monoxide treatment of microsomes reduced their activity as expected, showing that the cytochrome P-450 system was involved in the bioactivation of the cyclophosphamide. Further studies need to be performed with Aroclor 1254 induction for the appropriate test compounds. We plan to test other substances requiring bioactivation in the near future as a means of validating the use of both uninduced and Aroclor-induced microsomes.

In summary, uninduced rat liver microsomes can be successfully used as a bioactivating system for FETAX. The concentration of microsomes used must be limited to <0.12 mg/mL microsomal protein and 1 X generator system. Penicillin and streptomycin must be used to inhibit bacterial growth but do not deleteriously affect any of the FETAX endpoints. Induced microsomes show promise of increased activation using a lower amount of microsomal protein, but further experimentation is needed to show that Aroclor or its metabolites do not interact to alter assay results.

Other methods of bioactivating proteratogens need to be tried in order to find the best metabolic activation system for FETAX. Rat embryos have previously been cocultured with adult hepatocytes [*17*]. The hepatocytes were able to activate cyclophosphamide more efficiently than S9 supernatant and did not prove toxic to the embryos. The main problem with this technique is the difficulty of cell culture and the danger of toxicity to the hepatocytes. We have already succeeded in finding media in which frog embryos and frog hepatocytes can coexist and intend to study this system as well in the near future.

Acknowledgments

We wish to thank J. Steve Bell and Debbie Newell for their technical assistance. This work was supported by a grant from the Oklahoma State University Center for Water Research and Reproductive Hazards in the Workplace Grant No. 15-30 from the March of Dimes Birth Defects Foundation. Douglas A. Dawson was a recipient of a Presidential Fellowship in Water Resources from Oklahoma State University.

References

[1] Best, J. B. and Morita, M., "Planarians as a Model System for *in vitro* Teratogenesis Studies," *Teratogenesis, Carcinogenesis and Mutagenesis*, Vol. 2, 1982, pp. 277–291.

[2] Schuler, R., Hardin, B. D. and Niemer, R., "*Drosophila* as a Tool for the Rapid Assessment of Chemicals for Teratogenicity," *Teratogenesis, Carcinogenesis and Mutagenesis*, Vol. 2, 1982, pp. 293–301.

[3] Greenberg, J., "Detection of Teratogens by Differentiating Embryonic Neural Crest Cells in Culture: Evaluation as a Screening System," *Teratogenesis, Carcinogenesis and Mutagenesis*, Vol. 2, 1982, pp. 293–301.

[4] Sadler, T. W., Horton, W. E. and Warner, C. W., "Whole Embryo Culture: A Screening Technique for Teratogens," *Teratogenesis, Carcinogenesis and Mutagenesis*, Vol. 2, 1982, pp. 243–253.

[5] Dumont, J., Schultz, T. W., Buchanan, M., and Kao, G. in *Short-Term Bioassays in the Analysis of Complex Mixtures III*, Waters, Sandhu, Lewtas, Claxton, Chernoff, and Nesnow, Eds., Plenum, New York, 1983, pp. 393–405.

[6] Smith, M. K., Kimmel, G. L., Kochhar, D. M., Shepard, T. H., Spielberg, S. P., and Wilson, J. G., "A Selection of Candidate Compounds for *in vitro* Teratogenesis Test Validation," *Teratogenesis, Carcinogenesis and Mutagenesis*, Vol. 3, 1985, pp. 461–480.

[7] Kimmel, G. L., Smith, K., Kochhar, D. M., and Pratt, R. M., "Overview of *in vitro* Teratogenicity Testing.: Aspects of Validation and Application to Screening," *Teratogenesis, Carcinogenesis and Mutagenesis*, Vol. 2, 1982, pp. 221–229.

[8] Kitchin, K. T. and Woods, J. S., "2,3,7,8-Tetrachlorobenzo-p-Dioxin Induction of Aryl Hydrocarbon Hydroxylase in Female Rat Liver, Evidence for *de novo* Synthesis of Cytochrome P-450," *Molecular Pharmacology*, Vol. 14, 1978, pp. 890–899.

[9] Bradford, M. M., "A Rapid and Sensitive Method for the Quantitation of Microgram Quantities of Protein Utilizing the Principle of Protein-Dye Binding," *Analytical Biochemistry*, Vol. 72, 1976, pp. 248–254.

[10] Dent, J. G., Schnell, S., Graichen, M. E., Allen, P., Abernathy, D., and Couch, D. B., "Stability of Activating Systems for *in Vitro* Mutagenesis Assays: Enzyme Activity and Activating Ability Following Long-Term Storage at $-85°C$," *Environmental Mutagenesis*, Vol. 3, 1981, pp. 167–179.

[11] Nieuwkoop, P. D. and Faber, J., *Normal Table of* Xenopous Laevis *(Daudin)*, 2nd ed., North Holland Publishing Co., Amsterdam, 1975.

[12] Tallarida, R. J. and Murray, R. B., *Manual of Pharmacologic Calculations with Computer Programs*, Springer-Verlag, New York, 1980, pp. 1–150.

[13] Fantel, A. G., Greenaway, J. C., Juchau, M. R., and Shepard, T. H., "Teratogenic Bioactivation of Cyclophosphamide *in vitro*," *Life Sciences*, Vol. 25, 1979, pp. 67–72.

[14] Kitchin, K. T., Schmid, B. P., and Sanyal, M. K., "A Coupled Microsomal-Activating/Embryo Culture System: Toxicity of Reduced Betanicotinamide Adenine Dinucleotide Phosphate (NADPH)," *Biochemical Pharmacology*, Vol. 30, 1981, pp. 985–992.

[15] Mirkes, P. E., "Cyclophosphamide Teratogenesis: A Review," *Teratogenesis, Carcinogenesis and Mutagenesis*, Vol. 5, 1985, pp. 75–88.

[16] Braun, A. G., Harding, F. A., and Weinreb, S. L., "Teratogen Metabolism: Thalidomide Activation is Mediated by Cytochrome P-450," *Toxicology and Applied Pharmacology*, Vol. 82, 1986, pp. 175–179.

[17] Oglesby, L. A., Ebron, M. T., Beyer, P. E., Carver, B. D., and Kavlock, R. J., "Co-culture of Rat Embryos and Hepatocytes: *in vitro* Detection of a Proteratogen," *Teratogenesis, Carcinogenesis and Mutagenesis*, Vol. 6, 1986, pp. 129–138.

John Hadjinicolaou[1] and Gilles LaRoche[2]

Behavioral Responses to Low Levels of Toxic Substances in Rainbow Trout (*Salmo Gairdneri,* Rich)

REFERENCES: Hadjinicolaou, J. and LaRoche, G., "**Behavioral Responses to Low Levels of Toxic Substances in Rainbow Trout (*Salmo Gairdneri,* Rich),**" *Aquatic Toxicology and Hazard Assessment: 10th Volume, ASTM STP 971,* W. J. Adams, G. A. Chapman, and W. G. Landis, Eds., American Society for Testing and Materials, Philadelphia, 1988, pp. 327–340.

ABSTRACT: An avoidance apparatus was designed with a 10-m-long channel to obtain time-lapsed, three-dimensional display of fish positioning. With this new apparatus, effects of the toxicant dodecyl sodium sulfate (DSS) and two different polymers and monomers were tested on rainbow trout. The results show that four types of avoidance-preference, dose-response curves may be observed and that at low concentrations (0.01 to 0.08 mg/L for the DSS and in a range of 0.4 to 4 mg/L for different monomers and polymers) some toxicants may attract rather than repel certain organisms. This response is interpreted as representing behavioral extension of hormesis, the name given to the stimulatory effects caused by low levels of potentially toxic agents. An interpretation of unexplained literature data presents findings that endorse the concept of behavioral hormesis.

KEYWORDS: avoidance behavior, hormesis, rainbow trout (*Salmo gairdneri,* Rich), dose-response curve, toxic substances

In all animals including humans, exposure to toxic chemicals can lead to functional anomalies. These derangements may produce numerous effects, including alterations in behavior [1]. In fish, certain toxicant-induced behavioral changes have been shown to result from impaired sensory nervous system function [2–4].

Avoidance-preference reactions to a variety of toxicants can be related to effects that they may exert on chemoreceptors [5–7,] mechanoreceptors [4], or both. Chemoreception plays a dominant role in mediating behavior of aquatic animals, because vision and hearing have limited ranges and also because chemoreceptive membranes are directly exposed to the environment and are not protected by external barriers or internal detoxifying systems [6].

The emergence of behavioral toxicology is a relatively recent development. Behavioral responses to toxicant exposures may be critical for the survival of species and ecosystems. It may also provide an empirical basis for legislative regulations on acceptable concentrations of pollutants in the environment [8]. Experimental behavior toxicology can detect and measure behavioral effects of pollutants and offers sensitive evaluations of subtle changes on life expectancy.

Southam and Ehrlich [9] first proposed the term "hormesis" to describe "a stimulatory effect of subinhibitory concentrations of any toxic substance on any organism," and this definition is generally accepted. Specifically, hormesis is the name given to apparent or real stimulatory effects

---

[1] Assistant professor, Civil Engineering Department, McGill University, Montreal, Quebec, H3A 2K6, Canada.
[2] Research scientist, SNC Engineering Consulting Group, Montreal, Quebec H5B 1C8, Canada.

caused by low levels of potentially toxic agents [*10*]. Hormesis is a general phenomenon in which exposure of various organisms to traces or low levels of some toxic substances or agents actually stimulates physiological mechanisms in a manner that appears to stimulate growth and possibly improves chances of survival [*11*]. In recent years some interpretations have been given to stimulatory effects of low toxicant concentrations in a wide variety of organisms [*10*].

But what are the physiological mechanisms associated with hormesis? Stebbing, in his paper on the stimulation of growth [*10*], examines possible explanations: (a) that hormesis might be due to specific properties of the agents that enhanced growth; (b) that hormesis is a consequence of a response common to different organisms to some stimulus shared by different toxicants.

The purpose of this study was to examine if the physiological concept of hormesis can be extended to behavioral responses of fish at low toxicant concentrations. Using results of avoidance-preference tests and relating types of concentration-response curves to behavioral responses of fish to specific toxicants, the primary objective of this paper was to evaluate the possibility of endorsing the concept that at low concentrations some toxicants may attract organisms in a manner suggesting hormesis.

Materials and Methods

The entire design of the experimental setup of the flow through avoidance apparatus used in this study (Fig. 1) was a modification of the more commonly used steep gradient system to test avoidance preference reactions [*21–23*] and contained five basic components: (a) the water treatment facilities; (b) the holding facilities; (c) the temperature control system; (d) the channel; and (e) the data acquisition facilities [*12–14*].

The water treatment facilities contained an activated carbon unit and ultraviolet lights to reduce bacterial counts. The holding facilities contained one storage reservoir (Tank A), two holding tanks (Tanks B and C) for different fish sizes, and one acclimation Tank (Tank E) just before the channel. An automatic temperature control unit provided a continuous temperature of 15°C in the system with the help of submerged heating or cooling units in Tank A.

The experimental channel was 914.4 cm long, 30.5 cm wide, and had a continuous water elevation of 33.02 cm (max up to 45.7 cm). It was divided into five continuous sections of 182.9 cm each. The first section was constructed of Plexiglass to allow for jet holes and a variety of diffusers, while the other four were made of glass. In the first section of the channel, 152.4 cm from the front gate and 30.5 cm from the second section, provision was made for the injection of the effluents and dyes from the effluent tank.

A 0.6-cm-thick plexiglass barrier with holding devices was placed in the center of the channel, extending from 10.2 cm upstream of the first section through to the third section (426.7 cm) (Fig. 1). This separation unit was introduced to give the fish a chance to select either the contaminated or the noncontaminated side of the channel.

Because of the probable interest in a three-dimensional analysis of the positioning of the fish in the channel, mirrors were placed along one side of the channel. The mirrors were placed at an angle of 45° and, when photographed from above, both the horizontal and the vertical positioning were observed on the screen of television monitors. Consequently, both the avoidance and preference for the contaminated or noncontaminated side, and the distribution along the length of the channel or the depth of the individual fish from the surface, was recorded on videotape. Direct top-view coverage was provided by four overlapping video cameras along the length of the channel.

For rapid visual positioning of fish, both the bottom and one side of the entire channel were divided into 10.2 cm by 10.2 cm squares with black stripes. Each of the squares was color coded horizontally and vertically to aid in fish positioning over time.

Because natural light was inadequate for video recording and effective monitoring, fluorescent

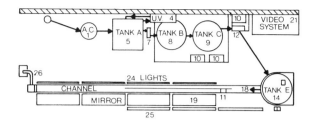

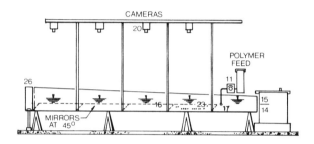

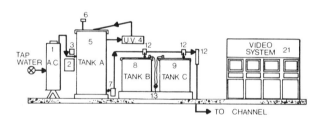

Number	Description	Number	Description
1	Activated Carbon Unit	14	Acclimation Tank E
2	Heating Control Unit	15	Front Gate to Channel
3	Automatic Temperature Control	16	Channel
		17	Injection Point
4	Ultraviolet Light	18	Centre Separation Unit
5	Storage Reservoir (Tank A)	19	Mirrors
6	Heating Coil	20	Video Cameras
7	Main Pump	21	Video System
8	Fish Holding Tank B	22	Thermometer
9	Fish Holding Tank C	23	Channel Position Indicators
10	Small Water Coolers (Optional)	24	Channel Lighting
		25	Sight and Sound Barriers
11	Polymer Storage & Pumping	26	End Gate and Drainage
12	Flowmeters		
13	Large Cooler for Acclimation Tank & Channel		

FIG. 1—*Plan view, side view of channel, and side view of fish-handling apparatus.*

lights were placed along the length of the channel, which resulted in a light intensity of 30 lx at the water surface. These lights were maintained on a 12-h photo period.

The four cameras covering the channel were mounted on moveable trolleys 289.6 cm above the water surface of the channel. Each camera could cover approximately 182.9 cm of the channel, resulting in total coverage of the area where the fish were able to swim.

Pictures from the four cameras were displayed on a separate video monitor for each section of the channel. Simultaneously with the pictures, long-playing video recorders recorded video signals for later analysis. Four video cassettes stored every experiment. For experiments of 2-h duration or less, the recording was continuous.

Following an experiment, the positions of fish at 5-min intervals were studied and evaluated. At these intervals, the tapes were stopped and static distributions of fish in the channel were quantitatively determined by a count of the number of fish within each 10.2-cm^2 section described earlier. The result of each count was recorded on counting sheets for subsequent statistical analysis.

The fish used were *Salmo gairdneri,* Rich (rainbow trout) one to two months old, purchased from two commercial hatcheries in the vicinity of Montreal. Preparatory experiments with different numbers of fish (20 to 200) were conducted to evaluate the optimum number of fish per experiment, taking into account the dimensions of the channel and the elimination of schooling or territorial bias factors. Ultimately, for every experiment 100 new fish were used and used only once. Three experiments were conducted for each pollutant concentration after an adequate number of control runs. Of the three experiments, two injected the pollutant from the left side and one from the right side of the channel to evaluate the presence of any side bias factor.

The control runs included: (a) the establishment of the general disposition of fish (baseline) without any pollutant (nine tests); and (b) the evaluation of the effect of the water jet (six tests).

Table 1 illustrates the results of these experiments, showing a 5% preference of the fish to align themselves with a small hydraulic disturbance in the flow field, which was incorporated in all subsequent analyses of the results so that the side bias factor was negligible.

The avoidance tests described herein contain multiple experiments with: (a) DSS (dodecyl sodium sulfate); and (b) polymers [polyacrylamide and copolymer and poly (dimethyldiallyl ammonium chloride)] and monomers [(polyacrylamide and poly (dimethyldiallyl ammonium chloride)].

DSS was first used in 1970 as a reference toxicant [15]. It has since been used widely in pollution bioassays because of its reproducible LC_{50} value with static toxicity tests (96-h LC_{50} = 5 to 7 mg/L for rainbow trout in fresh water) [20].

Polyelectrolytes combine the properties of polymers and electrolytes. Essentially, they are water soluble polymers with ionized groups [16]. Polyelectrolytes may be natural or synthetic and either inorganic or organic.

Synthetic polyelectrolytes—hereafter referred to as "polymers"—are in widespread use for

TABLE 1—*Results of control runs and side bias factors.*

		Time Spent on Respective Side of Apparatus as % of Total Time		
		Baseline	Acclimation Jet	
Preference-avoidance reaction	Side A	49.0	54.0	Injected Side A
	Side B	51.0	46.0	
	Side B	50.5	55.0	Injected Side B
	Side A	49.5	45.0	

treatment of water and wastewater. They are used as coagulant aids, as flocculants and flocculant aids, and as aids in sludge dewatering. Even though polymers have been used widely in effluent treatment, mineral processing, and other industrial processes, they are rarely used in the clarification of water for public consumption [*17*]. This was primarily due to concern about the toxicity of these compounds. As far as can be ascertained, the present study was the first instance where avoidance tests have been used on polymers and monomers.

Experimental Results
Tests with DSS

Twenty-seven tests were conducted with DSS at nine different concentrations. A behavior response curve appears in Fig. 2. Each point represents the average value of the avoidance reaction of a 2-h experiment in which 100 fish were used. Preference for the toxicant was observable until a concentration of 0.07 mg/L was reached.

At higher concentration an avoidance reaction appeared, indicating that the fish found the DSS objectionable even at concentrations much below the LC_{50} at 96 h (approximately 0.1 mg/L). At concentrations higher than 0.8 mg/L (Fig. 2), clear avoidance occurred. When the fish were exposed to the LC_{50} concentration, the avoidance behavior exhibited 65% (Fig. 2).

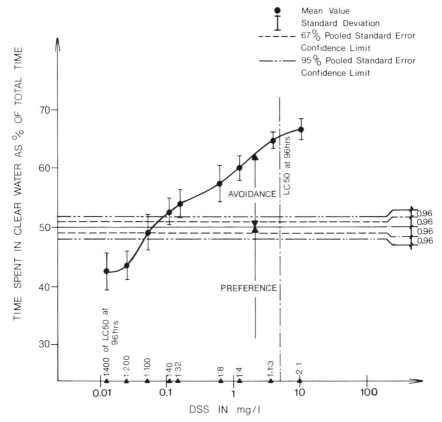

FIG. 2—*Avoidance reaction of DSS.*

TABLE 2—Results of the avoidance reaction of DSS.

Concentration, mg/L	Ratio of 96 h LC_{50}	Avoidance Response Time Spent in Clear Water as % of Total Time	Standard Deviation	Mean Standard Error	Pooled Standard Error
0.02	1:400	43	6 (± 3)	1.22	67% confidence limit
0.04	1:200	44	5 (± 2.5)	1.02	= 0.96 and 95%
0.07	1:100	49	5.5 (± 2.75)	1.12	confidence limit = 1.92
0.15	1:40	52.5	5 (± 2.5)	1.02	
0.2	1:32	54	5 (± 2.5)	1.02	
0.8	1:8	58	5.5 (± 2.75)	1.12	
2	1:4	60.5	4 (± 2)	0.81	
5	1:1.3	64	3 (± 1.5)	0.61	
10	2:1	66	3.5 (± 1.75)	0.71	

Considering the standard error of the avoidance values as shown in Table 2, a pooled standard error can plot probability parallel lines ±0.96% from the 50% line within a 67% confidence limit and ±1.92% within a 95% confidence limit.

Significant preference was considered to be below the interception of the 67 or 95% confidence limits lines (49.08 to 48.16%) with the dose-response curve (0.06 to 0.07 mg/L).

The finding that a significant preference occurred in very low concentrations can be considered as suggesting a hormetic response.

Tests with Polymers and Monomers

The polymers and monomers tested were:

1. Polymer A (63% by weight of dimethylaminoethyl acrylate quaternized with methyl chloride and 37% acrylamide).
2. Polymer C (dimethyldiallyl ammonium chloride).
3. Monomer A (polyacrylamide).
4. Monomer C poly (dimethyldiallyl ammonium chloride).

Table 3 shows the results of two static tests which were performed prior to the avoidance experiments in the federal laboratories of Environment Canada using *Salmo gairdneri*, Rich for the static 96-h LC_{50} and *Selenastrum capricornutum* for the algae tests. These results indicate significantly higher polymer acute toxicities than that observed for their constituent monomers.

Twenty-two avoidance-preference experiments were conducted with the two polymers and the two constituent monomers mentioned above using 100 fish each (*Salmo gairdneri*, Rich). Avoidance-preference curves for all four were derived.

The spectrum of the pollutant concentrations ranged from 0.3 to 20 mg/L, covering concentrations much lower than the lethal levels for the monomers and equal to or greater than the lethal levels for the polymers.

The results of the avoidance reaction of the polymers and monomers are shown in Table 4. The avoidance curves of Polymers A and C and Monomers A and C appear in Figs. 3 and 4 and 5 and 6, respectively.

Significant preference results were found for Polymer A and Monomers A and C for concentrations below the 67% confidence limit of the pooled standard error. In that case the preference threshold concentration was <0.4 mg/L for Polymer A and 0.42 and 0.33 mg/L for Monomers A and C,

TABLE 3—*Acute results for polymers and monomers.*

	Fish Static 96h—LC_{50}, mg/L	Algae 50% inhibition, mg/L
Monomer A	140	72
Polyelectrolyte A	0.4	2
Monomer C	300	220
Polyelectrolyte C	0.2	0.2

respectively. Using 95% standard error, significant preference appeared for Polymer A and Monomer A, with preference threshold concentrations 0.42 and 0.48 mg/L, respectively. In a recent review paper, "threshold" was defined in different ways [21]. Here, the term "preference threshold" refers to the first concentration causing preference.

TABLE 4—*Results of the avoidance reaction of Polymers A and C and Monomers A and C.*

						Pooled Standard Error	
	Dilution Ratio	Concentration, mg/L	Avoidance Reaction Time Spent in Clear Water as % of Total Time	Standard Deviation	Mean Standard Error	67% Confidence Limit	95% Confidence Limit
			POLYMERS				
A	1:2500	0.4	45	7.44 (±3.72)	2.15		
A	1:5000	0.66	42	3.54 (±1.77)	1.08		
A	1:1000	1.0	61	15.04 (±7.52)	3.54	2.9	5.8
A	1:750	1.33	68	13.9 (±6.95)	4.01		
A	1:300	2	72	13.2 (±6.6)	4.68		
A	1:250	4	85	7.44 (±3.72)	2.14		
C	1:2500	0.4	53.5	5.52 (±2.76)	1.47		
C	1:2000	0.5	58	7.12 (±3.56)	2.04		
C	1:1500	0.66	65	8.68 (±4.34)	2.24	2.0	4.0
C	1:1000	1.0	77	7.58 (±3.74)	2.28		
C	1:750	1.33	93	8.12 (±4.06)	1.96		
			MONOMERS				
A	1:3000	0.3	53	9.5 (±4.75)	2.74		
A	1:1500	0.66	45	7.0 (±3.50)	2.02		
A	1:750	1.33	45	5.2 (±2.61)	1.71	1.6	3.2
A	1:250	4	50	4.46 (±2.23)	1.28		
A	1:100	10	38	3.42 (±1.71)	0.98		
A	1:50	20	63	4.0 (±2.00)	1.15		
C	1:3000	0.3	50	6.62 (±3.31)	2.05		
C	1:2000	0.5	45	8.04 (±4.02)	2.32		
C	1:1000	1.0	55	9.58 (±4.79)	2.76	2.1	4.2
C	1:750	1.33	50	9.56 (±4.78)	2.38		
C	1:500	2	49	4.62 (±2.31)	1.15		
C	1:250	4	48	6.78 (±3.39)	2.4		

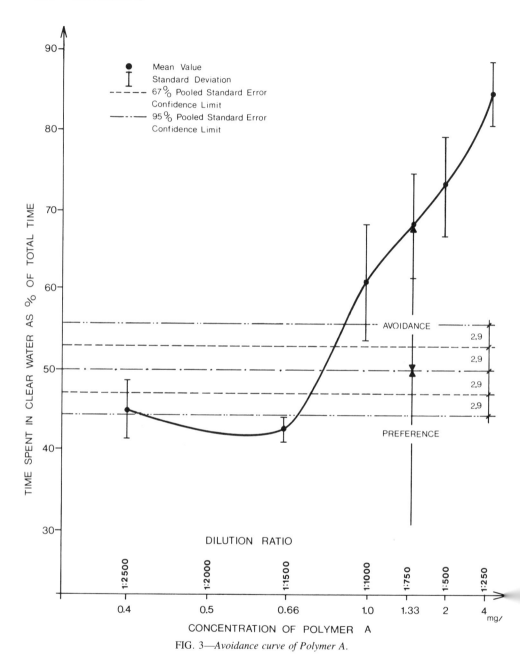

FIG. 3—*Avoidance curve of Polymer A.*

Discussion

With some substances known to be toxic at certain concentrations, current evidence shows that rainbow trout prefer waters containing lower levels of these substances. Why the fish are attracted by these lower concentrations is difficult to explain at present. Indeed, it is conceivable that slight

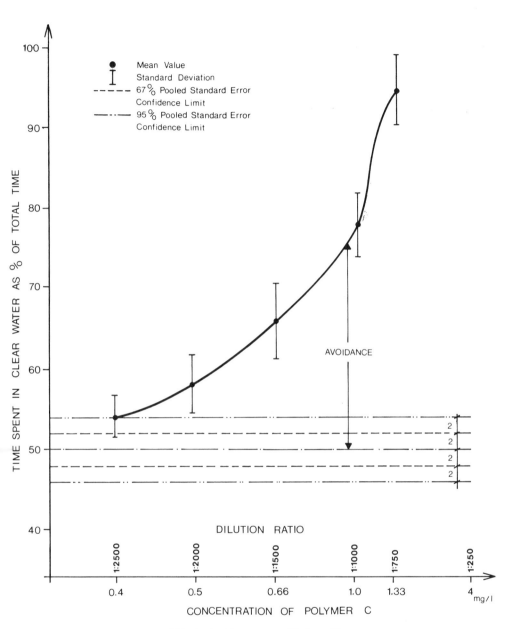

FIG. 4—*Avoidance curve of Polymer C.*

functional effects on chemo- and/or mechanoreceptors at lower levels could stimulate a positive attraction. The net result still remains that after repeated experiments young rainbow trout prefer low concentrations of some toxicants to purified, dechlorinated, and well-oxygenated Montreal drinking water. It may be that even purified Montreal drinking water used in both channels may contain unpalatable components to trout that are masked by low concentrations of potentially toxic

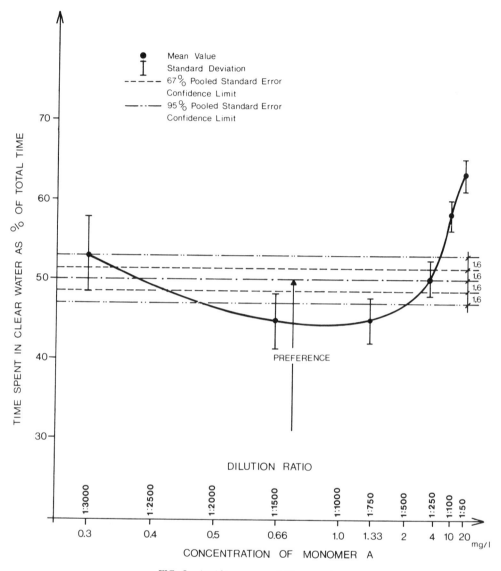

FIG. 5—*Avoidance curve of Monomer A.*

substances. In this instance, preference might be identified as unawareness if these low concentrations turn out to be toxic or damaging on a chronic basis.

Past observations [*10*] have shown that a wide variety of potentially toxic substances will, at certain levels, stimulate growth as well as a number of important vital functions in an equally varied number of species (hormesis). From these same observations it cannot be suggested that any of these levels are noxious or even suggest a degree of toxicity until responses are clearly associated with forms of biological impairments. At present, the only conclusion that may be drawn is that some products, toxic at certain concentrations, will stimulate growth, protein

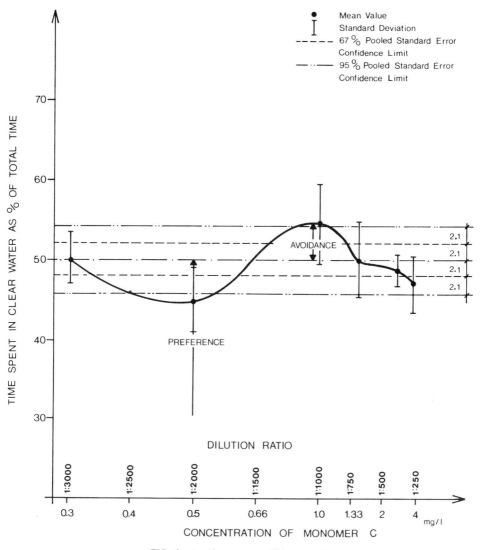

FIG. 6—*Avoidance curve of Monomer C*.

synthesis, etc. at other generally lower levels [*10,11*]. Further, it can be advanced that conditions inducing hormesis might attract organisms through appropriate stimulations or inhibitions of sensory receptors and ultimate neurophysiological responses.

To date, the concept of hormesis has not been extended to behavioral responses. However, since behavioral results from or is conditioned by a number of physiological reactions, it is probable that stimulatory effects may manifest as attraction at lower levels of some toxicants.

In 1975 Luckey, reviewing the types of effects caused by pollutants and toxic agents, identified four types of concentration-response curves (Fig. 7). The α-curve is the familiar pattern commonly observed for the effect of a toxic substance, showing no departure of the process or state from normal at low concentrations, followed by a progressive inhibition above a threshold concentration.

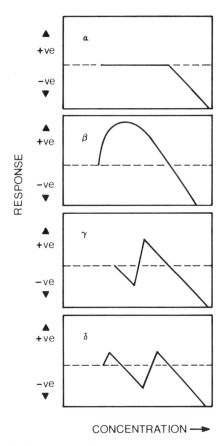

FIG. 7—*Various types of concentration-response curves identified by Luckey [11] in reviewing the types of effects of drugs and toxic agents.*

Curves β through δ show hormesis and describe responses most frequently observed when tests with low concentrations of various toxicants are analyzed. Curve β shows a single stimulatory peak at concentrations immediately below those that become progressively inhibitory.

For the Curves γ and δ, Luckey suggested that there are specific types of dose-response relationships and more data were needed to reconfirm them. Generally, it is well documented that hormesis is a response that occurs in living organisms in the form of a stimulation in biological activities at the cellular level [10].

The avoidance curve of DSS (Fig. 2) corresponded to a type α described in Fig. 7 and represented a typical response to a toxicant with a reproducible LC_{50}.

In Polymer A the avoidance curve (Fig. 3) resembled the dose-response Curve β described by Luckey (Fig. 7), Polymer C (Fig. 4) resembled the Curve α, and the avoidance curves of Monomers A and C (Figs. 5 and 6) looked like the Curves γ and δ, respectively.

In 1966 Sprague and Drury [18] published results of avoidance reactions of rainbow trout to representative pollutants. In these results the response of rainbow trout to solutions of free chlorine was difficult to explain. Original experiments were repeated by them with confirmation of the results [18].

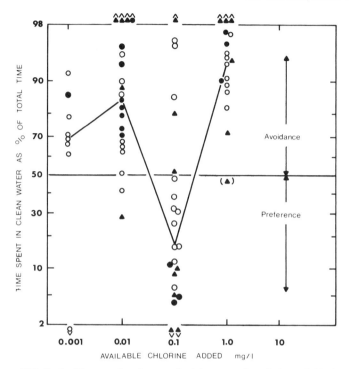

FIG. 8—*Avoidance and preference of rainbow trout for solutions of chlorine added as calcium hypochlorite* [*18*].

At the lowest concentration 0.001 mg/L, avoidance reaction was slight (Fig. 8). Most fish showed avoidance at a concentration of 0.01 mg/L of chlorine, lethal in twelve days according to laboratory tests [*18*]. Surprisingly, most trout appeared to prefer 0.1 mg/L of chlorine, which would kill them in four days [*18*]. These findings (Fig. 8) may correspond to a hormetic behavioral response corresponding to a dose-response Curve γ mentioned previously by Luckey (Fig. 7). This suggests that Curves γ and δ may be more frequently noticeable when a wider spectrum of toxicants at lower concentrations is tested. With chlorine and rainbow trout there is, however, a very distinct negative effect whereby fish appear to be attracted to a concentration of chlorine that is lethal within a few days. From this, it is therefore significant to point out that what rainbow trout prefer may not necessarily help in their survival. It can be further stated that such preference may not always be identified with optimization of behavior [*19*] or with the survival of desirable species and ecosystems.

Summary and Conclusions

1. Hormesis is a physiological concept that may be extended to behavioral responses of fish at low toxicant concentrations.

2. Results show that four types ($\alpha,\beta,\gamma,\delta$) of suggested dose-response curves appear possible (Figs. 3–6). Each dose-response curve gives a visual appreciation of the behavioral response of fish to specific toxicants at a range of concentrations (Figs. 3–6).

3. A greater understanding of the molecular biology of hormesis is needed before an understanding of avoidance and preference can be explained in functional terms.

4. Preference of toxicants at certain concentrations may represent a new dimension to the problem of environmental protection or management of species.

Acknowledgments

The authors are indebted to Terry Rigby, Norman Birmingham, Christian Blaise, Ronald Gehr, Robert Hutcheon, Joan Cornell, and Lesley-Ann Judge for their assistance.

References

[1] Beitinger, T. L. and Freeman, L., *Residue Revues,* Vol. 90, 1983, pp. 35–55.
[2] LaRoche, G., *Proceedings,* National Conference on Control of Hazardous Material Spills, 21–23 Mar. 1972, Houston, TX.
[3] LaRoche, G., Gardner, G. R., Eisler, R., Jackim, E. H., Yevich, P. P. and Zaroogian, G. E., ACS symposium, Sept. 1971, in *Bioassay Techiques and Environmental Chemistry,* Ann Arbor Science Publishers, Inc., 1973.
[4] Gardner, F. R., and LaRoche, G., *Journal of the Fisheries Research Board of Canada,* Vol. 30, 1973, pp. 363–368.
[5] Bardach, J. E., Fujiya, M, and Holl, A., *Science,* Vol. 148, No. 3677, 1965, pp. 1605–1607.
[6] Hara, T. J., Brown, S. B., and Evans, R. E., *Aquatic Toxicology,* J. O. Driagu, Ed., Wiley, New York, 1983.
[7] Hara, T. J., *The Behaviour of Teleost Fishes,* T. Pitcher, Ed., Croom Helm Ltd., Kent, England, 1984.
[8] Mello, N. K., *Federation Proceedings,* 58th Annual Meeting of the Federation of American Societies for Experimental Biology, Vol. 34, No. 9, 1975, pp. 1832–1854.
[9] Southam, C. M. and Ehrlich, J., *Phytopathology,* Vol. 33, 1943, pp. 517.
[10] Stebbing, A. R. D., *The Science of Total Environment,* Vol. 22, 1982, pp. 213–234.
[11] Luckey, T. D., *Heavy Metal Toxicity Safety and Hormology,* Stuttgart, Georg Thieme, Ed., 1975.
[12] Hadjinicolaou, J. and Spraggs, L. D., *Proceedings,* Eighth Aquatic Toxicity Workshop in Guelph, Canada, Canadian technical report of fisheries and aquatic science, 1982, pp. 68–82.
[13] Hadjinicolaou, J., "Water Pollution Control with Toxicant Avoidance," Ph.D. thesis, McGill University, 1983.
[14] Spraggs, L. D., Gehr, R. and Hadjinicolaou, J., *Water Science and Technology,* Vol. 14, 1982, pp. 1564–1567.
[15] LaRoche, G., Eisler, R. and Tarzwell, M., *Journal WPCF,* Vol. 42, No. 11, 1970, pp. 1982–1983.
[16] Packham, R. F., *Proceedings of the Society for Water Treatment and Examination,* Vol. 16, 1967, pp. 88.
[17] Hesner, B. D., *Journal of the Water Pollution Control Federation,* Vol. 113, 1975, pp. 13–16.
[18] Sprague, J. B. and Drury, D. E., *Advances in Water Pollution Research,* Vol. 1, proceedings of the fourth international conference held in Prague, J. H. Jenkins, Ed., Pergamon Press, New York, 1969.
[19] Howell, D. J., *American Zoology,* Vol. 23, 1983, pp. 257–260.
[20] Pessah, E., Wales, T. G. and Schnaider, G. R., *Proceedings,* 2nd Annual Canadian Aquatic Toxicity Workshop, Toronto, Canada, pp. 93–127.
[21] Giattina, J. D., and Garton, R. R., *Residue Reviews,* Vol. 87, 1983, pp. 44–90.
[22] Giattina, J. D., Cherry, D. S. Larrick, S. R. and Cairns, J., *Transactions of the American Fisheries Society,* Vol. 110, No. 576, 1981.
[23] Cherry, D. S. and Cairns, J., *Water Research,* Vol. 16, No. 263, 1982.

Laboratory and Field Comparisons

Mostafa A. Shirazi[1] and LeVaughn N. Lowrie[2]

An Approach for Integration of Toxicological Data

REFERENCE: Shirazi, M. A. and Lowrie, L. N., "**An Approach for Integration of Toxicological Data**," *Aquatic Toxicology and Hazard Assessment: 10th Volume, ASTM STP 971*, W. J. Adams, G. A. Chapman, and W. G. Landis, Eds. American Society for Testing and Materials, Philadelphia, 1988, pp. 343–360.

ABSTRACT: The practice of reducing a dose-response test into a single number such as LC_{50} is convenient but inadequate for distinguishing between dissimilar modes of biological response. When the mean curves of two groups of tests with dissimilar slopes at LC_{50} cross over at a 50% line, they give rise to equal LC_{50} cross over at a 50% line, they give rise to equal LC_{50} numbers for both groups of tests, and if used exclusively to represent the results, the single LC_{50} will mask this difference. The sensitivity of an organism to an incremental change in the dose is measured by the slope of the dose-response curve at any dose level. The slope for some modes of toxic response can be initially zero at small dose, then increase substantially at higher values, and for other modes it can be infinitely large initially, then level off at higher values of the dose. The management of different chemicals requires different strategies that can be supported with a more comprehensive analysis of existing test data.

Several similar or replicate tests produce slightly different dose-response curves due to uncontrollable experimental and biological test factors. Continued focus on summarizing these tests by averaging their LC_{50} numbers hinders progress toward discovering a biological response mode from the dose-response points of similar tests.

The paper presents an approach for aggregating a group of similar test data into a single mean dose-response curve with error bounds quantifying the variability of the original data. Groups of similar tests are then classified using generalized exponential dose-response functions whose constants define modes of biological response and facilitate calculation of toxicity and sensitivity. Groups of similar data and classes of different groups of tests are used to demonstrate the application and the utility of the approach.

KEYWORDS: comparative toxicology, bioassay end points, toxicology scalers, centroid, data aggregation

Nomenclature

a	Form factor
COVD	Coefficient of variation STD/D
COVR	STR/R
COVX	STX/X
COVY	STY/R
D	Dose, ppm, or mg/L, also T
D_{50}	Dose at 50% response, ppm
DBAR	Dose coordinate of mean data, ppm
DI	Dose intercept, ppm

---
[1] Senior research scientist, Environmental Protection Agency, Corvallis, OR 97333.
[2] Associate scientist, Northrop Services, Corvallis, OR 97333.

DM	Dose at midrange, ppm, also TM
DP	Dose pivot (centroid coordinate), ppm
DR	Dose range, ppm
Exp[ ]	Exponential function $e^{-(kD)^a}$
k	Scale factor, 1/ppm
$\log(k)$, a	Ranking factor
R	Response in fraction of control
R_{50}	Response at 50%
RBAR	Response coordinate of mean data
RI	Response intercept
RR	Response range
RT	Response pivot (centroid coordinate)
S	Slope, sensitivity, 1/ppm
S_{50}	Sensitivity at 50% response, 1/ppm
SM	Sensitivity at midrange, 1/ppm
STD	Standard deviation of data relative to DBAR, ppm
STR	Standard deviation of data relative to RBAR
STX	Standard deviation of data relative to Exp[ ]
STY	Standard deviation of data relative to R = Exp[−x]
T	Toxicity, ppm
T_{50}	Toxicity at 50% response, ppm
TM	Toxicity at midrange, ppm
x	Scaled dose = $(kD)^a$

Toxicological tests of a vast number of chemicals and organisms are being conducted throughout the government and in private laboratories to evaluate potential impacts of toxic chemicals on the environment and to humans. Individual test data must be condensed for interpretation and use, but often only a single toxicity number such as LC_{50} [1,2] is reported for a whole test. Other biological information useful for environmental management can be extracted from the test results, but a consistent, general approach is lacking.

Typically, a dose-response relationship is obtained from exposing a group of similar organisms to a series of different, but constant, levels of toxicant concentrations and recording the number of organisms surviving a test. The duration of exposure is fixed, for example, 96 h in some aquatic toxicity tests and five days in avian dietary tests. The concentration series is preselected from preliminary range-finding trial tests so that the highest levels produce total mortality of test organisms within the exposure duration. There are as few as four and as many as ten treatment levels to a test with concentrations spaced geometrically in an ascending order and as many observations of response within the exposure duration. The entire number of observation points (say 16 to 100) are often reduced to a single end point such as 96-h LC_{50} to express the concentration that kills 50% of test organisms at the end of 96 h.

The single LC_{50} number is inadequate for distinguishing between different biological sensitivities to chemicals. Some chemicals produce a steep dose-response and a steep time-response at 50% response, while others produce a shallow dose-time-response or a shallow dose-response and a steep time-response at 50% response. The management of different chemicals requires different strategies that cannot be determined based on a single 96-h LC_{50}. The steepness of the response surface at the 50% point determines the change in survival due to incremental change in LC_{50} (or Lt_{50}). The steepness is a measure of sensitivity and can be quantified by the value of the slope (S_{50}) at LC_{50} (or Lt_{50}). In this paper, we will be concerned exclusively with dose-response curves and defer the consideration of exposure duration to another paper.

The values of LC_{50} and the slope at LC_{50} (that is, S_{50}) cannot be used to calculate other concentrations, for example, near zero dose. Therefore, test data must be condensed to provide information on more than a single point on the dose-response curve.

Information on the near-zero dose level can indicate if an organism lacks or possesses a defense mechanism for the tested chemical, depending on the slope of the curve at that point. If there is little or no defense, there will be a measurable change in the response at small dose and the slope of the curve at zero is (infinitely) large. If there is an inherent defense, there will be no measurable change in response at small dose and the slope at zero is zero.

Information on large doses can tell if an organism is metabolizing a chemical or rapidly losing resistance to it depending on the value of the slope. In the first situation, the value of the slope is small and the dose-response curve levels off rather gradually; in the second, there is a more rapid decline with dose.

These toxicological insights can be captured directly from the original data but are lost when the data are condensed into a single end point. Since original dose-response data are seldom published, reliance on a conventional single toxicity number (LC_{50}, etc.) leads to a continuing loss of valuable information in archived notebooks.

Scalers of a Test

When dealing with large and diverse bioassay results it is useful to identify certain scalers within the data that facilitate biological explanation, integration, and classification. This emphasis on scaling as a vehicle for better understanding the structure of the data will be a focal point in this paper. Shirazi and Hart [3] used the centroid coordinates of individual dose-response tests as scalers and succeeded in collapsing extensive and widely divergent crop root bioassay tests into a band of closely clustered dose-response points. The dose-response curve was explained by means of a simple exponential curve. Centroid coordinates are calculated mean values that determine the centers of gravity of an experimental curve along the dose and response axes. Their effectiveness in unifying data is a measure of their suitability as scalers of a test. We extend the approach by relating the centroid coordinates of a test to a new set of scalers that offer a greater degree of explanation of biological data.

These new scalers are the constants a and k of the family of generalized exponential survival functions $R = \text{Exp}[-(kD)^a]$, where D is the dose, R is the response, $R = 1$ at $D = 0$, $a > 0$ and $k > 0$. The general forms of these curves for several values of a and k are shown plotted in Fig. 1. The curves include inverse sigmoidal or inverse "S"-shaped forms, "L"-shaped curves, and intermediate shaped curves. They can approximate, but are not restricted to, the shape commonly represented by integrated normal curve (probit). They have been used in the past for analyzing survival time in engineering and biomedical sciences [7]. Weibull [8] made the first application and provided interpretation of the constants a and k. We emphasize the biological interpretation and use the number of organisms surviving a test as a response variable.

The constant a defines the shape of various dose-response curves and we refer to it as the form factor. A family of curves represented by $a > 1$ has zero slope at zero dose and can be produced by organisms displaying defense mechanisms to small levels of toxic materials or by chemicals that are harmless at small dose levels. Figure 1 shows the plots of such curves for $a = 2, 4,$ and 8. The greater the value of the form factor, the stronger the defense mechanism and the delayed response.

The family of curves represented by $a < 1$ has infinite initial slope at zero dose and can be produced by organisms having no defense mechanism to a chemical. Two curves, one for $a = 0.25$ and a second for $a = 0.5$, are plotted in Fig. 1 as examples of this type of response.

The form factor $a = 1$ represents intermediate situations where the initial response is proportional to the small levels of the dose.

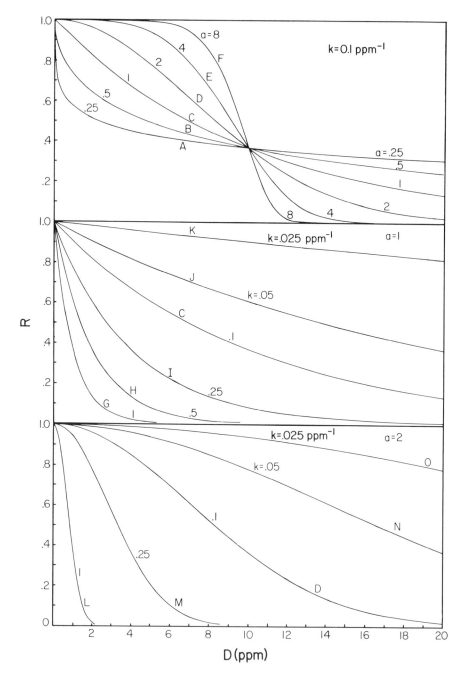

FIG. 1—*Generalized exponential dose-response functions* $R = Exp[-(kD)^a]$. *Three classes of response are defined (Fig. 1a) for values of* $a > 1$, $a = 1$, *and* $a < 1$, *corresponding respectively to flat, proportional, and steep responses at small values of the dose and to steep, gradual, and flat responses at large values of the dose. A continuous range of sensitivity and toxicity classes is definable for each response class by various values of* k *(Fig. 1b, 1c).*

The form factor influences the overall slope of the dose-response function at intermediate and large values of the dose as well. Very steep slopes (for example, $a = 8$) are produced by organisms that rapidly lose resistance with an increased dose. Very shallow slopes are produced in a dose-response of organisms that became adaptive to a chemical after an adverse initial reaction or by organisms that metabolize a chemical (for example, $a = 0.5$).

The scale factor k is a measure of strength of toxicity of a chemical. Organisms may be thought to magnify the scale of the dose axis when a substance is extremely toxic and reduce the scale when the substance is less harmful by the level of severity of their response. When comparing two dose-response curves, one produced from an extremely toxic chemical with another produced from one less harmful, the first has a contracted dose axis relative to the more expanded dose axis of the second type. Examples of these two types of response are shown in Fig. 1 for $a = 1$, $k = 1.0$, and $k = 0.25$, corresponding to the first and the second types, respectively. The differences between the two curves are produced solely by different scale factors.

At any point along the exponential dose-response curve, the following relationships hold among the quantities D, R, a, k, and S

$$D = [-\ln(R)]^{(1/a)} k^{-1} \quad (1)$$
$$S = a R [-\ln(R)]^{(a-1)/a} k \quad (2)$$
$$S = a R [-\ln(R)] D^{-1} \quad (3)$$

Where S is the (negative) slope and $\ln(R)$ is the natural logarithm of the response. If, in particular, we let $R = 0.5$ in Eqs 1–3, we obtain the 50 percentile values of the dose (D_{50}) from Eq 1 and the slope (S_{50}) from Eq 2. If these equations represented real test data and D_{50} and S_{50} were used to summarize the test results, D_{50} would equal the values of LC_{50}, LD_{50}, EC_{50}, etc., commonly used to quantify toxicity, and S_{50} would equal the sensitivity to incremental changes in toxicity. Since we intend to apply these relationships to a wide range of classes of tests, we will denote the toxicity at 50% of a test by T_{50}.

Real test data may be represented by the above equations once the scalers a and k of a test are calculated. We will extend the centroid approach of Shirazi and Hart [3] to obtain a and k of a test directly from data. The centroid coordinates do not generally fall on a dose-response curve. Shirazi and Hart [3] introduced the concept of midrange characteristics of a test. The values corresponding to the response, dose, and slope will be denoted by RM, DM, and SM, respectively. These characteristics are uniquely related to the centroid coordinates of a test (Fig. 2). They define a point on a dose-response curve, and they can be used with Eqs 1, 2, and 3. The calculation procedure leading to midrange values is summarized as follows.

Let (D0,R0), (D1,R1), ..., (Dn,Rn) represent n + 1 dose-response points representing a test. Let A1, A2, ..., An denote the areas of n trapezoidal segments A1 = 0.5(R0 + R1)(D1 − D0), A2 = 0.5 (R1 + R2) (D2 − D1), ..., etc. The total area under the dose-response curve is A = A1 + A2 + ... + An. Then, the centroid coordinates DP and RP on the dose and the response axes are, respectively,

$$DP = [0.5(D0 + D1)A1 + \ldots + 0.5(Dn - 1 + Dn)An]/A$$
$$RP = 0.5[0.5(R0 + R1)A1 + \ldots + 0.5(Rn - 1 + Rn)An]/A$$

Reference to Fig. 2 facilitates understanding the derivation of midrange quantities from the centroid. If we draw two lines parallel to the axes through the centroid to intercept the dose-response curve, we obtain the dose intercept DI and the response intercept RI. RR and DR will define the effective ranges corresponding to the response and the dose, respectively. They are obtained from RR = RI − RP and DR = DI − DP. RM and DM are points located approximately midrange within RR and DR segments, respectively. The slope SM is calculated at midrange and

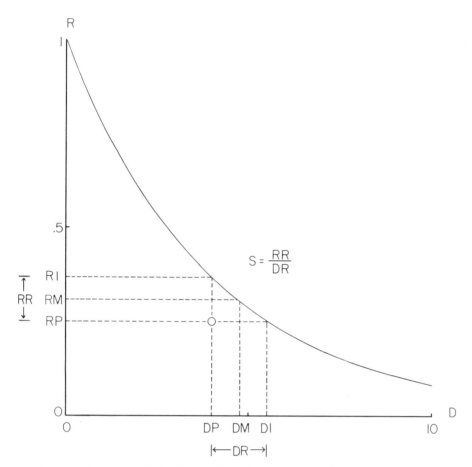

FIG. 2—*The midrange values RM and DM are located approximately halfway within the respective ranges RR and DR. The centroid coordinate DP defines the response intercept RI, and RP defines the dose intercept DI. The slope (or the sensitivity) S = RR/DR is defined for a test only when intercepts are obtainable without interpolation.*

approximated by SM = RR/DR or more precisely from the geometric mean of the slopes of three line segments between midrange and intercept. A test that fails to produce DI, that is, when the low dilution levels are terminated prematurely and a slope cannot be calculated from the data, that test will not be used.

Once the midrange values are obtained from the centroid, the scalers are calculated for a test from Eqs 1 and 3. These same scalers enable the calculation of toxicity (T_{50}) and sensitivity (S_{50}).

The representation of the mode of biological response with a and k improves when a and k are calculated from several similar tests, separately first for each test, then averaged to obtain the mean scalers for the group. A preferred approach is to aggregate the original data to discover the inherent structure of the dose-response curve by its scalers. The advantage is that the data can be fitted with any suitable curve if the above is found inadequate. Further, the experimental variability relative to the central tendency of the data can be calculated directly independent of any preimposed structure.

The calculation of variability along the dose-response curve is facilitated when evaluated at similar points for all tests. Shirazi and Hart [*3*] introduced a procedure that is applicable to aggregating a group of similar tests without scaling the data or a group of dissimilar tests after scaling the data. Their procedure leads to evaluation of experimental variability relative to mean aggregated curve without preimposing any structure (Steps 1–6 below) and relative to mean scaled (theoretical) exponential curve (Steps 7–10). The variability relative to the second curve consists of experimental variability and parameter estimation error.

1. Select an array Q1, Q2, ..., Qr on the response axis to be used as a common reference for interpolating among data of all tests. The intervals between consecutive Q's may be uneven but not so coarse as to smooth out the experimental fluctuation in the original data nor so fine and numerous as to add unnecessary detail and labor. The array 0.001, 0.005, 0.01, 0.05, 0.1, 0.2, 0.3, 0.4, 0.5, 0.6, 0.7, 0.8, 0.9, 1, ..., etc. may be used in many examples.

2. Interpolate between data points in a test to obtain dose values (D's) for each common response (Q's). Since neighboring data points in a test fluctuate relative to one another, the constant response lines Q1, Q2, ..., Qr may intercept the lines drawn between consecutive points more than once, each time producing one possible dose. For example, at Q1 there might be two dose values (D11, D12) from Test 1, one additional dose value (D13) from Test 2, ..., etc. for a total of q1 dose values at Q1 and q2 dose values at Q2, ..., etc.

3. Calculate the mean dose DBAR from all tests at each common response (Q's). That is,

$$DBAR1 = [D11 + D12 + \ldots + D1q1]/q1$$
$$DBAR2 = [D21 + D22 + \ldots + D2q2]/q2, \text{ etc.}$$

Corresponding to each mean dose coordinate DBAR, we calculate a mean response coordinate in Steps 4 and 5 below. The procedure in these two steps parallels the procedure of Steps 2 and 3 above.

4. Use the array [DBAR1, DBAR2, ..., DBARr] as common mean dose values to interpolate the original test data to obtain corresponding response values (R's). For example, there may be one response value (R11) at DBAR1 from Test 1, none from Test 2, ..., etc. for a total of p1 response values at DBAR1, p2 response values at DBAR2, ..., etc.

5. Calculate common mean response values (RBAR's) for all tests at each common mean dose. That is,

$$RBAR1 = [R11 + R12 + \ldots + R1p1]/p$$
$$RBAR2 = [R21 + R22 + \ldots + R2p2]/p2, \ldots, \text{etc.}$$

The mean curve defined by DBAR, RBAR represents the experimental data when there are more than one test in a group. The mean curve is used in our approach to calculate scalers a and k of a theoretical curve representing the data. It is also used to calculate the variability of the original data as in step 6 below.

6. Calculate the standard deviations STD1, STD2, at Step 3 for the dose and STR1, STR2 ... at Step 5 for the response as a means of describing the experimental variability for the group of tests relative to the mean dose-response curve (that is, DBAR, RBAR).

7. Calculate theoretical responses y11, y12, ..., etc. corresponding to x11, x12, ..., etc. at Step 3 from $y11 = \text{Exp}[-x11]$, $y12 = \text{Exp}[-x12]$, ..., etc. where, x11, x12, ..., etc. are scaled doses.

8. Calculate theoretical doses x11, x12, ..., etc. at $x11 = -\ln[y11]$, $x12 = -\ln[y12]$, ..., etc.

9. Calculate the mean theoretical response YBAR from Step 8 and the mean theoretical dose XBAR at Step 7.

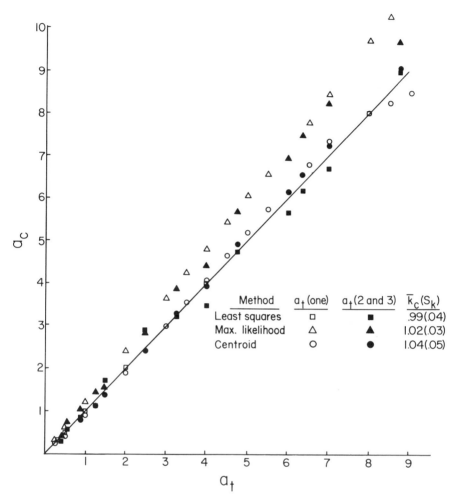

FIG. 3—*Comparison of calculated form factor and scale factor a_c, k_c using hypothetical tests with 15 evenly spaced data points and assumed theoretical a_t and k_t.*

10. Calculate the standard deviations along the dose axis (STX) from Step 7 and along the response axis (STY) from Step 8.

Parameter Estimation

The calculation of the scalers a and k is subject to parameter estimation error. The magnitude of the error depends on the estimation procedure used. The centroid approach is compared with the least squares approach and the maximum likelihood approach in Fig. 3. The latter two approaches are detailed in Lawless [7]. Hypothetical tests in Fig. 3 were generated from data points belonging to theoretical survival functions. The data points were spaced according to $R_i = 1 - (i - 1)/n$, $i = 1, \ldots, n$; n being the number of data points in a test. Figure 3 provides a particular example using n = 15. Other calculations were made for n = 5, 10, 20, and 25 points on a curve. In all cases, the least squares method was superior when only a single curve was represented.

Real data points are irregular and seldom all fall on a single curve. Neighboring data points in a real test belong to curves with slightly different a and k values. Hypothetical tests consisting of two and three form factors were constructed to simulate this situation and used to compare the three methods in Fig. 3. The centroid and the least squares method produced comparable results. The centroid approach was superior to the maximum likelihood approach in all but very small values of the form factor.

More than 50 calculations (m) were conducted with different values of n ranging from 5 to 25. The data were summarized by fitting them with a linear regression

$$a_t = b_0 + ba_c$$

where a_t is the theoretical form factor, a_c the calculated form factor, b_0 the zero intercept, and b the regression coefficient. We expect $b_0 = 0$ and $b = 1$ for a perfect estimation. The table below provides the comparison for the three methods and shows the centroid and

Method	b_0	b	r^2	m
Least squares	0.02	0.97	0.99	52
Maximum likelihood	−0.003	1.22	0.97	53
Centroid	−0.002	0.99	0.98	50

the least squares methods producing comparable results, and both are better than the maximum likelihood.

The Classification of Test Data

Figure 4 provides a graphic representation of Eqs 1, 2, and 3 showing the twin relationships among the scalers (a and k) and toxicity, sensitivity (T_{50} and S_{50}). A single point on the T-S plane of Fig. 4 represents a whole dose-response curve as indicated by the data points A, B, . . ., each locating a corresponding curve from Fig. 1.

A straight line in Fig. 4 represents a whole family of dose-response curves with a fixed form factor but varying values of scale factor. Theoretical curves representing responses to extremely toxic substances are represented by points along the upper left hand side of these lines. They define response curves having large scale factors. Curves representing less sensitive responses and low chemical toxicities fall on the lower right portion of the lines. These curves have small scale factors. Various sensitivity and toxicity classes are quantified by the loci of equal scale factors.

The upper family of curves ($a > 1$) represent theoretical responses with delayed reaction to low levels of toxicants, produced by organisms displaying initial defense mechanisms. The lower family of curves ($a < 1$) are responses where an organism offers no initial resistance.

It is worth noting that the integrated normal curve (probit) can be represented only by a single line in Fig. 4 corresponding to approximately $a = 5$ with the scale factor represented by the reciprocal of the standard deviation. The S_{50} and T_{50} coordinates of that line will not coincide with Fig. 4.

The sensitivity and toxicity values at 50 percentile can be used as a pair to classify relative rankings of biological effects of various chemicals using only a single point from the dose-response curve. However, the relative ranking will change for different points, for example, 90 percentile [9]. The use of the scalers as a pair enables classification independent of any particular percentile of the dose-response curve but relative to the overall form of the dose-response curve and the overall sensitivity of the organism to chemicals.

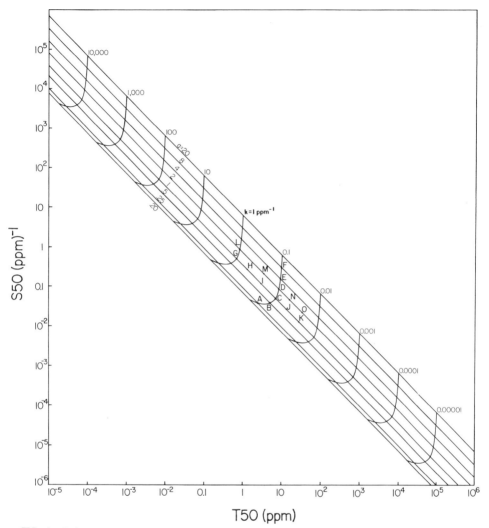

FIG. 4—*A dose-response curve appears as a single point (for example, A, B, . . . , from Fig. 1) on a toxicity-sensitivity plot. The values of the scalers k and a at a point are sufficient for reproducing the curves and are used to classify the biological modes of response.*

The scale factor may vary from 10^{-6} to 10^{+6} (1/ppm) while the form factor is generally less than 10 and often between 0.5 and 3.0. For this reason, it is convenient to use the logarithm of the scale factor while using the arithmetic value of the form factor. We will limit the class designation to two significant digits. As an example, the classification numbers of a few of the curves in Fig. 1 are: $[\log(k),a] = (-1.0,0.2)$ for A, $(0.0,1.0)$ for G, $(-2.0,1.0)$ for K, etc. Class designations are, of course, arbitrary.

Examples of Test Data

Ratsch [4] reported on extensive interlaboratory root elongation tests conducted on five crop roots using eight different chemicals. The data set was analyzed by Shirazi and Hart [3] using the centroid scaling approach. They did not classify the tests according to the scalers a and k. Our first example is a reexamination of the data using the classification and error analysis procedure of this paper. A second example is a data set analyzed by Callahan et al. [5] consisting of 62 chemicals and four species of earthworms. A third example consists of dietary tests using quail and 69 chemicals (Hill and Camardese [6]).

The solid curves passing through the data in Fig. 5 resulted from applying Steps 1 through 5 above. They represent the loci of scaled data mean DBAR, RBAR. The clustering is a graphic expression of the variability in the original data. The broken curve is the exponential function $R = \text{Exp}[-x]$, considered in our theory to be the limit of the scaled data.

The curves of the mean scaled dose-response data closely follow the simple exponential curve $R = \text{Exp}[-x]$. The agreement is good for a variety of tests, including a group of similar tests (Clover/2,4-D), several dissimilar groups of related tests (Crop Roots), or different classes of bioassay tests (Earthworm and Quail). This agreement of theory with data confirms the reasonableness of the scaling approach.

Figure 6 provides plots of standard deviations STX and STY for the aggregated data of Fig. 5. There are three curves on each plot, two standard deviation curves and one simple exponential dose-response curve. The numbers marked on the standard deviation curves refer to the sample size.

A coefficient of variation at a point is defined by the ratio of the standard deviation and the coordinate of the point on a dose-response curve. Figure 6 provides an overall mean coefficient of variation for each class of test. These numbers are useful for comparing the variability of various classes of tests on the same basis.

Tabulation and Ranking of Test Data

Table 1 summarizes the results, listing from right to left the overall mean coefficient of variation, the sensitivity S_{50}, the toxicity T_{50}, the ranking factor $(\text{Log}(k),a)$, number of tests in a group, the organism, the chemical, and a code for identifying the data on a T-S diagram of Fig. 7.

The data are ranked in the table first relative to the logarithm of the scale factor and then the form factor. Figure 7 displays this ranking for the data, better distinguishing the influence of different form factors. The middle straight line ($a = 1$) in the figure is the locus of all responses exhibiting an intermediate pattern between a distinct initial resistance to a toxicant ($a > 1$) and no initial resistance ($a < 1$).

A wide range of sensitivity and toxicity values are presented by the tests in Table 1. The most sensitive and toxic responses are obtained with earthworm contact tests and phenolic compounds (a6, b6, and c6). The same compounds and species of earthworm in soil tests produce substantially less sensitive and less toxic responses. On the other extreme is Carbowax, which produces the least toxic response with all crop roots tested.

The relative rankings of various responses may be modified when additional similar data are incorporated in the table. The number of tests in the table varies from a maximum of 24 for certain interlaboratory crop root tests to a single test for quail. The variabilities within a group of interlaboratory tests can be substantial (mean COVD up to 200%) raising concern about the exclusive reliance on limited number of tests to predict the response of an organism to a chemical. While this is an important consideration not to be overlooked when using such data, one can benefit from an aggregation of a class of similar tests whose variabilities have been quantified. An example detailing such application follows.

Table 1 contains single tests of various chemicals with quail, for example, test r8 with Phosmet. The aggregated scaled data for 69 chemicals using quail have been analyzed in Fig. 6d. The test r8 belongs to this population and our assumption is that it is subjected to the same level of experimental error as others in the population. Therefore, in the absence of more specific information, we can rely on the general data in Fig. 6d. Since we are interested in toxicity at 50%

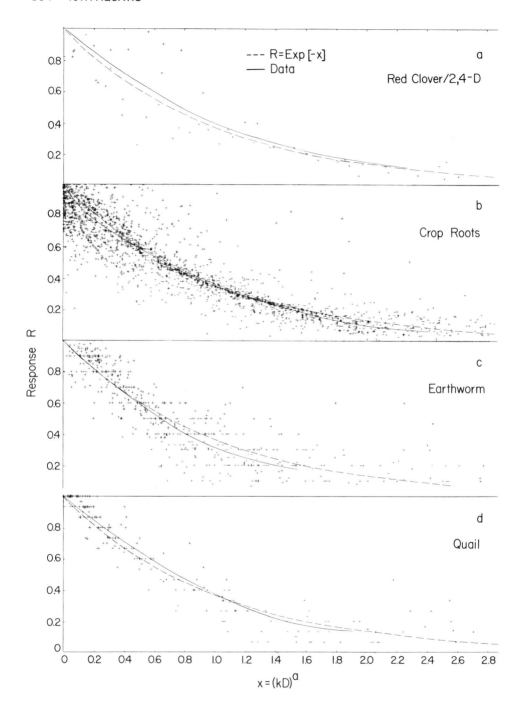

FIG. 5—*Experimental variability in bioassay tests of five crop roots, four species of earthworms, and quail using more than 100 organic compounds and metallic salts. Clover/2,4-D tests are subsets of crop root tests.*

response, we use the exponential curve of Fig. 5d at $x = -\ln 0.5 = 0.693$ to obtain the dimensionless standard deviation STX = 0.22. This number is converted to ppm by using the scale factors of the test with Phosmet. The ranking factor for this test (Table 1) is $\log(k) = -3.3$, $a = 2.4$, and $k = 0.0005$ (1/ppm). This leads to a dimensional standard deviation of $[(0.22)^{1/2.4}]/0.0005 = 1064$ ppm and a coefficient of variation at T_{50} of $(1064/1900)100 = 56\%$.

Discussion

The quantification of experimental variability and parameter estimation errors is facilitated by the use of models that can organize and classify different types of tests. Likewise, the interpretation and comparison of various classes of tests are facilitated by distinguishing various response patterns that can be simply modeled. The scaler approach introduced in this paper links together the quantification of errors and the mathematical interpretation of biological information.

The limited examples of data examined in this report show that toxicity numbers (T_{50}) for two or more tests can be numerically equal while sensitivities differ by a factor of 10 (for example, test q8 and j3 in Table 1) or in the extreme by a factor of 100. Therefore, it is useful to distinguish between equal toxicity numbers by reporting the sensitivity.

In the particular case of test q8, the ranking factor $(-3.2,4.7)$ classes the response curve as initially flat, then steep with increased dose ($a > 1$), while the test j3, with a ranking factor of $(-3.4,0.8)$ is just as clearly initially steep, then flat type of response. In other words, the curves of these two types of tests can be represented by their respective ranking factors, revealing even greater insight about each test.

The scalers of groups of similar tests are derived from the aggregation of whole tests in our approach and not merely their summaries as would be, for example, when averaging LC_{50} numbers of several tests. The use of whole test data facilitates the discovery of biological dose-response patterns and better explanation of cause and effect based on data.

The scalers of groups of similar tests enable the aggregation and comparison of experimental variabilities from different laboratories on equal mathematical basis using whole tests. The strengths and weaknesses of various laboratory techniques can be better identified.

The scalers make possible the aggregation of unlimited numbers of data, better facilitating classification of chemical effects and evaluating the feasibility of extending data within a class to predict the potential effect of untested chemicals. The conduct of experiments are also facilitated by reducing the need for range-finding, by anticipating the patterns of a new test from previous tests, and by reducing the need for many new tests.

There are some limitations in this paper: (1) A small number of classes of potential dose-response functions are examined. For example, the integrals of the generalized exponential functions used in this paper can lead to a workable procedure. The mathematics would be more complex if intractable. The examination of other suitable functions leading to smaller errors in calculating the scalers based on a limited number of data points may be fruitful. (2) Many types of growth tests, for example, root elongation, produce stimulatory response at low dose values. The problem is mathematically similar to situations where organisms in the control test die. We have removed these types of data from consideration using our approach. It is possible to modify the approach to overcome the problem, however. (3) Our emphasis is on understanding different classes of dose-response patterns. This requires tests that are complete individually and contain evenly spaced data points from 100% response to near zero response. However, many tests that otherwise can be useful for toxicity calculations and include data points up to and slightly past 50% response must be dropped in our analysis. This shortcoming cannot be removed without imposing arbitrary rules for extrapolation beyond tested values.

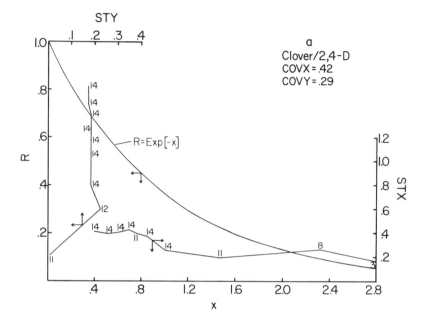

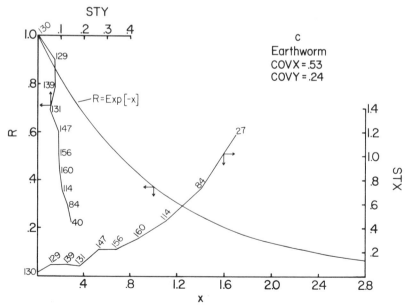

FIG. 6—*The standard deviations STX and STY of the scaled data from the simple exponential curve $R = Exp[-x]$ describe the prediction error for the population and can be applied to single tests in the population whose scale factors are known.*

SHIRAZI AND LOWRIE ON INTEGRATION OF DATA 357

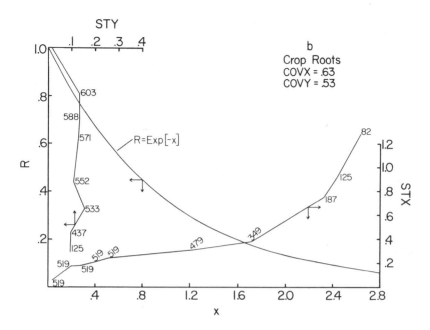

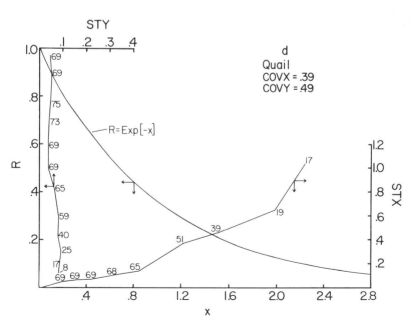

FIG. 6—Continued.

TABLE 1—*Classification of bioassay tests using the ranking factors: log*(k), *a.*

Code	Chemical	Organism	No. of Tests	Log(k), a	T_{50}, ppm	S_{50}, 1/ppm	COVD	COVR
a6	4-Nitrophenol	Earthworm	3	4.2, 1.9	0.580E−04	0.115E+05	0.11	0.23
b6	Diphenylnitrosamines	Earthworm	2	3.6, 1.7	0.183E−03	0.320E+04	0.21	0.12
c6	Phenol	Earthworm	4	3.1, 0.7	0.461E−03	0.521E+03	0.50	0.50
d6	Nitrobenzene	Earthworm	4	3.0, 2.1	0.884E−03	0.819E+03	0.32	0.34
e6	1,2-Dichloropropane	Earthworm	4	2.1, 1.7	0.598E−02	0.101E+03	0.28	0.25
g1	2,4-D	Clover	15	2.0, 0.2	0.105E−02	0.542E+02	0.80	0.21
f6	Dimethylphthalate	Earthworm	4	1.3, 1.6	0.399E−01	0.135E+02	0.31	0.16
g4	2,4-D	Cucumber	19	0.5, 0.3	0.771E−01	0.122E+01	1.54	0.59
g5	2,4-D	Radish	14	0.3, 0.4	0.196E+00	0.626E+00	2.08	0.42
g3	2,4-D	Wheat	7	−0.8, 0.4	0.224E+01	0.551E−01	0.74	0.34
i4	AgNO3	Cucumber	14	−1.1, 0.2	0.117E+01	0.456E−01	0.96	0.34
h2	CdCl2	Lettuce	16	−1.2, 0.8	0.101E+02	0.288E−01	0.76	0.42
h1	CdCl2	Clover	20	−1.5, 0.5	0.166E+02	0.106E−01	1.00	0.29
h4	CdCl2	Cucumber	17	−1.5, 0.3	0.707E+01	0.127E−01	0.80	0.33
i2	AgNO3	Lettuce	12	−1.6, 0.7	0.220E+02	0.114E−01	0.76	0.59
j5	MAA	Radish	12	−1.6, 0.3	0.134E+02	0.870E−02	0.55	0.21
j1	MAA	Clover	14	−1.6, 0.2	0.591E+01	0.120E−01	1.23	0.20
a6	4-Nitrophenol	Earthworm	4	−1.7, 2.1	0.469E+02	0.152E−01	0.17	0.39
k6	2,4,6-Trichlorophenol	Earthworm	4	−2.0, 1.9	0.858E+02	0.752E−02	0.31	0.38
l1	Monuron	Clover	6	−2.0, 0.8	0.643E+02	0.444E−02	0.58	0.22
l5	Monuron	Radish	13	−2.1, 0.8	0.866E+02	0.332E−02	0.80	0.20
i5	AgNO3	Radish	16	−2.1, 0.3	0.393E+02	0.299E−02	0.52	0.23
b6	Diphenylnitrosamines	Earthworm	4	−2.2, 3.2	0.143E+03	0.772E−02	0.19	0.36
m6	Carbaryl	Earthworm	4	−2.2, 2.2	0.144E+03	0.533E−02	0.57	0.65
j4	MAA	Cucumber	12	−2.2, 0.9	0.101E+03	0.302E−02	0.42	0.18
j2	MAA	Lettuce	11	−2.2, 0.6	0.977E+02	0.217E−02	0.75	0.20
i1	AgNO3	Clover	14	−2.3, 0.7	0.124E+03	0.189E−02	0.60	0.31
h5	CdCl2	Radish	16	−2.3, 0.6	0.121E+03	0.176E−02	0.94	0.20
c6	Phenol	Earthworm	4	−2.5, 1.8	0.269E+03	0.232E−02	0.38	0.67
n6	1,2,4-Trichlorobenzene	Earthworm	4	−2.5, 1.7	0.256E+03	0.226E−02	0.27	0.32
o4	NaF	Cucumber	24	−2.5, 1.0	0.226E+03	0.159E−02	0.26	0.32
o1	NaF	Clover	23	−2.6, 1.7	0.360E+03	0.166E−02	0.65	0.44
o5	NaF	Radish	22	−2.6, 1.5	0.320E+03	0.162E−02	0.57	0.41
o3	NaF	Wheat	18	−2.6, 1.1	0.296E+03	0.129E−02	0.65	0.39
i3	AgNO3	Wheat	6	−2.6, 0.7	0.211E+03	0.107E−02	0.67	0.18
o2	NaF	Lettuce	17	−2.8, 1.4	0.483E+03	0.102E−02	0.36	0.36
p2	Cineole	Lettuce	12	−2.8, 0.5	0.257E+03	0.615E−03	0.49	0.40
p3	Cineole	Wheat	16	−3.0, 0.5	0.444E+03	0.377E−03	0.67	0.41
q8	TEPP	Quail	1	−3.2, 4.7	0.147E+04	0.111E−02	0.06	...
p5	Cineole	Radish	12	−3.2, 0.6	0.890E+03	0.242E−03	0.81	0.48
r8	Phosmet	Quail	1	−3.3, 2.4	0.190E+04	0.438E−03	...	...
t8	Propanil	Quail	1	−3.4, 2.9	0.237E+04	0.431E−03	0.03	...
s8	Endosulfan	Quail	1	−3.4, 1.9	0.215E+04	0.314E−03	...	...
j3	MAA	Wheat	9	−3.4, 0.8	0.145E+04	0.180E−03	0.52	0.09
v8	Endosulfan (Thiodan E)	Quail	1	−3.5, 3.0	0.257E+04	0.407E−03	0.06	...
u8	Cadmium Succinate	Quail	1	−3.5, 1.8	0.266E+04	0.235E−03	0.03	...
p1	Cineole	Clover	8	−3.5, 1.0	0.200E+04	0.170E−03	0.34	0.42
w8	Malathion	Quail	1	−3.6, 2.3	0.325E+04	0.242E−03	0.05	...
e6	1,2-Dichloropropane	Earthworm	4	−3.7, 3.6	0.466E+04	0.266E−03	0.20	0.61
x8	Phosalone	Quail	1	−3.7, 3.1	0.439E+04	0.246E−03	0.06	...
y8	Rotenone	Quail	1	−3.8, 2.7	0.590E+04	0.159E−03	0.03	...
z5	Carbowax	Radish	13	−5.2, 1.2	0.122E+06	0.344E−05	0.52	0.33
z4	Carbowax	Cucumber	16	−5.2, 0.8	0.930E+05	0.306E−05	0.43	0.38
z3	Carbowax	Wheat	14	−5.3, 2.0	0.180E+06	0.383E−05	0.62	0.40
z1	Carbowax	Clover	15	−5.3, 1.5	0.173E+06	0.291E−05	0.55	0.33
z2	Carbowax	Lettuce	17	−5.3, 1.2	0.157E+06	0.254E−05	0.38	0.47

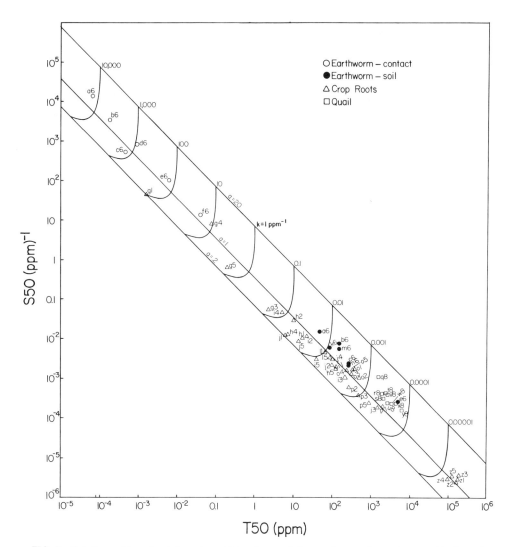

FIG. 7—*Relative ranking of toxicity and sensitivity of several classes of tests using scalers* a *and* k. *Strongly toxic and sensitive responses appear on the upper left hand side of the plot; responses with initial defense appear above the* a = 1 *line.*

References

[1] "Methods for Measuring the Acute Toxicity of Effluents to Freshwater and Marine Organisms, No. EPA/600/4-85/013, Peltier, W. H., and Weber, C. I., Eds., U.S. Environmental Protection Agency, Environmental Monitoring and Support Laboratory, Cincinnati, OH, 1985.
[2] Hamilton, V. A., Russo, R. C., and Thurston, R. V., "Trimmed Spearman-Karber Method for Estimating Median Lethal Concentration in Toxicity Bioassays," *Environmental Science and Toxicology*, Vol. 11, pp. 714–717, 1977. Correction: Ibid., Vol. 12, p. 417, 1978.
[3] Shirazi, M. A., and Hart, J. W., "A Unifying Scaler for Bioassay Tests," *ISEM Journal*, Vol. 6, Nos. 3–4, 1984, pp. 25–53.

[4] Ratsch, H. C., "Interlaboratory Root Elongation Testing on Toxic Substances on Selected Plant Species," No. EPA-600/3-83-05, U.S. Environmental Protection Agency, Environmental Research Laboratory, Corvallis, OR, PB 83-226 126, National Technical Information Service, 5285 Port Royal Road, Springfield, VA, 1983.

[5] Callahan, C. C., Shirazi, M. A., and Newhauser, E. F., "A Unifying Scaler for the Evaluation of Relative Toxicity to Earthworms," unpublished.

[6] Hill, E. F. and Camardese, M. B., "Lethal Dietary Toxicities of Environmental Contaminants and Pesticides to Coturnix," U.S. Department of Interior, Fish and Wildlife, Technical Report 2, Washington, DC, 1986.

[7] Lawless, J. F., *Statistical Models and Methods for Lifetime Data*, John Wiley and Sons, Inc., New York, 1982.

[8] Weibull, W., "A Statistical Distribution Function of Wide Application," *Journal of Applied Mechanics*, Vol. 18, 1951, pp. 293–297.

[9] Shirazi, M. A., Bennett, R., and Lowrie, L., "An Approach to Environmental Risk Assessment Using Avian Toxicity Tests," *Archives of Environmental Contamination and Toxicology*, Vol. 17, 1988, pp. 263–271.

John Cairns, Jr.[1]

What Constitutes Field Validation of Predictions Based on Laboratory Evidence?

REFERENCE: Cairns, J., Jr., "**What Constitutes Field Validation of Predictions Based on Laboratory Evidence?**" *Aquatic Toxicology and Hazard Assessment: Tenth Volume, ASTM STP 971,* W. J. Adams, G. A. Chapman, and W. G. Landis, Eds., American Society for Testing and Materials, Philadelphia, 1988, pp. 361–368.

ABSTRACT: The term *validation* has appeared in the literature with great regularity in the last few years, but the process has yet to be explicitly stated. Most of the predictions of environmental hazard are based primarily, and sometimes entirely, on laboratory toxicity tests involving single species. While this may change over the years, it will remain true for the immediate future. Therefore, the validation process must start with an explicit statement of the types of predictions being made on the basis of single species laboratory toxicity tests. Unfortunately, the terms *no-adverse-biological-effects concentration* or *biologically safe concentration*, or *no-observable-effects level* might lead the unwary to believe that all possible adverse effects were unlikely to occur at concentrations below the one stated. In fact, I have never seen a body of evidence in a scholarly professional journal that would convince reviewers that these assertions had been confirmed in a scientifically sound way. Some more modest but more explicit predictions and validation criteria, for illustrative purposes only, follow.

1. Only the test species is expected to be protected fully in natural systems, and validation means confirming this assumption in the field.

2. Other species than the one(s) actually tested are thought to be protected at the no-observable-effects concentration, and a list of these species inhabiting the natural systems in question is included. Validation is carried out with field observations on all of these species.

3. No adverse effects at the community level of biological organization will occur, and the characteristics used to validate this assumption are listed. Validation may be carried out in microcosms or mesocosms under certain circumstances but should be based on field observations whenever possible.

4. No adverse effects will occur at the ecosystem level of organization, and the characteristics at the ecosystem level used to validate this assumption are listed. Validation in field enclosures or natural systems should probably be mandatory.

By coupling the explicit predictions being made with the explicit end points being used to validate these predictions, a more systematic and orderly process of validation will ensue. Until more hypothesis testing is carried out in the field of hazard evaluation than is in place presently, it is unlikely that the field will get the recognition it deserves.

KEYWORDS: ecotoxicology, toxicity testing, hazard evaluation, bioassays, field studies, validation

Sanders [1], in his excellent analysis of the problem of field validation, discusses scientific field evaluation and hypothesis testing. These are important concepts and have not been applied as Sanders recommends. He also discusses such factors as sensitivity analysis for parameter screening and modeling and analysis. Readers interested in these matters should refer to the Sanders chapter.

---

[1] University Distinguished Professor, Department of Biology, and director, University Center for Environmental Studies, Virginia Polytechnic Institute and State University, Blacksburg, VA 24061.

One of the first attempts to demonstrate a scientific basis for using laboratory tests to predict field effects is discussed by Geckler et al. [2], and a more recent attempt is found in Suter et al. [3]. *ASTM STP 865, Validation and Predictability of Laboratory Methods for Assessing the Fate and Effects of Contaminants in Aquatic Ecosystems,* also provides some preliminary examples, although the process does not go as far as recommended in my discussion here.

The book *Testing for Effects of Chemicals on Ecosystems,* representing the report of a committee on ecotoxicology which I chaired for the National Research Council [4], made me think very seriously about the notable failure to cope with the critical issue of validation. The issue itself was defined a number of years ago in a book on hazard evaluation [5]. There were several places in that book where the need for confirmative or validating studies was stated. One of the most notable was in a chapter by Kimerle et al. [6], where the sequence of screening studies, predictive studies, confirmative studies (for example, validating), and monitoring studies were called for as a logical progression in the development of hazard evaluation criteria. A colleague and I experimented with the validation of predictive models developed in laboratory test systems by checking these predictions in the field [7] at a single level of biological organization—namely, single species. The correlation between the field and laboratory data was much better than anticipated, despite the dissimilarity in habitat complexity between laboratory and field systems. The major reasons for the close correspondence appeared to be the proximity of the laboratory and field study sites that reduced the damage to organisms due to transport, and so on. Equally important was that river water was pumped directly from the New River into the laboratory, and its quality was not significantly altered in the process.

In addition, some tests with multispecies systems suggest that, at least for some, the validation process can be carried out with less difficulty than was once thought [8,9]. However, there is essentially no substantive evidence validating predictions based on single species toxicity tests that were used to predict responses at higher levels of biological organization (for example, communities and ecosystems). In June 1985 [10], I discussed the process of validation at different levels of biological organization. These preliminary thoughts were further refined in a manuscript devoted solely to the subject of validation [11]. Although there is no reason to repudiate these early ideas, it is quite evident that they need further refinement—this is the purpose of this manuscript.

I now believe there should be at least three different levels of validation: (*a*) scientific validation, (*b*), site-specific operational validation, and (*c*) monitoring validation. Each of these will call for strikingly different types and quantities of information and different levels of professional capability in providing this information.

As I write this discussion in late March 1986, the hearings on the Challenger space shuttle explosion are in full swing. There are some interesting parallels in the reasoning between predicting the effects of potential toxicants in a complex, highly variable system from data generated in a simple, comparatively uncomplicated system and the problem of predicting how a space shuttle will perform in a variety of conditions, some or many of which have not been simulated in the laboratory tests and all of which are performed in a more complex set of circumstances than those in tests of various components carried out individually. While we can hope in both circumstances to reduce the number of failures, it is not likely that we can eliminate them entirely. The Challenger hearings indicate that the reasoning process that lead to the decision to launch that fatal day should have been more thoroughly documented before the launch occurred. In that way, one could determine expeditiously whether the failure was due to a lack of information, misuse of information (including failure to communicate the information), or both. In both cases, the question is whether an artificial system (the space shuttle) or a natural system (an ecosystem) will perform as predicted when subjected to stresses that have been simulated in test units generally less complex individually than the actual exposure conditions. Predictions must be validated in each case as well as present methodology permits. Since present methodology is not yet so extensive, sophisticated, or widely used as to totally eliminate error in judgment, it seems mandatory to show the data base used for making a decision as well as the decision process that led to a particular decision. Consequently, this discussion on validation is presented in those terms. First, an orderly, systematic process for validating predictions of harm or risk to ecosystems must be developed, and, concomitantly, a procedure must be developed for documenting the reasoning process that led to one of three decisions: (a) probability of environmental damage minimal—no further evidence needed; (b) environmental damage highly probable—prohibit intrusion of the chemical into the environment or severely limit use according to generally accepted risk management practices; or (c) probability of environmental damage uncertain—more evidence is needed (and the type of evidence needed would be explicitly stated, together with an indication of the degree to which the estimate of hazard or risk would be improved).

Statement of Predictions of Natural System Response that Can Be Made from Test Results

Most of the field of environmental toxicology rests on a series of implicit, but rarely stated, assumptions infrequently backed by scientifically justifiable evidence. As noted in "Water Quality Criteria of 1972" [12] and elsewhere, regulatory standards need not be scientifically justifiable. Regulatory criteria, however, should be. If regulatory agencies are using single species laboratory toxicity tests with the assumption that the results from these can be successfully extrapolated to predict the response of more complex and more highly variable natural systems, evidence should be provided in the peer-reviewed scholarly literature. Since the response of a natural system to a toxicant might theoretically be predicted from a laboratory toxicity test, it is clear that judgment must be used in selecting particular attributes that are crucial to the well being of natural systems in general for the validation process. In so doing, one would identify those ecosystems in which the particular test would be most useful, since these attributes would be essential to their well being. An illustrative list of predictions for discussion purposes only follows:

1. That all species inhabiting natural systems are less sensitive to the toxicant than the test species used.
2. That the response of the test species is highly correlated with the response of species that are key regulators of natural systems.
3. That key functional attributes of natural systems, such as nutrient spiraling and energy flow, will not be disrupted or impaired at concentrations of toxicant predicted to have no-adverse-biological effects by the single species or other laboratory toxicity tests.
4. That key structural attributes of natural communities, such as the proportion of different trophic levels, will be unaltered at concentrations predicted to have no-adverse effects by the laboratory toxicity tests.

Three Levels of Validation

As suggested earlier, there should be three levels to the validation process. The first level of scientific validation is by far the most important to the credibility of predictions of no-adverse effects based on laboratory data and has already been discussed. The second level, site-specific operational validation, merely confirms that the results of the more general scientific validation apply to a particular site. All ecosystems have some attributes in common, but these may be strikingly modified at a particular site. Level 2 validation is designed to determine the degree of difference between the general model and a specific site. The third level, monitoring, is designed to detect errors in Levels 1 and 2. It is particularly important where there are still uncertainties after Levels 1 and 2 validations. The uncertainty might be due, for example, to the fact that the ecosystem is known to be highly variable and the effects of this variability had not been fully explored in the Levels 1 and 2 validations. Meeting the rigorous scientific requirements of peer review and publication in a professional journal and demonstrating the general applicability of a test method would require such validation if the laboratory toxicity tests are used to protect the various components of natural systems just mentioned. This validation would serve two important purposes: (*a*) to provide evidence that the laboratory data generated can be used to predict responses in natural systems either in a variety of conditions beyond those actually measured or at levels of biological organization other than those measured or for species other than those studied, or all of these, and (*b*) to provide guidelines on future site-specific validation efforts. The proliferation of unvalidated new laboratory toxicity testing methods is adding more to the confusion than to the capabilities of the field of environmental toxicology. At the present time, keeping track of all the new methods is difficult. Even if all can be followed, choosing among them is difficult because the results are rarely given a detailed comparison with already existing or commonly used methods and practically never is there validation of the type described. It seems intuitively reasonable that all of these methods are not equal in their utility for predicting responses in more complex systems, and the validation process will serve to eliminate those that are not, keeping the number that user groups must cope with at a more reasonable level. An economic objection to the validation process just described will be its cost. There is no question that the additional evidence gathered will increase costs by an order of magnitude or more. On the other hand, validation of the predictions is as much a part of method development as the round robins designed to determine replicability, utility of instructions for carrying out the method, and the like. The validation process may well demonstrate that all of the assumptions made about the ability of single species laboratory toxicity tests to predict concentrations of potential toxicants that will cause no-adverse effects in natural systems are correct. If so, this should provide the scientific evidence for stronger support of their use. If the predictions are not as verifiable as assumed, correcting both the means of making the predictions and perhaps even the environmental realism of the tests is necessary so that the accuracy of the predictions is improved. As Sanders [1] notes, we cannot afford to err badly in either direction.

At the outset, it is difficult to estimate how many end points should be chosen for validation, how many ecosystems should be used in the process, and how many different levels of biological organization are appropriate. Initially, it would be well to pick extremes in order to determine the boundary conditions beyond which the predictions are not valid. Using these for range-finding purposes, one might then become more specific to show those conditions under which predictions can be expected to be reliable. Since it is clearly impossible to do every test or even a small fraction of the total, scientific judgment will be particularly useful in this regard. The reasoning used in the selection process should be explicitly stated, since it is as important as the actual validation itself.

Level 1—Scientific Validation

There is considerable discussion these days about the appropriate content of standard methods (this term is used in the legal sense—that is, formal professional endorsement through a specified process by such organizations as the American Society for Testing and Materials or the consortium of organizations that produces the book *Standard Methods for the Examination of Water and Wastewater*). Standard methods for toxicity testing of aquatic organisms provide great detail on equipment to be used, procedures in carrying out the tests, and the like. They do not regularly explicitly state how the results of the test should be and should not be used. One would not find explicit directions on the types of predictions that can be made from laboratory tests regarding the response of natural systems. Furthermore, there is no discussion of the ways in which one can confirm or validate these predictions or even whether one can use end points in natural systems other than the end points used in the laboratory toxicity tests.

Standard methods for freshwater toxicity testing have very explicit instructions on how the tests should be carried out, which are further improved by round robins and various other quality assurance processes. Curiously, practically nothing is said either about how the information generated should be used or, equally important, how it should not be used. The Ohio River Valley Water Sanitation Commission (ORSANCO) 24-h toxicity test [13] is virtually unique in explicitly stating what the test is and what it is not. Even the unusually candid statements in the ORSANCO test could easily be improved. Nevertheless, ORSANCO should be commended for delimiting the utility of the test much more explicitly than is common even today, and the ORSANCO test was developed many years ago.

Recent issues of professional journals reveal six new toxicity testing methods involving a variety of organisms. In some cases, no attempt is made to contrast the new method with existing methods or statements to the effect that the quality of information was unique and did not replicate that of existing methods. Not only are data contrasting the sensitivity, reliability, and so on with more commonly accepted methods lacking, but the way in which the predictions made from the information can be validated is also not given. In a purely scientific publication, this is understandable since the practical utility and relevance is not essential to a scientific paper. On the other hand, since most of the papers did not represent major conceptual breakthroughs in toxicology or science, the absence of such information is no longer advisable.

The United States is clearly a litigatious society, and, as such, it is noteworthy that developers of standard methods in aquatic toxicology are not taken to court frequently for not including warning statements (truth in packaging) on misuse and more explicit directions for proper use. For the record, I did none of these things in papers on toxicity testing that were published about 20 to 30 years ago. Even recent papers are notably deficient in terms of the criteria just stated. Charitable individuals will probably forgive these sins of omissions committed by early practitioners in any developing field. However, ecotoxicology has now developed to a point where such omissions may shortly not be viewed as permissively and charitably.

Stages. The first stage of scientific validation should demonstrate that an ecologically important indigenous species in the receiving system has a response threshold reasonably close to or more tolerant than the laboratory test species. This is extremely important when the laboratory test species is not indigenous. Even when the same species is used in both the laboratory and field, one should have evidence that the field response is as predicted by the laboratory tests. The second stage should demonstrate that an array of species from a variety of taxonomic groups and trophic levels do not suffer deleterious effects (for example, lethality, reproductive impairment, or reduced growth rate) at concentrations predicted to have no-adverse effects based on laboratory tests. The third stage should demonstrate that ecologically important community or ecosystem attributes, such as energy flow or nutrient cycling, are not impaired at concentrations predicted to have no effect by the laboratory toxicity tests.

Statement of Purpose of the Test. Ideally, a newly developed toxicity test, in addition to providing information on how the test is carried out and other customary data, should provide the following information.

1. What is particularly valuable and unique about this particular method? Details might be provided on comparison with other somewhat similar methods; a description of how the information generated fills a need not filled by existing methods; and how the method is less expensive, more reliable, and more suitable for making predictions to larger and more complex systems.

2. Is the test designed for rangefinding or screening, predicting responses in natural systems,

validating predictions made with another system, or monitoring to see if quality control conditions are being met?

3. Is this method designed to be used in both rivers and lakes? If applicable to only a limited number of ecosystems, this should be explicitly stated. If there are some ecosystems for which the information has no predictive or other value, these should be stated as well. If the organism is not indigenous to the receiving system, in what ways is it superior to indigenous organisms?

4. What statistical methods and procedures should be used to determine the reliability of test results? What is the rationale for choosing these particular methods, and are there alternative methods that should be used in certain circumstances?

5. What level of professional competence is necessary in general for successful utilization of the test results? What level of professional competence is necessary for generating basic data?

Level 2—Site-Specific Validation Process

The site-specific validation process will be easier when there are more publications on the validation process published in peer-reviewed professional journals [9]. Since the validation is for a specific ecosystem or habitat, a means is needed to determine whether the critical attributes of a specific receiving system are being protected as predicted. If the laboratory toxicity test method is accompanied by illustrations of the validation process, the method would not have been selected unless the end points of the toxicity test are important to the ecosystem or unless they can be used to predict the effect on these attributes in the natural system even though they are not present in the laboratory test system. If the correspondence between the laboratory predictions and the field response is high, one need go no further; if it is not, an alternative method might well be selected for making these predictions.

Level 3—Monitoring

There are some cases where the validation process may be unsatisfactory because the method does not provide predictions that are validated well in the natural system but no superior method seems to be available. There are other reasons why the validation process might not work well either, such as extensive natural variability, episodic events such as floods, and the like. In such cases, some form of monitoring of selected characteristics of the natural system would be advisable. The monitoring should serve two purposes: (*a*) to determine why the validation was not better than it turned out to be, and (*b*) to alert management to deleterious effects upon the system that occur due to failure of the predictive model. *Monitoring* is used in this context as defined by Hellawell [*14*] as "surveillance undertaken to ensure that previously formulated standards are being met." If after a reasonable period of monitoring, the predictive model is found to be better than the validation process suggested, the monitoring can cease. Otherwise, the monitoring should provide some evidence regarding the reason for the failure of the validation process, and correction can be made.

Summary

Environmental toxicology or ecotoxicology has many of the attributes of traditional science since species laboratory toxicity tests are often published in peer-reviewed literature. The results can be replicated by others, various statistical analyses can be applied to determine the confidence one can have in the numbers generated as reflecting the response pattern being tested, and a variety of other attributes. The departure from the characteristics of the scientific method develops when the results from single species and other laboratory tests are used to predict responses of natural systems and these predictions are not validated in a scientifically justifiable manner. The problem is complicated and expensive. Because ecology itself is nearly as new a field as environmental toxicology, there are many uncertainties about how things work. Only relatively recently did Robert A. MacArthur and others develop predictive models for ecological systems and validate them. The same developmental process is essential for environmental toxicology and involves many of the same processes as validating a theoretical model for the field of "pure" ecology. As is the case for theoretical ecology, there will always be some doubt about the transferability of results from one ecosystem to another, the determination of the most important functional attributes of the system, and identifying key or regulating species. In the field of theoretical ecology, one can afford to wait years before deciding whether or not a method should be used. In the field of applied ecology, which includes environmental toxicology, the luxury of waiting until one has full confidence in the methods is not possible. Nevertheless, this is no excuse for not improving the methods while they are being used. It is my opinion that the chief deficiency in the field of environmental toxicology is the failure to validate widely the assumptions upon which the predictions are based when single species toxicity tests and other laboratory toxicity tests are used to predict the effects of various concentrations of chemical and other potential toxicants in natural systems.

The validation process still requires considerable thought, but even the most limited confirmatory tests will provide very useful information. There are even heartening indications that redundancy of natural systems is such that extrapolations from one attribute of a system to another may be less difficult than it now seems. On the other hand, the complexity and variability of natural systems is sufficiently well documented to urge caution on the use of simple laboratory models for predicting events in complex natural systems.

Acknowledgments

I am deeply indebted to James R. Pratt and Barbara R. Niederlehner for comments on an early draft of this manuscript. Darla Donald did her usual fine job of putting the manuscript in the form required by ASTM. The manuscript was typed in draft form by Becky Allen and in final form by Betty Higginbotham.

References

[1] Sanders, W. M., III in *Fundamentals of Aquatic Toxicology*, G. M. Rand and S. R. Petrocelli, Eds., Hemisphere Publishing Corp., New York, 1985, pp. 601–618.
[2] Geckler, J. R., Horning, W. B., Neiheisel, T. M., Pickering, Q. H., and Robinson, E. L., "Validity of Laboratory Tests for Predicting Copper Toxicity in Streams," EPA-600/3-76-116, National Technical Information Service, Springfield, VA, 1976.
[3] Suter, G. W., II, Barnthouse, L. W., Breck, J. E., Gardner, R. H., and O'Neill, R. V. in *Aquatic Toxicology and Hazard Assessment: Seventh Symposium, ASTM STP 854*, R. D. Cardwell, R. Purdy, and R. C. Bahner, Eds., American Society for Testing and Materials, Philadelphia, 1985, pp. 400–413.
[4] National Research Council, *Testing for Effects of Chemicals on Ecosystems*, National Academy Press, Washington, DC, 1981.
[5] *Estimating the Hazard of Chemical Substances to Aquatic Life, ASTM STP 657*, Cairns, J., Jr., Dickson, K. L., and Maki, A., Eds., American Society for Testing and Materials, Philadelphia, 1978.
[6] Kimerle, R. A., Gledhill, W. E., and Levinskas, G. J. in *Estimating the Hazard of Chemical Substances to Aquatic Life, ASTM STP 657*, J. Cairns, Jr., K. L. Dickson and A. Maki, Eds., American Society for Testing and Materials, Philadelphia, 1978, pp. 132–146.
[7] Cairns, J., Jr. and Cherry, D. S., *Water Science and Technology*, Vol. 15, 1983, pp. 31–58.
[8] *Multispecies Toxicity Testing*, J. Cairns, Jr., Ed., Pergamon Press, New York, 1985.
[9] Livingston, R. J., Diaz, R. J., and White, D. C., "Field Validation of Laboratory-Derived Multispecies Aquatic Test Systems," EPA/600/S4-85-039, U. S. Environmental Protection Agency Project Summary Report, Cincinnati, OH 1985.
[10] Cairns, J., Jr., "Politics, Economics, Science—Going Beyond Disciplinary Boundaries to Protect Aquatic

Ecosystems," in *Great Lakes Ecosystem Health: Effects of Toxic Substances,* J. Gannon and M. Evans, Eds., Wiley, New York, in press.

[11] Cairns, J., Jr., "What is Meant by Validation?" *Hydrobiologia,* Vol. 137, 1986, pp. 271–278.

[12] "Water Quality Criteria of 1972," National Academy of Sciences, National Academy of Engineering, Washington, DC, 1972.

[13] ORSANCO 24-Hour Bioassay, ORSANCO Biological Water Quality Committee, Ohio River Valley Water Sanitation Commission, Cincinnati, OH, 1974.

[14] Hellawell, J. M. *Biological Surveillance of Rivers,* Water Research Centre, Stevenage, England, 1978.

Robert J. Livingston[1]

Field Verification of Multispecies Microcosms of Marine Macroinvertebrates

REFERENCE: Livingston, R. J., "**Field Verification of Multispecies Microcosms of Marine Macroinvertebrates,**" *Aquatic Toxicology and Hazard Assessment: 10th Volume, ASTM STP 971,* W. J. Adams, G. A. Chapman, and W. G. Landis, Eds., American Society for Testing and Materials, Philadelphia, 1988, pp. 369–383.

ABSTRACT: A five-year study was carried out to determine the feasibility of using multispecies microcosms of infaunal macroinvertebrates to predict the responses of estuarine systems to toxic substances. Criteria were developed to evaluate the field validation of laboratory microcosms. Experiments carried out in the Apalachicola Bay system in Florida demonstrated that infaunal macroinvertebrates can be established for short periods (five to six weeks) and that the microcosms can be used to simulate certain features of natural field assemblages.

Water quality in the microcosms essentially paralleled that in the field, although variation of certain water features and sediment characteristics was noted. These laboratory artifacts were apparently caused by the isolation of the microcosms from natural phenomena of the estuarine environment that were not replicable in the laboratory. Physical habitat features and biological responses in the study area were extremely complex and highly variable in space and time. Factors such as water and sediment quality, predator-prey relationships, recruitment, and dominance relationships among infaunal populations, influenced the community structure of benthic organisms in the laboratory and the field. The relative influence of physical and biological factors varied considerably between habitats and through time. Consequently, the extent to which the laboratory microcosms paralleled field conditions depended to a considerable degree on the time of testing and dominance/recruitment features of the system in the source area. Species richness in the laboratory microcosms appeared to be a good indicator of field conditions. Multispecies tests have certain advantages over single-species tests in terms of the expanded scope of evaluation. However, results from such experiments should be interpreted within established limits of natural environmental variation. Each microcosm should undergo a thorough calibration with field conditions to qualify the influence of laboratory artifacts and to identify the potential problems associated with isolation from key controlling factors in the field.

KEYWORDS: field verification, multispecies microcosms, marine macroinvertebrates, water quality, recruitment, dominance relationships

Microcosms have become an effective tool in the evaluation of the environmental effects of toxic substances [1]. The need to develop and use multispecies tests to determine the impact of toxic waste disposal has been clearly stated [2]. However, there is an associated need to verify such methods in the field [3] as a prerequisite for the identification of those tests that "most accurately represent the natural prototype" [2]. Relatively little has been done to determine the relationship of laboratory to field relationships even though the criteria for such an evaluation are relatively clear [4,5].

The basic question that underlies the research results reported here involves what is required to predict the environmental effects of a toxicant or stimulatory substance on a given ecological

---

[1] Professor and director, Center for Aquatic Research and Resource Management, Florida State University, Tallahassee, FL 32306-2043.

system. Field verification of multispecies microcosms involves the testing of the predictive capability of such laboratory test systems relative to the actual environmental response of natural populations to xenobiotic substances. This process involves the selection of a particular microcosm and the acquisition of sufficient field data to account for natural variation and the dynamics of the field populations on which the test is based. In this case, natural soft-bottom (unvegetated), estuarine macroinvertebrates were used because of their importance to the food webs of shallow coastal waters, the close association of such organisms with sediments (which tend to concentrate various toxic substances), and the considerable body of background scientific literature that now exists concerning such organisms. In addition, because of the ability to replicate, realism, generality, and the potential application to protocol development, sediment cores as multispecies microcosms represent a potentially useful method in the intermediate and advanced stages of a hazardous waste assessment [2].

The principal objective of this study was to evaluate the capability of the sediment core microcosm as a realistic analog to the simulation of the natural estuarine community from which it was derived. Such evaluation included verification at three levels of order: physicochemical differences, similarity of population/community structure, and differences in various processes noted under full-field and semi-field treatments and laboratory (microcosm) conditions.

Materials and Methods

Study sites, located in the Apalachicola Bay system (East Bay, St. George Sound, Fig. 1), are shallow (1 to 2 m), unvegetated, soft-bottom areas located in oligohaline (Stations 3,5A) and polyhaline (Station ML) conditions. Sediments in such areas were composed of silty sand (3,5A) and fine sand (ML). Previous studies (Livingston, unpublished data) indicated that the East Bay macroinvertebrate biota (3,5A) was characterized by high dominance (polychaete worms such as *Mediomastus ambiseta* as dominants), low to moderately low species richness and diversity, and moderate to high productivity. The polyhaline area (ML) was characterized by moderate dominance (polychaetes such as *Paraprionospio pinnata*), high species richness and diversity, and low productivity.

The field component of the experimental program was run using movable (anchored) aluminum frames (2 by 2 by 3 m in height). A coring assembly (platform with 100 numbered holes) was placed on the frame to allow a collecting team to stand on the platform and take core samples at exact, predetermined locations in the sediments. Multiple (12) random samples were taken with a polyvinyl chloride (PVC) corer (7.5 cm diameter, 10.0 cm depth) and run through nested 0.25 and 0.50-μm sieves. Each sieve fraction was preserved in 10% formalin and stained with rose bengal to facilitate picking. Picked organisms were identified to species and counted.

Four field-laboratory experiments were carried out over a two-year period. Such tests were conducted during the spring and fall seasons of peak biological activity at the study sites. Physical-chemical water measurements included temperature (°C), salinity (parts per thousand), dissolved oxygen (parts per million), pH, color (platinum-cobalt units), and turbidity (NTU). Sediments were analyzed for organic content and grain size. A background sampling program was carried out weekly over the entire study period; physical-chemical and biological samples were taken as described above. For each experiment, a randomized block design (Fig. 2) of multiple (3) samples was carried out for each treatment (unscreened platforms, primary control; screened platforms, predator exclusion; screened platforms, predator inclusion). The inclusion-exclusion cages were made by attachment of 6-mm^2 nylon mesh to the frame with monofilament line. Galvanized metal flashing was driven 9 to 10 cm into the sediment to anchor the mesh. Predator inclusions were made using species type and densities comparable to those taken in the field at the time of experimentation.

Laboratory microcosms were constructed in boxes (0.8 m^2) set on seawater tables (flowing

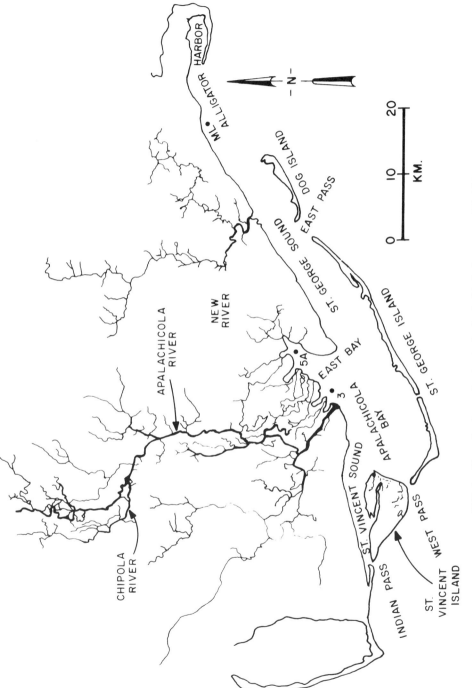

FIG. 1—*Chart showing study sites in the Apalachicola Bay system (Florida).*

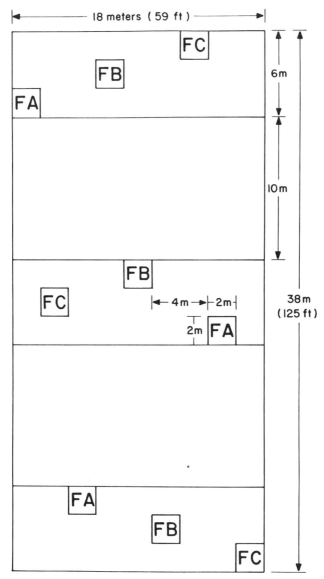

FIG. 2—*Diagram of a typical field array showing placement of inclusion/exclusion cages and cage controls.*

seawater; 95% overturn every 3 to 4 h). A series of cores was taken with hand-operated box corers (10 by 20 by 10 cm deep). The core samples were placed in the laboratory boxes in the same arrangement in which they were taken in the field to form a given microcosm. Light, temperature, and salinity regimes were comparable to those in the field in each test. Synoptic samples (twelve cores per treatment) were taken from the microcosms in a manner similar to that described above for the field sampling program.

Laboratory-field samples were taken weekly or biweekly, depending on the experiment. To prevent slumping in the microcosm, replacements of azoic sediments were made in the core holes. Variables analyzed during the experimental series included numerical abundance, number of species, Brillouin species diversity and evenness. All analyses were carried out with and without $\log_{10} (x + 1)$ transformations. A nested ANOVA model was used to test for differences between the laboratory microcosms and field collections by treatment. The null hypothesis that there was no significant difference among the field and laboratory treatments with respect to the above variables was tested with selected ANOVA models. A one-way ANOVA was run on all treatments by sampling period over each six-week experiment. A randomized-block, repeated-measures ANOVA was used for the field data with locations as the blocking factor and time as the repeated measure. Tukey's method of multiple comparisons was used to test the differences between all possible pairs of means. An analysis of qualitative changes in macroinvertebrate assemblages was carried out with Czekanowski (or Bray-Curtis; 6) similarity coefficients and the flexible grouping strategy with beta = -0.25 [7].

Results and Discussion

Background field information

Overall, seasonal changes in macroinvertebrate assemblages at the oligohaline and polyhaline sites were somewhat similar although dominance hierarchies remained distinctive. During the study period, oligohaline dominants included *Paranais littoralis, M. ambiseta, S. benedicti, Cerapus* sp., *Hobsonia florida,* and *Grandidierella bonnieroides.* Polyhaline dominants were *M. ambiseta,* tubificid worms, and *P. pinnata.* Community indices were seasonally and annually variable at the oligohaline site (3), with peak numerical abundance during the winter of 1981–82 (Fig. 3). Species richness was high at the polyhaline station (ML), with seasonal peaks during winter-spring periods. Most of the dominant species taken at the oligohaline site reached the highest numerical abundance during winter-spring periods. At the polyhaline site, there were complex dominance cycles with winter-spring peaks for *M. ambiseta* and spring-fall peaks for *Brania wellfleetensis*. Others, such as *Aricidea fauvelli,* showed no distinct seasonal patterns. Winter recruitment patterns were most common at both stations for species such as *M. ambiseta, Paranais littoralis,* and immature tubificids.

Field Experiments

Sediments from the oligohaline and polyhaline sites tended to differ in terms of the relative proportions of sand, silt, and clay. The polyhaline site was characterized by sandy sediments with relatively small components (1 to 2%) of silt and clay. Mean grain size ranged from 2 to 3 phi units. No seasonal trends in sediment characteristics were noted at the polyhaline site. Sediments in the oligohaline study area tended to have higher silt and clay fractions (usually ranging from 40 to 80% sand, 13 to 48% silt, and 3 to 10% clay). Average grain size was usually smaller (mean grain size, 3 to 4 phi units). Seasonal trends were apparent in the oligohaline system, with generally higher percentages of sand noted during spring periods of high river flooding.

An analysis of sediment characteristics (mean grain size, standard deviation, skewness, kurtosis; percent sand, silt, clay, and percent organics) as a function of treatment-related effects for each of the four experiments indicated no qualitative treatment effect. Analysis of variance for mean grain size and percent sand indicated no significant ($p < 0.05$) treatment-related effects on sediment composition of field samples or laboratory microcosms. No time-based differences were noted in sediments taken from field treatments or laboratory microcosms over the course of an individual experiment.

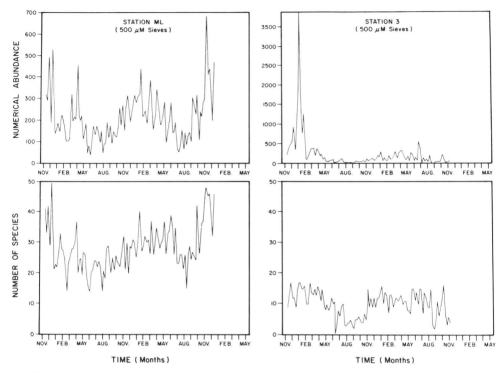

FIG. 3—*Numerical abundance and species richness of infaunal macroinvertebrates taken weekly at Stations 3 and ML in the Apalachicola Bay system over the study period.*

Field interactions of the infaunal macroinvertebrates were complex and highly variable in spatial as well as temporal terms. The benthic macroinvertebrate assemblages were influenced by a complex series of physical, chemical, and biological stimuli and processes. In polyhaline areas during the spring of 1982, exclusion of predators led to increased numbers of the dominant polychaete *Mediomastus ambiseta* (Fig. 4). Although species richness was not affected by predator exclusion, species diversity and evenness were significantly reduced, in large part because of the increased dominance of *Mediomastus*. The relationship to predation was corroborated by the predator-inclusion treatments, which were not statistically different from the cage controls and uncaged areas. This predation effect occurred during the usual spring decline of the macroinvertebrate community (Fig. 4). Randomized block ANOVA tests showed significant ($p < 0.05$) exclusion treatment effects for species diversity indices and log numbers of *Mediomastus* and total numbers of individuals. Time-based (week-to-week) effects were significant ($p < 0.05$) for all variables except species richness and numbers of individuals. The treatment $\times$ week (T $\times$ W) interaction was significant, which indicated that different treatments had diverging temporal trends. Less significant effects ($p > 0.05$) were noted because of the disturbance of sampling. The main treatment-induced effects were shown for organisms taken in the top 2 cm of sediment, where the highest numbers of invertebrates are located. There were also indications of treatment-related effects on recruitment of young organisms in the field, with lower numbers of smaller organisms in the uncaged field controls. Recruitment of the cosmopolitan *Mediomastus* was an important factor in the relative abundance patterns of the macroinvertebrate assemblages. This species was a major colonizer of azoic sediments during the spring experimental period.

Subsequent experiments in oligohaline and polyhaline areas during different seasons of the year

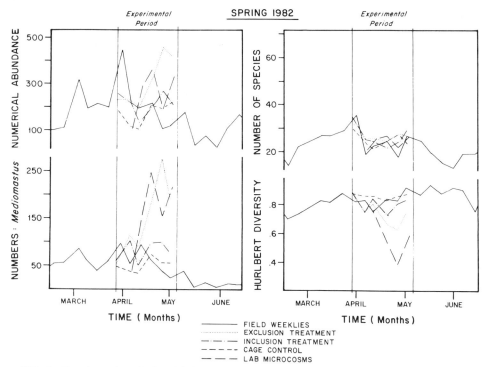

FIG. 4—*Experimental results in polyhaline area (Station ML) showing long-term field changes and short-term (during spring, 1982, experiment) treatment effects involving changes in numbers of individuals, numbers of* Mediomastus, *species richness, and diversity of infaunal macroinvertebrate assemblages.*

revealed no similar treatment-related effects in the field. The oligohaline tests indicated no treatment-related effects during fall or late winter-spring periods (Figs. 5 and 6). These results confirmed previous findings [8]. Tests during the spring indicated significant storm-induced effects on the macroinvertebrate assemblages although the effects were noted across all treatments. The lack of a predation effect in the field was tentatively attributed to the high dominance values, recruitment trends, and habitat variability in oligohaline portions of the study area. Analysis of six-week experiments in the polyhaline area of study during the fall showed no treatment effects concerning either the quantitative or the qualitative distribution of organisms over the study period. There was no clear evidence of significant ($p < 0.05$) predation or sedimentation effects on the major population and community indices of infaunal macroinvertebrates. Time-based differences were noted for important species with high recruitment values (that is, *Mediomastus* and immature tubificids).

Overall, results from the field tests indicated spatial and temporal gradients of habitat (for example, salinity) on predation effects, dominance characteristics, and community indices. Predation was an important controlling variable in high-salinity areas during spring periods of low general recruitment and high predation pressure on infaunal species.

Laboratory Microcosms

The establishment of microcosms sometimes led to changes in the laboratory assemblages relative to the field assemblages from which they were derived. Such changes included a general reduction of numerical abundance and species richness due to the loss of certain populations (Figs.

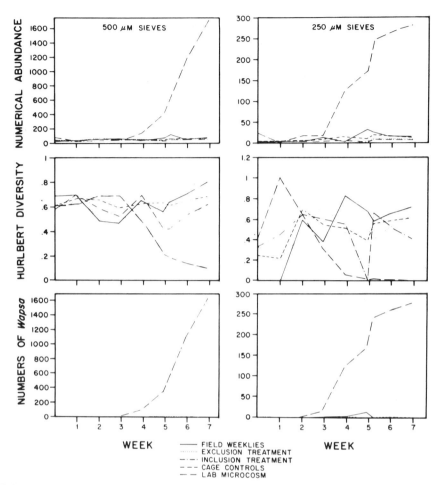

FIG. 5—*Experimental results in the oligohaline area (Station 3) showing changes in numerical abundance, diversity, and top dominant (500- and 250-μm sieve fractions) in various treatments during the fall of 1982.*

7 and 8). However, the pattern of reduced abundance and loss of populations varied from experiment to experiment depending on the seasonally variable dominance relationships. Throughout each 6 to 9 week experimental period, the macroinvertebrate microcosms changed relative to the field. This change was thought to be caused by sampling activities in the laboratory microcosms. Significant ($p < 0.05$) differences were noted between sampled and unsampled microcosms after weekly sampling over a six-week period. Such differences were also noted between the unsampled microcosms and the field samples. Sampling effects were noticeably reduced in the fall of 1982 when weekly coring was changed to biweekly coring. Subsequent experiments (Livingston, unpublished data) indicated that the size of the microcosm can be reduced and multiple subsamples taken to eliminate the sampling factor.

During the second week (T_1) of the Spring 1982 experiment at the polyhaline site (ML), inclusion and exclusion cages were grouped together in a cluster analysis of treatments by week, as were cage controls and laboratory microcosms (Fig. 9). By Week 3 (T_2), the laboratory microcosms were somewhat different from the other treatments. However, for the balance of the

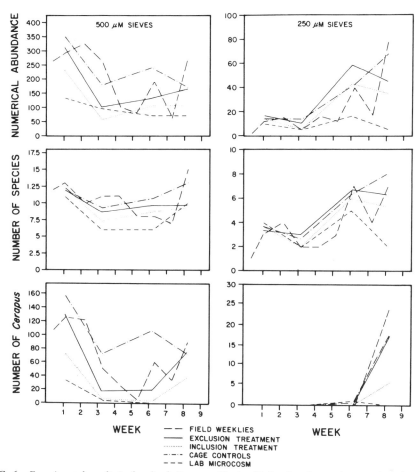

FIG. 6—*Experimental results in the oligohaline area (Station 3) showing changes in numerical abundance, species richness, and a top dominant (500- and 250-µm sieve fractions) in various treatments during the spring of 1983.*

experiment (Weeks 4 to 6), field exclusions were grouped with laboratory microcosms. The inclusion treatment was intermediate between the above treatments and the two field controls. Although the laboratory microcosms differed from the field controls, the changes (that is, numbers of *Mediomastus* with time) resembled the exclusion treatment, which indicates that, under the circumstances of the experiment, the lack of predation may have affected the laboratory microcosms (Fig. 4). Subsequent experiments with predators in the laboratory microcosm indicated predation-related effects on the laboratory test systems. The effects varied according to the system (no effects were noted under oligohaline conditions) and the time of experimentation (no release of dominant species was noted during the fall experiment in the polyhaline system). Thus, predation was a complex but potentially important factor in the establishment and maintenance of laboratory microcosms.

Experiments in polyhaline areas during the fall showed significant differences in the total numerical abundance and species richness in the laboratory microcosms relative to the field (Figs. 7 and 8). Differences between the lab and field were also noted in the oligohaline tests (Figs. 5

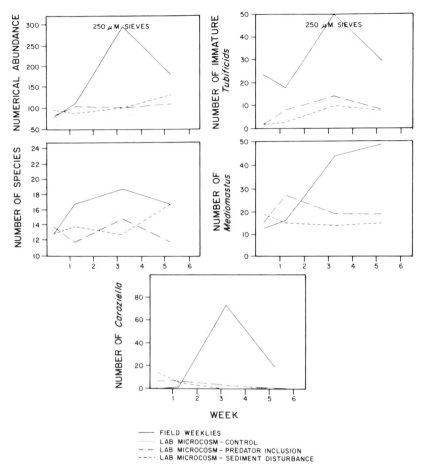

FIG. 7—*Comparison of field weeklies with laboratory microcosms (predator inclusions, sediment disturbance); organisms were taken from Station ML during the fall of 1983 (250-µm sieve).*

and 6). The differences were caused by occasional increases in a given population or reduced numbers of organisms in the microcosms at the start of the experiment (Fig. 6). Natural disturbances in the field led to responses of the infaunal macroinvertebrate assemblages that did not occur in the laboratory microcosms. The general increase in recruitment in the field following the storm in the spring of 1983 was not evident in the laboratory microcosms (see 250-µm sieve data, Fig. 6). Specific changes in physical and biological features of the natural environment had complex effects on the laboratory test systems, which were largely dependent on the timing of the test and the nature of the associated macroinvertebrate assemblages. However, indices such as species richness often showed similar, if not parallel, trends in the lab microcosms and field treatments. Such indices were not appreciably affected by changes in a single population.

Recruitment is one of the important processes of benthic communities. Recruitment of juveniles in the Apalachicola Bay system peaked from November through February (Fig. 10). Recruitment was apparent for *P. littoralis, M. ambiseta, S. benedicti,* and *G. bonnieroides* at Station 3; at station ML, such dominants included *M. ambiseta, Brania wellfleetensis, Exogone dispar, Carrazziella hobsonae,* immature tubuficids, and *Tellina texana.* Recruitment in the laboratory

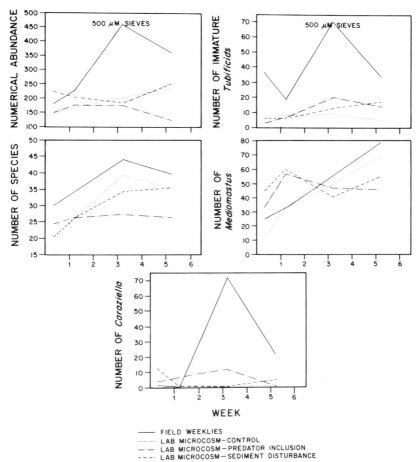

FIG. 8—*Comparison of field weeklies with laboratory microcosms (predator inclusions, sediment disturbance); organisms were taken from Station ML during the fall of 1983 (500-µm sieve).*

microcosms did not follow natural field conditions. Often, significant differences occurred in the recruitment of dominant populations in the microcosms. These differences may have led to increased recruitment in the laboratory for species such as *Mediomastus* (as a response to the disturbance of establishing the microcosm and sampling) or reduced recruitment of other species such as tubificid worms. Despite precautions taken with the seawater input from control areas, recruitment was particularly low in the laboratory microcosms during the fall experiments under polyhaline conditions (Fig. 8). This result contrasted directly with results of the spring experiments, which were dominated by enhanced *Mediomastus* recruitment in the laboratory treatments relative to field controls (Fig. 4). Other species such as *Wapsa grandis* showed higher recruitment in the laboratory during the fall experiments with oligohaline systems (Fig. 5). Significant ($p < 0.05$) differences existed between laboratory microcosms and field conditions for various population and community features (numerical abundance, species richness, dominant populations) of the organisms taken in the 250-µm sieves. Such results indicated that recruitment of juveniles was affected by laboratory conditions, but that the exact effect was influenced by the nature of the source area and the time of sampling.

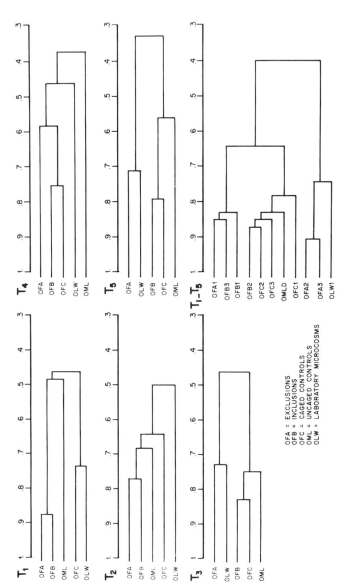

FIG. 9—*Cluster analysis of changes in species composition by treatment (summed cage results) for field and laboratory data over a six-week period in tests with polyhaline macroinvertebrates during spring 1982. Also shown is the cluster analysis by treatment for all replicates summed over the experimental periods (T_1–T_5).*

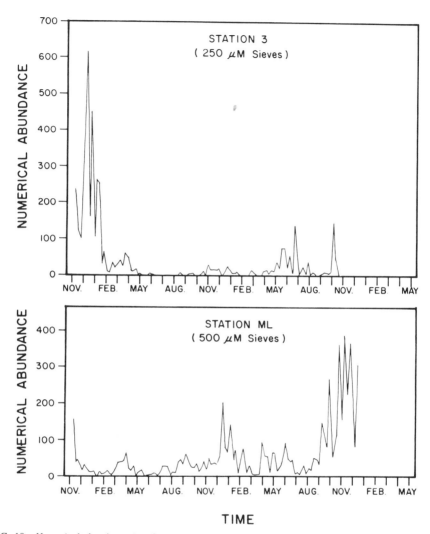

FIG. 10—*Numerical abundance (numbers per core sample) of macroinvertebrates on 250-μm and 500-μm sieves taken weekly at the study sites in the Apalachicola system from November 1981, through December 1983.*

Summary of Findings

It is not necessary to understand every factor that contributes to a given population/community trend when developing a laboratory microcosm. However, certain factors play an important role in the evaluation of the predictive capabilities of a given microcosm.

1. Duplication of a specific result from a microcosm test is difficult because of ongoing natural variation of the field components.
2. The limits of natural variation of field assemblages should be known in any verification process so that the altered response of the microcosm can be evaluated.

3. Various features of the laboratory microcosms are difficult to control. Isolation from natural disturbance is an important artifact in the development of microcosms. Species response to such isolation will vary depending on the time of the year and the natural history (tolerance, reproduction, recruitment, vulnerability to predation) of a given population.

4. Habitat- and time-dependent biological processes such as predation pressure and recruitment may be altered under laboratory conditions. Gradient effects of key habitat features such as salinity may control or alter such impact, which, in turn, may lessen the predictive capacity of the microcosm. The elimination of predation, under certain conditions, may lead to the release of dominant populations in the microcosm. Such release can then affect the composition of the microcosm community to a point where little direct connection to the field is possible. Such altered predator-prey effects may also change complex competitive interactions, again favoring certain species under laboratory conditions. Thus, laboratory microcosms may, under specific conditions, favor certain adventitious species that can then, with time, grow to dominate the microcosm.

5. Specific community features of the microcosm such as species richness, when qualified by known changes caused by laboratory artifacts, may be representative of field conditions. However, field verification of the microcosm is only possible within the bounds of our knowledge of the associated level of variation in the field. Critical factors that determine such qualification include response of dominant populations to specific laboratory artifacts, recruitment processes, predator-prey interactions, and time-based changes relative to the isolation of the microcosm from natural disturbance.

6. Specific populations of infaunal macroinvertebrates in the microcosms sometimes differed from field populations. These differences were often associated with known artifacts in the laboratory and could be qualified as such. The establishment of the microcosm was a crucial factor in the continuity and success of the test system. Microcosms of infaunal macroinvertebrates differed from field conditions depending on the timing of the experiment and recruitment features of the subject populations. Thus, the predictive capacity of the laboratory microcosm depended to a considerable degree on habitat characteristics and recruitment trends in the source area at the initiation of the microcosm.

7. Microcosms of benthic organisms, although sometimes characterized by changes in individual populations, provided a broad spectrum of biological information for use in evaluating the impact of toxic wastes. Community indices such as species richness approximated field conditions quite closely in most of the tests.

Conclusions

The use of multispecies microcosms adds a new, more sophisticated dimension to the science of bioassay evaluations. A valid argument that has been used to justify such use is that the multispecies approach is closer to the natural environment and thus represents an expansion of the utility of such tests to predict the actual impact of xenobiotics in the field. The results of our experiments indicate that even relatively simple microcosms such as sediment cores of estuarine infaunal macroinvertebrates present certain problems in the extrapolation of experimental results to field conditions. There are indications that such microcosms can be used effectively for such analysis but only after a complete evaluation of the particular artifacts associated with the microcosm in question. Field validation of the core assemblages showed that indices such as species richness were robust when used in conjunction with short-term tests. Because of the specific nature of proven laboratory artifacts, it is suggested that multispecies bioassay tests would be more useful if they were field tested and calibrated so that the most applicable indices can be used for predictive purposes. Field verification of microcosms would thus allow more effective use of such methods in the analysis of the impact of toxic substances on natural aquatic systems. Analyses are continuing for the determination of useful indices of impact that coincide with demonstrated field response. Such tests include transformation of species-specific data into trophic modes and guild systems.

Acknowledgments

Project officers for this research were Thomas W. Duke and James H. Gilford. Other project personnel included Robert J. Diaz and David C. White. Macroinvertebrate identifications were made by Gary L. Ray and Kenneth Smith. Computer operations were run by Loretta E. Wolfe. Laboratory operations were maintained by Duncan Cairns, Robert L. Howell, IV, Michael Kuperberg, and William Greening. Statistical advice was provided by Duane A. Meeter. This project (CR 810292-01-0) was funded by the U.S. Environmental Protection Agency.

References

[1] *Microcosms in Ecological Research,* J. P. Giesy, Jr., Ed., Technical Information Center, U.S. Department of Energy, Springfield, VA, 1980.
[2] Hammons, A. S., Giddings, J. M., Suter, G. W., II, and Barnhouse, L. W., *Methods for Ecological Toxicology: A Critical Review of Laboratory Multispecies Tests,* Publication No. 1710, Environmental Sciences Division, U.S. Environmental Protection Agency, Washington, DC, 1981.
[3] Cairns, J., Jr., Alexander, M., Cummings, K. W., Edmondson, W. T., Goldman, C. R., Harte, J., Hartaung, R., Isensee, A. R., Levins, R., McCormick, J. F., Peterle, T. J., and Zar, J. H., *Testing for Effects of Chemicals on Ecosystems,* National Academy Press, Washington, DC, 1981.
[4] Livingston, R. J. and Meeter, D. A., "Correspondence of Laboratory and Field Results: What Are the Criteria for Verification?" in *Multispecies Toxicity Testing,* J. Cairns, Jr., Ed., Pergamon Press, New York, 1985, pp. 76–88.
[5] Livingston, R. J., Diaz, R. J., and White, D. C., *Field Validation of Laboratory-derived Multispecies Aquatic Test Systems,* Project Summary, U.S. Environmental Protection Agency, Gulf Breeze, FL, 1985.
[6] Bloom, S. A., *Marine Ecology Progress Series,* Vol. 5, 1981, pp. 125–128.
[7] Lance, G. N. and Williams, W. T., *Computer Journal,* Vol. 9, 1967, pp. 323–380.
[8] Mahoney, B. M. S. and Livingston, R. J., *Marine Biology,* Vol. 69, 1982, pp. 207–213.

Frieda B. Taub,[1] *Andrew C. Kindig,*[2] *and Loveday L. Conquest*[3]

Interlaboratory Testing of a Standardized Aquatic Microcosm[4]

REFERENCE: Taub, F. B., Kindig, A. C., and Conquest, L. L., **"Interlaboratory Testing of a Standardized Aquatic Microcosm,"** *Aquatic Toxicology and Hazard Assessment: 10th Volume, ASTM STP 971*, W. J. Adams, G. A. Chapman, and W. G. Landis, Eds., American Society for Testing and Materials, Philadelphia, 1987, pp. 384–405.

ABSTRACT: The "Standardized Aquatic Microcosm" (SAM) Protocol is being tested in three laboratories as well as in the laboratory where it was developed (University of Washington). Each laboratory has used a control and three concentrations of copper sulfate and has completed at least one experiment.

The successional sequence in the controls has consistently included nitrate depletion and an early algal bloom terminated by increases in grazer populations, especially *Daphnia*. In all experiments analyzed to date, low copper sulfate concentrations were associated with temporary reduction in *Daphnia* and concurrent algal blooms; at higher concentrations, the algae were inhibited to a greater extent and for a longer duration.

The concept of a "days-weighted-by-variable" statistic (DWV) is introduced to assess variation between experiments. The DWV, as a measure of the center of gravity of a microcosm time trace, is useful to detect shifts in timing under increased copper concentrations.

KEYWORDS: microcosms, interlaboratory testing, bioassay, toxicity testing, copper, *Daphnia*, algae

Multispecies toxicity tests are being considered as potential tools for evaluating the ecological effects of chemicals [1,2]. To gain acceptability for regulatory use, multispecies test protocols must demonstrate ecologically meaningful responses to treatment effects in a statistically testable manner. Acceptability would also be increased if the degree of reproducibility between laboratories as well as repeatability within laboratories were known. Regulatory agencies would have more confidence if decisions were based on results that could be verified by other laboratories.

The current study is testing the hypothesis that the "Standardized Aquatic Microcosm" (SAM) Protocol can be performed in different laboratories and that similar control and treatment effects will result. The protocol describes the procedures for initiating, monitoring, and analyzing the results of the microcosm experiments [3]. The term "standardized" refers to the microcosms being initiated with chemically defined medium and of the same species (ten species of algae and five species of animals) with similar densities and physiological states, to the degree feasible. The purpose of this study includes describing, identifying, and reducing sources of variation between experiments.

Earlier reports displayed the development of the microcosms [4], their responses to toxic chemicals [5–7], sensitivity as a result of successional stage [8], and comparison to field studies [9].

---

[1] Professor, School of Fisheries, WH-10, University of Washington, Seattle, WA 98195.
[2] Fisheries biologist, School of Fisheries, WH-10, University of Washington, Seattle, WA 98195.
[3] Associate professor, Center for Quantitative Science, University of Washington, Seattle, WA 98195.
[4] School of Fisheries Contribution No. 737.

The principles of interlaboratory (round-robin) testing of microcosm tests were discussed earlier [10], and results of the first two tests in the host and first participating laboratory have been reported [11]. Another laboratory, Corvallis-EPA, has also used an earlier prototype of the SAM with atrazine as the test chemical and has compared the microcosm responses with pond and single species responses [12,13].

The purpose of this paper is to present the results of SAM experiments in two additional laboratories and an additional experiment in our laboratory and to compare these to earlier experiments. A new statistical method is demonstrated by computing a "days-weighted-by-variable" (DWV, details below) on each microcosm. Briefly, the DWV yields a single number representing the center of gravity for a given time trace. This was done on the controls and 500 μg L^{-1} and is described for two variables on six experiments. Both intra- and interlaboratory comparisons were obtained. A major source of between-laboratory variation was discovered to be associated with the medium in which the *Daphnia* were reared; recommendations for correction are suggested.

Methods

The interlaboratory test consists of experiments performed at the host (University of Washington) and three participating laboratories: (1) EPA–Duluth (under a cooperative agreement with the University of Minnesota at Duluth, MN); (2) U.S. Army, Aberdeen Proving Grounds, MD; and (3) Marine Bioassay Laboratories, Carlsbad, CA. The status of the experimental design is shown in Table 1. The participating laboratories made all of the medium and reared all of the organisms

TABLE 1—*Status of interlaboratory SAM experiments.*

Laboratory	First Experiment	Second Experiment
U of W		
Initiated	2 Dec. 1983	24 Jan. 1986
	(to be redone[a])	
Experiment no.[b]	ME 64	ME 82
Cu deviation[c]	−13%	+5%
EPA-Duluth		
Initiated	2 Dec. 1983	6 Apr. 1984
Experiment no.	ME 65	ME 67
Cu deviation	−3%	+6%
Aberdeen		
Initiated	3 May 1985	24 Jan. 1986
Experiment no.	ME 74	ME 87
Cu deviation	. . .[d]	−9%
MB Lab		
Initiated	15 Nov. 1986	
Laboratory	ME 79	
Cu deviation	−1%	

[a] This experiment will be redone because *Daphnia* populations were low at the end of the experiment (see text) and Cu concentration was low.

[b] Experiment numbers (ME ##) indicate the Model Ecosystem number under which the original data are stored in the host laboratory.

[c] Cu deviation indicates the difference between the nominal concentration of Cu and the measured total copper on Day 7.

[d] For this experiment only, the treatments consisted of control, 127, 255, and 509-μg L^{-1} Cu; all other experiments consisted of control, 500, 1000, 2000-μg L^{-1} Cu.

used in their experiments; initial organism stocks were supplied by the host laboratory, as well as a copy of the Protocol [3]. A four-day training course was provided at the host laboratory for researchers from the participating laboratories. Apart from a site visit by Dr. Kindig prior to the initiation of studies, no host staff was involved in the experiments by Laboratories 2 and 3. Raw data from the participating laboratories were sent to the host laboratory and were statistically analyzed on the data-handling package (CYBER computer). The participating laboratories audited the results and agreed that the reported results represent the data sent.

Each experiment consists of 24 microcosms (3 L each), 6 controls, and 6 each of 3 concentrations of $CuSO_4$. The medium, T82MV, contained 0.5-mM nitrate, 0.04-mM phosphate, other inorganic nutrients, and vitamins. The sediment was sand, chitin, and cellulose. Ten algal species were inoculated on Day 0 at initial concentrations of 10^4 cells mL^{-1}. Five species of animals, including *Daphnia*, were inoculated on Day 4. Copper sulfate was added to randomly assigned microcosms on Day 7 to provide six each of controls, 500, 1000, and 2000 µg L^{-1} Cu. Samples were removed on Days 7 and 63 for Cu analyses at the host laboratory. Species counts (algae and animals, not bacteria), pH, oxygen (night/day/night), optical density, absorbance, and in vivo fluorescence were measured twice weekly; nutrients were measured twice weekly during the first month, once weekly thereafter. The measured variables were used to calculate others. For example, the biovolume of algae was calculated by summing the product of the algal counts times nominal cell volume for each species. This allowed us to estimate total algal volume from cell counts. Data were collected directly on computer coding sheets for data entry. Reinoculation of very small numbers of microscopic organisms was done weekly (Friday); larger organisms, for example, *Daphnia*, were reinoculated only if required to maintain three animals per microcosm. The reinoculation simulated immigration, permitted recovery of species that were killed during temporary toxicity phases, and prevented random extinctions. The numbers of organisms reinoculated on Friday, after sampling, were below the level of detection of the sampling technique; they would not influence the data collected unless they reproduced.

Within-experiment statistics included separate univariate analyses of variance (ANOVA) and a priori t-tests on means for each sampling day. Results are expressed as the "Interval of Nonsignificant Difference" (IND) around the control means [14], calculated by the formula

$$x_c \pm t\sqrt{s^2\left(\frac{1}{n_c} + \frac{1}{n_t}\right)} \tag{1}$$

where
x_c = control mean,
t = 95% tabled value from a "t" probability distribution,
s^2 = pooled variance from ANOVA,
n_c = number of control replicates, and
n_t = number of treatment replicates.

Treatment means outside of these bounds are statistically distinguishable from the control means. Between-experiment statistics include computation of a "days-weighted-by-variables" (DWV), the "centroid" of a time trace for each microcosm. This is followed by two-way ANOVAs on the DWVs to compare the relationship from controls to treatments among the various laboratories.

Variations in techniques between laboratories did occur. The medium in which the *Daphnia* were reared differed: University of Washington (UW) used the microcosm medium T82MV in the first experiment [Model Ecosystem (ME) 64] and in an earlier second experiment (ME 69, [11]); ME 69 has been replaced in this report with a subsequent experiment in which the Lake Washington water (Seattle, WA) was the rearing medium; Duluth used Lake Superior water; Aberdeen used treated well water; Marine Bioassay used aged microcosm medium T82MV. Aberdeen omitted amphipods and did not measure in vivo fluorometry; in their second experiment they also omitted *Ulothrix*, one of the filamentous green algae. Nutrients were analyzed by different methods in some of the laboratories. Stock cultures of protozoa and rotifers had different animal densities, and different amounts of culture medium were added during inoculation and reinoculation.

Variations in the amounts of Cu differed; with the exception of Aberdeen No. 1, all were -13 to $+6\%$ of the nominal 500, 1000, 2000 μg L^{-1} Cu (Table 1). Aberdeen No. 1 used the weight of CuSO$_4$, and the corresponding Cu concentrations were control, 127, 255, and 509 μg L^{-1} Cu. Although this varied from the intended commonality of experimental treatments, the lower concentrations do allow us to explore the sensitivity of the microcosms. The control and nominal 500 μg L^{-1} Cu treatments will be compared across all experiments.

Results

Biological

Given the limited size of this publication, algal biovolume and *Daphnia* abundance have been selected to display the effects of the Cu treatment (Figs. 1–3), although there were also significant effects on algal species dominance, other invertebrates, nitrogen chemistry, pH, in vivo fluorescence, absorbance, and oxygen dynamics, which will sometimes be referred to in the text. Treatment means of three of the experiments are published [*11*] and are repeated here for comparison. Treatment means and intervals of nonsignificance are shown in Fig. 1. Replicate values for algal biovolume and *Daphnia* control and 500-μg L^{-1} Cu are shown for all seven experiments in Figs. 2 and 3.

Controls. The control microcosm events followed the sequence of nitrate (the limiting nutrient) depletion concurrent with increased algal abundance; algal biovolume increased until the *Daphnia* population bloomed. Two species of algae, *Ankistrodesmus* and *Scenedesmus*, were the dominant and codominants during much of the experiment, although other species were sometimes more abundant for a few sampling days. Contaminant algal species never became established in the controls. The algal peaks (means) ranged from circa 150 to 600 $\times$ 10^5 μm^3/mL and occurred from Days 7 to 14; subsequently, algal biovolumes were generally less. The *Daphnia* populations peaks occurred at abundances ranging from 100 to 280 animals per 100 mL from Days 18–28. A *Daphnia* bloom always followed an algal bloom, and *Daphnia* peaks were associated with declines in algal biovolume. Other grazers were present in measurable populations but did not appear to control algal abundance to the extent of the *Daphnia*.

Daphnia reared in the chemically defined medium T82MV displayed lower peaks, poorer replication during the initial peaks (were not synchronous), and failed to maintain >20 animals per 100 mL during the last part of the experiment (UW No. 1 and Marine Bioassay Laboratory No. 1) (Fig. 3). This pattern of poor *Daphnia* population development was consistently shown in an earlier UW experiment (ME 69, [*11*]) and in other technique experiments (ME 72, ME 75). Because the Cu treatment effects were so dramatic, statistical treatment effects were still shown in copper experiments; however, it is undesirable to have unhealthy *Daphnia* populations. The lack of control *Daphnia* populations at the end of the experiment makes it impossible to show statistical differences to poisoned treatments. *Daphnia* reared in lake water or treated tap water had better characteristics in these respects. Some preliminary suggestions will be discussed in the Discussion Section.

500-μg L^{-1} Copper Treatments. After the addition of Cu on Day 7, algal biovolume was reduced for one or more sampling days, although the reduction was not statistically significant in

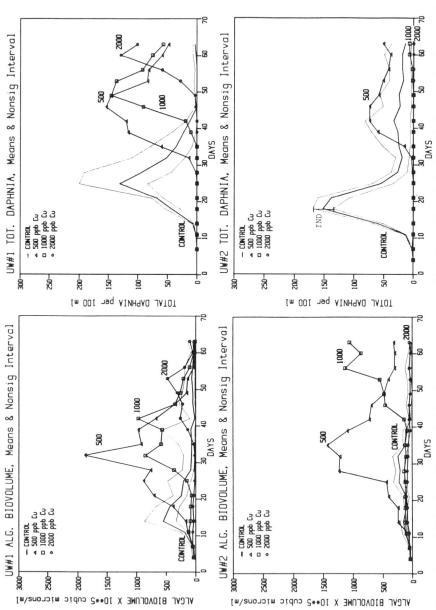

FIG. 1—*Means of control (thick line without symbols) and treatment microcosms (thick lines with symbols). The thin lines are the "interval of nonsignificant difference" (IND), a type of least significant difference; it is calculated from the log transformed, pooled variance estimate and is therefore nonsymmetrical around the control mean. If a mean is outside of the IND area, it is statistically different than the control. All experiments show Control, 500, 1000, 2000-µg L^{-1} Cu, except Aberdeen No. 1, which shows Control and 500-µg L^{-1} Cu.*

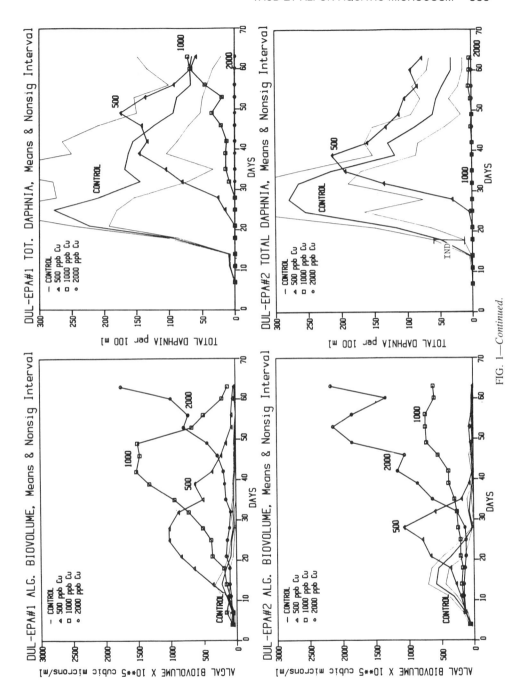

FIG. 1—*Continued.*

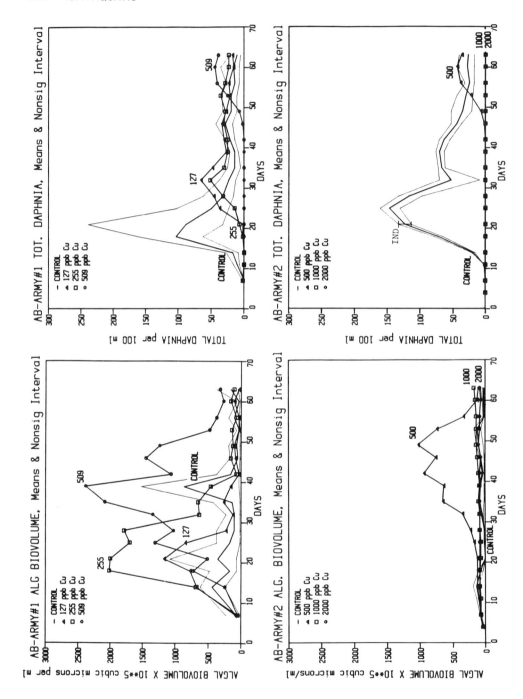

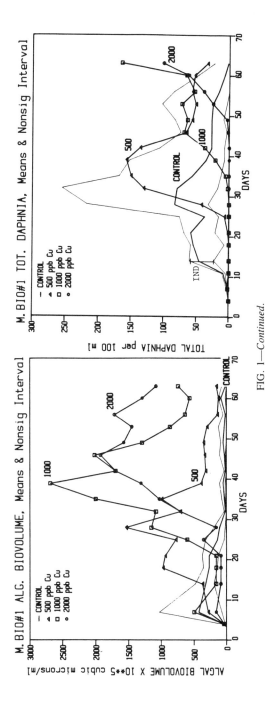

FIG. 1—Continued.

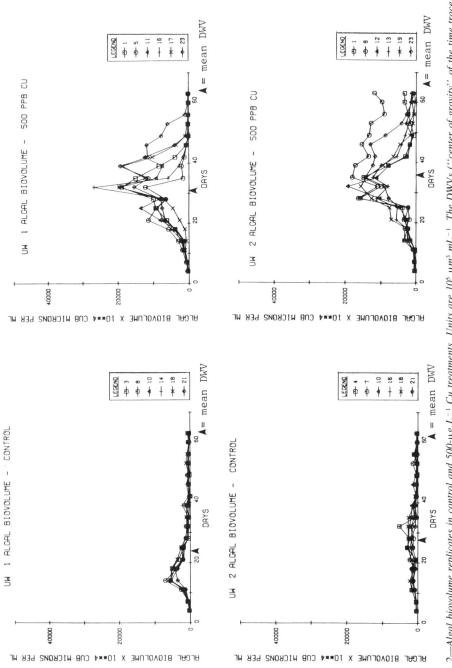

FIG. 2—*Algal biovolume replicates in control and 500-μg L^{-1} Cu treatments. Units are 10^5 μm^3 mL^{-1}. The DWVs ("center of gravity" of the time trace) are shown by an arrow on the time axis.*

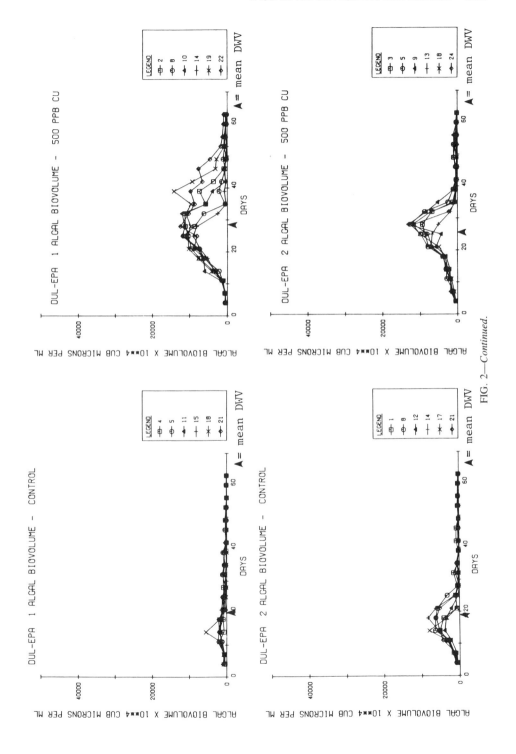

FIG. 2—*Continued.*

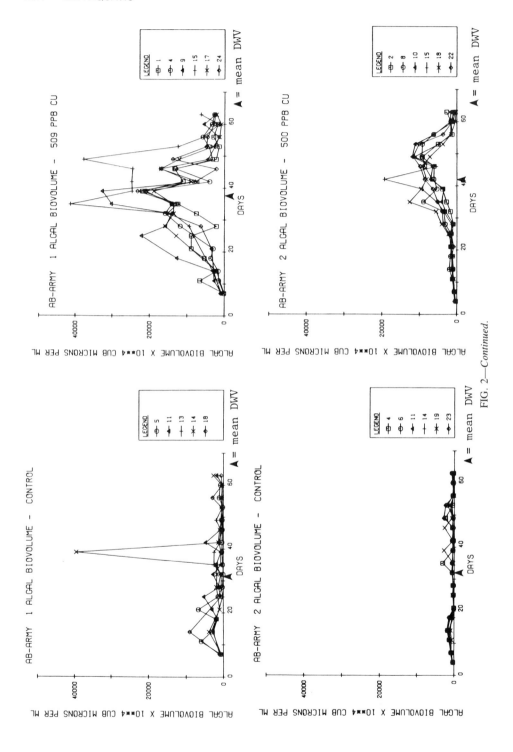

FIG. 2—Continued.

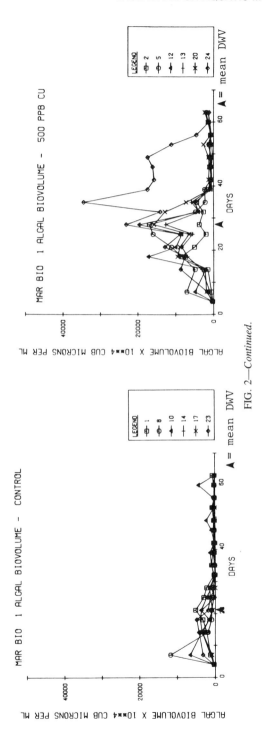

FIG. 2—Continued.

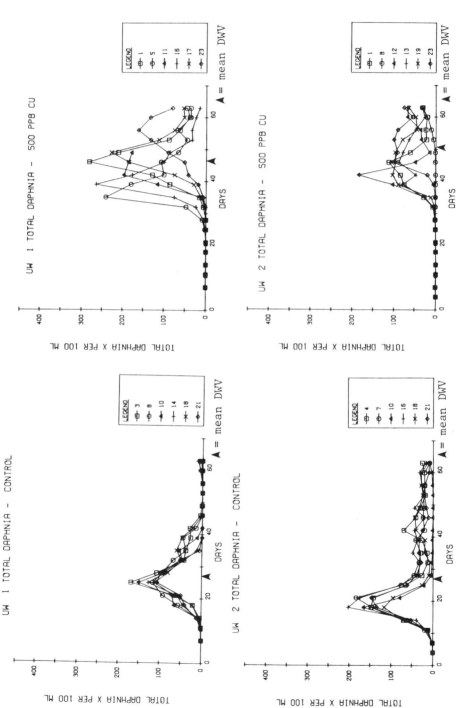

FIG. 3.—Daphnia abundance in control and 500-μg L^{-1} Cu treatments. The microcosm symbols are the same as in Fig. 2; high Daphnia abundances terminate algal blooms. The DWVs ("center of gravity" of the time trace) are shown by an arrow on the time axis. Units are animals per 100 mL.

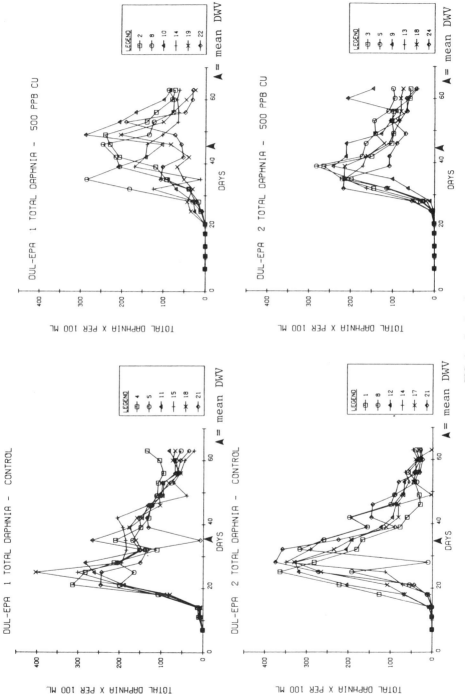

FIG. 3—*Continued.*

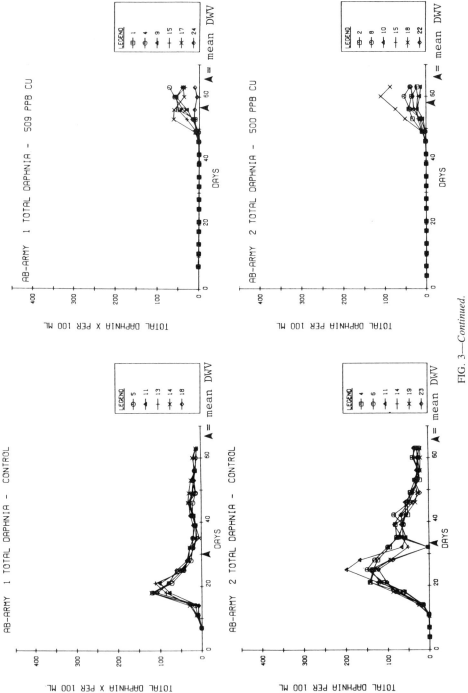

FIG. 3—*Continued.*

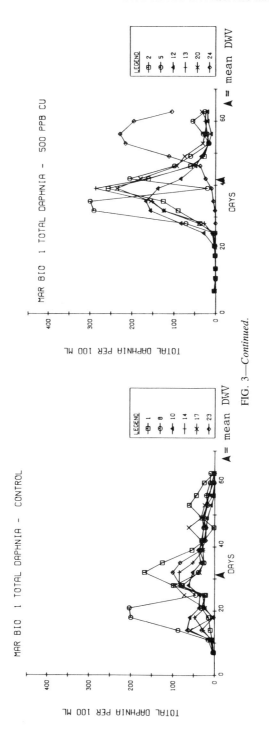

FIG. 3—*Continued.*

some experiments; oxygen gain, pH, and algal species diversity were usually significantly reduced, supporting the hypothesis that primary production was inhibited. Thereafter, algal biovolume increased above controls during the temporary absence of *Daphnia*. In all experiments, algal biovolume was significantly greater than controls for a time. The algal biovolume peak was later than the controls (Days 25 to 39) and exceeded the earlier control peaks (1030 to 2370 × 10^5 $\mu m^3/mL$). Because the peaks occurred after the controls had lower algal abundance, the 500-μg L^{-1} Cu treatment peaks were vastly higher than the controls on those days. The algal blooms consisted of relatively Cu-resistant species *Ankistrodesmus*, *Scenedesmus*, *Chlorella*, and *Selanastrum*; sensitive species such as *Lyngbya*, *Anabaena*, and *Nitzschia* were virtually eliminated by the Cu. *Daphnia* were eliminated from the microcosms by the Cu for several sampling days; the differences were statistically significant. *Daphnia* eventually established populations from reinoculated animals. The time and magnitude of the peaks varied from 39 to 60 days and 40 to 210 animals per 100 mL. Because the control *Daphnia* populations were declining at these times, the Cu-treated *Daphnia* were significantly more abundant than controls at the time of their bloom.

Comparison of the control replicates with the 500-μg L^{-1} Cu replicates (Fig. 2) shows the delay in the development of the algal blooms and the delay in the establishment of *Daphnia* population (Fig. 3). The timing of recovery was variable among the Cu-treated replicates. Recovery could have been caused by algae that survived the Cu treatment or reinoculated cells; all of the *Daphnia* present at the time of the Cu treatment died, and recovery was from reinoculated animals.

1000-μg L^{-1} Copper Treatments. This treatment resulted in greater inhibition and for a longer duration. Algal biovolume was significantly lower than the control in all experiments but did eventually recover and bloomed. The delayed blooms were sometimes, but not always, of greater abundance than the control or low Cu treatments and were composed of fewer algal species. In the Duluth–EPA experiments, a contaminant alga *Oocystis* was the dominant bloom species. It is the only contaminant alga that became established in these seven experiments, and it did so only in this and the higher Cu treatments. In Duluth-EPA and Aberdeen-Army experiments, *Selenastrum* was much more dominant in this treatment than in controls (where they were out competed and maintained very low populations) and the lower Cu treatment. Similar results occurred to some degree in all experiments.

Daphnia populations became established near the end of the 63-day experiment in all but one experiment (Aberdeen-Army No. 2). Recovery might have occurred in that experiment had the experiments continued beyond 63 days.

2000-μg L^{-1} Copper Treatments: Extreme and prolonged inhibition of algae occurred. Late algal blooms occurred in Duluth No. 1 and 2 and Marine Bioassay; these were virtually unialgal cultures of *Oocystis* (a contaminant) in Duluth experiments and *Selenastrum* in the Marine Bioassay experiments. Algal inhibition of most species occurred in all experiments, and algal biovolume was very reduced in some of the experiments.

Daphnia populations were reestablished only in UW No. 1 and Marine Bioassay No. 1 experiments, later than in the 1000-μg L^{-1} Cu treatment. Termination of the experiment at 63 days may have prevented eventual reestablishment in the other experiments.

In summary, the delay in the algal and *Daphnia* blooms (relative to controls, and the progressively later blooms at the higher Cu concentrations) and the display of algal blooms of Cu-resistant algal species during the *Daphnia* absence were consistently shown in all experiments. Variation in the timing and magnitude of the blooms became progressively greater at higher Cu concentrations.

Statistical Treatment

Statistical analyses of repeated measurements of related variables present special problems. The microcosm data share these attributes with pond and lake studies. After exploring a variety of statistics, results are shown here for control and 500-μg L^{-1} Cu replicates of the seven experiments (Table 2). Complete statistics for all treatment groups and all experiments will be published when the interlaboratory testing is complete. Our current approach is to compare the days-weighted-by-variable (DWV) between control replicates and treatment replicates. This acknowledges that we are comparing the effect within each experiment and comparing these for the different experiments. The DWV for each microcosm is computed as follows

$$\text{DWV} = \sum (\text{day} \times \text{abundance}) / \sum (\text{abundance})$$

where the summation is performed over all (fixed) sampling days for a variable. Table 2 shows the means and standard deviations of the DWV for the control and 500-μg L^{-1} Cu for seven experiments for algal biovolume and *Daphnia*. The DWVs are also noted in Figs. 2 and 3. For both variables, there was a significant delay ($P < 0.00005$) in the peak (mean DWV in the 500-μg L^{-1} Cu significantly greater than in the controls) over all seven experiments. The magnitudes of the difference were not significantly different ($p > 0.25$) for algal biovolume over the seven experiments; they were for *Daphnia* ($P < 0.001$). By looking at the differences between *Daphnia* mean DWV (500-μg L^{-1} Cu minus the controls), the experiments performed within a laboratory tend to be more similar than those performed in other laboratories. However, the direction of the shift (delay in the *Daphnia* recovery) was the same across all laboratories.

Using the DWV to produce a single value for each microcosm, the interlaboratory comparisons used a two-way analysis of variance (ANOVA) with treatment group as one factor and experiment as the other factor (Table 2). By weighting each sampling day with its associated count, the DWV can reflect differences between an early, low bloom (low DWV) versus a period of inhibition followed by an increased abundance (high DWV).

TABLE 2—*Mean DWV for controls and 500-ppb Cu treatments.*

	Control	ALGAL BIOVOLUME Difference (Delay)	500-ppb Cu
UW No. 1	24.03(1.41)	6.87	30.90(3.75)
UW No. 2	28.31(1.32)	7.63	35.94(3.59)
Duluth No. 1	20.11(3.40)	8.27	28.38(3.12)
Duluth No. 2	18.91(1.59)	6.23	25.14(1.72)
Aberdeen No. 1	31.40(6.59)	6.25	37.65(2.68)
Aberdeen No. 2	31.57(3.61)	9.76	41.33(1.85)
Marine Bioassay	21.49(3.61)	7.24	28.73(3.75)
	Control	DAPHNIA Difference (Delay)	500-ppb Cu
UW No. 1	27.44(1.22)	19.64	46.98(4.36)
UW No. 2	27.82(3.88)	22.38	50.20(4.53)
Duluth No. 1	35.56(0.74)	10.22	45.78(1.36)
Duluth No. 2	34.41(2.57)	9.67	44.08(2.12)
Aberdeen No. 1	29.65(1.05)	27.17	57.12(1.71)
Aberdeen No. 2	33.21(0.78)	24.38	57.59(0.88)
Marine bioassay	31.59(5.08)	10.33	41.92(6.42)

NOTES: Units are days. DWVs are also marked in Figs. 2 and 3. Sample sizes for each "cell" are $n = 6$, except Aberdeen No. 1 control ($n = 5$). Standard deviations are in parentheses. F tests ($df = 1,69$) to compare treatments: 105.44 (algal biovolume, $P < 0.001$), 714.3 (*Daphnia*, $P < 0.001$). F tests ($df = 6,69$) for treatment-experiment interaction: 0.42 (algal biovolume, $P > 0.25$), 18.3 (*Daphnia*, $P < 0.001$).

Averaging over either the entire experiment or a distinct time period (where "averaging" here is simply the summed counts/number of sampling days) has been used [13] to compare the end points of algal production in ponds and microcosms treated with atrazine.

In doing a sequence of one-way ANOVAs that compared means for each variable on each sampling day, we found that differences in magnitude among controls and slight differences in timing made all the experiments appear statistically different. Even experiments that appeared qualitatively similar were statistically different when events were offset by a single period between samplings (three to four days). This issue seems to have been resolved by the use of the DWV, which characterizes the "center of gravity" in terms of day for each time trace.

An earlier effort which compared fitted regression parameters via ANOVA was useful as long as the data could be expressed by a simple equation. Quadratic models of the log transformed *Daphnia* and algal biovolume, and a reciprocal linear model for the nitrate data provided high correlation coefficients for several experiments [11]. Some variables, such as oxygen gain and pH, were too complex for this treatment. The method depended greatly on the goodness of fit of the equation to the original data. Also, if some experiments had statistically higher quadratic coefficients but lower linear coefficients than others, the biological meaning was not clear. This method does have the advantage of comparing the shape of the data, rather than the absolute points, on a day-by-day comparison.

Discussion

General trends in controls and treatments of the seven experiments were similar; the same conclusions, which follow, could be drawn in all experiments. (1) The addition of Cu selectively inhibited algal species and was acutely toxic to *Daphnia*. (2) During the absence of *Daphnia*, algal blooms occurred of the relatively Cu-resistant species *Ankistrodesmus*, *Scenedesmus*, *Chlorella*, *Selenastrum*, and *Oocystis*. Sensitive species such as *Lyngbya*, *Anabaena*, and *Nitzschia* were virtually eliminated by the Cu, in spite of reinoculation. (3) The magnitude and duration of the effects were greater at higher concentrations. (4) Algal species diversity was reduced at higher Cu concentrations. (5) Biological communities eventually became established in all of the 500-μg L^{-1} Cu, most of the 1000-μg L^{-1} Cu, and some of the 2000-μg L^{-1} Cu treatments. Had the experiments lasted longer than 63 days, recovery might have been shown in all the experiments and treatments.

Between-Experiment Control Behavior

Some of the between-experiment variation has identified causes and can probably be reduced in future experiments. These include *Daphnia* rearing techniques, algal inoculation, and avoidance of excessive rotifer and protozoan medium additions during reinoculations.

Variation among controls in different experiments was likely to have been caused by differences in initial conditions and slight differences in temperature and light intensity during the experiment. We had shown that altering the initial algal density altered the timing and magnitude of algae and *Daphnia* blooms [15]. There are tradeoffs between increasing the cost by requiring greater standardization in setting up the experiments and tolerating the differences displayed. The results emphasize the need to compare treatment groups with simultaneous controls.

The magnitude and persistence of the control *Daphnia* populations were related to the rearing medium of the *Daphnia* stocks. Experiments in which natural waters were used to culture *Daphnia* (Duluth, Aberdeen) had tighter replication, higher peaks, and sustained populations throughout the experiment; experiments for which T82MV was used to rear *Daphnia* (UW No. 1, other UW experiments, ME 72, 75, and Marine Bioassay No. 1) had poorer replication during the initial peaks, lower peak populations, and low *Daphnia* populations toward the end of the experiment.

Keating and others have shown that selenium is a requirement for *Daphnia*, especially for healthy cuticle development [*16–18*]; T82MV lacks Se and several other trace metals that are included in Keating's medium. We have since shown that the addition of these metals (in μg L^{-1}, Se, 1; Br, 50; Sr, 100; Rb, 100; Li, 100; I, 5; V, 0.5) to T82MV (renamed T85MVK) provides a better rearing medium for microcosm test animals (unpublished data).

Based on the experiments presented here, the quality control criteria for *Daphnia* should be that replicate means of at least 85 animals per 100 mL be obtained during the first 28 days and that at least 20 animals per 100 mL be maintained during the entire experiment. These are slightly lower than recommended in the draft protocol [*3*] and an earlier paper [*10*]. The only experiment presented in this paper that would fail to meet these criteria is UW No. 1, which was initiated with animals reared in T82MV. We plan to redo this experiment to correct that problem and to correct the low Cu (-13% of nominal) treatment.

Differences in algal dominance, especially with filamentous algae, may have been caused by differences in the initial numbers. When clumps of filamentous algae were not eliminated from the inoculated material, overinoculation of filamentous forms occurred because clumps were underestimated in the microscopic cell counts. This problem was minor in the experiments reported here, but was a problem in some earlier experiments. Some of the laboratories had excessive accumulations of phosphate and delayed depletion of nitrate during the experiment, caused by the excess volume of high-phosphate rotifer and protozoa medium added during reinoculation. The phosphate additions prevented the demonstration of phosphate depletion that usually was concurrent with nitrate depletion and may have caused some differences in ammonia metabolism. The slightly increased nitrate may have allowed denser algal blooms in those experiments. These problems were not serious and are easily avoided.

Between-Experiment Differences in Copper Treatments

All of the experiments displayed a toxicity to algae and *Daphnia*; the major differences were the timing of the recovery of the more Cu-resistant algae and the later recovery of the *Daphnia*. Recovery was always more delayed in the higher Cu concentrations. Timing of recovery differed among experiments and also among replicates within an experiment. Variation in the recovery of the organisms indicated that the rate at which the Cu was removed from the active form varied. We know that much of the $CuSO_4$ added interacts with other ions in the medium, especially carbonate, and precipitates. Measurement of total and dissolved Cu a few minutes after introduction indicates that most of the added Cu becomes associated with the particulate fraction; over time, much of the Cu becomes associated with the detritus and algal mat on the sand. The detoxification probably is related to the speciation of the Cu rather than to the dissolved concentration. A larger proportion of the dissolved Cu will be in the cupric ion form at low pH [*19*]. Shortly after the Cu is added, mortality is high and photosynthesis is low; the pH remains low until a Cu-resistant algae starts to grow. Its photosynthesis increases Ph and may detoxify the system for other organisms. We have been modeling the ecological effects of Cu on the microcosms [*20,21*] and are exploring the roles of dissolved organic carbon and alkalinity.

Within each experiment, recovery in the low Cu treatment occurred before recovery at the higher Cu concentrations. If recovery in low treatment occurred later, recovery at higher treatments may not occur during the 63-day test. This variability in the recovery of the 1000 and 2000-μg L^{-1} Cu treatments could be misinterpreted as disagreements between experiments.

The recovery of Cu from subsamples during the experiments has varied from close to nominal to less than 50% of the initial Cu added. Much of the unrecovered Cu has been shown to be associated with detritus mixed into the sand. The Cu associated with the deep detritus can be recovered only if stirring is very vigorous, or if the microcosms are dissected at the end of the experiment. The "missing" Cu is therefore an artifact of the sampling technique. Sorption on the sand and glass are under study and will be reported when the interlaboratory study is complete.

Additional evidence that control behavior is reproducible between laboratories is also supported by experiments done at Corvallis-EPA [12] and the Aberdeen-Army Brass experiment reported in this volume. At least one additional experiment will be conducted at our laboratory to further test the reproducibility of the methods.

The availability of a reproducible microcosm protocol may lend credibility to the use of multispecies bioassays by ecologists and by regulatory organizations. Standardization has substantial advantages by speeding data analyses, aiding data interpretation, and allowing verification. By examining ecological processes common to all aquatic communities (primary and secondary production and nutrient recycling with competition within trophic levels), the microcosm bioassay provided insights into responses of some natural systems. Since different aquatic systems responded differently to Cu, it would not have been possible for the microcosm results to have agreed with all of them [9]. In the future, we hope to use the microcosms and a simulation model (MICMOD [20,21]) to explore why different aquatic communities respond differently to the same chemical stress.

Acknowledgments

We wish to thank James Meador for Cu analyses, Thomas Sibley for supervision of the Cu chemistry studies, Rob Rieger for technical assistance, and Sharon Roloff for data storage/retrieval, manuscript preparation, and graphics. Special thanks are due the staffs of the participating laboratories: Duluth–EPA, David Yount, Lyle Shannon, Michael Harrass (during these studies), and Chuck Walbridge; U.S. Army-Aberdeen, Wayne Landis, Mark Haley, and Dennis Johnson; Marine Bioassay Laboratory, Allan Thum, Skip Newton, Sherrie Hamer, and John Hardin.

The project has been supported by the U.S. Food and Drug Administration under contract 223-83-7000. We also acknowledge the help of the Project Officers Buzz Hoffmann and John Matheson, who contributed to the general development of the project and helped to stimulate progress.

References

[1] *Multispecies Toxicity Testing*, J. Cairns, Jr., Ed., Pergamon Press, New York, 1985, pp. 261.
[2] Environmental Studies Board, Committee to Review Methods for Ecotoxicology, National Academy of Science, *Testing for Effects of Chemicals on Ecosystems*, National Academy Press, Washington, DC, 1981, pp. 103.
[3] Taub, F. B. and Read, P. L., "Standardized Aquatic Microcosm Protocol," Vol. II, Final Report, Food and Drug Administration Contract No. 223-80-2352, University of Washington, Seattle, WA, 1982.
[4] Taub, F. B. and Crow, M. E., "Synthesizing Aquatic Microcosms," in *Microcosms in Ecological Research*, J. P. Giesy, Jr., Ed., Symposium Series 52, CONF-781101, Department of Energy, Washington, DC, 1980, pp. 69–104.
[5] Taub, F. B., Read, P. L., Kindig, A. C., Harrass, M. C., Hartmann, H. J., Conquest, L. L., Hardy, F. J. and Munro, P. T., *Aquatic Toxicology and Hazard Assessment (Sixth Symposium), ASTM STP 802*, American Society for Testing and Materials, Philadelphia, 1983, pp. 5–25.
[6] Taub, F. B. in *Algae as Ecological Indicators*, L. E. Shubert, Ed., Academic Press, New York, 1984, Chap. 13, pp. 363–394.
[7] Taub, F. B. in *Concepts in Marine Pollution Measurements*, H. H. White, Ed., University of Maryland, College Park, MD, 1984, pp. 159–192.
[8] Kindig, A. C., Conquest, L. L., and Taub, F. B., *Aquatic Toxicology and Hazard Assessment (Sixth Symposium), ASTM STP 802*, American Society for Testing and Materials, Philadelphia, 1983, pp. 192–203.
[9] Harrass, M. C. and Taub, F. B. in *Validation and Predictability of Laboratory Methods for Assessing the Fate and Effects of Contaminants in Aquatic Ecosystems, ASTM STP 865*, American Society for Testing and Materials, Philadelphia, 1985, pp. 57–74.

[10] Taub, F. B. in *Multispecies Toxicity Testing*, J. Cairns, Jr., Ed., Pergamon Press, New York, 1985, Chap. 13, pp. 165–186.
[11] Taub, F. B., Kindig, A. C., and Conquest, L. L., in *Community Toxicity Testing, ASTM STP 920*, American Society for Testing and Materials, Philadelphia, 1986, pp. 93–120.
[12] Stay, F. S., Larsen, D. P., Katko, A., and Rohm, C. M. in *Validation and Predictability of Laboratory Methods for Assessing the Fate and Effects of Environmental Contaminants in Aquatic Ecosystems, ASTM STP 865*, American Society for Testing and Materials, Philadelphia, 1985, pp. 75–90.
[13] Larsen, D. P., DeNoyelles, F., Jr., Stay, F., and Shiroyama, T., *Environmental Toxicology and Chemistry*, Vol. 5, 1986, pp. 179–190.
[14] Andrews, H. P., Snee, R. D., and Sarner, M. H., *American Statistician*, Vol. 34, No. 4, 1980, pp. 195–199.
[15] Taub, F. B., Harrass, M. C., Hartmann, H. J., Kindig, A. C., and Read, P. L., *Internationale Vereinigung für Theoretische und Angewandte Limnologie*, Vol. 21, 1981, pp. 197–204.
[16] Keating, K. I. and Dagbusan, B. C., *Proceedings of the National Academy of Sciences*, Vol. 81, 1984, pp. 3433–3437.
[17] Keating, K. I., *Water Research*, Vol. 19, No. 1, 1985, pp. 73–78.
[18] Winner, R. W., *Bulletin of Environmental Contamination and Toxicology*, Vol. 33, 1984, pp. 605–611.
[19] Shott, G. J., "Predicting Copper(II) Speciation During Single Species and Microcosm Bioassays," Master of Science thesis, University of Washington, Seattle, WA, 1984.
[20] Rose, K. A., "Evaluation of Nutrient-Phytoplankton-Zooplankton Models and Simulation of the Ecological Effects of Toxicants Using Laboratory Microcosm Ecosystems," Doctoral dissertation, University of Washington, Seattle, WA, 1985.
[21] Swartzman, G. L. and Rose, K. A., *Ecological Modelling*, Vol. 22, 1983/1984, pp. 123–134.

Foster L. Mayer, Jr.,[1] *Bengt-Erik Bengtsson,*[2] *Steven J. Hamilton,*[1] *and Ake Bengtsson*[3]

Effects of Pulp Mill and Ore Smelter Effluents on Vertebrae of Fourhorn Sculpin: Laboratory and Field Comparisons

REFERENCE: Mayer, F. L., Jr., Bengtsson, B.-E., Hamilton, S. J., and Bengtsson, A., "**Effects of Pulp Mill and Ore Smelter Effluents on Vertebrae of Fourhorn Sculpin: Laboratory and Field Comparisons,**" *Aquatic Toxicology and Hazard Assessment: 10th Volume, ASTM STP 971,* W. J. Adams, G. A. Chapman, and W. G. Landis, Eds., American Society for Testing and Materials, Philadelphia, 1988, pp. 406–419.

ABSTRACT: Vertebral quality of fourhorn sculpin (*Myoxocephalus quadricornis*) exposed to pulp mill or ore smelter effluents was investigated in the laboratory and in contaminated sites near the Swedish coast of the Gulf of Bothnia. Actual effluent samples from pine and birch pulp processes (chlorine bleaching) and a simulated effluent of the ore smelter effluent were tested in the laboratory. In the field, fish were collected from both reference (control) and contaminated sites. Laboratory exposures of pulp mill effluent significantly affected biochemical composition of vertebrae, but no statistically significant effects were observed in fish from the field. Mechanical properties were significantly affected in fish from both the laboratory and field; spinal anomalies ranged from 19 to 38%. Major effects on mechanical properties of vertebrae were observed in the ore smelter study, but the properties affected in laboratory fish differed from those affected in the field. Incidence of spinal anomalies was 47 to 58% in the laboratory and 29% in the field. Effects on fish in the laboratory were related to those observed in the field for pulp mill effluents, but the simulated effluent for the ore smelter elicited vertebral responses different from those in fish exposed to actual effluents in the field.

KEYWORDS: Fourhorn sculpin, *Myoxocephalus quadricornis,* vertebral anomalies, pulp mill effluent, chlorinated organics, ore smelter effluent, arsenic, heavy metals

Several potential adverse effects on essential biological functions may occur from vertebral anomalies in fish [*1*]. Vertebral lesions that impair swimming ability, increase vulnerability to predation, or alter feeding behavior could be expected to decrease survival and growth. Other adverse effects include a decreased ability to defend a territory and to compete for sexual partners, and a general physiological vulnerability to environmental stress. Presently, fish vertebrae (biochemical composition and mechanical properties) appear promising for predicting contaminant effects from laboratory studies to the field, assessing fish health in the field, and determining whether problem contaminants are organic or inorganic. The main difficulty in assessing the utility

---

[1] U.S. Fish and Wildlife Service, Columbia National Fisheries Research Laboratory, Columbia, MO 65201, (Dr. Mayer is now at the U.S. Environmental Protection Agency, Environmental Research Laboratory, Gulf Breeze, FL 32561).

[2] National Swedish Environmental Protection Board, Brackish Water Toxicology Laboratory, Studsvik, S-611 82 Nykoping, Sweden,

[3] University of Umea, Department of Animal Ecology, S-901 87 Umea, Sweden.

of biochemical and mechanical measurements is in finding field situations where fish population effects and history of contamination are at least partly known, so that these approaches can be validated.

Certain areas in Sweden in the Gulf of Bothnia (Baltic Sea) appeared to provide opportunities for validation. Skeletal abnormalities observed in fourhorn sculpin (*Myoxocephalus quadricornis*) have been as high as 40% near ore smelters and 58% near pulp industries [2]. The Husum kraft pulp industry is located on the coast of the Gulf of Bothnia in northern Sweden, about 15 km northeast of Ornskoldsvik. About 33×10^6 m^3 of wastewater are produced by the plant annually, and considerable amounts of chlorinated organics occur in the effluent from the chlorine bleaching process. For more than 50 years, Skelleftea Bay and adjacent areas of the Gulf of Bothnia have been contaminated by arsenic and metals in effluents from a complex of sulphide ore smelters (Ronnskarsverken) near the town of Skelleftea in northern Sweden. Elevated concentrations of arsenic, cadmium, copper, lead, mercury, zinc, and other metals have been detected in sediments and biota collected more than 100 km from the industry [3].

In addition to having a high incidence of vertebral anomalies in apparently polluted areas, the fourhorn sculpin was the species of choice for four other reasons:

1. It is a rather sedentary species and its condition should therefore reflect localized pollution conditions.
2. It has few predators in this area of the Baltic Sea, increasing the potential for affected sculpins to survive for sampling.
3. Populations in the Gulf have been genetically characterized [4], thus facilitating the selection of reference or control fish.
4. Food types and quantities consumed are identical in normal and affected fish, thereby eliminating dietary interactions [5].

Fourhorn sculpin were exposed to effluents from the pulp industry and simulated effluents from the smelter complex in the laboratory. Fish were also collected in the field near both industries and from reference (control) areas. The purpose of our study was to determine the effects of effluents on vertebral composition and mechanical properties of sculpin held in the laboratory, and to compare them with those of sculpin collected in the field. We anticipated that the combined results would help explain the etiology of vertebral damage in the fish resulting from pollution.

Methods and Materials

Laboratory Tests

All laboratory exposures were conducted at the Brackish Water Toxicology Laboratory, National Swedish Environmental Protection Board, Nykoping, Sweden. The fish were obtained by artificial fertilization of eggs from adults caught during the spawning season (January-February) in the Gulf of Bothnia. The eggs hatched after about three months of incubation at 2 to 4°C, and the pelagic fry were fed live *Artemia* nauplia for one month. During the next two months, the fish were also fed progressively increasing amounts of live limnetic zooplankton of larger size (*Cyclops, Daphnia, Corethra,* and *Culex*). When the fry became benthic (circa three months), the diet was supplemented with commercially available live *Tubifex*. From August until the experiments started, *Tubifex* was the only food offered.

Sediment was collected from the Baltic Sea at a depth of 30 m and distributed on the bottom of each test aquarium. In the study of pulp mill effluent, 30 to 35 sculpin (ten months old and weighing circa 1 g) were placed in each of seven 70-L glass aquaria. Two aquaria were used as controls and the others for various concentrations of effluents: two for a low (0.12%) effluent concentration from the pine pulp process, one for high (0.6%) pine, one for birch (0.6%), and one for pine plus birch (0.3% each). The fish were exposed for nine months. Fish in the birch effluent exposure were accidently lost after four months, and a new group of fish were added for the remaining five months. The replacement fish were from the same batch of fish that the original exposures were initiated with so that weights and age would be the same as in the controls and fish in the other exposures. For the ore smelter effluent study, 20 fish (six months old and weighing circa 0.2 g) were placed in each of six 60-L aquaria arranged to provide duplicate controls and 0.1 and 1.0% simulated effluent exposures. Fish were exposed for 11.5 months. In both studies, live *Tubifex* was added to the aquaria in surplus amounts during the first two months and maintained by weekly additions. During the rest of the exposures, *Tubifex* was supplemented with live opossum shrimp (*Neomysis*). At the end of the studies, fish were removed from each aquarium and frozen for X-ray and vertebral analyses. The hepatosomatic index (liver weight/body weight × 100) was also determined on 15 to 17 fish from the control and each exposure.

Samples of effluents from the pulp mill (pine and birch) were collected in January, June, and August 1983 (Table 1). Effluent samples were kept frozen (-20 to $-25°C$) before use. Thawed 25-L stock solutions of effluent were kept chilled (4°C) to reduce bacterial activity. Stock solutions were continuously pumped by a peristaltic pump (SP1A-20 Stalprodukter, Sweden), which delivered a flow of 15 (± 1) mL/h during the first two months and 30 (± 2) mL/h during the rest of the experiment, depending on water flow. The effluent was mixed with natural brackish water from a depth of circa 40 m and pumped into the laboratory. The incoming water was evenly distributed to each of the test aquaria by a water flow system described by Granmo and Kollberg [6]. The water flow was 40 (± 2) mL/min for the first two months and then increased to 80 (± 4) mL/min. Total volume of effluent from the mill decreased during collections for the pine effluent, but dilutions were adjusted to maintain the initial concentrations.

The effluent from the Ronnskarsverken ore smelter enters the Gulf of Bothnia through two main outlets. The effluents are analyzed for arsenic and metal content by the company. The mean discharge concentrations (mg/litre) in 1980–1981, based on an average wastewater effluent of 600 m^3/h, were as follows: arsenic 48, cadmium 0.4, copper 0.88, lead 1.0, mercury 0.06, and zinc 2.4.

Stock solutions were prepared to simulate exposures of the smelter effluent (Table 2). The arsenic concentration was halved to better reflect a recently reduced industrial discharge. Solutions were made from analytical grade $NaAsO_2$, $CdCl_2 \cdot 2\ 1/2H_2O$, $CuCl_2 \cdot 2H_2O$, $PbCl_2$, $HgCl_2$, and $Zn(NO_3)_2 \cdot 4H_2O$ dissolved in distilled water and acidified to pH 2.5 with concentrated hydrochloric acid. Stock solutions were continuously pumped by a peristaltic pump, which delivered a flow of 13.5 (± 1.5) mL/h per channel. The flow was mixed with filtered natural brackish water and pumped into the laboratory as described for pulp mill effluent. Incoming water was evenly distributed to each of the test aquaria (215 $\pm$ 10 mL/min). Water analyses for arsenic and metals were performed on pooled weekly samples. Water samples were kept frozen and were acidified before analysis. Arsenic was determined by flameless atomic absorption spectrophotometry (AAS) after UV-irradiation; cadmium, copper, and lead by flame or flameless AAS with Zeemanncorrect; mercury by cold vapor AAS; and zinc by flameless AAS.

Water temperature in the aquaria was checked daily, and salinity, pH, and dissolved oxygen were measured weekly. No attempts were made to adjust water temperature; it followed the natural fluctuations in Tvaren Bay, outside the laboratory. Average physicochemical characteristics [standard deviation (SD) in parentheses] of the test water for pulp mill and ore smelter effluents, respectively, were as follows: pH, 7.5 (0.1) and 7.8 (0.1); salinity, 7.4 (0.3) and 7.1 (0.4) 0/00; and temperature, 8.8 (0.7) and 7.0°C (1.1). Oxygen ranged from 70 to 90% of saturation in both studies. Artificial illumination (fluorescent tubes; L25 Universalvit, Osram, Sweden) was

TABLE 1—Chemical characterization of the pulp mill effluent used in laboratory exposures.

Characteristic	Pine			Birch		
	January	June	August	January	June	August
Washing loss Na_2SO_4 (kg/metric ton of pulp)[a]	16	13	15	16	15	15
Biological oxygen demand (BOD_7, mg/litre)	270	330	350	250	235	270
Chemical oxygen demand (COD, mg/litre)	1130	1200 to 1400	1200	1060	960	1050
Kappa-number	20 to 22	13 to 16	20	17 to 18	14 to 17	18
Suspended matter (mg/litre)	25	13	11	42	42	34
pH	2.5	2.1	2.1	3.9	4.3	2.5
$EOCl$[b] (mmol/litre)	3.3	2.4	1.0	0.4	0.4	0.94
Waste water volume (m^3/metric ton of pulp)[a]	70	31	45	60	60	60
Nitocra spinipes, 96-h LC_{50} (% w/w)	8.7	7.4	8.8	28	19	19
Microtox, 5-min EC_{50} (% w/w)	18	12	11	12	17	8.7

[a] Metric tons at 90% dryness.
[b] Extractable organic chlorine (total chlorinated organics = $EOCl \times 13.3$).

TABLE 2—*Mean measured concentrations (µg/litre; SD in parentheses) of arsenic and heavy metals in the exposure water of test aquaria for the simulated effluent of the ore smelter.*

Element	Simulated Exposure as Estimated Percent of Effluent		
	Control	0.1%	1.0%
Arsenic	0.75 (0.05)	32 (2.3)	323 (22)
Cadmium	<0.10	0.46 (0.23)	2.9 (0.11)
Copper	0.63 (0.23)	0.75 (0.26)	6.8 (0.37)
Lead	<0.70	1.2 (0.6)	8.2 (2.0)
Mercury	0.06 (0.02)	0.11 (0.04)	1.1 (0.36)
Zinc	3.9 (3.0)	5.3 (3.1)	39 (14)

used throughout the experiment. Day length was adjusted regularly by a timer to follow the natural photoperiod; light intensities 1 m from the aquaria were 50 lx for day and <0.1 lx for night.

Further details were given by Andersson et al. [7] for pulp effluents and by Lithner and Bengtsson [8] for ore smelter effluents.

Field Collections

All sampling sites were in northern Sweden, on or near the coast of the Gulf of Bothnia. Reference sites (controls) were selected to the north of the exposure areas, since the prevailing current moves in a southerly direction, and the fish from each reference site did not differ genetically from those in the respective exposure areas [4]. Collections were made with gill nets having mesh sizes (stretched measure) of 6 to 100 mm, set on the bottom at depths of 10 to 25 m. The fish were carefully removed from the nets to prevent physical induction of vertebral damage. The fish were then frozen for later X-ray and vertebral analyses.

For the pulp mill effluent study, 158 sculpin were collected 1.5 km from the mill (Husum) from 3 to 13 Nov. 1981, and 149 were collected from the reference area (Degerfjarden) 14 km north of the mill on 23 July 1981 [9]. Field sampling sites for the ore smelter study were in the Skelleftea and Byske bays [10]. The sculpin samples (195 fish) were collected on 25 Sept. 1982 from Skelleftea Bay near the ore smelter. In addition to arsenic and metals, which are the predominant pollutants, Skelleftea Bay is somewhat polluted by organics (for example, resinous substances) discharged from a wallboard factory at the Skelleftea River about 4 km from the mouth, and from a mechanical pulp mill at Burea in southern Skelleftea Bay. Byske Bay (circa 30 km north of Skelleftea Bay) was used as a reference control area; 196 fourhorn sculpin were collected there on 17 Sept. 1982. Because the Byske area may be slightly influenced by pollutants from Ronnskarsverken, as was revealed by sediment investigations in 1973 [3], it was not a true control.

Vertebral Analyses

Vertebral quality was characterized by evaluating biochemical composition, density, and mechanical properties of vertebrae. All fish from the laboratory exposures and field collections were X-rayed, and radiographs were examined for alterations in vertebral structures, as described by Bengtsson and Bengtsson [11]. Eight vertebrae (the third through tenth) from each fish were individually dissected and mechanically tested. Mechanical properties of bone were tested by compressive loading of individual vertebrae to determine strength, elasticity, and toughness [12]. Strength characteristics measured were rupture (the force causing failure of bone structural integrity) and elastic limit (the force above which permanent structural damage occurs in bone tissue). Elasticity characteristics measured were strain (the amount of deformation incurred by bone tissue at failure) and modulus of elasticity (the index of bone stiffness). We also measured toughness (the energy-absorbing capacity of bone tissue). Vertebral density was calculated as mean vertebral dry weight (mg) per mean vertebral volume (cm^3) per fish. After mechanical testing, we determined biochemical composition of the vertebrae. Vertebrae were dried for 2 h in a forced-air oven at 100°C, split into two fractions, and weighed. Collagen was isolated by the method of Flanagan and Nichols [13]. The isolated collagen was weighed and hydrolyzed at 115°C in 2 mL of 6 N HCl for 16 h and analyzed for hydroxyproline [14] and proline [15]. Calcium was determined by the method of Gitelman [16] and phosphorus by the method of Fiske and Subbarow [17].

Statistical Analyses

Data were analyzed by analysis of variance and covariance (adjusting for size by weight) and the least significant difference mean comparison test [18]. Analysis of covariance did not result in an improvement in statistical significance over analysis of variance, and therefore analysis of variance was used in all tests. The level of statistical significance was $P \leq 0.05$.

The data were also subjected to discriminate analysis to determine if a vector of variables could distinguish between site-exposure combinations. Differences in vertebral constituents due to age negated laboratory-to-field comparisons; however, graphical multivariate (radial) profiles [19] were constructed from the data expressed as percent of controls to allow for such comparisons.

Results

Pulp Mill

Laboratory exposures of pulp mill effluent significantly affected the biochemical composition of vertebrae (Table 3). Collagen and the proline:hydroxyproline ratio significantly decreased, and hydroxyproline increased. Vertebral density tended to increase with exposure to pulp mill effluent in both laboratory and field exposures, but the increase was not statistically significant. Mechanical properties were significantly affected in both laboratory and field exposures. The vertebrae became more elastic (increased strain) and required more force to break (rupture and elastic limit). However, the laboratory exposure with birch pulp effluent differed from that with the pine or pine plus birch effluents; the vertebrae required less force to break and were less stiff (decreased rupture and decreased modulus of elasticity). Those changes caused by pine or pine plus birch effluents resulted in an increased energy-absorbing capacity (toughness) of the vertebrae. Percentage occurrence of spinal anomalies after laboratory exposures [9] were as follows: control 11, low pine 33, high pine 37, birch 38, and pine plus birch 19; in the field, spinal anomalies were observed in 9.4% of the fish from the control and 31% of those from exposed sites.

The multivariate response profiles varied (Fig. 1). However, they were almost identical for fish from the low pine effluent exposure and fish collected from the field (Fig. 2), indicating similar exposure types and concentrations. This similarity was supported in part by like incidences of spinal anomalies. The profile for birch was very different from that for either pine effluent concentration. The pine-birch combination profile was very comparable to that found in the high pine effluent exposure, but the combination appeared antagonistic in regard to vertebral anomalies (19%). This antagonism was also observed with the hepatosomatic index (%) as follows: control 2.69, low pine 3.25, high pine 3.52, birch 3.17, and pine plus birch 2.51.

Ore Smelter

Vertebral phosphorus decreased in sculpins exposed in the laboratory, whereas it increased in fish exposed to metal smelter effluents in the field (Table 4). The proline:hydroxyproline ratio also

TABLE 3—Mean vertebral characteristics (SD in parentheses) of fourhorn sculpin exposed to pulp mill effluents in the laboratory and field.

Vertebral Characteristic	Laboratory Exposures					Field Sites	
	Control	Low Pine	High Pine	Birch	Pine + Birch	Degerfjarden[a]	Husum
Biochemical composition							
Collagen, mg/g bone	216 (17)	190 (32)[b]	176 (33)[b]	156 (22)[b]	154 (22)[b]	184 (10)	187 (10)
Hydroxyproline, mg/g collagen	55 (5)	57 (5)	60 (2)[b]	62 (8)[b]	68 (6)[b]	57 (4)	58 (3)
Proline, mg/g collagen	85 (7)	92 (8)	89 (6)	94 (15)	93 (10)	120 (9)	122 (7)
Proline:hydroxyproline	1.57 (0.08)	1.60 (0.07)	1.48 (0.07)[b]	1.53 (0.10)	1.38 (0.08)[b]	2.15 (0.14)	2.09 (0.13)
Calcium, mg/g bone	310 (25)	321 (85)	323 (38)	302 (35)	288 (17)	217 (20)	206 (28)
Phosphorus, mg/g bone	128 (7)	128 (7)	127 (8)	124 (5)	124 (7)	101 (7)	101 (8)
Density, mg/cm³	313 (59)	322 (54)	352 (37)	337 (41)	347 (26)	1003 (118)	1046 (144)
Mechanical properties							
Rupture, g-force/mm²	227 (49)	272 (48)[b]	299 (59)[b]	180 (51)[b]	249 (25)	866 (146)	962 (233)[b]
Elastic limit, g-force/mm²	189 (39)	234 (41)[b]	238 (46)[b]	151 (46)[b]	200 (20)	643 (112)	730 (163)[b]
Strain, %	8.3 (1.5)	9.2 (1.3)	11 (1.6)[b]	10 (2.5)[b]	11 (1.8)[b]	13 (2.9)	14 (3.1)
Modulus of elasticity, kg-force/mm²	3.5 (0.7)	3.7 (0.4)	4.0 (0.7)	2.3 (0.9)[b]	3.2 (0.5)	9.3 (2.0)	9.5 (2.2)
Toughness, g-mm/mm³	11 (4)	14 (4)	19 (6)[b]	10 (4)	17 (4)[b]	66 (23)	79 (31)
Number analyzed	12	10	10	10	10	32	34

[a] Control reference site.
[b] Significantly different from respective controls ($P \leq 0.05$).

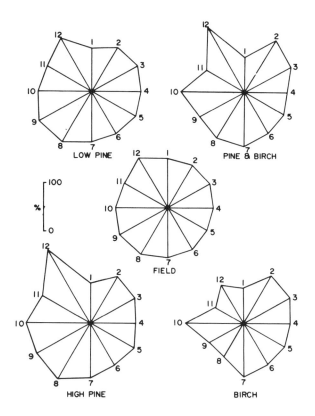

FIG. 1—*Multivariate response profiles of pulp mill effluent effects on vertebral characteristics of fourhorn sculpin in the laboratory and collected from the field. The data are expressed as percent of controls, with the center of the profile beginning at 0% (1 = collagen, 2 = hydroxyproline, 3 = proline, 4 = proline; hydroxyproline ratio, 5 = calcium, 6 = phosphorus, 7 = density, 8 = rupture, 9 = elastic limit, 10 = strain, 11 = modulus of elasticity, and 12 = toughness).*

increased in fish collected from areas near metal smelter effluents. These changes, although statistically significant, were very small. Vertebral density tended to increase after exposure to metals in both the laboratory and field, but the increases were not statistically significant. Some mechanical properties of vertebrae in fish from both the laboratory and field exposures were also significantly affected. The vertebrae were more elastic (modulus of elasticity decreased and strain increased), and more force was required to break them (rupture and elastic limit increased). These changes resulted in an increased energy-absorbing capacity (toughness) of vertebrae and an increase in spinal anomalies. The average incidences of spinal anomalies after laboratory exposures [*10*] were 24% in the control, 58% in the 0.1% effluent, and 47% in the 1.0% effluent, compared with 22% in fish from the control site and 29% in those from the exposed field sites.

Residue analyses were performed on liver and whole fish samples from both the laboratory study and the field sites (G. Lithner, unpublished observations). Since the analyses are not yet complete, our comparisons here are only preliminary. The residues of arsenic, cadmium, copper, lead, mercury, and zinc in liver and whole fish were roughly similar in the low laboratory exposure

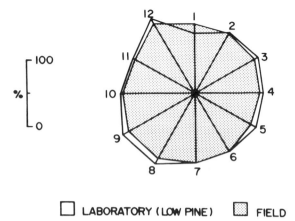

FIG. 2—*Comparison of multivariate response profiles of fourhorn sculpin from the low laboratory exposure of pine pulp effluent and fish from the field (see caption of Fig. 1 for explanation).*

(0.1% simulated effluent) and the contaminated field site (Skelleftea Bay). In the high laboratory exposure (1.0%), residues of cadmium, lead, and mercury were 6 to 20 times those of fish collected from Skelleftea Bay. However, the similarity in inorganic residues in fish from the low laboratory exposure and contaminated field site were difficult to compare in terms of trends in vertebral characteristics. Most biochemical and vertebral characteristics that were significantly affected were not identical in the laboratory and field (Fig. 3). The differences may have been due to different exposure conditions (metal composition and speciation) and the possible presence of other contaminants, as indicated previously. Also, the incidence of vertebral anomalies was high in both the laboratory and field controls.

Discussion

In the pulp mill effluent study, vertebral characteristics were significantly affected in fish from both the laboratory and field. Spent bleach liquor of pine kraft pulp mills contains chlorinated components similar to those of toxaphene [20], and toxaphene causes vertebral anomalies in fish [21–23]. The similarity of the characteristics measured was high between the fish exposed to the low pine effluent in the laboratory and those collected in the field.

The mode of action proposed for chlorinated organic chemicals on vertebral quality in fish involves a possible contaminant-induced competition for ascorbic acid between collagen metabolism in bone and microsomal mixed function oxidases (MFO) that detoxify or metabolize a broad range of organic chemicals in liver [22,24,25]. Increased hepatosomatic index as an indicator of induced MFO activity was observed in all laboratory exposures except for the pine plus birch. The competition for ascorbic acid may decrease the vitamin in bone and alter hydroxyproline and collagen content, with a concomitant alteration in bone integrity that renders the bone more fragile. This concept is further supported by the vertebral responses we observed in the laboratory exposure with birch effluent, but not with the other laboratory exposures or fish from the field.

Hamilton et al. [26] showed that bone density and the hydroxyproline concentration in bone collagen are important factors in assessing toxaphene effects on mechanical properties of fish vertebrae. In our study, vertebral strength increased in fish exposed to pine and pine plus birch effluents and in fish collected in the field; elasticity increased in the laboratory exposures only. These changes in mechanical properties are the same as those observed with toxaphene [26], but

TABLE 4—Mean vertebral characteristics (SD in parentheses) of fourhorn sculpin exposed to ore smelter effluents in the laboratory and field.

Vertebral Characteristic	Laboratory Exposures			Field Sites	
	Control	0.1%	1.0%	Byske Bay[a]	Skellefteå Bay
Biochemical composition					
Collagen, mg/g bone	206 (12)	207 (7)	216 (6)	163 (10)	156 (7)
Hydroxyproline, mg/g collagen	53 (8)	66 (16)	55 (10)	70 (6)	68 (5)
Proline, mg/g collagen	137 (15)	130 (10)	129 (6)	128 (15)	135 (10)
Proline:hydroxyproline	2.6 (0.6)	2.0 (0.5)	2.4 (0.6)	1.8 (0.2)	2.0 (0.1)[b]
Calcium, mg/g bone	266 (9)	276 (30)	255 (28)	202 (24)	198 (27)
Phosphorus, mg/g bone	135 (7)	138 (6)	128 (7)[b]	115 (7)	123 (7)[b]
Density, mg/cm^3	285 (23)	292 (40)	323 (50)	891 (138)	948 (202)
Mechanical properties					
Rupture, g-force/mm^2	438 (50)	508 (97)	487 (91)	708 (122)	865 (225)[b]
Elastic limit, g-force/mm^2	353 (40)	426 (92)	381 (72)	493 (78)	672 (145)[b]
Strain, %	9.5 (2.3)	9.7 (1.3)	13 (2.0)[b]	14 (2.8)	15 (4.5)
Modulus of elasticity, kg-force/mm^2	6.2 (1.1)	6.9 (1.3)	5.4 (1.4)	7.3 (1.7)	8.0 (1.9)
Toughness, g-mm/mm^3	25 (9)	29 (7)	36 (10)[b]	59 (16)	79 (38)
Number analyzed	8	8	8	16	16

[a] Control reference site.
[b] Significantly different from respective controls ($P \leq 0.05$).

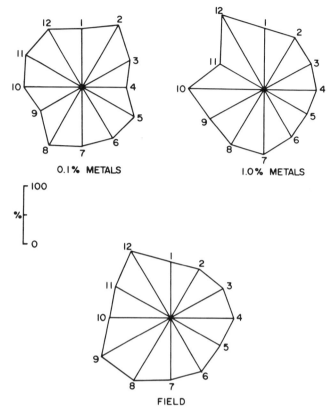

FIG. 3—*Multivariate response profiles of simulated ore smelter effluent effects on vertebral characteristics of fourhorn sculpin in the laboratory and collected from the field (see caption of Fig. 1 for explanation).*

are contrary to those expected if only the reduction in collagen (that is, more fragile) is considered in predicting altered mechanical properties of bone. The present study and the study by Hamilton et al. [26] indicate that increased vertebral strength is associated with increased hydroxyproline content. Consequently, increased hydroxyproline content in collagen may provide a more meaningful measure of vertebral strength. This is supported by the work of Lees and Davidson [27] and Carter and Spengler [28], who found that collagen cross-link (hydroxyproline) density can increase bone strength.

Obviously, sufficient alterations in bone composition can lead to decreased strength and weakened vertebrae. It is less obvious that increased bone strength would result in vertebral anomalies. Currey [29] stated that the mechanical properties of bone, and the histology that produces them, are mainly determined by the selective balance between stiffness (elasticity), and resistance to fracture; there must always be a compromise between stiffness and resistance to fracture, since these two properties oppose each other. Enough irritation could arise between stronger and denser vertebrae to stimulate repair mechanisms that could result in vertebral fusion and other anomalies. In a histopathology study with scorbutic channel catfish [30], spinal deformity was evoked by three types of vertebral abnormality: osteoporosis, abnormally increased ossification (repair), and dysplasia.

The chronic effects of cadmium [31,32], lead [33], and zinc [32,34] on vertebral damage in fish, and our results with sculpin, indicate that the smelter effluent might be contributing to the high incidence of vertebral damage in wild sculpins in Skelleftea Bay. The capability of metals such as cadmium, lead, and zinc to induce this kind of damage in fish has been demonstrated in laboratory studies [31,33–36], but has not been previously demonstrated in fourhorn sculpin. Unlike an earlier study [37] relating decreased strength to increased contaminant exposures, the present work associated vertebral damage in sculpin with increased elasticity and toughness (laboratory) and increased strength (field). Bengtsson et al. [2] suggested that cadmium, lead, and zinc affect the neuromuscular function of fish, causing stress on the vertebrae that in turn results in skeletal fractures. However, increased strength, when other mechanical properties remain relatively constant, can result in increased wearing and erosion on adjoining vertebral surfaces. This increase can lead to constant irritation and eventually vertebral deformation, as we discussed for pulp mill effluent.

Although not specifically addressed in this study, the mode of action of inorganic contaminants on vertebral composition and structure can be elucidated to some degree. A study by Srivastava et el. [38] on fetal rat bone showed that collagen, as well as overall protein synthesis, was considerably reduced by cobalt, gold, and rhodium; slightly reduced by nickel; and unaffected by chromium, molybdenum, titanium, and tungsten. Bhatnager and Hussain [39] reported that cadmium, mercury, palladium, and platinum decreased collagen synthesis in mammalian lung tissue by binding essential sulfhydryl groups on prolylhydroxylase. Arsenic also binds certain enzymatic sulfhydryl groups [40] and inhibits intercellular transport of proline from membrane vesicles in rats [41]. Cadmium inhibits the cross-linking (that is, stiffening) of collagen and causes pathological bone changes that probably lead to altered structural integrity of bone [42]. Lead has also been shown to reduce proline hydroxylation and biosynthesis of collagen [43]. These studies suggest that lesions induced by some inorganic chemicals may be due to a decrease in collagen or cross-linking of collagen and can be a result of inorganic binding of active sites on prolylhydroxylase. This hypothesis was supported since sculpin collected from the field had an increased ratio of proline:hydroxyproline.

The measurement of skeletal deformities in fish has been proposed as a means of monitoring pollution in marine environments [11,32]. Likewise, measurements of biochemical composition and mechanical properties of vertebrae have been shown to be indicators of bone development in fish exposed to contaminants [21,26,37]. Effects of organic and inorganic contaminants on bone integrity are similar in that vertebral anomalies are produced—although they may develop through different modes of action. This similarity makes the use of biochemical composition and mechanical properties of vertebrae, as well as vertebral deformities, conducive to assessing the effect of an array of contaminants on fish health.

Because the interactions of biochemical constituents are complex, a sufficient deviation from the norm of few or many of those constituents can apparently result in vertebral damage, as evidenced in the present study. Therefore, the etiology of vertebral anomalies observed in fish from contaminated areas can be appropriately assessed only by an integrative approach. Vertebral X-ray examinations [11] should be accompanied by biochemical and mechanical analyses of vertebrae, as well as by histopathological examination [30,44,45].

Acknowledgments

This project was sponsored by the Research Committee of the National Swedish Environmental Protection Board and by the Fish and Wildlife Service, U.S. Department of the Interior, and supported in part by the Missouri Research Assistance Act and Monsanto Co. through the University of Missouri-Columbia. The Brackish Water Toxicology Laboratory staff is acknowledged for conducting the chronic laboratory studies. We thank M. R. Ellersieck for statistical recommendations, F. J. Dwyer and S. M. King for bone analysis, and A. C. Akerstedt-Bengtsson and M.

L. Lofgren for X-ray work. We especially acknowledge the technical assistance of J. S. Thomas and statistical analyses by C. H. Deans and L. F. Sappington.

References

[1] Hickey, C. R., "Common Abnormalities in Fishes, Their Causes and Effects," Technical Report No. 0013, New York Ocean Sciences Laboratory, Montauk, NY, 1972.
[2] Bengtsson, B.-E., Bengtsson, A., and Himberg, M., *Ambio,* Vol. 14, 1985, pp. 32–35.
[3] Lithner, G., "Ronnskarsundersokningen 1973," National Swedish Environment Protection Board, Brackish Water Toxicology Laboratory, Studsvik, S-611 82 Nykoping, Sweden, 1974.
[4] Gyllensten, U. and Ryman, N., *Ambio,* Vol. 14, 1985, pp. 29–31.
[5] Hansson S., Bengtsson, B.-E., and Bengtsson, A., *Marine Pollution Bulletin,* Vol. 15, 1984, pp. 375–377.
[6] Granmo, A. and Kollberg, S., *Water Research,* Vol. 6, 1972, pp. 1597–1599.
[7] Andersson, T., Bengtsson, B.-E., Forlin, L., Hardig, J., and Larsson, A., "The Long-Term Effects of Bleached Pulp Mill Effluent on Carbohydrate Metabolism and Hepatic Xenobiotic Biotransformation Enzymes in Fish," National Swedish Environment Protection Board, Brackish Water Toxicology Laboratory, Studsvik, S-611 82 Nykoping, Sweden, 1986.
[8] Lithner, G. and Bengtsson, B.-E., "Long-Term Accumulation of Metals and Arsenic in Juvenile Fourhorn Sculpin (*Myoxocephalus quadricornis* L.)," National Swedish Environment Protection Board, Brackish Water Toxicology Laboratory, Studsvik, S-611 82 Nykoping, Sweden, 1986
[9] Bengtsson, A., "Slutrapport fisk/cellulosa ryggradsskador," National Swedish Environment Protection Board, Brackish Water Toxicology Laboratory, Studsvik, S-611 82 Nykoping, Sweden, 1986.
[10] Bengtsson, A., Bengtsson, B. E., and Lithner, G., *Journal of Fish Biology,* 1986, in press.
[11] Bengtsson, A. and Bengtsson, B.-E., *Aquilo Ser Zoologica,* Vol. 22, 1983, pp. 61–64.
[12] Hamilton, S. J., Mehrle, P. M., Mayer, F. L., and Jones, J. R., *Transactions of the American Fisheries Society,* Vol. 110, 1981, pp. 708–717.
[13] Flanagan, B. and Nichols, G., *Journal of Biological Chemistry,* Vol. 237, 1962, pp. 3686–3692.
[14] Woessner, J. F., *Archives of Biochemistry and Biophysics,* Vol. 93, 1961, pp. 440–447.
[15] Troll W., and Lindsley, J., *Journal of Biological Chemistry,* Vol. 215, 1955, pp. 655–660.
[16] Gitelman, H. J., *Analytical Biochemistry,* Vol. 18, 1967, pp. 521–531.
[17] Fiske, C. H. and Subbarow, Y., *Journal of Biological Chemistry,* Vol. 66, 1925, pp. 374–400.
[18] Snedecor, G. W. and Cochran, W. G., *Statistical Methods,* Iowa State University Press, Ames, 1967.
[19] Eisler, R. in *Marine Pollution and Sea Life,* Fishing News (Books) Ltd., London, 1972, pp. 229–233.
[20] Pyysalo, H. and Antervo, K., *Chemosphere,* Vol. 14, 1985, pp. 1723–1728.
[21] Mayer, F. L., Mehrle, P. M., and Schoettger, R. A. in *Recent Advances in Fish Toxicology,* Ecological Research Series, No. EPA 600/3-77-085, Environmental Protection Agency, Corvallis, OR, 1977, pp. 31–54.
[22] Mayer, F. L., Mehrle, P. M., and Crutcher, P. L., *Transactions of the American Fisheries Society,* Vol. 107, 1978, pp. 326–333.
[23] Mehrle, P. M. and Mayer, F. L., *Journal of the Fisheries Research Board of Canada,* Vol. 32, 1975, pp. 593–598.
[24] Street, J. C., Baker, R. C., Wagstaff, D. J., and Urry, F. M., *Proceedings,* Second International Union of Pure and Applied Chemistry International Congress on Pesticide Chemistry, Tel Aviv, Israel, 1971, Gordon and Breach, New York, pp. 281–302.
[25] Wagstaff, D. J. and Street, J. C., *Toxicology and Applied Pharmacology,* Vol. 19, 1971, pp. 10–19.
[26] Hamilton, S. J., Mehrle, P. M., Mayer, F. L., and Jones, J. R., *Transactions of the American Fisheries Society,* Vol. 110, 1981, pp. 718–724.
[27] Lees, S. and Davidson, C., *Journal of Biomechanics,* Vol. 10, 1977, pp. 473–486.
[28] Carter, D. R. and Spengler, D. M., *Clinical Orthopedics and Related Research,* Vol. 135, 1978, pp. 192–217.
[29] Curry, J., *American Zoologist,* Vol. 24, 1984, pp. 5–12.
[30] Miyazaki, T., Plumb, J. A., Li, Y. P., and Lovell, R. T., *Journal of Fish Biology,* Vol. 26, 1985, pp. 647–655.
[31] Bengtsson, B.-E., Carlin, C. H., Larsson, A., and Svanberg, O., *Ambio,* Vol. 4, 1975, pp. 166–168.
[32] Bengtsson, B.-E., *Philosophical Transactions of the Royal Society of London,* Vol. B286, 1979, pp. 457–464.
[33] Davies, P. H. and Everhart, W. H., *Effects of Chemical Variations in Aquatic Environments, Vol. 3, Lead Toxicity to Rainbow Trout and Testing Application Factor Concept,* EPA-R3-73-011c, U.S. Environmental Protection Agency, Washington, DC, 1973.

[34] Bengtsson, B.-E., *Oikos,* Vol. 25, 1974, pp. 134–139.
[35] Hodson, P. V., Milton, J. W., Blunt, B. R., and Slinger, S. R., *Canadian Journal of Fisheries and Aquatic Sciences,* Vol. 37, 1980, pp. 170–176.
[36] Holcombe, G. W., Benoit, D. A., Leonard, E. N., and McKim, J. M., *Journal of the Fisheries Research Board of Canada,* Vol. 33, 1976, pp. 1731–1741.
[37] Mehrle, P. M., Haines, T. A., Hamilton, S. J., Ludke, J. L., Mayer, F. L., and Ribick, M. A., *Transactions of the American Fisheries Society,* Vol. 111, 1982, pp. 231–241.
[38] Srivastava, R., Lefebvre, N., and Onkelinx, C., *Proceedings of the European Society of Toxicology,* Vol. 17, 1976, pp. 191–197.
[39] Bhatnager, R. S. and Hussain, M. A., *Proceedings,* Fourth Joint Conference on Sensing Environmental Pollutants, American Chemical Society, Washington, DC, 1977, pp. 527–531.
[40] Luh, M.-D., Baker, R. A., and Henley, D. F., *Science of the Total Environment,* Vol. 2, 1973, pp. 1–12.
[41] Kanner, B. I. and Sharon, I., *Biochemistry and Biophysics Acta,* Vol. 60, 1980, pp. 185–194.
[42] Iguchi, H. and Sano, S., *Toxicology and Applied Pharmacology,* Vol. 62, 1982, pp. 126–136.
[43] Visticia, D. T., Ahrens, F. A., and Ellison, W. R., *Archives of Biochemistry and Biophysics,* Vol. 179, 1977, pp. 15–23.
[44] Couch, A. J., Winstead, T. J., and Goodman, R. L., *Science,* Vol. 197, 1977, pp. 585–586.
[45] Couch, A. J., Winstead, T. J., Hansen, D. J., and Goodman, R. L., *Journal of Fish Diseases,* Vol. 2, 1978, pp. 35–42.

Aquatic Toxicology

Thomas F. Parkerton,[1] Susan M. Stewart,[1] Kenneth L. Dickson,[1] John H. Rodgers, Jr.,[1] and Farida Y. Saleh[1]

Evaluation of the Indicator Species Procedure for Deriving Site-Specific Water Quality Criteria for Zinc

REFERENCE: Parkerton, T. F., Stewart, S. M., Dickson, K. L., Rodgers, J. H., Jr., and Saleh, F. Y., "**Evaluation of the Indicator Species Procedure for Deriving Site-Specific Water Quality Criteria for Zinc,**" *Aquatic Toxicology and Hazard Assessment: 10th Volume, ASTM STP 971*, W. J. Adams, G. A. Chapman, and W. G. Landis, Eds., American Society for Testing and Materials, Philadelphia, 1988, pp. 423–435.

ABSTRACT The indicator species procedure enables modification of a national water quality criterion to account for differences in the biological availability/toxicity of a chemical introduced into site waters. To evaluate this procedure, a case study was conducted with zinc using the Trinity River near Denton, Texas as an example site.

Acute and chronic toxicity tests were performed with *Pimephales promelas* and *Daphnia pulex in* both natural river and reference waters. Toxicity results were expressed as a function of both total and soluble metal.

An acute water effects ratio (WER) was used to calculate the site-specific criterion maximum concentration. The site-specific value showed excellent agreement with the national water quality criterion adjusted for the site water hardness when based on total zinc. These findings were altered when toxicity results from a different reference water were used in calculations, but were unaffected if test organisms were not acclimated in site water prior to testing. Three different methods were used to calculate the site-specific criterion continuous concentration, which was found to be up to four times lower than the national criterion. Toxicological differences observed between lab and site waters were reduced when expressed in terms of mean soluble metal.

Additional experiments demonstrated that particulate zinc was not biologically available under test conditions and that soluble zinc may vary in its toxicity. Data also suggest that kinetic factors influence zinc toxicity. The implications of this research for the development of site-specific metal criteria are discussed.

KEYWORDS: site-specific water quality criteria, zinc, Daphnia pulex, *Pimephales promelas,* bioavailability, kinetics, toxicity

As point source dischargers comply with technology-based effluent controls, increased priority is being directed toward water bodies that exceed water quality standards. The U.S. Environmental Protection Agency (EPA) has provided guidance for assessing whether designated uses are attainable as well as for procedures for modifying national water quality criteria to reflect local site-specific conditions [*1,2*]. These procedures were developed by EPA to ensure the preservation of national water quality without the burdensome costs of unreasonable and scientifically unjustifiable pollution control restraints.

Site-specific criteria may be warranted if (1) the sensitivity of local species differs appreciably from those used in deriving the national criterion or (2) site water-toxicant interactions significantly

---

[1] Institute of Applied Sciences, North Texas State University, Denton, TX 76203-3078.

mitigate or enhance toxicant effects [1,2]. The recalculation and indicator species procedures are two methods developed for deriving site-specific criteria that address these situations, respectively. A third alternative, the resident species procedure, is recommended if both species sensitivity and differences in toxicant bioavailability are anticipated.

Literature has indicated that toxicity of pollutants, especially certain metals, is highly dependent on physical/chemical characteristics of a water [3–5]. In the case of many metals, total concentrations upon which metal criteria are based do not appear to be related to bioavailability. Often toxicity appears to reside primarily in the soluble fraction, although the toxicity of sediment-bound metals has not yet been well elucidated [6,7]. Consequently, the U.S. Environmental Protection Agency is encouraging states to apply the indicator species procedure so that metal standards which account for site-specific conditions that influence toxicity may be developed. This study was conducted in order to evaluate the utility of this procedure for deriving site-specific water quality criteria for zinc. Guidelines for the use of this procedure are found in the *Water Quality Standards Handbook* [6]. This procedure is applicable in cases where a discharge contains a few well-quantified contaminants whose effects and interactions with other contaminants are understood. Whole effluent rather than chemical specific techniques have recently been advocated for water quality-based toxics control of complex effluent matrices or multiple source discharges [8].

The primary objectives of this study were to:

1. Compare national criteria with site-specific criteria as derived via the indicator species procedure. All three methods for calculating a site-specific criterion continuous concentration were used by employing either the national or site-specific acute/chronic ratio or a chronic water effects ratio.

2. Assess the impact of two procedural modifications on results, namely, choosing an alternate reference laboratory water or acclimating test organisms in site water prior to testing.

3. Evaluate differences in observed toxicity in terms of both point (LC_{50}) as well as concentration response (slope) estimates.

4. Determine if observed toxicological differences between waters could be reduced when expressed in terms of soluble rather than total metal concentrations.

5. Examine the effect of zinc-site water interaction time (kinetics) on zinc toxicity.

Zinc was chosen as a model metal due to its widespread occurrence in both municipal and industrial effluents [9] and the reported importance that physiochemical variables exert on its fate and ultimate effect [10–17]. These include pH and hardness [11–13], complexation with inorganic and organic ligands [5,14,15], and adsorption to solids [16–17].

The Trinity River near Denton, Texas was selected as a study site because this water provided common water quality characteristics that are typical of many U.S. rivers [18].

Materials and Methods

Site Water

A single grab sample of site water was collected in clean 30-gallon plastic containers in late March 1985. Water was immediately returned to the laboratory and stored at 4°C until needed. Toxicity experiments were conducted over the next seven months. Temporal changes in water quality that might have occurred during this time due to sample storage were assumed to be negligible in comparison to natural seasonal water quality fluctuations reflected by monitoring data. In addition, periodical monitoring of pH, hardness, alkalinity, and dissolved oxygen demonstrated that these parameters did not change appreciably over the course of this study.

Stock Cultures

Daphnia pulex cultures were maintained in growth chambers providing constant temperature (20 ± 2°C) and light conditions (550 lx, 16:8 h light/day). Ten animals were reared in moderately hard reconstituted water [19] in 600-mL glass beakers and were fed approximately 3.75×10^6 cells of *Selenastrum capricornutum* daily. *Pimephales promelas* were cultured in aquaria in a flow-through system of carbon-filtered, dechlorinated tap water, hereafter referred to as dechlorinated tap water, at approximately 25°C and constant light conditions (325 lx, 16:8 h light/day). Spawning fish were fed *ad libitum* twice daily, once with frozen brine shrimp and once with dry fish food. Eggs were removed daily and hatched in aerated 500-mL separatory funnels. Hatched fry were raised in aquaria and fed *ad libitum* <24-h-old brine shrimp two or three times daily.

Zinc/Water Quality Analysis

Zinc solutions were prepared by dissolving reagent grade zinc chloride (CAS Reg. 7646-85-7) into deionized water after it had been passed through a millipore Mill-Ro4 ultrapurification system. Zinc was determined using a Perkin Elmer model 2380 atomic absorption spectrophotometer. The lower detection limit for this method was approximately 10 μg/L. Water quality parameters used to characterize river and reference waters were determined in accordance with the American Public Health Association (APHA) [19] and are provided in Table 1.

Acute Toxicity Tests

Static acute toxicity tests were conducted in general accordance with recommended test procedures [19,20] using five concentrations and a control. Daphnids were tested in 250-mL glass beakers using five replicates for each treatment level with ten <48-h-old neonates per replicate. Organisms of this age class were chosen instead of <24-h-old individuals, as usually recommended, after

TABLE 1—*Comparison of water quality parameters in site and lab reference waters.*

Parameter	Reconstituted Water	Dechlorinated Tap Water	Elm Fork Trinity River
Calcium, mg/L	19.1[a]	15.2	54.0
Magnesium, mg/L	12.1[a]	4.4	6.8
Sodium, mg/L	26.3[a]	41.0	43.0
Potassium, mg/L	2.10[a]	5.0	8.7
Manganese, mg/L	...	.03	0.07
Iron, mg/L	...	.05	0.10
Silicate, mg/L	...	.20	0.96
Sulfate, mg/L	93.6[a]	51.9	47.6
Chloride, mg/L	1.9[a]	20.8	50.3
pH	7.5 to 7.9	7.8 to 8.1	7.7 to 8.0
Dissolved oxygen, mg/L	7.4 to 8.5	7.9 to 8.5	6.9 to 8.5
Alkalinity, mg/L as $CaCO_3$	44 ± 6	96 ± 6	176 ± 10
Hardness, mg/L as $CaCO_3$	97 ± 16	109 ± 6	201 ± 16
Noncarbonate hardness, mg/L as $CaCO_3$	0	53	38
Conductivity, μmhos	323 ± 47	321 ± 14	546 ± 23
Total suspended solids, mg/L	...	...	32.5 ± 3.3
Volatile suspended solids, mg/L	...	...	6.4 ± 0.7

[a]Values calculated from nominal composition of reconstituted water.

preliminary experiments in reconstituted water indicated that 24 and 48-h-old neonates were equally sensitive to zinc challenges. Three replicate exposures were used for fish tested in gallon glass jars with ten one-month-old fry measuring approximately 1 cm in length per replicate. Toxicity tests were conducted simultaneously for both site and reference waters so that a water effects ratio (WER) could be determined. Reconstituted water and dechlorinated tap water served as the reference water for daphnid and fish tests, respectively. Daphnid (48 h) and fish (96 h) tests were maintained under similar conditions as the stock cultures except without feeding.

In order to assess the importance of organism acclimation to site water prior to testing, daphnids were cultured for three generations in site water under identical conditions as the stock culture. Acute tests with neonates from the site water culture were tested in conjunction with unacclimated neonates tested in site water. Fish were also acclimated by holding fry in site water two weeks prior to testing, but, unlike daphnids, tests were not run simultaneously with unacclimated fish but instead were run separately. An accompanying reference test in dechlorinated tap water was conducted during each of these experiments. Similarly, to assess the impact that alternate reference waters might impose on results, toxicity tests were conducted in which both organisms were tested concurrently in the two reference waters. Lastly, the effect of kinetics on zinc toxicity was explored by allowing the toxicant to interact with either site or reference water for 24 h prior to organism exposure. During this period, spiked test solutions were agitated at 50 rpms with the aid of a shaker table.

At the termination of each experiment, daphnids were counted using a dissecting microscope (X15) by pouring the contents of each replicate beaker through a counting chamber constructed of petri plates and 250 μm Nitex TM screening. Daphnid response was judged in terms of immobilization and lack of internal movement. Samples for total zinc analysis were taken at the start of both daphnid and fish tests immediately after spiking and were then acidified with 1 mL of concentrated HNO_3 per 50-mL sample. Samples for total zinc at the end of daphnid tests were taken by acidification of an additional replicate carried through experiments without animals. Total samples for fish tests were taken by directly acidifying one replicate from each treatment at the termination of the experiment after fish mortality had been recorded. Soluble zinc, defined as zinc that passed through a 0.45-μm filter, was also measured by taking composite samples from replicates at 0, 24, and 48 h for daphnid acute tests and at approximately 0, 1, 3, 6, 12, 24, 48, and 96-h intervals for fish acute tests. After filtration, samples were then acidified to pH <2.0 with concentrated HNO_3. Dissolved oxygen and pH were monitored at the start and termination of each test.

Chronic Experiments

Static renewal 21-day daphnid tests and fathead minnow 7-day embryo larval early life stage tests were performed in both natural and reference waters. Tests were conducted simultaneously for daphnids and one week apart for fish. Procedures followed those outlined by U.S. EPA [21,22]. Daphnids <24 h old were individually tested in 50-mL glass test chambers using ten organisms per treatment under the same environmental conditions described for acute tests. Animals were fed approximately 1×10^7 algal cells three times each week. Fertilized fish embryos <12 h old were introduced into 250-mL incubation cups that were attached to a rocker arm apparatus driven by a 4-rpm motor to keep water circulating over embryos [23]. Two replicate plexiglass chambers each containing approximately 40 embryos/replicate and 2.0 L of test solution were randomly positioned in a circulating water bath maintained at (25 ± 1°C) and held at constant light conditions (425 lx, 16:8 light/day).

Test solutions were renewed daily for fish and three times each week for daphnids. Samples for total zinc were taken at the beginning of each renewal period for both organisms. Composite samples for soluble zinc were taken at 0, 1, 3, 8, and 24-h intervals during each renewal period for fish but only at renewal termination for daphnid experiments.

Statistical Analysis

Data were first explored via graphical displays and univariate statistics as recommended [20–22]. Mortality data were analyzed using the probit procedure available in the SPSS-X statistical package [24]. Statistical differences between LC_{50}s or EC_{50}s were determined by the method suggested by Sprague [25]. Arcsine transformed and untransformed chronicity data were examined to determine if normality and homogeneity of variance assumptions could be met. Parametric or nonparametric analysis of variance was then used depending upon appropriateness. If statistical significance was demonstrated, a Dunnett's one-sided multiple comparison test was then used to compare the mean of each level of toxicant with mean control response to delineate statistical differences [26]. Analysis of covariance was employed to test if the rate of change of soluble zinc concentrations during toxicity tests was significantly different between the waters tested. SAS [27] programs were used for these analyses. Significance in all statistical tests was judged at the $\alpha = 0.05$ level.

Results

The results of acute toxicity tests are presented in Table 2. Fathead minnow LC_{50}s were within the range of previous studies while *D. pulex* EC_{50}s were higher but were generally within a factor of five from those reported in the literature [11,28]. Control mortality in all experiments was <10% except in Experiment 2 with fathead minnows at 0 h interaction time in which 13% mortality occurred in both site and lab waters. However, since control mortalities were equal, results are reported because relative comparisons between waters could still be made. The ranges of pH and dissolved oxygen measured during the experiments are included in Table 1. Percent recovery of total zinc calculated by dividing initial and final total zinc measurements for each level of zinc spike was nearly 100% in all tests. Sorption of zinc to test chambers was found to be negligible (<10%). Replicate experiments (Table 2) were combined and subjected to probit analysis. These results are summarized in Table 3 and were used to compare point and slope estimates of zinc toxicity between site and lab waters.

Analysis of covariance revealed that the rate of change of aqueous zinc concentrations during tests was highly significantly different ($P < 0.0001$) between the two reference waters and one site water tested. Aqueous zinc decreased during toxicity tests conducted in both dechlorinated tap and Trinity River waters. In contrast, aqueous zinc changed little in reconstituted water and was approximately equal to the total concentration for total concentrations as high as 8 mg/L.

Estimates of chronic toxicity are reported in Table 4. On the 18th day of the daphnid 21-day test in reconstituted water, three controls died for unknown reasons, hence exceeding recommended requirements for control mortality [21]. A concomitant increase in mortality was not observed at any level of toxicant, and in fact only 20% mortality was observed at the first two levels of zinc after 21 days. Therefore, we chose to proceed with statistical analysis based on the seven remaining controls, although these data should be judged with appropriate caution. In addition, in an effort to compare chronic toxicity between lab and site waters, 21-day EC_{50}s corrected for control mortality for both waters are also presented in Table 4. Maximum allowable toxicant concentration (MATC) values for daphnids agreed well with previous work, while fathead minnow MATCs were higher than reported in other studies but within a factor of approximately five [11,28,29].

Discussion

National water quality criteria are currently expressed in terms of a 1-h average criterion maximum concentration (CMC) and a four-day average criterion continuous concentration, neither of which is to be exceeded more than once every three years [11]. Providing these limits are not exceeded, aquatic life and its uses should be protected from the acute and chronic effects of zinc.

TABLE 2—Results of acute toxicity experiments.

Zinc/Water Interaction Time, h	Organism	Experiment	Water[1]	Total Zinc, mg/L			WER[2]	Mean Aqueous Zinc, mg/L			WER[2]
				LCL	LC$_{50}$	UCL		LCL	LC$_{50}$	UCL	
0	Daphnia	1	EFT	2.18	2.44	2.74	2.01	1.54	1.73	1.95	1.43
			RHW	0.33	1.21	4.29	[2.16]*[a]	0.33	1.21	4.29	[1.53]*[a]
				[0.62	1.13	2.03][a]		[0.62	1.13	2.03][a]	
		2	EFT	2.08	2.95	5.77	1.79*	1.96	2.08	2.32	1.26
			RHW	1.42	1.65	2.03		1.42	1.65	2.03	
		3	EFT	1.47	2.45	5.25	2.09	...	1.80	...	1.54
			RHW	...	1.17	...		...	1.17	...	
		1	RHW	...	1.30	...	2.00	...	1.30	...	2.55
			DECL2	0.42	0.65	0.85		0.37	0.51	0.60	
		2	RHW	0.95	1.15	1.37	2.67*	0.95	1.15	1.37	2.95*
			DECL2	0.31	0.43	0.53		0.29	0.39	0.46	
	Fathead minnow	1	EFT	2.48	3.19	3.70	1.61*	1.50	1.85	2.10	1.13
			DECL2	1.55	1.98	2.35		1.33	1.63	1.88	
		2	EFT[c]	1.91	2.84	4.07	0.82	1.01	1.58	2.17	1.13
			DECL2	2.94	3.48	4.17		1.15	1.40	1.74	
		1	RHW	1.03	1.36	1.64	0.55*	0.92	1.22	1.47	0.94
			DEL 2	2.04	2.49	2.98		1.10	1.30	1.51	

24	Daphnia	1	EFT	... — 1.92 — ...	2.23	... — 1.38 — ...	1.64
			EFT^c	2.29 — 2.79 — 3.50	(3.24)*^c	1.74 — 2.06 — 1.31	(2.45)*^c
			RHW	0.43 — 0.86 — 1.33		0.40 — 0.84 — 1.31	
		2	EFT	1.48 — 1.70 — 1.92	3.33*	1.19 — 1.37 — 1.54	2.85*
			EFT^c	1.43 — 1.73 — 2.11	(3.40)*^c	1.16 — 1.41 — 1.72	(2.94)*^c
			RHW	0.18 — 0.51 — 0.72		0.17 — 0.48 — 0.69	
		3	EFT	1.07 — 2.23 — 3.25	2.56*	0.98 — 1.61 — 2.22	1.92*
			EFT^c	1.96 — 2.23 — 2.51	(2.56)*^c	1.43 — 1.61 — 1.81	(1.92)*^c
			RHW	0.46 — 0.87 — 1.29		0.44 — 0.84 — 1.24	
	Fathead minnow	1[b]	EFT	... — 9.52 — ...	2.52	... — 3.10 — ...	1.95
			DECL2	2.78 — 3.78 — 5.13		1.38 — 1.59 — 1.84	
		2	EFT	8.80 — 9.49 — 10.11	1.99*	2.97 — 3.20 — 3.42	1.78*
			DECL2	3.10 — 4.76 — 8.73		1.57 — 1.81 — 2.15	

[1] EFT = Elm Fork of the Trinity River.
DECL2 = Dechlorinated and carbon filtered tap water.
RHW = Reconstituted hard water.
[2] Water effects ratio calculated as site water LC_{50}/lab water LC_{50}.
[3] Previous work in our lab using zinc as a reference toxicant has shown that aqueous concentration does not change in 48-h Daphnia tests. Therefore we assumed that EC_{50}'s based on total and mean aqueous concentrations would be identical in 0-h zinc/water interaction time experiments.
[a] Result after outlier replicate was excluded from calculation.
[b] Experiment conducted with only 3 concentrations and a control.
[c] Organisms acclimated in site water prior to exposure.
* Significantly different from one.

TABLE 3—*Comparison of zinc toxicity between site and lab water for Daphnia pulex and Pimephales promelas based on point and slope estimates.*

TOTAL ZINC, MG/L

Zinc/Water Interaction Time, h	Organism	Water[1]	LCL — LC_{50} — UCL	WER[2]	LCL — RMP[3] — UCL	χ^2 Parallel
0	Daphnia	EFT	2.27 — 2.67 — 3.06	1.95*	1.51 — 2.06 — 3.23	1.47
		RHW	1.11 — 1.37 — 1.64			
	Fathead minnow[a]	EFT	2.48 — 3.19 — 3.70	1.61*	1.17 — 1.52 — 2.23	0.60
		DECL2	1.55 — 1.98 — 2.35			
24	Daphnia	EFT	1.89 — 2.09 — 2.31	2.61*	1.90 — 2.48 — 3.53	6.13**
		RHW	0.65 — 0.80 — 0.93			
	Fathead minnow	EFT	4.39 — 9.24 — 12.87	2.15*	1.05 — 2.10 — 13.04	8.80**
		DECL2	3.36 — 4.29 — 5.59			

MEAN AQUEOUS ZINC, MG/L

Zinc/Water interaction Time, h	Organism	Water[1]	LCL — LC_{50} — UCL	WER[2]	LCL — RMP — UCL	χ^2 Parallel
0	Daphnia	EFT	1.66 — 1.88 — 2.16	1.42*	1.71 — 1.44 — 2.05	13.04**
		RHW	1.11 — 1.32 — 1.64			
	Fathead[a] minnow	EFT	1.50 — 1.85 — 2.10	1.13	0.91 — 1.09 — 1.34	0.43
		DECL2	1.33 — 1.63 — 1.88			
24	Daphnia	EFT	1.40 — 1.54 — 1.71	2.00*	1.53 — 1.90 — 2.50	3.08
		RHW	0.63 — 0.77 — 0.90			
	Fathead minnow	EFT	2.97 — 3.15 — 3.34	1.83*	1.37 — 1.84 — 3.01	8.94**
		DECL2	1.56 — 1.72 — 1.91			

[1] EFT = Elm Fork of the Trinity River.
  RHW = Reconstituted hard water.
  DECL2 = Dechlorinated and carbon filtered tap water.
[2] WER = Water effects ratio (site water LC_{50}/lab water LC_{50}).
[3] RMP = Relative mean potency.
* Significantly different from one.
** Chi-square test for parallelism is significant indicating that regression slopes are not equivalent for site and lab water toxicity tests.
[a] Based on results from Experiment 1 only (Table 2).

TABLE 4—*Chronic toxicity of zinc in site and lab waters based on total and mean aqueous (in parenthesis) concentrations[1] (mg/L).*

	DAPHNIA								
	Reproduction			Survival			LC_{50} Survival[5]		
Water	NOEC[2] —	MATC[3] —	LOEC[4]	NOEC —	MATC —	LOEC	LCL —	EC_{50} —	UCL
EFT	0.09 —	0.12 —	0.14	0.14 —	0.17 —	0.21	0.19 —	0.25 —	0.32
	(0.09 —	0.11 —	0.13)	(0.13 —	0.15 —	0.18)	(0.16 —	0.20 —	0.25)
RHW	0.14 —	0.16 —	0.18	0.18 —	0.24 —	0.32	0.10 —	0.15 —	0.20
	(0.13 —	0.15 —	0.18)	(0.18 —	0.22 —	0.27)	(0.08 —	0.13 —	0.16)
	FATHEAD MINNOW								
	Viable Hatch[6]			Fry Survival			LC_{50} Survival[7]		
Water	NOEL —	MATC —	LOEC	NOEL —	MATC —	LOEC	LCL —	EC_{50} —	UCL
EFT	0.72 —	0.92 —	1.18	0.72 —	0.92 —	1.18	0.62 —	1.11 —	1.79
	(0.44 —	0.56 —	0.72)	(0.44 —	0.56 —	0.72)	(0.33 —	0.49 —	0.65)
DECL2	0.47 —	0.58 —	0.72	0.27 —	0.36 —	0.47	0.33 —	0.49 —	0.65
	(0.38 —	0.45 —	0.54)	(0.22 —	0.29 —	0.38)	(0.25 —	0.38 —	0.52)

[1] Mean soluble concentration was calculated in fish tests by averaging aqueous measurements taken during each renewal period. Mean soluble concentration was estimated in Daphnid tests by averaging initial total measurements taken at the beginning of each renewal with aqueous measurements determined at the end of the renewal.
[2] No observed effect concentration.
[3] Maximum acceptable toxicant concentration.
[4] Lowest observed effect concentration.
[5] Refers to 21-day EC_{50} corrected for control mortality.
[6] Refers to the MATC based on the percent of embryos nondeformed at hatching.
[7] Refers to LC_{50}s based on embryo and fry survival at test termination (8 days).

Calculation of the Site-Specific Criterion Maximum Concentration

Total zinc was approximately two times less toxic to daphnids in site water for tests run immediately after spiking (Table 2). The water effects ratios for the two fish experiments were variable, primarily due to the change in LC_{50} values that was observed between the two dechlorinated tap water reference tests. Apparent discrepancies were not observed if mean aqueous results were considered. Additional fish reference tests demonstrated that the results of the first experiment were reproducible and more accurately reflected acute zinc toxicity in this water. The second test was anomalous due to the slower rate of zinc precipitation that occurred during exposures. Thus a final acute WER of 1.77 was calculated by taking the geometric mean of the significant daphnid (1.95) and fathead minnow (1.61) WERs. The site-specific criterion maximum concentration was then calculated by multiplying the WER by the national criterion maximum concentration adjusted for laboratory water hardness using the equation provided in the national water quality criteria document for zinc [11]. Since two reference waters of comparable hardness were used, an average hardness of 103 mg/L as $CaCO_3$ was used in calculations. The resulting site-specific maximum concentration was found to be 1.77 × 98.0 μg/L or 173.5 μg/L. This value shows excellent agreement with the value of 169.6 μg/L that would have been predicted using the national equation adjusted for the hardness of the collected site water (Table 1). These results indicate that the differences in toxicity results between site and reference waters were attributable to differences in water hardness and that the national criterion, appropriately adjusted for site water hardness, would therefore account for these differences. However, the observed and predicted criterion values would be strictly applicable only at the hardness value of the site water used in toxicity tests. U.S. Geological Survey data at this site from 1968 to 1985 indicated that hardness varied considerably (213 ± 65, mean ± standard deviation, n = 113). Thus, a conservative estimate of receiving water hardness would be needed to ensure that the system would be protected. This would be analogous to the use of a critical design flow such as the 7Q10. Alternatively, dynamic modeling techniques could be employed to account for system-dependent hardness variation [30]. This points out the need for incorporating monitoring data with the results of site-specific criteria derivations.

Effect of Acclimation

Data presented in Table 2 suggest that acclimation of test organisms prior to acute testing does not influence calculated WERs. Sublethal stress may result, which would not be detected in controls, when unacclimated organisms are directly transferred from culture to site water. However, the resulting WER would tend to be conservative since only organisms exposed in site water would experience this additional stress. Acclimation of fish to site water presents no major difficulty, but developing separate daphnid cultures in site water may be impractical and cost ineffective. Such an approach in which organisms are not acclimated has generally been accepted in effluent toxicity testing and might be considered for site-specific applications.

Effect of Reference Water

The WERs calculated from parallel toxicity tests using the two different reference waters were used to evaluate the impact that choosing alternate pairs of reference waters might have on site-specific criterion maximum concentration derivations. Table 5 indicates that the site-specific criterion value can vary by a factor of two for reference waters of similar hardness. This indicates that the reference water chosen can significantly affect the site-specific criterion derivation procedure despite adjustments made to account for differences in reference water hardness. The development and implementation of standardized laboratory reference waters in site-specific studies would provide a basis for comparisons between site and reference waters and would ultimately provide a better understanding of site-specific toxicity as the guidelines are applied in the future with various receiving systems.

Effects of Concentration-Response Relationships

One of the underlying assumptions of the indicator species procedure is that the toxicity of a chemical differs between lab and field by a constant factor and that this difference may be

TABLE 5—*The effect of choosing different laboratory reference waters on the derivation of zinc site-specific maximum concentrations for total recoverable zinc, $\mu g/L$.*

		Daphnia	
		Reconstituted water	Dechlorinated tap water
Fathead minnow	Reconstituted water	234	383
	Dechlorinated tap water	174	284

adequately estimated by the ratio of point (LC_{50}) toxicity estimates in lab and site waters. However, this assumption appears tenuous if concentration response curves are not parallel, as indicated by significant chi-square tests for parallelism in some of our experiments (Table 3). In such cases, a WER is not constant but is concentration dependent. Failure to consider the nature of concentration response curves when conducting site-specific studies may result in the use of erroneous WERs and the development of inappropriate site-specific criteria.

Effects of Precipitation and Kinetics

Generally, the differences in zinc toxicity between waters were small and were reduced when results were expressed in terms of mean soluble metal. Apparently, zinc that precipitated during toxicity tests was less bioavailable than soluble zinc. Results also demonstrated that the toxicity of total zinc was significantly reduced when zinc was allowed to interact with test waters prior to test initiation. Again this was attributed to precipitation reactions and illustrates the effect that kinetics may impose on the observed toxicity. Although adsorption of zinc to suspended solids present in site water may have reduced aqueous concentrations, previous work suggests that this contribution would be minor due to the low level of solids present [17]. The relative sensitivity of the two organisms to zinc in the two reference waters may also be a consequence of precipitation. Daphnids were less sensitive in reconstituted water presumably due to the protective effect afforded by the higher levels of calcium and magnesium and/or the formation of less toxic aqueous zinc species. However, at the higher zinc levels used in fish tests, the greater extent of oversaturation enhanced zinc precipitation in dechlorinated tap water, partially offsetting the protective effect noted for daphnids in reconstituted water. Table 2 indicates that significant toxicological differences may occur between waters for zinc even if normalized for soluble metal. Furthermore, WERs based on mean soluble zinc tended to increase after 24 h of interaction time, suggesting that the kinetics of aqueous speciation may influence zinc toxicity.

Recently, thermodynamic equilibrium models [31] have been applied to aid in the identification of individual toxic metal species [32]. This information coupled with model predictions of site-specific speciation and fate should provide a useful tool in the refinement of metal criteria and in the establishment of water quality-based permit limits. However, since these models do not consider the kinetics of speciation, they should be applied with caution. Further research on the relationship between the kinetic aspects of metal speciation and bioavailability is warranted.

Calculation of the Site-Specific Criterion Continuous Concentration

The national criterion continuous concentration for zinc is also specified as a function of water hardness [11] and was calculated to be 153.6 µg/L for the Trinity River sample collected. Since a final residue value or final plant value is unavailable for zinc, the site-specific final chronic value calculated by the three methods provided in the guidelines was equivalent to the site-specific criterion continuous concentration.

The first procedure for calculating a site-specific criterion continuous concentration, obtained by dividing the final acute site-specific value (site-specific maximum concentration × 2) by the national acute/chronic ratio of 2.208 yielded a criterion of 157 µg/L. Site-specific acute/chronic ratios for both species were calculated by taking the LC_{50} value at 0 h interaction time (Table 3) and dividing these values by the lowest MATC value reported for total zinc in site water (Table 4). Since these ratios were within a factor of 10 (22.2 for daphnids, 3.5 for fish), the geometric mean of these two values was used to calculate a site-specific final acute/chronic ratio of 8.8. The site-specific criterion continuous concentration of 39.5 µg/L was determined by dividing 8.8 into the site-specific final acute value of 348 µg/L. From Table 4 the chronic WER was not significant for daphnids and was less than a factor of three for fish. Therefore, application of the last method resulted in a site-specific criterion continuous concentration equivalent to the national criterion of

87 µg/L, based on the reference water hardness of 103 mg/L as $CaCO_3$, since the final chronic WER was equal to 1.0. The preceding derivations were intended to illustrate the use of these methods, and no attempt was made to account for seasonal changes in site water quality, which has been shown to affect procedure results [33,34].

Conclusions

This study illustrates the application of the indicator species procedure for deriving site-specific zinc criteria and identifies several factors which should be considered when applying the procedure. The site-specific maximum criterion concentration was shown to generally agree well with the national criterion based on total zinc but depended on the reference water(s) chosen. This indicates that standardized laboratory waters should be used in this procedure so that a suitable baseline can be established for comparing differences in bioavailability observed at other sites. It has been pointed out that evaluating toxicological differences exclusively in terms of LC_{50} estimates may be misleading if concentration-response slopes are not parallel. Data presented suggest that acclimation to site water did not affect derived criteria and from a practical viewpoint may not be necessary in such studies. The site-specific criterion continuous concentration derived was found to vary by a factor of 4 based on total zinc depending upon the calculation method used.

These findings suggest that differences between national criteria and site-specific criteria derived via the indicator species procedure would be small when applied to zinc. This information may provide guidance in deciding cost tradeoffs for applying this procedure versus going beyond technology-based controls to comply with national criteria. Although toxicological differences between waters may be reduced if soluble rather than total zinc is used, the effects of kinetics on equilibrium model predictions and bioavailability must be considered if meaningful site-specific metal criteria are to be developed.

Acknowledgments

We would like to thank Willard Gibbons for providing water quality monitoring data, Ed Price for technical advice, and Carson Harrod and Rich New for their assistance in the laboratory.

References

[1] Water Quality Standards Regulation, U.S. Environmental Protection Agency, *Federal Register*, Nov. 1983, Vol. 48, p. 51400.

[2] Gostomski, F. E., "Water Quality Criteria: Protection of Use Perspective," *Aquatic Toxicology and Hazard Assessment: (Eighth Symposium), ASTM STP 891*, R. C. Bahner and D. J. Hansen, Eds., American Society for Testing and Materials, Philadelphia, 1985, pp. 45–47.

[3] Black, J. A., Roberts, R. F., Johnson, D. M., Mincucci, D. D., Nancy, K. H., Allen, H. E., "The Significance of Physiochemical Variables in Aquatic Bioassays of Heavy Metals," in *Bioassay Techniques and Environmental Chemistry*, Ann Arbor Science Publishers, Inc., 1973, pp. 259–275.

[4] Sprague, J. B., "Factors that Modify Toxicity," in *Fundamentals of Aquatic Toxicology*, G. M. Rand and S. R. Petrocelli, Eds., Hemisphere Publishers, New York, 1985, pp. 122–176.

[5] Borgman, V., "Metal Speciation and Toxicity of Free Metal Ions to Aquatic Biota, *Aquatic Toxicology*, J. O. Nriagu, Ed., Wiley, New York, 1983, pp. 47–72.

[6] *Water Quality Standards Handbook*, U.S. Environmental Protection Agency, Washington, DC, Dec. 1983.

[7] O'Donnell, J. R., Kaplan, B. M., and Allen, H. E., "Bioavailability of Trace Metals in Natural Waters," *Aquatic Toxicology and Hazard Assessment (Seventh Symposium), ASTM STP 854*, R. D. Cardwell, R. Purdy, and R. C. Bahner, Eds., American Society for Testing and Materials, Philadelphia, 1985, pp. 485–501.

[8] Technical Support Document for Water Quality Based Toxics Control, U.S. Environmental Protection Agency, Office of Water, Washington, DC, Sept. 1985.

[9] Forestner, V. and Wittman, G. T. W., *Metal Pollution in the Aquatic Environment*, Springer-Verlag Heidelberg, New York, 1979, pp. 397.

[10] Callahan, M. A., Slimak, M. W., Gabel, N. W., May, I. P., Fowler, C. F., Freed, J. R., Jennings, P., Durfee, R. L., Whitmore, F. C., Maestri, B., Mabey, W. R., Holt, B. R., and Gould, C., "Water-Related Fate of 129 Priority Pollutants," Vol. 1, EPA 440/4-79-029a, U.S. Environmental Protection Agency, Washington, DC, 1979.
[11] "Ambient Aquatic Life Water Quality Criteria for Zinc," EPA-440/5-80-79, April 10th draft, U.S. Environmental Protection Agency, Office of Water Regulations and Standards Division, Washington, DC.
[12] Mount, D. I., "The Effect of Total Hardness and pH on the Acute Toxicity of Zinc to Fish," *International Journal of Air Water Pollution,* 1966, Vol. 10, pp. 49–56.
[13] Bradley, R. W. and Sprague, J. B., "Accumulation of Zinc by Rainbow Trout as Influenced by pH, Water Hardness and Fish Size," *Environmental Toxicology and Chemistry,* Vol. 4, 1985, pp. 685–694.
[14] Allen, H. E., Hall, R. H., and Brisban, T. D., "Metal Speciation, Effects on Aquatic Toxicity," *Environmental Science & Technology,* 1980, Vol. 14, pp. 441–443.
[15] Gauss, J. D. and Winner, R. W., "The Interactive Effects of Water Hardness and Humic Acid on the Acute and Chronic Toxicity of Zinc to *Daphnia magna,*" Poster Session, 6th Annual Meeting, Society of Environmental Toxicology and Chemistry, 10–13 Nov. 1985, St. Louis, MS.
[16] Nienke, G. E. and Lee, G. F., "Sorption of Zinc by Lake Michigan Sediments. Implications for Zinc Water Quality Criteria Standards," *Water Research,* 1982, Vol. 16, pp. 1373–1378.
[17] Hall, W. S., Dickson, K. L., Saleh, F. Y., Rodgers, J. H., Wilcox, D., and Entatezami, A., "Effects of Suspended Solids on the Acute Toxicity of Zinc to *Daphnia magna* and *Pimephales promelas,*" *Water Resources Bulletin,* Vol. 22, No. 6, 1986.
[18] Britton, L. J., Goodard, D. E. and Brioggs, J. C., "Quality of Rivers of the United States, 1976 Water Year—Based on Open File Report, the National Stream Quality Accounting Network (NASQAN)," U.S. Geological Survey, Washington, DC, 1976, pp. 80–594.
[19] "Standard Methods for the Examination of Water and Wastewater," 15th ed., American Public Health Association, Washington, DC, 1980.
[20] Peltier, W., "Methods for Measuring the Acute Toxicity of Effluents to Aquatic Organisms," EPA 600/4-78-012, U.S. Environmental Protection Agency, Washington, DC, 1978.
[21] "Interim Procedures for Conducting the *Daphnia magna* Toxicity Assay," Draft 3rd version, Office of Research and Development, U.S. Environmental Protection Agency, Washington, DC, 1984.
[22] Horning, W. B., and Weber, C. I., "Short Term Methods for Estimating the Chronic Toxicity of Effluents and Receiving Waters to Freshwater Organisms," EPA 600/4-85-014, U.S. Environmental Protection Agency, Washington, DC, 1985.
[23] Mount, D. I., "Chronic Toxicity of Copper to Fathead Minnows *Pimephales promelas* Rafinesque," *Water Research,* Vol. 2, 1968, pp. 215–223.
[24] SPSS-X Inc. Users Guide, 2nd ed., McGraw-Hill Co., Chicago, IL, 1985.
[25] Sprague, J. B. and Fogels, A., "Watch the Y in Bioassay," *Proceedings,* 3rd Aquatic Toxicity Workshop, Halifax, NS, 2–3 Nov. 1976, Environmental Protection Technical Report No. EPA-5-AR-77-1, Halifax, Canada, 1977, pp. 107–118.
[26] Zarr, J. H., *Biostatistical Analysis,* Prentice-Hall, Inc., Englewood Cliffs, NJ, 1974, p. 620.
[27] SAS Institute, SAS Statistics Guide, Cary, NC, 1982.
[28] Mount, D. I. and Norberg, T. J., "A Seven Day Life Cycle Cladoceran Life Cycle Test," *Environ. Toxicol. Chem.,* Vol. 3, 1984, pp. 425–434.
[29] Norberg, T. J. and Mount, D. I., "A New Fathead Minnow (*Pimephales promelas*) Subchronic Toxicity Test," *Environ. Toxicol. Chem.,* Vol. 4, 1985, pp. 711–718.
[30] Parkerton, T. F., Stewart, S. M., Dickson, K. L., Rodgers, J. H., Jr., and Saleh, F. Y., "Derivation of Site-Specific Water Quality Criteria for Zinc: Implications for Wasteload Allocation" presented at the 59th Water Pollution Control Federation Conference, Los Angeles, CA, 7 Oct. 1986.
[31] Felmy, A. R., Brown, S. M., Onishi, Y., Yabusaki, S. B. and Argo, R. S., "MEXAMS—The Metals Exposure Analysis Modeling System," U.S. Environmental Protection Agency, Office of Research and Development, Athens, GA, 1984.
[32] Cowan, C. E., Jenne, E. A. and Kinnison, R. R., "A Methodology for Determining the Relationships Between Toxicity and the Aqueous Speciation of a Metal," presented at the Aquatic Toxicology and Environmental Fate: 9th Symposium, American Society for Testing and Materials, Philadelphia, PA, 14–16 April 1985.
[33] Sphear, R. L. and Carlson, A. R., "Derivation of Site-Specific Water Quality Criterion For Cadmium and the St. Louis River Basin, Duluth, Minnesota," *Environ. Toxicol. and Chem.,* Vol. 3, 1984, pp. 651–665.
[34] Carlson, A. R. and Roush, T. H., "Site-Specific Water Quality Studies of the Straight River, Minnesota: Complex Effluent Toxicity, Zinc Toxicity and Biological Survey Relationships," EPA 600/3-85-005, Environmental Protection Agency, Washington, DC, 1985.

Raymond N. Yong[1] and Ralph D. Ludwig[1]

A Toxicity Assessment of Tar Sands Tailings

REFERENCE: Yong, R. N. and Ludwig, R. D., "**A Toxicity Assessment of Tar Sands Tailings,**" *Aquatic Toxicology and Hazard Assessment: 10th Volume, ASTM STP 971,* W. J. Adams, G. A. Chapman, and W. G. Landis, Eds., American Society for Testing and Materials, Philadelphia, 1988, pp. 436–446.

ABSTRACT: Toxicity assessment was conducted on the liquid fractions of waste tailings generated from the hot water extraction process employed in recovering oil from the Athabasca tar sands. Indices of toxicity employed included a mixed culture of microorganisms indigenous to a major river flowing in the vicinity of containment ponds used to store the waste tailings. Toxicity assessment employing the mixed culture of microorganisms was conducted by monitoring the growth of microbial suspensions through turbidity measurements. Other indices of toxicity employed were the freshwater green alga *Selenastrum capricornutum* and the bacterium resembling *Photobacterium phosphoreum*. The liquid fractions considered for toxicity assessment included those of both fresh tailings and aged tailings. In addition, toxicity assessment was conducted on the fluid emanating from accumulated sludge in the containment ponds.

Fresh tailings were observed to be highly toxic to all indices of toxicity employed. At a concentration of 40% by volume, fresh tailings were observed to support on average less than 38% of the growth observed in controls for the mixed culture of river microorganisms, while at a concentration of 25% by volume the fresh tailings brought about an average light emission reduction of greater than 65% in the bacterium resembling *Photobacterium phosphoreum*. Toxicity assessment with *Selenastrum capricornutum* gave an interpolated EC_{50} value of 24% by volume.

Aged tailings were observed to exhibit only slight toxicity to *Selenastrum capricornutum* and the bacterium resembling *Photobacterium phosphoreum* and no toxicity to the mixed culture of river microorganisms. In fact, aged tailings at a concentration of 40% by volume appeared to stimulate growth in the mixed culture of river microorganisms over that observed in the controls. At 100% concentration by volume, an average light emission reduction of 16% was observed for the bacterium resembling *Photobacterium phosphoreum,* and an average 33% inhibition of growth relative to controls was observed for *Selenastrum capricornutum.*

Emanated sludge fluid was observed to be highly toxic only to the alga *Selenastrum capricornutum* with an interpolated EC_{50} value of 29% by volume. At a concentration of 40% by volume, the emanated sludge fluid was observed to apparently stimulate growth in the mixed culture of river microorganisms over that observed in the controls. At a concentration of 100% by volume, an average light emission reduction of only 13% was observed for the bacterium resembling *Photobacterium phosphoreum.*

The results suggest that, although initially very toxic, the waste tailings undergo a self-detoxification process with time. This may have important implications with regard to measures which might be implemented in the containment and treatment of the tailings.

KEYWORDS: oil, oil sands, wastes, tailings, toxicology, assessments, plankton, microorganisms, algae, bacteria, *Selenastrum capricornutum, Photobacterium phosphoreum*

The Athabasca tar sands of northern Alberta have in recent years developed into an increasingly important source of fuel oil. The tar sands contain in excess of 700 billion barrels of bitumen,

---

[1] Director and research associate, respectively, Geotechnical Research Centre, Environmental Geotechnology Division, McGill University, Montreal, Quebec, Canada H3A 2K6.

which is presently being extracted using the "hot water extraction process" as described by Camp [1]. The "hot water extraction process," involving the addition of large volumes of hot water to achieve separation of the bitumen from the sand, has resulted in the production of large volumes of wastewater or "tailings." These tailings, containing sand and clay-sized particles as well as unextracted bitumen, are released into large holding ponds or "tailings ponds" at temperatures approaching 80°C (176°F). The 2 to 10% fines solids content of the tailings and its inherent settling properties has resulted in the formation of a large volume of slowly dewatering sludge. Previous work conducted by Hrudey et al. [2] has indicated that dyke drainage from these tailings ponds is lethal to *Salmo gairdneri* (rainbow trout) with an extrapolated 96-h LC_{50} value of 11% by volume. The study described in this paper strives to provide greater insight into the toxicity of tar sands tailings by employing indices of toxicity from lower trophic levels in the aquatic food chain.

Procedure

Test Samples

Toxicty assessment was conducted on three different samples originating from two sources. The two sources of samples were: (1) tar sands tailings produced as a result of the "hot water extraction process" destined for release into the holding ponds; and (2) pore fluid released in the natural "in situ" dewatering process of accumulated sludge in the holding ponds. Two samples from the first source were considered, namely "fresh" tailings (that is, tailings resulting from extraction process and no more than three days old) and "aged" tailings (that is, tailings resulting from extraction process and approximately one year old). Fresh tailings samples consisted of tailings collected prior to disposal into the holding ponds and temporarily held in five-gallon covered (nonairtight) plastic pails at room temperature for three days to allow for settling of much of the solids content. One-litre aliquots of clarified sample were subsequently collected and stored at $-20°C$ (that is, frozen) in plastic (linear polyethylene) bottles until required for testing. Aged tailings consisted of tailings collected prior to disposal into the holding ponds and stored in covered (nonairtight) 60-gallon plastic barrels at room temperature for an entire year. Only then were 1-L aliquots collected and stored at $-20°C$ (that is, frozen) in plastic (linear polyethylene) bottles until required for testing.

One sample from Source 2 was considered and consisted of original sludge pore fluid having emanated over a period of three months from approximately 8 m^3 of untreated sludge which had been removed from a tailings pond and stored in a large epoxy-lined covered (nonairtight) concrete storage tank at 20°C. One-litre aliquots from the supernatant, formed from the emanating pore fluid, were collected and subsequently stored at $-20°C$ (that is, frozen) in plastic (linear polyethylene) bottles until required for testing.

All samples in the study were, therefore, stored at $-20°C$ (that is, frozen) from the time of collection until the time at which any test was to be conducted. Samples were frozen in an effort to minimize chemical, physical, and biological changes which might occur in the samples over the time during which the study was conducted.

All samples in the study were, furthermore, prefiltered with 0.45-μm membrane filters prior to any test being conducted. The suspended solids content of the samples was visibly very high (particularly in the tailings samples) and would thus have significantly interfered with many of the tests conducted.

Indices of Toxicity

Three indices of toxicity were employed in the study. They were: (1) a mixed culture of microorganisms indigenous to the Athabasca River (a major river flowing in the vicinity of some tailings ponds near Ft. McMurray, Alberta); (2) *Selenastrum capricornutum,* a freshwater, green alga (Chlorophyceae) of the order Chlorococcales; and (3) a marine bacterium resembling *Photobacterium phosphoreum*.

Toxicity assessment employing microorgansims indigenous to the Athabasca River was conducted in the following manner. A 2-L sample of Athabasca River water containing a mixed population of microorganisms (to be used as inoculum) was collected upstream of a tar sands processing plant and temporarily stored in a 4-L plastic container. Within 48 h of collection, the river sample was placed in a refrigerator at 4°C and stored there until further use (48 h). Given volumes of test sample (either 5 or 10 mL) to be assessed for toxicity were pipetted into preautoclaved (sterilized) growth flasks (Fig. 1) using standard aseptic techniques. (Distilled water rather than test sample was pipetted into control flasks.) Ten millilitres of preautoclaved (sterilized) liquid minimal medium containing salts and glucose were then added to appropriate flasks. (Glucose served as the substrate for decomposition by the microorganisms.) The liquid minimal medium was prepared according to Pelczar and Chan [3] by combining a sterilized minimal salt solution with a sterilized 20% glucose solution. The minimal salt solution consisted of 20-g NH_4Cl, 4-g KH_2PO_4, 4-g NH_4NO_3, 0.4-g $MgSO_4 \cdot 7H_2O$, 8-g Na_2SO_4 (anhydrous), 8-g K_2HPO_4 (anhydrous), and distilled water for

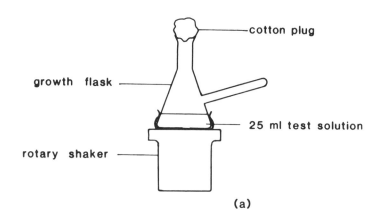

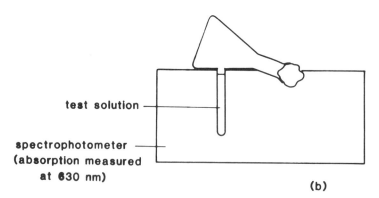

FIG. 1—*Microbial growth experimental setup; growth flask in inverted position for absorption reading.*

a total of 1000 mL of solution. The 20% glucose solution consisted of 200-g D-glucose plus distilled water for a total of 1000 mL of solution. The liquid minimal medium was then prepared by combining 1 mL of 20% glucose solution, 25 mL of minimal salt solution, and sterilized distilled water to a total volume of 100 mL.

After addition of the test sample and the liquid minimal medium, 5 mL of Athabasca River water (inoculum) were added to all flasks. Five milliliters of sterilized distilled water were aseptically added to those flasks containing only 5 mL of test sample such that all flasks before testing contained a total of 25 mL of solution. All flasks were subsequently placed on a rotary shaker for continuous shaking over a 48-h period of time at a temperature of 25°C. Increases in the total biomass of the microorganisms were monitored by means of a Bausch & Lomb Spectronic 20 Spectrophotometer set at a wavelength of 630 nm. Increased absorption at this wavelength was interpreted as being a linear reflection of an increase in total microbial biomass.

Toxicity assessment employing the green alga *Selenastrum capricornutum* was conducted in accordance with the "*Selenastrum capricornutum* Printz Algal Assay Bottle Test" [4]. Erlenmeyer flasks containing various sample dilutions were inoculated with a given number of algae (40 000 cells) plus a given quantity of macro- and micronutrients [4] and then incubated (to facilitate free gas exchange and to provide consistent light intensity) for eight days. The quantity of algae present after the eight-day incubation period was then determined employing a Coulter Electronics, Inc. Model TA II Coulter Counter. Biomass yields of algae exposed to sample dilutions were then compared to biomass yields of algae obtained in controls, and the relative toxicity of each sample dilution was thus determined (controls after eight days contained approximately 4 000 000 cells). Selection of an alga belonging to the genus *Selenastrum* as an index of toxicity was based on the fact that algae of this particular genus are found in waters of the most diversified composition [4] and have a wide tolerance towards environmental conditions [5]. Reasons for employing *Selenastrum capricornutum* in particular include ease of handling and ease of enumeration.

Toxicity assessment employing the marine bacterium resembling *Photobacterium phosphoreum* was conducted in accordance with the Microtox toxicity test [6]. A Microtox Model 2055 Toxicity Analyzer System was employed, and the test is based on the concept that the production of light by the bacteria resembling *Photobacterium phosphoreum* is a result of the total metabolic processes of the cells and is thereby a reflection of their overall state of health. Thus, any inhibition of the metabolic processes would be expected to manifest itself in a net reduction in light productivity. A decrease in light productivity of the microorganisms is therefore interpreted as an indication of toxicity. In all cases considered, the bacteria were exposed to the test samples for a period of 5 min. Sodium chloride was added to all samples to provide the appropriate saline environment, and light productivity after the 5-min exposure time was then compared to light productivity in controls. Controls consisted of saline solution and the bacteria only.

All tests conducted at any given test sample concentration employing the three indices of toxicity were in this study conducted in triplicate.

Results and Discussion

Mixed Culture of Microorganisms

The results of the toxicity assessment for the three samples employing microorganisms indigenous to the Athabasca River as indices of toxicity are provided in Figs. 2 and 3. Figures 2 and 3 depict growth curves for the microorganisms over a 24-h period after their inoculation into the samples. Growth of the microorganisms was reflected by increased absorption at a wavelength of 630 nm. In all cases, absorption readings were observed to vary by no more than 5% within any given set of triplicates at any given time that readings were taken. Glucose served as the substrate (food

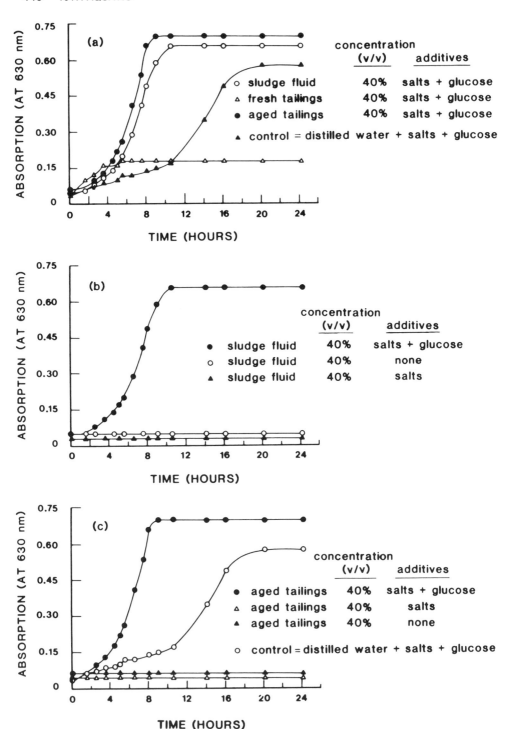

FIG. 2—*Growth curves for Athabasca River microorganisms.*

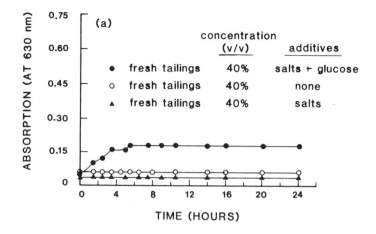

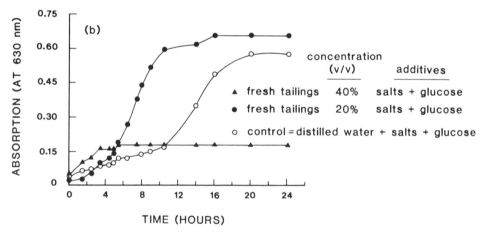

FIG. 3—*Growth curves for Athabasca River microorganisms.*

source) for the microorganisms, and salts were added to ensure that the essential elements required for growth were available. Sample concentrations of 20% by volume and 40% by volume were considered. Figure 2a provides growth curves for fresh tailings, aged tailings, and emanated sludge fluid at a concentration of 40% by volume in the presence of added salts and glucose. Also provided in Fig. 2a is the growth curve for a control consisting of distilled water, salts, and glucose. The growth curves obtained appear to indicate that, of the three samples, only the fresh tailings were toxic to the mixed culture of microorganisms (at a concentration of 40% by volume). The fresh tailings (at a concentration of 40% by volume) yielded on average no more than 37% of the growth obtained in the control after 24 h. Emanated sludge fluid and aged tailings, however, not only appeared to be nontoxic to the microorganisms at a concentration of 40% by volume based on their 24-h growth yields but, moreover, appeared to stimulate the initial growth rate of the microorganisms over that observed in the control. This perhaps suggests that other nutrients may be available to the microorganisms in these two samples, thereby enhancing their initial growth rate. Figures 2b and 2c, however, indicate that the initial enhanced growth in the emanated sludge fluid and the aged tailings at a concentration of 40% by volume is not likely a result of additional biologically available carbon sources present in the samples. This was evidenced by the fact that, in the absence of glucose and in both the presence and absence of additional salts, no growth was obtained in either emanated sludge fluid or aged tailings. If additional carbon sources were biologically available, then at least some growth should have been observed in these samples, particularly in the presence of salts. Thus, the enhanced initial growth rate observed in the emanated sludge fluid and the aged tailings over that observed in the control is probably due rather to additional growth-stimulating elements already present in these samples. Figure 3a indicates that the minimal growth observed in the fresh tailings at a concentration of 40% by volume was dependent on the presence of glucose and not any other biologically available carbon sources. Figure 3b indicates that the fresh tailings at a concentration of 20% by volume were nontoxic to the glucose-degrading microorganisms. Moreover, the fresh tailings at a concentration of 20% by volume also appeared to stimulate initial growth of the microorganisms over that observed in the control. Thus, it is likely, as in the case of the emanated sludge fluid and the aged tailings, that additional growth-stimulating elements are present in the fresh tailings.

It should be noted that tests conducted on all three samples in the presence of both salts and glucose but in the absence of Athabasca River inoculum resulted in no observed growth after 24 h. This indicates that the growth curves presented in Figs. 2 and 3 reflect the growth of microbial populations originating from the Athabasca River and not the test samples themselves.

Selenastrum capricornutum *Algae*

The results of the *Selenastrum capricornutum* Printz Algal Assay Bottle Test are provided in Fig. 4. The results are presented in the form of histograms in which the height of an individual histogram for any given sample is equal to the percent algae growth observed in the given sample relative to the algae growth observed in the controls. (Algae growth refers to the biomass (ppm) of *Selenastrum capricornutum* accumulated after an eight-day incubation period.) Controls consisted of distilled water plus nutrient supplements only. In all cases, algae growth within any given set of triplicates was observed to vary by less than 5%. Test sample concentrations ranging from 100% by volume to 25% by volume were considered. The effects of two supplements were considered. The supplements were: (1) nutrient medium (a solution consisting of macro- and micronutrients); and (2) EDTA (ethylenediaminetetraacetic acid). The nutrient medium was added to samples to ensure the availability of sufficient nutrients for algae growth.

The EDTA was added as a complexing agent for purposes of rendering any metals in a potentially toxic (bioavailable) state into a nontoxic (nonbioavailable) state. Thus EDTA was added in an effort to determine whether metals were a source of toxicity in the samples. It can be observed from Figs. 4b and 4c that addition of EDTA had little, if any, effect on algae growth. Aged tailings in the presence of nutrient solution alone supported on average 66% of the algae growth observed in controls, while the addition of EDTA to the aged tailings containing nutrient solution yielded on average 71% of the algae growth observed in controls. Because the difference in algae growth in the presence and absence of EDTA is relatively small (less than 5%), it would be precarious to conclude that metals in the aged tailings contribute to toxicity. Nevertheless, if metals do contribute to toxicity, their contribution appears to be relatively insignificant.

From Fig. 4a, it is clear that all samples (that is, fresh tailings, aged tailings, and emanated sludge fluid) were not able to support any significant algae growth in the absence of nutrient supplements. Only the aged tailings were able to support significant algae growth after addition of nutrient supplements (see Fig. 4b), indicating that fresh tailings and emanated sludge fluid are characterized by toxic properties, factors, and/or components to a greater degree than in the aged

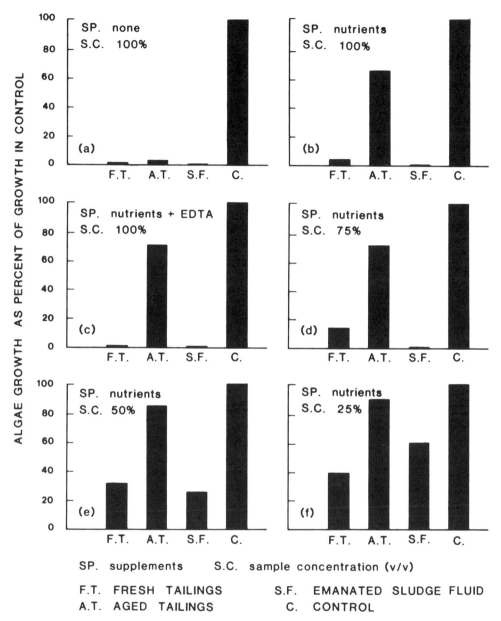

FIG. 4—*Growth histograms for the alga* Selenastrum capricornutum.

tailings. At a sample concentration of 75% by volume, the fresh tailings appeared to be less toxic than the emanated sludge fluid (see Fig. 4d). The fresh tailings supported on average 14% of the algae growth observed in controls, while emanated sludge fluid supported on average less than 1%. The aged tailings were the least toxic of the samples at this concentration (that is, 75% by volume) and supported on average 72% of the growth observed in controls. (The increase in algae growth observed in aged tailings at a concentration of 75% by volume represents an increase of only 6 and 2% over the algae growth observed in aged tailings at a concentration of 100% by volume in the absence and presence of EDTA, respectively.) At a sample concentration of 50% by volume, samples in order of increasing toxicity were aged tailings, fresh tailings, and emanated sludge fluid with algae growths on average of 85, 31, and 25%, respectively, relative to the growth observed in controls (see Fig. 4e). At a sample concentration of 25% by volume, samples in order of increasing toxicity were aged tailings, emanated sludge fluid, and fresh tailings with algae growths on average of 91, 61, and 40%, respectively, relative to the growth observed in controls (see Fig. 4f). Thus, the results indicate that at a sample concentration of 50% by volume the emanated sludge fluid is more toxic than the fresh tailings, while at a sample concentration of 25% by volume, the fresh tailings are more toxic than the emanated sludge fluid.

The EC_{50} values for the three samples considered in this study were determined by plotting the percent inhibition in algae growth against the log of the sample concentration. Regression analysis was then applied in order to find the "best fitting line" for the given set of points. The EC_{50} values were then extrapolated from the equations describing the "best fitting line." The EC_{50} values obtained for fresh tailings and emanated sludge fluid were 24 and 29%, respectively. Because aged tailings at a concentration of 100% by volume in the presence of nutrient supplements supported more than 50% of the algae growth observed in controls, extrapolation of an EC_{50} value was not applicable. Thus, based strictly on EC_{50} values, the samples in order of increasing toxicity to *Selenastrum capricornutum* were aged tailings, emanated sludge fluid, and fresh tailings.

Bacteria Resembling Photobacterium phosphoreum

The results for the toxicity test employing the marine bacterium resembling *Photobacterium phosphoreum* are provided in Fig. 5. In this toxicity test, the quantity of light emitted by the bacteria 5 min after inoculation into the samples relative to the quantity of light emitted by the bacteria immediately after inoculation into the samples is considered to be a reflection of the toxicity of the samples. Thus, a decrease in light production by the bacteria 5 min after inoculation into the samples is an indication of an induced poorer state of health in the bacteria (a poorer state of health does not necessarily imply death of the bacteria). Sample concentrations of 100, 50, 25, and 12.5% by volume were considered. Light productivity within any given set of triplicates was observed to vary by less than 5%. Figure 5 is presented in the form of histograms where the height of each individual histogram for any given sample is equal to the average percent light emission observed in the sample relative to the light emission observed in controls. Controls consisted of the bacteria suspended in a nontoxic saline diluent, while samples consisted of the bacteria suspended in test sample dilutions adjusted to the appropriate sodium chloride concentration.

Figure 5a indicates that at a sample concentration of 100% by volume only fresh tailings were highly toxic to the bacteria. An average decrease in light emission of no less than 95% was observed after 5 min in the fresh tailings. Emanated sludge fluid and aged tailings produced average light emission reductions of no more than 13 and 17%, respectively, at 100% concentration by volume. At a sample concentration of 50% by volume, fresh tailings produced a light emission reduction of no less than 84%, while emanated sludge fluid and aged tailings produced average light emission reductions of no more than 7 and 9%, respectively (see Fig. 5b). At a sample concentration of 25% by volume, fresh tailings produced an average light emission reduction of no less than 65%, while emanated sludge fluid and aged tailings produced no light emission reductions (see Fig. 5c). Even at a concentration of 12.5% by volume, fresh tailings produced an average light emission reduction of no less than 31% (see Fig. 5d). Thus, fresh tailings were highly toxic to the bacterium resembling *Photobacterium phosphoreum,* while emanated sludge fluid and aged tailings were relatively nontoxic.

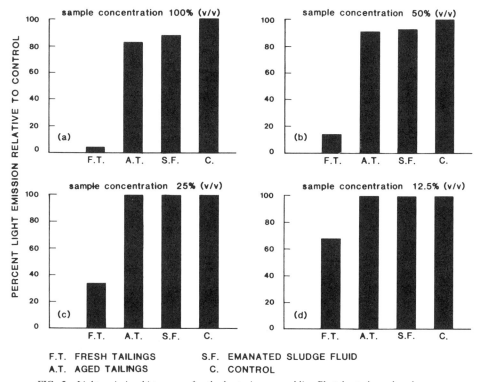

FIG. 5—*Light emission histograms for the bacterium resembling* Photobacterium phosphoreum.

Conclusions

The results of the toxicity tests employing the three indices of toxicity indicate that of the three samples considered, aged tailings exhibited the least toxicity while fresh tailings exhibited the greatest toxicity. Fresh tailings were highly toxic to all three indices of toxicity employed, while aged tailings were relatively nontoxic. The emanated sludge fluid, although highly toxic to the alga *Selenastrum capricornutum,* exhibited no toxicity to microorganisms indigenous to the Athabasca River (for the sample concentrations considered) and relatively little toxicity to the bacterium resembling *Photobacterium phosphoreum* compared to the fresh tailings. The results appear to indicate that the tailings produced in the hot water bitumen extraction process are initially highly toxic but that with time and under appropriate conditions, a gradual self detoxification process occurs. The rate at which the detoxification process occurs has not been established, although it would appear that a one-year period of time is sufficient. These results may have important implications with regard to any treatment and/or storage procedures which may be employed to reduce the potential environmental impact of disposed tar sands tailings.

Acknowledgments

The authors are grateful to the Natural Sciences and Engineering Research Council of Canada (NSERC) for providing the funding necessary to conduct the study (Research Grant No. A-882). The authors would also like to thank Christian Blaise, Norman Birmingham, and the laboratory staff of Environment Canada (Longueuil) for providing their facilities and invaluable contributions throughout the course of the study. Gratitude is also extended to Diana Mourato for her technical assistance.

References

[1] Camp, F., *The Tar Sands of Alberta, Canada,* 2nd ed., Cameron Engineers, Inc., Denver, 1974.
[2] Hrudey, S., Sergy, G., and Thackeray, T., "Toxicity of Oil Sands Plant Wastewaters and Associated Organic Contaminants," *Proceedings. 11th Canadian Symposium on Water Pollution Research Canada,* pp. 34–44.
[3] Pelczar, M. and Chan, E., *Laboratory Exercises in Microbiology,* 4th ed., McGraw-Hill, 1977.
[4] Miller, W., Greene, J., and Shiroyama, T., "The *Selenastrum capricornutum* Printz Algal Assay Bottle Test: Experimental Design, Application, and Data Interpretation Protocol," EPA-600/9-78-018, Environmental Protection Agency, Corvallis, OR, 1978.
[5] Rodhe, W., "Algae in Culture and Nature," *Mitteilungen des Internationalen Vereins der Limnolnologie,* in press.
[6] Qureshi, A. A., Coleman, R. N., and Paran, J. H., "Evaluation and Refinement of the Microtox Test for Use in Toxicity Screening," in *Toxicity Screening Procedures using Bacterial Systems,* B. J. Butka and D. Liu, Eds., *Toxicology Series Vol. 1,* Marcel Dekker, Inc., New York, 1983, pp. 89–118.

Michael J. Machuzak[1] and Thomas K. Mikel, Jr.[1]

Drilling Fluid Bioassays Using Pacific Ocean Mysid Shrimp, Acanthomysis sculpta, a Preliminary Introduction

REFERENCE: Machuzak, M. J. and Mikel, T. K., Jr., "**Drilling Fluid Bioassays Using Pacific Ocean Mysid Shrimp, *Acanthomysis sculpta*, a Preliminary Introduction,**" *Aquatic Toxicology and Hazard Assessment: 10th Volume, ASTM STP 971,* W. J. Adams, G. A. Chapman, and W. G. Landis, Eds., American Society for Testing and Materials, Philadelphia, 1988, pp. 447–453.

ABSTRACT: The toxicity evaluation of drilling fluids using Gulf Coast mysid shrimps has become well developed during the past several years and is used by the Region II EPA to evaluate discharges from offshore platforms in the mid-Atlantic. Development of a method to utilize native mysid shrimps for Pacific Ocean platforms, however, is still in its infancy. A new preliminary acute toxicity method has been developed using *Acanthomysis sculpta*, a mysid shrimp commonly found associated with central California kelp forests. The new method is similar to mid-Atlantic procedures but has been modified to accommodate the life-style of the Pacific Coast species and to be in compliance with the specifications of EPA, Region IX. A slurry is prepared by combining one part of drilling fluid with nine parts of filtered natural seawater. This solution is mixed on a stir plate while being vigorously aerated for 30 min. The slurry is then allowed to settle for 1 h, and the supernatant is decanted and used directly in the toxicity test. Six 1-L concentrations of the extract solution (usually 100, 56, 32, 18, 10, and 0%) to seawater are prepared in triplicate and placed in a cooling bath maintained at 15.0 ± 1.0°C. Concentrations are aerated as needed to maintain 65% of oxygen saturation, and then 20 previously collected and similar-sized (6 to 8 mm) mysid shrimp are randomly placed in each container. The test continues for 96 h, after which the live organisms are counted and the LC_{50} interpolated from averaged percent mortality.

KEYWORDS: *Acanthomysis sculpta*, mysid shrimp, bioassay, acute toxicity, 96-h LC_{50}, drilling fluid, aquatic toxicity

In recent years, several standard methods have been published utilizing mysid shrimps for static acute toxicity bioassays [1,2]. Consequently, the U.S. Environmental Protection Agency has found mysids to be useful for testing the toxicity of drilling platform discharges in the Atlantic Ocean [3]. EPA Region IX, requiring a drilling discharge bioassay for Pacific Coast platforms, adopted the Atlantic Coast method several years ago. In 1984, the American Petroleum Institute published a very similar mysid shrimp/drilling fluid bioassay [4], and an official procedure was finally published in the Federal Register in 1985 [5].

Although generally effective, a criticism of the method might be that it lacks relevance to Pacific Coast ecosystems. The organism specified is *Mysidopsis bahia*, a Gulf Coast mysid that is commonly found in southern United States estuaries [2] but does not occur on the West Coast. Also, specified test temperatures (20°C) would seldom be encountered in the open Pacific Ocean. With the assistance of the EPA Region IX staff [6], we modified the accepted procedures to

---

[1] Chief biologist and laboratory director, respectively, CRL Environmental, a division of Chemical Research Laboratories, Ventura, CA 93001.

accommodate the colder-living *Acanthomysis sculpta,* a mysid shrimp commonly found among central California kelp beds.

Materials and Methods

Test Organism Collection and Handling

We found no procedures for culturing *Acanthomysis sculpta,* so wild populations were collected in the field. Our populations were obtained from kelp beds adjacent to several of the Central Coast Channel Islands, although these animals are also found along mainland coastal beds. The mysids can be collected easily with a bucket dipped just below the kelp bed canopy. During transit, the holding waters are aerated and kept cool (15 ± 2°C) with external ice. Upon return to the laboratory, the mysids are gently introduced into a partially filled 200 gal aquarium containing precooled and aerated natural seawater. Seawater for both animal acclimation and test dilution are collected at least one mile offshore and away from river and harbor mouths or known ocean outfalls. In order to reduce potential for disease, all water is filtered through 0.45-μm membrane filters before use [5]. Mysids are acclimated for a period of at least 48 h in water gently aerated and maintained at 15 ± 1°C. Mysids are daily fed approximately 100 brine shrimp nauplii each [7]. During the acclimation period, natural death should not exceed 5%, and all dead organisms are removed immediately. As a general procedure, all glassware, collection containers, aquaria, etc. are thoroughly washed, rinsed, and acid-soaked prior to use [8,9].

Sample Collection and Preparation

Drilling fluids are collected from active field systems. Care is taken to exclude as much air as possible from the sampling containers (five-gallon polypropylene buckets with snap-on lids work well), and the samples are kept on ice during transit. Holding times in the laboratory greater than two weeks are acceptable as long as the sampling containers are not opened and samples are maintained at 4°C [5]. Drilling fluids which are discolored, have a foul odor, or have a pH value less than 9.0 have likely been degraded by bacteria and are discarded [5].

The testing slurry is prepared from the combination of one part drilling fluid to nine parts of filtered seawater (see above). The slurry is rapidly stirred with a magnetic stir plate and vigorously aerated for 30 min. The slurry pH is adjusted (if necessary) to ± 0.2 pH units of the filtered seawater with hydrochloric acid and is allowed to settle for 1 h. The supernatant is decanted (not siphoned) and used directly in the acute toxicity test. Storage of the supernatant is not permitted [5].

Acute Toxicity Testing

The supernatant is diluted with filtered seawater to obtain desired toxicity test concentrations [most commonly: 100, 56, 32, 18, 10, and 0% (control)] [1]. One litre of each concentration is poured into three replicate 1-gal aquaria (total of 18 test containers), placed in a cooling bath, and gently aerated via a small aquarium pump. Aeration is supplied at a rate of approximately 50 to 140 cm^3/min [5]. Once the dissolved oxygen concentrations have reached 65% of saturation, pH, dissolved oxygen, temperature, and salinity are measured and recorded. Twenty similarly sized mysids (6 to 8 mm long) are then randomly allocated to each of the 1-gal aquaria [5].

Toxicity tests continue for a period of 96 h. During that time, mysids are fed daily to prevent cannabalism (see above), and all initially recorded chemical and physical variables and survival are measured at 4, 8, 24, 48, 72, and 96 h. A continuous monitor graphically records all temperature fluctuations during the course of the test. All data are reported, and the survival counts after 96 h are used to calculate final results. If 10% average mortality or greater occurs in any control replicate (0%); the bioassay is repeated [8]. A summary of the recommended test conditions are included in Table 1.

LC_{50}s are calculated from a semilog graphic interpolation between the averaged results from the two successive concentrations surrounding 50% survival [8]. When more than two concentrations provide other than 0 and 100% survival, the Probit Method [7] can be used to generate a more accurate estimate of LC_{50}. The final report includes all water quality data and survival results throughout the test, semilog interpolation graphs, continuous temperature recording charts, and all dates, times, and descriptions of samples, seawater, organisms, etc. [8,9]. An example of water quality test conditions from one test is included as Table 2.

Results and Discussion

Since 1984, we have conducted 18 acute toxicity tests with *Acanthomysis sculpta* using suspended-phase extracts from various different drilling fluids (Table 3). In all cases, control survivals were high (85 to 100%), mortality trends were linear with percent concentration of extract, and within-concentration replicates were usually very similar. Thus, *A. sculpta* are hearty enough to survive well under laboratory conditions but sensitive enough to respond to a wide range of drilling fluid toxicities.

This paper is presented as a first-time introduction of *Acanthomysis sculpta* as a West Coast analogue to the extensively utilized Gulf Coast mysid, *Mysidopsis bahia*. In order to make East and West Coast methods comparable, a considerable amount of future testing is anticipated. First, the life cycle and culturing methods must be established so that homogeneous populations can be tested. Perhaps some modifications of Nimmos's methods [2] would be appropriate. Second, it

TABLE 1—*Recommended test conditions for Mysids* Acanthomysis sculpta.

Temperature	15 ± 1°C
Light quality	Ambient laboratory illumination
Light intensity	50 to 100 ft cd (ambient laboratory levels)
Photoperiod	16-h light/24 h
Size of test vessel	3.8 L (1 gal)
Volume of test solution	1 L
Age of test animals	6- to 8-mm long (exclude gravid females)
No. animals per test vessel	20
No. of replicate test vessels per concentration	3
Total no. organisms per concentration	60
Feeding regime	Two drops of concentrated brine shrimp nauplii suspension twice daily (approximately 100 nauplii/mysid)
Aeration	Gentle, to maintain at least 65% oxygen saturation. Supplied at a rate of approximately 50 to 140 cm^3/min. via small aquarium pump
Dilution water	Filtered (0.45μm) natural seawater, or synthetic salt water adjusted to 33 ppt salinity
Test duration	96-h (static test)
Effect measured	Mortality—no movement of body or appendages on gentle prodding
Endpoint	LC_{50} calculated from semilog graphic interpolation between the average results from the two successive concentrations surrounding 50% survival. Probit Method analysis when conditions are met (see Ref 7)

TABLE 2—Example of water quality data from acute toxicity tests using Acanthomysis sculpta.

Conc.	0 Hours (12/19/85)				4 Hours (12/19/85)				8 Hours (12/19/85)				24 Hours (12/20/86)			
	pH	D.O., mg/L	Temp., °C	Sal., ppt.	pH	D.O., mg/L	Temp., °C	Sal., ppt.	pH	D.O., mg/L	Temp., °C	Sal., ppt.	pH	D.O., mg/L	Temp., °C	Sal., ppt.
100%	8.2	7.4	15.7	30	8.2	7.4	15.7	30	8.2	7.4	15.7	30	8.2	7.4	15.7	30
	8.1	7.3	15.7	30	8.2	7.3	15.7	30	8.1	7.3	15.7	30	8.1	7.3	15.7	30
	8.2	7.3	15.7	31	8.1	7.3	15.7	31	8.1	7.3	15.7	31	8.1	7.3	15.7	31
56%	8.1	7.4	15.7	31	8.1	7.4	15.7	31	8.1	7.4	15.7	31	8.1	7.4	15.7	31
	8.1	7.3	15.7	31	8.1	7.3	15.7	31	8.1	7.3	15.7	31	8.1	7.4	15.7	31
	8.1	7.5	15.7	31	8.1	7.5	15.7	31	8.1	7.5	15.7	31	8.1	7.5	15.7	31
32%	8.0	7.6	15.7	32	8.0	7.6	15.7	32	8.0	7.6	15.7	32	8.0	7.6	15.7	32
	8.0	7.6	15.7	31	8.0	7.6	15.7	31	8.0	7.6	15.7	31	8.0	7.6	15.7	32
	8.0	7.6	15.7	32	8.0	7.6	15.7	32	8.0	7.6	15.7	32	8.0	7.6	15.7	32
18%	7.9	7.4	15.7	33	7.9	7.4	15.7	33	7.9	7.4	15.7	33	7.9	7.5	15.7	33
	7.9	7.6	15.7	33	7.9	7.6	15.7	33	7.9	7.5	15.7	33	7.9	7.6	15.7	33
	7.9	7.5	15.7	33	7.9	7.5	15.7	33	7.9	7.5	15.7	33	7.9	7.6	15.7	33
10%	7.9	7.8	15.7	33	7.9	7.8	15.7	33	7.9	7.8	15.7	33	7.9	7.8	15.7	33
	7.9	7.8	15.7	33	7.9	7.8	15.7	33	7.9	7.8	15.7	33	7.9	7.8	15.7	33
	7.8	7.7	15.7	33	7.8	7.7	15.7	33	7.8	7.7	15.7	33	7.8	7.7	15.7	33
Control	7.8	7.9	15.7	33	7.8	7.9	15.7	33	7.8	7.9	15.7	33	7.8	7.9	15.7	33
	7.8	7.8	15.7	33	7.8	7.8	15.7	33	7.8	7.8	15.7	33	7.8	7.9	15.7	33
	7.8	7.8	15.7	33	7.8	7.8	15.7	33	7.8	7.8	15.7	33	7.8	7.9	15.7	33

Conc.	48 Hours (12/21/85)			72 Hours (12/22/85)			96 Hours (12/23/85)		
	D.O.	Temp	Sal ppt	D.O.	Temp	Sal ppt	D.O.	Temp	Sal ppt
100%	8.2	7.4	15.7	8.2	7.4	15.7	...	...	...
	8.1	7.3	15.7	8.2	7.3	15.7	8.2	7.3	32
	8.1	7.3	15.7	8.1	7.3	15.7	8.1	7.3	32
56%	8.1	7.4	15.7	8.1	7.4	15.7	8.1	7.4	32
	8.1	7.3	15.7	8.2	7.3	15.7	8.2	7.3	33
	8.1	7.5	15.7	8.1	7.5	15.7	8.1	7.5	32
32%	8.0	7.6	15.7	8.0	7.6	15.7	8.0	7.6	33
	8.0	7.6	15.7	8.0	7.6	15.7	8.0	7.6	32
	8.0	7.6	15.7	8.0	7.7	15.7	8.0	7.7	32
18%	7.9	7.5	15.7	8.0	7.5	15.7	8.0	7.5	32
	7.9	7.6	15.9	7.8	7.6	15.7	7.8	7.6	34
	7.9	7.6	15.7	7.8	7.6	15.7	7.8	7.6	34
10%	7.9	7.8	15.7	7.9	7.8	15.7	7.9	7.8	34
	7.8	7.8	15.7	7.9	7.8	15.7	7.9	7.8	34
	7.8	7.8	15.7	7.8	7.8	15.7	7.8	7.8	34
Control	7.8	7.9	15.7	7.8	7.9	15.7	7.8	7.9	33
	7.8	7.9	15.7	7.8	7.9	15.7	7.8	7.9	34
	7.8	7.8	15.7	7.8	7.9	15.7	7.8	7.8	34

NOTE: Conc. = concentration; D.O. = dissolved oxygen; Sal. ppt = salinity, parts per thousand.

TABLE 3—*Percent survival results of acute toxicity tests using Acanthomysis sculpta in suspended-phase extracts of various drilling fluids—20 animals tested per replicate.*

Percent Conc. of Extract	Replicate	Apr. 1984	Apr. 1984	Apr. 1984	Aug. 1984	Dec. 1984	May 1985	Oct. 1985	Oct. 1985	Nov. 1985	Dec. 1985	Feb. 1986	Jul. 1986	Aug. 1986	Oct. 1986	Oct. 1986	Oct. 1986	Oct. 1986	Oct. 1986
100	1	0	0	20	65	0	5	65	65	0	90	0	0	0	0	45	0	0	0
	2	0	0	25	80	0	5	70	65	0	100	0	0	0	0	55	0	0	0
	3	0	0	20	60	0	10	75	45	0	95	0	0	0	0	50	0	0	0
56	1	30	10	30	100	0	25	85	85	0	100	0	0	35	25	60	0	0	5
	2	35	5	30	100	0	35	90	70	10	100	0	0	35	5	60	0	0	5
	3	25	5	30	95	0	20	85	70	0	100	0	0	25	15	60	0	0	0
32	1	60	40	55	100	55	55	95	90	50	100	0	5	65	75	80	0	0	15
	2	60	35	65	100	45	60	85	75	35	100	0	10	75	80	65	0	0	40
	3	70	50	50	100	50	35	95	85	25	100	0	15	65	80	55	0	0	50
18	1	70	70	60	100	70	95	90	90	50	100	65	45	85	80	85	5	65	60
	2	65	70	70	100	80	50	90	85	60	100	60	90	65	80	90	0	60	60
	3	65	75	65	100	85	55	90	90	55	100	45	70	85	85	70	5	50	80
10	1	75	75	80	100	100	85	95	95	75	100	75	90	80	90	90	60	90	70
	2	90	75	80	100	90	90	90	95	70	100	75	65	90	100	65	55	90	85
	3	85	70	85	100	90	85	85	80	70	100	75	70	85	95	70	60	85	75
0	1	95	100	95	100	100	100	95	95	100	100	100	100	90	90	95	95	95	100
	2	90	90	90	100	100	100	95	85	95	100	100	90	100	100	95	95	95	90
	3	90	90	90	100	100	100	95	95	95	100	100	90	100	100	100	100	100	100
LC$_{50}$	=	39.9	27.0	37.0	100	32.0	32.0	100	100	21.0	100	19.4	19.3	42.4	41.0	100	10.8	19.5	24.5

NOTES: Conc. = concentration.

will be necessary to conduct some replicate side-by-side testing with *Mysidopsis bahia* utilizing selected drilling fluids and perhaps standard toxicants such as dodecyl sodium sulfate.

Despite these research needs, we feel confident that *Acanthomysis sculpta* can be used as an effective acute toxicity test organism for Pacific Coast drilling fluid discharges. It appears to be responsive, well adapted to laboratory conditions, and has the advantage of being locally abundant and thus more likely to be potentially impacted by offshore drilling activities.

References

[1] Borthwick P. W., "Methods for Acute Static Toxicity Tests with Mysid Shrimp/*Mysidopsis bahia*," *Bioassay Procedures for the Ocean Disposal Permit Program*, EPA 600/9-78-010, U.S. Environmental Protection Agency, Washington, DC, March 1978.

[2] Nimmo, D. R., Hamaker, T. L., and Somers, C. A., "Culturing the Mysid (*Mysidopsis bahia*) in Flowing Seawater or a Static System," *Bioassay Procedures for the Ocean Disposal Permit Program*, EPA 600/9-78-010, U.S. Environmental Protection Agency, Washington, DC, March 1978.

[3] "Drilling Mud Bioassay Test Procedures," EPA (Region II), U. S. Environmental Protection Agency, Washington, DC, 1980.

[4] *Standard Procedures for Drilling Fluid Bioassays*, American Petroleum Institute, Washington, DC, May 1984.

[5] United States Environmental Protection Agency, "Proposed Regulations for Offshore, Oil and Gas Extraction Point Source Category—Appendix 3—Drilling Fluids Toxicity Tests," *Federal Register*, Vol. 50, No. 165, Washington DC, 26 Aug. 1985.

[6] Bromley, E., Personal communication, U.S. EPA Region IX, San Francisco, CA, 16 Oct. 1985.

[7] "Methods for Measuring the Acute Toxicity of Effluents to Freshwater and Marine Organisms," 3rd ed., Environmental Monitoring and Support Laboratory, U.S. EPA 600/4-85/013, U.S. Environmental Protection Agency, Washington, DC, March 1985.

[8] Kopperdahl, F. R., "Guidelines for Performing Static Acute Toxicity Fish Bioassays in Municipal and Industrial Wastewaters," Fish and Wildlife Water Pollution Control Laboratory, California Department of Fish and Game, Rancho Cordova, CA, 1976.

[9] *Standard Methods for the Examination of Water and Wastewater*, 16th ed., American Public Health Association, Washington, DC, 1985.

Harvey Babich[1] *and Ellen Borenfreund*[1]

In Vitro Cytotoxicity of Polychlorinated Biphenyls (PCBs) and Toluenes to Cultured Bluegill Sunfish BF-2 Cells

REFERENCE: Babich, H. and Borenfreund, E., "**In Vitro Cytotoxicity of Polychlorinated Biphenyls (PCBs) and Toluenes to Cultured Bluegill Sunfish BF-2 Cells,**" *Aquatic Toxicology and Hazard Assessment: 10th Volume, ASTM STP 971,* W. J. Adams, G. A. Chapman, and W. G. Landis, Eds., American Society for Testing and Materials, Philadelphia, 1988, pp. 454–462.

ABSTRACT: An *in vitro* cytotoxicity methodology, initially developed for use with mammalian cells to evaluate the relative acute toxicities of chemical agents, has been adapted for aquatic ecotoxicity studies by using cultured fish cells as the bioindicator system. This methodology, termed the neutral red assay, was applied to evaluating the comparative *in vitro* cytotoxicities of a series of polychlorinated biphenyl mixtures and substituted toluenes to bluegill sunfish BF-2 cells in culture. Although comparatively small differences were noted in the *in vitro* cytotoxicities of the polychlorinated biphenyl mixtures, their differential potency to the BF-2 cells was Aroclor 1248 $\simeq$ 1016 $\simeq$ 1254 $\simeq$ 1242 > 1221 $\simeq$ 1232 > 1260 > 1262. For the toluene series, the sequence of comparative potency was $\alpha,\alpha,2,6$-tetrachlorotoluene > 2,3-dinitrotoluene > 2,4-dichlorotoluene, 6-chloro-3-hydroxytoluene > 2,6-dinitrotoluene, 2,4-dinitrotoluene, *o*-, *m*-, and *p*-chlorotoluene >>> toluene. The *in vitro* cytotoxicities of the chlorinated toluenes, but not of the nitro-containing toluenes, were correlated with their log octanol/water partition coefficients. These data are in accord with published studies on the *in vivo* acute toxicities (LC_{50} assays) to fish of chloro- and nitro-containing toluenes. The greater *in vitro* cytotoxicity of 2,3-dinitrotoluene than of the two other dinitrotoluene analogs was also in agreement with published *in vivo* acute toxicity studies with freshwater fish. This present study, in conjunction with previous *in vitro* cytotoxicity assays with BF-2 cells, has demonstrated the potential usefulness of the neutral red assay for predicting the *in vivo* response of fish to acute exposures of chemicals and for establishing structure-activity relationships among groups of related chemicals.

KEYWORDS: bluegill sunfish BF-2 cells, neutral red cytotoxicity assay, PCBs, toluenes, aquatic toxicology, *in vitro* alternatives

In response to the need for alternative approaches in toxicity screenings of chemicals that use live animals as the test systems, several laboratories, both in the United States and in Europe [1,2], have been concerned with developing *in vitro* cytotoxicity assays. Of the various approaches and techniques being evaluated, the neutral red *in vitro* cytotoxicity assay [3] has been used to analyze a large series of chemical samples and has been shown to correlate rather well with the *in vivo* response to acute exposure to chemical toxicants. Developed and standardized initially with mammalian cells, this assay has been used to evaluate the comparative cytotoxicities of surfactants [4], inorganic metal salts [5], and a spectrum of chemicals, supplied as coded samples by the Fund for the Replacement of Animals in Medical Experiments [6]. Recently, the neutral

---

[1] Senior research associate and adj. associate professor, respectively, Laboratory Animal Research Center, The Rockefeller University, New York, NY 10021.

red assay has been adapted for use with cultured fish cells for evaluating the ecotoxicity to the aquatic biota of chemical pollutants [7–9] (Table 1).

This paper further evaluates the neutral red *in vitro* cytoxicity assay using cultured bluegill sunfish (*Lepomis macrochirus*) cells to ascertain the relative cytotoxicities of a series of polychlorinated biphenyl (PCB) mixtures and toluene compounds. This assay is based upon the incorporation of the supravital dye neutral red into lysosomes of viable, uninjured cells after incubation of the cell culture with toxic chemicals. This weakly-cationic dye penetrates membranes by nonionic diffusion and binds intracellularly to carboxylic and/or phosphate groups of the lysosomal matrix. Changes in the accumulation and retention of neutral red occur when cell membranes (plasma, as well as the very sensitive lysosomal membranes) are injured or damaged. Dead cells cannot retain neutral red. The amount of neutral red, after extraction from the lysosomes, is quantitated spectrophotometrically and compared with the level of dye recovered from nontreated control cultures. Quantification of extracted neutral red by spectrophotometric analysis has been correlated with cell numbers, both by direct counting and by protein determinations of cell numbers.

Materials and Methods

Cells

The established cell lines, BF-2 (bluegill sunfish fry), RTG-2 [rainbow trout (*Salmo gairdneri*) gonads], BALB/c mouse 3T3 (embryo), and N_2A (mouse neuroblastoma), and low-passage primary cells, SHE (Syrian hamster embryo), and RNB (rat neonatal brain) were maintained and grown in Dulbecco's modified Eagle medium (DMEM), supplemented with 10% fetal calf serum, 100 units/mL penicillin G, 100 µg/mL streptomycin, and 1.25 µg/mL Fungizone. All cells were grown and maintained in a 5.5% carbon dioxide (CO_2), humidified atmosphere, with the RTG-2 cells at 20°C, the BF-2 cells at 26°C, and the mammalian cells at 37°C.

Solutions and Test Agents

Cells were trypsinized with a solution consisting of 0.8-g sodium chloride (NaCl), 0.058-g sodium bicarbonate ($NaHCO_3$), 0.04-g potassium chloride (KCl), 0.1-g glucose, 0.05-g trypsin, and 0.02-g versene in 100 mL of deionized, distilled water. Aroclors 1016, 1221, 1232, 1242, 1248, 1254, 1260, and 1262, $\alpha,\alpha,2,6$-tetrachlorotoluene, 2,4-dichlorotoluene, *o*-, *m*-, and *p*-chlorotoluene, 6-chloro-3-hydroxytoluene, 2,4-dinitrotoluene, 2,6-dinitrotoluene (U.S. Environmental Protection Agency Repository for Toxic and Hazardous Substances), 2,3-dinitrotoluene (Aldrich), and toluene (Fisher) were dissolved in methanol. In comparative studies on the cytotoxicity of Aroclor 1254 to the fish and mammalian cells, the Aroclor (Analab) was dissolved in dimethylsulfoxide (DMSO); initial studies showed that the cytotoxicity of Aroclor 1254 was equivalent whether solubilized in methanol or in DMSO, providing that the solvent concentration was below 2 and 0.5%, respectively.

Neutral Red Assay

For studies on the response of the BF-2 cells to a 24-h exposure to toxicants, individual wells of a 96-well tissue culture microtiter plate were inoculated with 0.2 mL of complete medium containing 3.2×10^4 cells; for the study on the response of the BF-2 cells to a 7-day exposure of Aroclor 1254, individual wells were inoculated with 1.5×10^3 cells. Thereafter, in either the 24-h or 7-day exposure studies, the plate was incubated for 24-h, after which the medium was

TABLE 1—*Midpoint cytotoxicity values ($NR_{50}s$) of organic and inorganic chemical pollutants towards bluegill BF-2 cells.*[a]

Compound	NR_{50}, mmol/L	Compound	NR_{50}, mmol/L
Chlorinated Pesticide		**Organometal**	
chlordane	0.082	methylmercury (CH_3HgCl)	0.004
heptachlor	0.126	diethyltin [$(C_2H_5)_2SnCl_2$]	0.07
aldrin	0.192	dimethyltin [$(CH_3)_2SnCl_2$]	0.37
4,4'-DDD	0.199		
4,4'-DDT	0.234	**Cationic Inorganic Metal**	
4,4'-DDE	0.240	silver ($AgNO_3$)	0.006
endrin	0.756[b]	mercury ($HgCl_2$)	0.02
		cadmium ($CdCl_2$)	0.08
Polycyclic Aromatic Hydrocarbon		zinc ($ZnCl_2$)	0.19
acenaphthylene	0.243	copper ($CuCl_2$)	0.55
3-methylcholanthrene	0.344[b]	cobalt ($CoCl_2$)	0.95
benzo(a)pyrene	0.365[b]	nickel ($NiCl_2$)	2.0
		lead [$Pb(NO_3)_2$]	2.7
Phenolic		tin ($SnCl_2$)	4.7
pentachlorophenol	0.17	manganese ($MnCl_2$)	6.2[b]
2,3,5,6-tetrachlorophenol	0.27	chromium ($CrCl_3$)	27.8[b]
2,3,5-trichlorophenol	0.33		
2,4,6-trichlorophenol	0.48	**Anionic Inorganic Metal**	
2,4-dichlorophenol	0.51	arsenite ($NaAsO_2$)	0.05
4-chlorophenol	1.6	dichromate ($K_2Cr_2O_7$)	0.08
2-chlorophenol	2.4	chromate (K_2CrO_4)	0.18
2,4-dimethylphenol	2.4	arsenate (Na_2HAsO_4)	0.59
2,4-dinitrophenol	4.1	selenite (Na_2SeO_3)	2.2
4-nitrophenol	4.3	permanganate ($KMnO_4$)	5.3
3-methylphenol	5.3	selenate (Na_2SeO_4)	11.2[b]
phenol	6.9		

[a] BF-2 cells were exposed to the toxicants for 24 h at 26°C; NR_{50} values from Refs 7 and 9 and from unpublished data.
[b] Extrapolated values.

removed and the cells were refed with medium unamended (control) or amended with varied concentrations of toxicant. After another 24 h, or 7 days, of exposure, the medium was removed and replaced with 0.2 mL of complete medium containing 50 μg/mL neutral red (NR). The NR-containing medium had been preincubated overnight at 37°C and was centrifuged prior to use to remove fine precipitates of dye crystals. The plate was then returned to the incubator for another 3 h to allow for the uptake of the vital dye into the lysosomes of viable, uninjured cells. Thereafter, the medium was removed and the cells washed quickly with a fixative [1% formaldehyde–1% calcium chloride ($CaCl_2$)], and then 0.2 mL of a mixture of 1% acetic acid–50% ethanol was added to each well to extract the dye. After an additional 10 min at room temperature and rapid agitation on a microtiter plate shaker, the plates were transferred to a microplate reader equipped with a 540-nm filter to measure absorbance of the extracted dye [3].

All experiments were performed at least three times, and the relative cytotoxicities of the various test agents were compared to control cultures by computing by linear regression analysis the concentration of toxicant needed to reduce absorbance by 50% (that is, NR_{50} value). For graphic representations, the toxicity curves were constructed with individual data points presented as the arithmetic mean ± standard error of the mean.

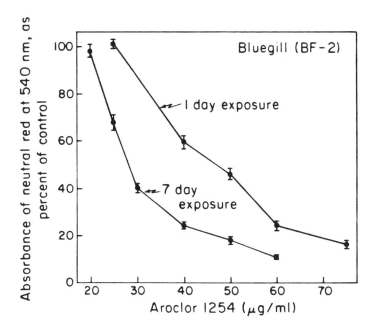

FIG. 1—*Comparative* in vitro *cytotoxicity of Aroclor 1254 to bluegill sunfish BF-2 cells after 24-h and 7-day exposures. The individual data points are expressed as the arithmetic mean ± the standard error of the mean.*

Results

Polychlorinated Biphenyls

A representative dose-response cytotoxicity curve is illustrated in Fig. 1, which shows the response of bluegill BF-2 cells to Aroclor 1254. Although the neutral red assay has been used primarily for 24-h exposures to chemical toxicants, by initially seeding fewer cells, this assay can be adapted for longer exposure times. As seen in Fig. 1, the 24-h and 7-day midpoint cytotoxicity values (that is, NR_{50}) for exposure to Aroclor 1254 were 45.4 and 29.4 µg/mL, respectively.

The BF-2 cell line was selected as the representative cell line as in previous studies with inorganic metals [8]; this cell line was shown to be more sensitive than the RTG-2 cell line, which is derived from rainbow trout and which has been used in cytotoxicity assessments of aquatic toxicants [10]. Figure 2 shows the comparative 24-h cytotoxicity of Aroclor 1254 to the BF-2 and RTG-2 cell lines, as well as to the N_2A mouse neuroblastoma and BALB/c mouse 3T3 cell lines and to low-passage primary cells derived from Syrian hamster embryo (SHE) and neonatal rat brain (NRB) tissues. The NR_{50} values for a 24-h exposure to Aroclor 1254 are 33.7, 54.1, 54.3, 70.0, and 94.9 µg/mL for the N_2A, NRB, BALB/c 3T3, SHE, and RTG-2 cells, respectively.

The relative 24-h midpoint cytotoxicity values for the eight different Aroclor mixtures towards the bluegill BF-2 cells are presented in Table 2. The Aroclor mixtures exhibiting the greater cytotoxicity were those with a chlorine content (by weight) of between 41 and 54% (that is, Aroclors 1016, 1242, 1248, 1254), with lesser cytotoxicity being noted for those Aroclor mixtures with a chlorine content either ≤32% (that is, Aroclor 1221, 1232) or ≥60% (that is, Aroclor 1260, 1262).

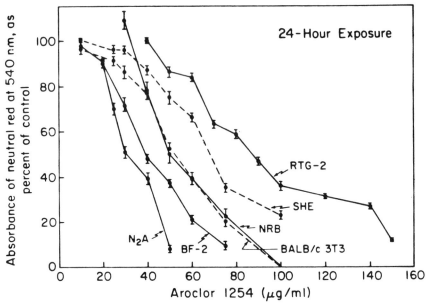

FIG. 2—*The comparative* in vitro *cytotoxicity of Aroclor 1254 to bluegill BF-2, rainbow trout RTG-2, BALB/c mouse 3T3, and N_2A mouse neuroblastoma cell lines and to low passage primary cells obtained from Syrian hamster embryos (SHE) and neonatal rat brain (NRB) tissues. The seeding concentrations of cells/well were 0.9×10^4 3T3, 1×10^4 RTG-2 or RNB, 1.1×10^4 SHE, 1.5×10^4 N_2A, and 3.2×10^4 BF-2. The exposure was for 24-h and the individual data points are expressed as the arithmetic mean $\pm$ the standard error of the mean.*

Toluene Compounds

The relative 24-h cytotoxicity values for toluene, chlorinated toluenes, and nitro-containing toluenes are presented in Table 3. Toluene was the least toxic, with toxicity increasing by incorporation of chlorine- or nitro-groups into the parent molecule. For the chlorinated toluenes the sequence of cytotoxicity towards the BF-2 cells was $\alpha,\alpha,2,6$-tetrachlorotoluene > 2,4-dichlorotoluene > o-, m-, and p-chlorotoluene. This sequence of cytotoxicity was correlated with their log octanol/water partition coefficients (log P values), that is, increasing the lipophilicity of

TABLE 2—*Midpoint cytotoxicity values ($NR_{50}s$) of different Aroclor mixtures toward BF-2 cells.*

Aroclor	Percent Chlorine, by Weight	NR_{50}, µg/mL
1221	21[a]	58.4
1232	32[a]	59.3
1016	41[b]	43.7
1242	42[a]	46.3
1248	48[a]	43.3
1254	54[a]	45.4
1260	60[a]	62.4
1262	62[a]	72.1

[a] Reference *11*.
[b] Reference *12*.

TABLE 3—*Midpoint cytotoxicity values ($NR_{50}s$) of toluene compounds towards BF-2 cells.*

Test Agent	NR_{50}, mmol/L	Log P
toluene	7.4	2.69[a]
o-chlorotoluene	1.6	3.42[a]
m-chlorotoluene	1.7	3.42[a]
p-chlorotoluene	1.7	3.42[a]
2,4-dichlorotoluene	0.64	3.98[a]
α,α,2,6-tetrachlorotoluene	0.21	4.64[b]
6-chloro-3-hydroxytoluene	0.70	3.10[a]
2,4-dinitrotoluene	1.6	1.98[a]
2,6-dinitrotoluene	1.4	1.98[a]
2,3-dinitrotoluene	0.41	1.98[a]

[a] Reference *13*.
[b] Reference *14*.

the chlorinated molecule increased its cytotoxic potency. The addition of an hydroxy-group to the monochlorinated toluene molecule further enhanced its cytotoxicity, as noted by the greater cytotoxicity of 6-chloro-hydroxytoluene than of *o*-, *m*-, and *p*-chlorotoluene. The cytotoxicity of the nitro-containing toluenes, however, was not correlated with their log P values, as noted by the greater cytotoxicity of 2,3-dinitrotoluene as compared to 2,4- or 2,6-dinitrotoluene, all of which have equivalent log P values.

Discussion

Polychlorinated Biphenyls

The midpoint cytotoxicity values of Aroclor 1254 towards the cells derived from fish and mammalian species, as determined by the neutral red assay, were comparable to midpoint cytotoxicity values obtained by others using different *in vitro* cytotoxicity endpoints and cell types. For example, using Chinese hamster ovary (CHO-KL) cells, Rogers et al. [*15*] noted that 27, 44, and 45 μg/mL Aroclor 1254 caused a 50% decrease in colony formation, cell numbers, and protein determinations of cell populations, respectively. Litterst and Lichenstein [*16*] noted that 63 and 110 μg/mL Aroclor 1254 caused a 50% reduction in *in vitro* growth of HeLa cells (a malignant human cell line) and of human skin fibroblasts, respectively. The lesser sensitivity of the rainbow trout RTG-2 cells to Aroclor 1254 was consistent with other studies, in which RTG-2 cells were more tolerant than BF-2 cells to a 24-h exposure to inorganic metals [*8*] or a 72-h exposure to organic mutagens/carcinogens [*10*]. It should be noted that the population doubling time of the RTG-2 trout cells is by far the slowest of all the cells investigated herein, extending to about 4 to 5 days, which could account for reduced sensitivity.

Using reductions in the lysosomal uptake and retention of neutral red as the cytotoxicity endpoint, the relative 24-h cytotoxicities of the different Aroclor mixtures to the BF-2 cells was 1248 ≃ 1016 ≃ 1254 ≃ 1242 > 1221 ≃ 1232 > 1260 > 1262 (Table 2). Rogers et al. [*15*], using reduction in cell numbers as the parameter of cytotoxicity towards CHO-Kl cells, noted a potency sequence of 1260 > 1016 > 1242 > 1254, whereas when reduction in colony formation was the parameter of cytotoxicity, the sequence was 1254 ≃ 1260 > 1242 ≃ 1016. When considering the suppression of replication of Chinese hamster (CH-461) cells as the cytotoxicity endpoint, Hoopingarner et al. [*17*] noted Aroclor 1016 to be more cytotoxic than Aroclor 1221, 1248, 1254, and 1260 (all of which exhibited approximately equivalent potencies). Apparently, the *in vitro* response of cultured cells to the different Aroclor mixtures was dependent not only on the specific cell type (Fig. 2) and exposure time (Fig. 2), but also on the cytotoxicity endpoint being considered [for example, *15*]. However, in the studies reported herein with the BF-2 cells (Table 2), as well as those with mammalian cells [*15*], the comparative *in vitro* cytotoxicities of the various Aroclor mixtures did not, in general, markedly differ from each other. When formulating the Water Quality Criteria to protect freshwater aquatic life from adverse exposures to PCBs, the U.S. Environmental Protection Agency treated all the different Aroclor mixtures as a single entity and set the criterion at 0.014 µg/L as a 24-h average [*12*].

Toluenes

There was a strong correlation ($r = -0.99$) between the log midpoint cytotoxicity values (log NR_{50}s) for toluene, *o*-, *m*-, and *p*-chlorotoluene, 2,4-dichlorotoluene, and $\alpha,\alpha,2,6$-tetrachlorotoluene and their log P values. The negative value of the correlation coefficient reflects the inverse relationship between log P and log NR_{50}, that is, as the octanol/water partition coefficient was increased, midpoint cytotoxicity occurred at a lower concentration of toxicant. 6-Chloro-3-hydroxytoluene exhibited greater cytotoxicity than the other monochlorinated toluenes, although its lipophilicity is less (that is, a log P value of 3.10 compared to 3.42 for *o*-, *m*-, and *p*-chlorotoluene). Apparently, the incorporation of the hydroxy-group into the monochlorotoluene structure enhanced the overall cytotoxicity of the test agent. Inclusion of 6-chloro-3-hydroxytoluene into the log P versus log NR_{50} correlation reduced the correlation coefficient to -0.87.

The *in vitro* cytotoxicities of the three dinitrotoluenes were, apparently, not related to their log octanol/water partition coefficients, as emphasized by their equivalent log P values but differing NR_{50} values. The greater *in vitro* cytotoxicity of 2,3-dinitrotoluene than of 2,4- and 2,6-dinitrotoluene was also evident in published studies of the acute *in vivo* toxicities of these compounds to freshwater fish and invertebrates [*18*]. There are only limited data on the acute *in vivo* toxicities of toluenes to bluegill sunfish. However, the greater 24-h cytotoxicity of 2,3-dinitrotoluene (NR_{50} = 73.8 mg/L) than of toluene (573 mg/L) that was noted with the cultured BF-2 cells was also evident *in vivo*. Thus, for bluegill sunfish the 96-h LC_{50}, in mg/L, was 0.33 for 2,3-dinitrotoluene and 13.00 for toluene [*13*].

Related Studies

In the studies reported herein, the lack of sufficient acute *in vivo* toxicity (LC_{50}) data for bluegills exposed either to PCBs [*12*] or to toluenes [*18*] prevented correlations between the *in vitro* response of the BF-2 cells with the *in vivo* acute response of bluegills to similar toxicants. However, in previous studies with BF-2 cells and with cytotoxicity being determined by the neutral red assay, good correlations were noted between the *in vitro* cytotoxicities of cationic metals ($r = 0.83$) and of phenolics ($r = 0.98$) with published data of their 96-h LC_{50} toxicity to bluegill sunfish [*7,9*]. Furthermore, Bols et al. [*20*], using inhibition of cell attachment to a substratum as the cytotoxicity endpoint, correlated the *in vitro* cytotoxicity of phenolics to RTG-2 trout cells with their published 96-h LC_{50} acute toxicity to rainbow trout. *In vitro* cytotoxicity assays with cultured fish cells may prove to be a meaningful tool for initial screenings of chemicals for their potential acute *in vivo* toxicity to fish.

In vitro cytotoxicity assays may also be useful for elucidating structure-activity relationships among groups of chemicals. The correlation between the log P values of the chlorinated toluenes with their *in vitro* cytotoxicity potencies demonstrated in this study with the BF-2 cells, is in agreement with similar *in vivo* studies with guppies (*Poecilia reticulata*) [*21*]. Previous *in vitro* cytotoxicity studies with BF-2 [*9*] and RTG-2 [*20*] cells noted a strong correlation between the cytotoxicity of phenolic compounds and their log P values; again, these findings were in accord with *in vivo* acute toxicity studies with guppies [22,23]. Babich et al. [7] correlated the *in vitro* cytotoxicities to BF-2 cells of a series of divalent cationic metals with their chemical softness parameters (σ_p); a similar correlation was noted in *in vivo* acute toxicity (LD$_{50}$) studies with mice [24].

General Conclusion

The data presented here, as well as that reported elsewhere [7,9,25,26], demonstrate the potential application of the neutral red cytotoxicity assay with cultured fish cells, both with respect to its usefulness in predicting the *in vivo* response of fish to acute exposures to chemical toxicants and for establishing structure-activity relationships among groups of chemicals. The volume of chemicals already in commerce but without sufficient ecotoxicity data, coupled with the relatively higher costs of *in vivo* assays for preliminary screenings, strongly suggests the consideration of *in vitro* assays for initial screening of chemical toxicities.

Acknowledgments

This research was supported, in part, by Revlon, Inc. The authors thank Robert L. Lipnick, Office of Toxic Substances, U.S. Environmental Protection Agency, Washington, DC for providing log P calculations obtained using the CLOGP3.3 computer program and to James A. Puerner for his technical support.

References

[1] Borenfreund, E. and Shopsis, C., "Toxicity Monitored With a Correlated Set of Cell-Culture Assays," *Xenobiotica*, Vol. 15, 1985, pp. 705–711.
[2] *In Vitro Toxicology: A Progress Report from the Johns Hopkins Center for Alternatives to Animal Testing. Alternative Methods in Toxicology*, A. M. Goldberg, Ed., Vol. 3, Liebert, Inc., New York, 1986.
[3] Borenfreund, E. and Puerner, J. A., "A Simple Quantitative Procedure Using Monolayer Cultures for Cytotoxicity Assays (HTD/NR-90)," *Journal of Tissue Culture Methods*, Vol. 9, 1984, pp. 7–9.
[4] Borenfreund, E. and Puerner, J. A., "Toxicity Determined *in Vitro* by Morphological Alterations and Neutral Red Absorption," *Toxicology Letters*, Vol. 24, 1985, pp. 119–124.
[5] Borenfreund, E. and Puerner, J. A., "Cytotoxicity of Metals, Metal-Metal and Metal-Chelator Combinations Assayed *in Vitro*," *Toxicology*, Vol. 39, 1986, pp. 121–134.
[6] Stark, D. M., Shopsis, C., Borenfreund, E., and Babich, H., "Progress and Problems in Evaluating and Validating Alternative Assays in Toxicology," *Food and Chemical Toxicology*, Vol. 24, 1986, pp. 449–455.
[7] Babich, H., Puerner, J. A. and Borenfreund, E., "*In Vitro* Cytotoxicity of Metals to Bluegill (BF-2) Cells," *Archives of Environmental Contamination and Toxicology*, Vol. 15, 1986, pp. 31–37.
[8] Babich, H., Shopsis, C., and Borenfreund, E., "*In Vitro* Cytotoxicity Testing of Aquatic Pollutants (Cadmium, Copper, Zinc, Nickel) Using Established Fish Cell Lines," *Ecotoxicology and Environmental Safety*, Vol. 11, 1986, pp. 91–99.
[9] Babich, H. and Borenfreund, E., "*In Vitro* Cytotoxicity of Organic Pollutants to Bluegill Sunfish (BF-2) Cells," *Environmental Research*, Vol. 42, 1987, pp. 229–237.
[10] Kocan, R. M., Landolt, M. L. and Sabo, K. M., "*In Vitro* Toxicity of Eight Mutagens/Carcinogens for Three Fish Cell Lines," *Bulletin of Environmental Contamination and Toxicology*, Vol. 23, 1979, pp. 269–274.
[11] Safe, S. "Polychlorinated Biphenyls (PCBs) and Polybrominated Biphenyls (PBBs): Biochemistry, Toxicology, and Mechanism of Action," *CRC Critical Reviews in Toxicology*, Vol. 13, 1984, pp. 319–395.
[12] *Ambient Water Quality Criteria for Polychlorinated Biphenyls*, U.S. Environmental Protection Agency, Washington, DC, 1980.
[13] Hansch, C. and Leo, A., *Substituent Constants for Correlation Analysis in Chemistry and Biology*, Wiley, New York, 1979.

[14] Leo, A. and Weininger, D., *Medchem Software Release 3.3, Medicinal Chemistry Project,* Pomona College, Claremont, CA, 1985.
[15] Rogers, C. G., Heroux-Metcalf, C., and Iverson, F., "*In Vitro* Cytotoxicity of Polychlorinated Biphenyls (Aroclors 1016, 1242, 1254, and 1260) and Their Effect on Phospholipid and Neutral Lipid Composition of Chinese Hamster Ovary (CHO-Kl) Cells," *Toxicology,* Vol. 26, 1983, pp. 113–124.
[16] Litterst, C. L. and Lichenstein, E. P., "Effects and Interactions of Environmental Chemicals on Human Cells in Tissue Culture," *Archives of Environmental Health,* Vol. 22, 1971, pp. 454–459.
[17] Hoopingarner, R., Samuel, A., and Krause, D., "Polychlorinated Biphenyl Interactions with Tissue Culture Cells," *Environmental Health Perspectives,* Vol. 1, 1972, pp. 155–158.
[18] *Ambient Water Quality Criteria for Dinitrotoluene,* U.S. Environmental Protection Agency, Washington, DC, 1980.
[19] Buccafusco, R. J., Ells, S. J., and LeBlanc, G. A., "Acute Toxicity of Priority Pollutants to Bluegill (*Lepomis macrochirus*)," *Bulletin of Environmental Contamination and Toxicology,* Vol. 26, 1981, pp. 446–452.
[20] Bols, N. C., Boliska, S. A., Dixon, D. G., Hodson, P.V., and Kaiser, K. L. E., "The Use of Fish Cell Cultures as an Indication of Contaminant Toxicity to Fish," *Aquatic Toxicology,* Vol. 6, 1985, pp. 147–155.
[21] Konemann, H., "Quantitative Structure-Activity Relationships in Fish Toxicity Studies. Part 1: Relationships for 50 Industrial Pollutants," *Toxicology,* Vol. 19, 1981, pp. 209–221.
[22] Konemann, H. and Musch, A., "Quantitative Structure-Activity Relationships in Fish, Part 2: The Influence of pH on the QSAR of Chlorophenols," *Toxicology,* Vol. 19, 1981, pp. 223–228.
[23] Lipnick, R. L., Bickings, C. H., Johnson, D. E., and Eastmond, D. A., "Comparison of QSAR Predictions with Fish Toxicity Screening Data for 110 Phenols," in *Aquatic Toxicology and Hazard Assessment, ASTM STP 891,* American Society for Testing and Materials, Philadelphia, 1985, pp. 153—176.
[24] Williams, M. W., Hoeschele, J. D., Turner, J. E., Jacobson, K. B., Christie, N. T., Paton, C. L., Smith, L. H., Witschi, H. R. and Lee, E. H., "Chemical Softness and Acute Metal Toxicity in Mice and *Drosophila,*" *Toxicology and Applied Pharmacology,* Vol. 63, 1982, pp. 461–469.
[25] Babich, H. and Borenfreund, E., "Structure-Activity Relationship (SAR) Models Established *in vitro* with the Neutral Red Cytotoxicity Assay," *Toxicology In Vitro,* Vol. 1, 1987, pp. 3–9.
[26] Babich, H. and Borenfreund, E., "Cultured Fish Cells for the Ecotoxicity Testing of Aquatic Pollutants," *Toxicity Assessment,* Vol. 2, 1987, pp. 119–133.

Norman O. Crossland[1]

A Method for Evaluating Effects of Toxic Chemicals on Fish Growth Rates

REFERENCE: Crossland, N. O., "**A Method for Evaluating Effects of Toxic Chemicals on Fish Growth Rates**," *Aquatic Toxicology and Hazard Assessment: 10th Volume, ASTM STP 971*, W. J. Adams, G. A. Chapman, and W. G. Landis, Eds., American Society for Testing and Materials, Philadelphia, 1988, pp. 463–467.

ABSTRACT: Provided that environmental factors and the food supply are not limiting, the growth rate of fish is more or less constant from the fry stage until the onset of maturity. It is independent of the initial size of the fish and the duration of the period between measurements. This paper describes a method which uses the growth rate of juvenile rainbow trout as a sensitive response for evaluating chronic toxicity.

Groups of 16 fish per 40-L aquarium were anesthetized, weighed, measured, and individually marked using freeze brands. Fish were fed a ration of pelleted food equivalent to 4% of wet body weight per day, given in two feeds per day. They were exposed under flow-through conditions to 3,4-dichloroaniline (DCA) at concentrations of 0, 19, 39, 71, 120, and 210 µg/L for a period of 20 days. Effects on length, weight, and growth rates were determined after 14 and 28 days exposure to DCA.

After 14 days there were no significant effects ($P > 0.05$) on length or weight, but there was a highly significant depression of growth rate ($P < 0.01$) among fish exposed to the highest concentration of DCA (210 µg/L). After 28 days there were statistically significant effects on length and growth rate at all concentrations between 39 and 210 µg/L, and there was a clear relationship between concentration and response. Growth rate was affected to a greater extent than length, and overall it proved to be a much more sensitive response than either weight or length.

KEYWORDS: toxicity, growth rate, rainbow trout, subchronic, 3,4-dichloroaniline

Fish convert their food into body tissue much more efficiently than warm-blooded vertebrates. The dietary energy requirement of fish is less than that of mammals because they do not have to maintain a constant body temperature and because they expend less energy in locomotion. There is also less energy expenditure in protein metabolism since they excrete nitrogenous waste as ammonia instead of urea. Provided that environmental factors and the food supply are not limiting, the growth rate of fish is more or less constant from the fry stage until the onset of maturity. Thus fish growth can be defined in terms of a rate that remains relatively constant for a substantial part of the life cycle.

It is well established that the growth rate of various organisms is affected by sublethal concentrations of toxic chemicals. The growth rate of fish is therefore potentially useful as a reliable and sensitive indicator of toxicological effects. This may not be generally recognized because standard procedures used in chronic and subchronic fish toxicity tests are insufficiently sensitive to measure effects on growth rates. For example, in widely used early life stage (ELS) tests, fish are reared in batches from embryo to the juvenile stage, typically 28 days after hatching. Under these conditions, individuals that are first to hatch grow more quickly than others, leading

---

[1] Principal scientist, Shell Research, Ltd., Sittingbourne, Kent, ME9 8AG, United Kingdom.

to high variability among the weights of individual fish at 28 days and consequently to lack of sensitivity in detecting differences in growth rates between batches. By comparison, in the method described here, relatively high sensitivity of the growth response is achieved by marking individual fish, grading them to reduce variability in starting weights, and using a diet and environmental conditions to ensure a relatively fast and constant rate of growth.

In preliminary work, described elsewhere [1], experiments were carried out to investigate the effects of water flow rate, ration level, and frequency of feeding on the growth rates of juvenile rainbow trout, *Salmo gairdneri*, Richardson. Growth rates were reported in terms of the relative growth rate (RGR), expressed as the increment of growth relative to the initial size, in units of mg/g. RGR is dependent on the initial size of the fish and on the duration of the period between measurements. Alternatively, since growth is exponential, the growth rates, r, may be defined as

$$r = \frac{\log_e W_{t_2} - \log_e W_{t_1}}{t_2 - t_1} \times 100 \qquad (1)$$

where W_{t_1} and W_{t_2} are the weights at times t_1 and t_2, respectively. Expressed in this way, in per cent per day, r is independent of the initial size of the fish and the duration of the period between measurements. In the light of these obvious advantages for toxicological work, r has been preferred to RGR in the present paper.

Materials and Methods

Juvenile rainbow trout were obtained from Parkwood Trout Farm, Harrietsham, Kent, United Kingdom. Each batch of fish was obtained from a single spawning, and the fish in each batch were reared together from hatching to the juvenile stage. This procedure was adopted to minimize aggressive behavior between individuals, which tends to occur if fish that are reared in separate batches are subsequently mixed. Fish obtained from the farm were graded and acclimatized to conditions in the laboratory for two weeks before the start of the experiment. During this period they were fed a maintenance ration of 1% of their body weight per day.

During the experimental period the fish were fed a daily ration equivalent to 4% of their mean initial wet body weight. After 14 days the fish were reweighed and the ration level was increased to 4% of mean wet body weight on Day 14. The daily ration was divided into two equal portions and given in two feeds per day.

Throughout the acclimatization and experimental periods the fish were fed with British Petroleum Mainstream salmon fry diet, particle size 2.0 mm (range 1.4 to 2.4 mm), recommended by the manufacturers for feeding juvenile salmonids in the range of 2.6 to 15 g. According to manufacturer's specifications, this diet contains three different types of marine fish meal, mainly of Scandinavian origin. Other minor ingredients are blood meal, fish oil, soya oil, dried yeast, wheat flour, vitamins, and mineral supplements.

Fish were held in six 40-L glass aquaria, the sides of which were covered by black polythene to minimize disturbance. Filtered, dechlorinated tap water was supplied continuously to each aquarium via a reservoir, where the water level was maintained at a constant height with the aid of a solenoid switch. The flow rate of water from the reservoir to each aquarium was maintained constant at 200 L/day using a series of six diaphragm metering pumps which incorporated control values for regulation of flow rates. The water in each aquarium was aerated, and the rate of air flow was maintained at 2 L/min. Water temperatures were measured at intervals of 4 h using a thermistor coupled to a data logger. The pH and dissolved oxygen were measured daily, and the total hardness was measured weekly.

Fish were individually marked using a freeze branding technique. Branding irons were made from stainless steel wire, bent into the shape of a letter or arabic numeral, and attached to a wooden handle. Prior to branding, fish were lightly anesthetized in 0.1 g/L tricaine methanesulfonate (MS222). They were then placed on a moist pad of tissue, and excess moisture was removed from the area to be branded. A brand, cooled by immersion in liquid nitrogen (temperature $-196°C$), was pressed onto the fish for a period of 3 s. Soon after branding, the mark was almost invisible but gradually appeared during the next two days and was then clearly visible for the duration of the experiment. With a little experience it was possible to mark the fish very rapidly without causing them any apparent harm and with less than 1% mortality associated with handling and marking.

Before starting the experiment, a batch of about 200 juvenile fish were starved for 24 h. Then they were carefully graded to reduce the variability between individuals to a minimum. Fish weighing 4.5 to 6.0 g were used for the experiment. About half of the original batch of 200 were outside of this weight range and were rejected. Groups of 16 fish, mean weights varying from 4.9 to 5.4 g [Standard Deviation (SD) 0.39 to 0.46] were placed in each aquarium. After anesthetizing and marking them, the fish were weighed on a top pan balance to an accuracy of ± 0.1 g, and their length, from the snout to the fork of the tail, was measured to the nearest millimeter. Fourteen days later the fish were again starved for 24 h and then reweighed and remeasured. The fish in each aquarium were examined daily, mortalities were recorded, and any dead fish were removed and replaced with individuals of a similar weight to maintain a constant density of fish in the aquaria. However, any replacements were excluded from subsequent analysis of the data.

A reference compound, 3,4-dichloroaniline (DCA), was used to assess the sensitivity of the method. Nominal concentrations of DCA were 0, 30, 55, 100, 170, and 300 μg/L, and these were maintained in the aquaria for 28 days after introducing the fish. A primary stock solution of 17 g/L of DCA in analytical grade acetone was prepared and stored in the dark at 4°C. Secondary stock solutions were prepared every two or three days using the primary stock solution and acetone. The secondary stock solutions were delivered to the aquaria from syringes through stainless steel tubes at a rate of 66 μL/h using a continuous infusion syringe pump. DCA solutions and tap water were mixed together at a junction of toxicant and water flows situated immediately above each aquarium. Dosing of the aquaria was started 48 h before introducing the fish to ensure that equilibrium conditions existed at the start of the exposure period. The concentrations of DCA in the aquaria were determined by reverse-phase, high-performance liquid chromatography (HPLC) at intervals of three to four days throughout the experiment. On each sampling occasion, 20-mL water samples were drawn from the center of each aquarium. Samples were passed through disposable cartridges packed with a reversed phase (C18) absorbent, which were first prepared by washing with 20 mL of acetonitrile and 20 mL of distilled water. The DCA retained by the cartridges was eluted with acetonitrile and diluted with distilled water.

Growth rates were calculated using Eq 1. Data for weights, lengths, and growth rates were analyzed using a one-way analysis of variance, and the treatment means were compared with the controls using Dunnett's test.

Results

The range of values obtained for water quality parameters monitored in this experiment was as follows: temperature 13.5 to 16.8°C; pH 7.4 to 8.0; dissolved oxygen 7.4 to 10.1 mg/L; total hardness 240 to 270 mg/L calcium carbonate ($CaCO_3$).

There were no treatment-related mortalities of fish. Only one of the 96 fish died, the others surviving the full 28-day period of exposure. Concentrations of DCA, as determined by HPLC, were approximately 30% lower (28 to 35%) than the nominal values. The data are summarized in Table 1.

After 14 days exposure to DCA there were no significant effects ($P > 0.05$) on lengths or

TABLE 1—*Effects of DCA on lengths, weights, and growth rates of rainbow trout after 14 days exposure.*

Concentration of DCA, µg/L (±S.D.)	<1.0	19 (±3.2)	39 (±4.7)	71 (±11.7)	120 (±24.6)	210 (±30.5)
n	16	15	16	16	16	16
Mean lengths, cm (±S.D.)	8.6 (0.33)	8.8 (0.38)	8.6 (0.51)	8.5 (0.36)	8.3 (0.36)	8.4 (0.27)
Mean wet weights, g (±S.D.)	8.0 (0.81)	8.4 (1.13)	7.9 (1.49)	7.8 (0.86)	7.7 (1.01)	7.4 (0.68)
Growth rates, % per day (±S.D.)	3.2 (0.81)	3.2 (0.83)	3.0 (1.3)	3.2 (0.28)	3.2 (0.72)	2.4[a] (0.53)

NOTE: n = number of fish.
[a] Significantly different from the control at the 1% level.

weights at any concentration. However, at 210 µg/L the growth rate was depressed by 25%, and this effect was significant at the 1% level.

After 28 days exposure to DCA, wet weights of fish exposed to the two highest concentrations (120 and 210 µg/L) were significantly less ($P < 0.01$) than the control group. Lengths of fish exposed to concentrations in the range 39 to 210 µg/L were significantly less ($P < 0.05$ or 0.01) than the control group. There was no effect on growth rate in the 19 µg/L treatment, but growth rate was significantly less than in the control ($P < 0.05$ or 0.01) at DCA concentrations of 39, 71, 120, and 210 µg/L. Furthermore, there was a clear relationship between dose and response, that is, the higher the concentration of DCA, the lower the growth rate of fish. Expressed as a percentage of the control group, growth rates were reduced by 16, 19, 52, and 61% in the 39, 71, 120, and 210 µg/L treatments, respectively.

Discussion

Flow rates of 2 to 3 L/g of fish per day have been recommended for carrying out continuous flow toxicity tests [2]. In work reported elsewhere [1] it was found that there was no significant effect ($P > 0.05$) on growth rate for flow rates in the range 0.5 to 3 L/g of fish per day. Ideally,

TABLE 2—*Effects of DCA on lengths, weights, and growth rates of rainbow trout after 28 days exposure.*

Concentration of DCA, µg/L (±S.D.)	<1.0	19 (±3.2)	39 (±4.7)	71 (±11.7)	120 (±24.6)	210 (±30.5)
n	16	15	16	16	16	16
Mean lengths, cm (±S.D.)	9.9 (0.59)	9.9 (0.59)	9.5[a] (0.70)	9.4[a] (0.47)	8.9[b] (0.57)	8.8[b] (0.42)
Mean wet weights, g (±S.D.)	12.4 (2.18)	12.6 (2.34)	11.4 (2.68)	11.0 (1.84)	9.5[b] (1.88)	8.7[b] (1.19)
Growth rates, % per day (±S.D.)	3.1 (0.96)	2.9 (0.78)	2.6[a] (0.79)	1.5[b] (0.77)	1.5[b] (0.74)	1.2[b] (0.61)

[a] Significantly different from the control at the 5% level.
[b] Significantly different from the control at the 1% level.

fish should weigh 4 to 5 g at the start of an experiment. At this weight fish are big enough to brand without difficulty but small enough to be accommodated at a density of 16 fish in a 40-L aquarium. After 28 days, assuming average growth rates reported here, the average weight of fish in a control aquarium will be 10 to 12 g. Assuming a constant flow of water of 200 L per day to each aquarium, the flow rate per gram of fish per day will vary from approximately 3 L on Day 0 to 1 L on Day 28.

On the basis of results reported here, this method is sufficiently sensitive to detect a 10% depression of growth rate, with a statistical probability of 95%. In order to achieve an acceptable degree of precision, the density of fish should not be less than about 16 per treatment, the flow rate of water should be maintained within the limits of 1 to 3 L per gram of fish per day, and the ration level should be equivalent to 4% of wet body weight per day, given in two feeds per day.

Cannibalism within experimental groups of fish can be a problem. It has been reported [3] that 4-g rainbow trout were far too aggressive for use in growth experiments. However, in the work described here it was found that cannibalism can be avoided by careful grading of fish to obtain a narrow range of sizes at the start of an experiment and by ensuring that fish are raised as a single batch from egg to juvenile stage.

From the results of this experiment with DCA, the maximum acceptable toxicant concentration (MATC) to juvenile rainbow trout lies between 19 and 39 μg/L. This estimate is in good agreement with other estimates of MATC values for DCA. In laboratory and field tests with *Daphnia* spp. [4], the MATC was estimated to be 10 to 20 μg/L. In 32-day ELS tests with the fathead minnow, *Pimephales promelas* Rafinesque, it was found that the MATC was 5.1 to 7.1 μg/L [5].

In previously reported work with methyl parathion [6], significant effects ($P < 0.01$) on growth of rainbow trout were observed at 400 μg/L but not at 200 μg/L. This result was in good agreement with the results of an ELS test [7] in which it was shown that methyl parathion did not affect the growth or mortality of sensitive life stages of *P. promelas* at concentrations less than 310 μg/L.

References

[1] Crossland, N. O., *Chemosphere*, Vol. 14, No. 11/12, 1985, pp. 1855–1870.
[2] Alabaster, J. S. and Lloyd, R., *Water Quality Criteria for Freshwater Fish*, Butterworths, London, 1980.
[3] Rowe, D. W., Sprague, J. B., Heming, T. A., and Brown, I. T., *Aquatic Toxicology*, Vol. 3, 1983, pp. 161–169.
[4] Crossland, N. O. and Hillaby, J. M., *Environmental Toxicology and Chemistry*, Vol. 4, 1985, pp. 489–499.
[5] Call, D. J., Poirier, S. H., Knuth, M. L., Harting, S. L., and Lindberg, C. A., *Bulletin of Environmental Contamination and Toxicology*, Vol. 38, 1987, pp. 352–358.
[6] Crossland, N. O., *Ecotoxicology and Environmental Safety*, Vol. 8, 1984, pp. 482–495.
[7] Jarvinen, A. W. and Tanner, D. K., *Environmental Pollution Series A*, Vol. 27, 1982, pp. 179–195.

Mark V. Haley,[1] *Dennis W. Johnson,*[1] *William T. Muse, Jr.,*[1] *and Wayne G. Landis*[1]

The Aquatic Toxicity and Fate of Brass Dust

REFERENCE: Haley, M. V., Johnson, D. W., Muse, W. T., Jr., and Landis, W. G., "**The Aquatic Toxicity and Fate of Brass Dust,**" *Aquatic Toxicology and Hazard Assessment: 10th Volume, ASTM STP 971,* W. J. Adams, G. A. Chapman, and W. G. Landis, Eds., American Society for Testing and Materials, Philadelphia, 1988, pp. 468–479.

ABSTRACT: A series of environmental bioassays and fate studies have been performed on brass dust. The standardized aquatic microcosm (SAM) was used as the long term, ecosystem assay.

Short-term bioassays using *Daphna magna* and green algae were conducted. The 48-h EC_{50} of the brass dust to *D. magna* ranged from 0.016 to 0.026 mg/L. Microscopic investigations indicated that the daphnia filter and ingest the material. Green algae were not as affected by the brass dust as the daphnids. The 96-h IC_{50} for *Ankistrodesmus falcatus* is 0.316 mg/L and for *Selenastrum capricornutum*, 0.056 mg/L. At very low concentrations of the brass, algal growth was stimulated.

Fate studies indicate that brass disassociation in water varies with pH. At low pHs the material rapidly disassociates into Cu^{++} and Zn^{++} ions. At pHs above neutrality the disassociation is markedly slower.

The short-term bioassays and fate studies were used to set the concentrations of brass for the SAM (the SAM is standardized protocol containing algae, invertebrates, bacteria, and a protozoan inoculated in 3 L of medium). A study using $CuSO_4$ served as a positive control for the brass microcosm. In both microcosms the concentration of available copper slowly decreased until an equilibrium was reached. In addition, both microcosms demonstrated a dose response relationship with an increase in algal biomass correlated with an increase in toxicant dose. The cause is likely to be the elimination of the primary grazers, daphnia, because of the test chemical's toxicity. Recovery as measured by positive population growth of the daphids also demonstrated a dose-response relationship.

KEYWORDS: aquatic toxicology, brass dust, copper, zinc, *Daphnia magna,* algal bioassays, standard aquatic microcosm, metal toxicity, trophic interactions

Extensive literature exists on the aquatic toxicity of trace metals [1]. Traditionally the toxicant is delivered to the test system as a salt and rarely in combination with other metals. Often overlooked is the fact that many metals are introduced into the environment as alloys such as brass, pewter, solder, and stainless steel. The constitutents of the alloys may leach into the environment at different rates and interactions between the metals may alter the toxic response of the biological community.

Several recent papers have examined the interactions among metals on the toxic effects expressed by an organism. Prevot and Soyer-Gobillard [2] have reported recently on the combined action of selenium and cadmium to dinoflagellates. In the dinoflagellates studied, *Prorocentium micans* and *Crypthecodinium cohnii*, the interaction was antagonistic if both toxicants were present simultaneously in low concentrations. Surprisingly, a synergistic interaction occurs if only one of the elements is present in a high concentration. Mixtures of cadmium, copper, mercury, and zinc delivered to cyclopoid copepods demonstrated a toxicity higher than any individual component

---

[1] Environmental Toxicology Branch, Toxicology Division, Chemical Research Development and Engineering Center, Aberdeen Proving Ground, MD 21010-5423.

[*3*]. Another important finding from this study was that populations from areas of previous trace metal contamination demonstrated an increase in resistance. Cultures of *Selenastrum capricornutum* have also been demonstrated to adapt to higher concentrations of trace metals [*4*]. Whether this increase in resistance to metal intoxication is due to physiological adjustments or natural selection remains to be determined.

This article reports the toxicity of an alloy, brass, on aquatic communities. Single species daphnid and algal assays were conducted as well as standard aquatic microcosm (SAM) [*5*] studies using $CuSO_4$ and brass dust as toxicants. The SAM was chosen as the multispecies assay since it is a defined, generic system that has been demonstrated to mimic larger ecosystems in many ways [*6,7*].

In addition to toxicity studies, we followed the fate of copper sulfate and brass dust in the two separate microcosms. Both substances contain copper as the predominate element of interest. Although necessary as an essential nutrient, low concentrations (0.1 to 1.0 mg/L) of copper can cause toxic effects to many species of algae and other lower aquatic organisms [*6*]. Hence, the disassociation of copper from both test compounds is an important toxicological consideration. In these experiments, soluble copper is referred to as ionized copper and includes Cu^{++}, $CuCO_3$, $CuOH$, and $Cu(CO_3)_2$ that can pass through a 0.45-μm filter. Measurement of this form of copper is a better approximation of its potential toxicity to aquatic organisms than the measurement of total copper [*8*].

In this report we concentrate on the evaluation of the acute toxicity data as a predictor of the impacts seen on the daphnid and algal populations. The results of the acute daphnid and algae assays have recently been reported and are included for completeness [*9,10*]. Evaluation of the brass SAM experiments in regards to species diversity, dynamics, energy flow, and physical characteristics is to be incorporated into subsequent publications.

Methods and Materials

The composition and physical specifications of the brass particles are detailed in Table 1. Because both particulate and dissolved metal was of concern, a suspension of brass was made for distribution to sample vessels. The brass was weighed on a Cahn-28 electrobalance and dispensed into disposable polycarbonate tubes. A suitable diluent, depending upon the assay to be performed, was added to make a 1 mg/mL suspension. The sealed tube was placed in an ultrasonic water bath and manipulated until all the particles were suspended as uniformly as possible. Appropriate dilutions were made from this stock suspension. Dissolved $CuSO_4$ was added in appropriate amounts in the microcosm assays.

TABLE 1—*Properties of brass dust* [*11*].

Average diameter		1.72 μm
Thickness range		320 to 800 nm
Composition:	Cu	68.5%
	Zn	27.5%
	Al	0.20%
	Pb	0.10%
	Sb	0.10%
	Palmetic acid	0.52%
	Steric acid	0.74%

NOTE: Purchased as Richgold Superfine from Atlantic Powdered Corp., New York, NY.

Bioassays

Short-term static assays using the algae *S. capricornutum* and *Ankistrodesmus falcatus* and the cladoceran *Daphnia magna* have been described previously [9,10]. The assays conformed to applicable ASTM and EPA standards. In the daphnid assays, water hardness ranged between 50 to 65 ppm $CaCO_3$ and pH varied from 6.8 to 7.0. A 16:8 light/dark cycle was maintained with a temperature of 20 ± 1°C. Conditions in the algal assays were similar, pH ranging from 6.6 to 6.9, hardness of 62.0 ± 5 ppm total $CaCO_3$, with a temperature of 25.0 ± 0.5°C. Algae were incubated with a light/dark cycle of 16:8 at 339/lm/m². (Ethylenedinitrilo) tetraacetic acid, a powerful chelator, was deleted from the culture media.

The 64-day SAM protocol followed has been described (Fig. 1) [5,6]. The microcosms were prepared by the introduction of ten algal, four invertebrate, and one bacterial species (Table 2) into 3L of sterile defined medium. Test containers were 4-L glass jars. An autoclaved sediment consisting of 200-g silica sand and 0.5-g ground chitin was added after the separate sterilizations of the medium and vessel. The separate sterilizations of the sediment, medium, and vessel help prevent cracking of the glass vessel during routine handling. Algae were added weekly (500 cells/microcosm). Invertebrates were added if less than three of a particular species were observed per jar.

Numbers of organisms, dissolved oxygen, and pH were determined twice weekly. Nutrients (nitrate, nitrite, ammonia, and phosphate) were sampled and measured weekly for the first four weeks, then once weekly thereafter. Room temperature was 20 ± 2° and illumination 79.2 $\mu Em^{-2} s^{-1}$ with a range of 78.6 to 80.4.

Copper sulfate was added at calculated concentrations of 0.5, 1.0, and 2.0 mg/L. Brass dust concentrations were determined by the results of the short-term bioassays so that a no-effect

FIG. 1—*Timeline for the standard aquatic microcosm as developed by Taub* [5,6].

TABLE 2—*Species composition of the SAM.*

ALGAE

Anabaena cylindrica
Ankistrodesmus
Chlanydomonas reinhard
Chlorella vulgarius
Lynghya sp.
Nitzchia kutzigiana
Scenedesmus obliquus
Selenastrum capricornutum
Stigeoclonium sp.
Ulothrix sp.

INVERTEBRATES

Daphnia magna
Cyprinotus sp.
Philodina sp.
Hypotrichus ciliate, species unknown

BACTERIA

Klebsiella pnuemoniae

---

concentration and a concentration exhibiting toxic effects would be bracketed (0.01, 0.5, and 1.0 mg/L).

Chemical Fate

Samples for dissolved metal analysis were collected weekly from each microcosm vessel. After stirring the vessel, a 7-mL sample was withdrawn from the upper water layer into a 10-mL plastic B&D syringe. The syringe was then connected to a Swinex-25-mm filter holder and the sample filtered through a prewashed 0.45-μm filter (Millipore) to separate dissolved from particulate metals. The filtrate was then put into a 25-mL polyethylene bottle which contained enough nitric acid (HNO_3) to bring the final sample concentration to 0.05 N HNO_3.

Sample analysis for dissolved metals was conducted on an atomic absorption spectrophotometer. The Perkin Elmer-460 was set at the following parameters, wavelength 324.7 nm, slit width 0.7 mm, air-acetylene plume, and a lamp of copper (HCL) 18 mo. Both samples and standards were matched for HNO_3 concentration. A linear standard curve was produced using both commercially prepared (1000 mg/L stock) and EPA quality assurance standards. Standards were also prepared in the sample media, without the addition of the test compound, to verify that the sample matrix did not affect metal quantitation. The concentration of dissolved copper and zinc was determined by comparing the absorbance of the samples to the standard curve. The analytical sensitivity using this technique was 0.02 mg/L for copper.

Results

Short-Term Bioassays

Table 3 summarizes the acute toxicity test results from organisms exposed to brass dust. The algal species were least affected. The 96-h IC_{50} for *A. falcatus* was 0.316 mg/L and for *S.*

TABLE 3—*Summary of short-term toxicity tests with brass.*

	EC_{50} mg/L	Hours of Exposure
Daphnia magna	0.023	48
	IC_{50} mg/L	Hours of Exposure
Selenastrun capricornutum	0.056	96
Ankistrodesmus falcatus	0.316	96

capricornutum was 0.056 mg/L. Algal growth was stimulated when exposed to brass concentrations below 0.010 mg/L.

The 48-h EC_{50} for *D. magna* was 0.023 mg/L. Microscopic observation demonstrated that the daphnia ingest the brass. The size of the brass particulate is the approximate size of the algae used to culture the daphnia (*A. falcatus, S. capricornutum, C. vulgaris,* and *C. reinhardi*). Titanium dioxide and quartz (similar in size to brass) were examined for toxicity to daphnia (acute tests). These materials were also ingested by daphia but were not toxic up to 1000 mg/L.

SAM Bioassays

The copper sulfate and brass SAMs exhibited good internal repeatability among replicates. These results are also comparable to those performed with $CuSO_4$ by Taub et al. (this volume) as part of the round-robin validation of the SAM protocol. The following discussion will concentrate on the daphnia-algae interaction within these systems.

Table 4 presents the relationship between initial concentration of the toxicant, measured Cu, and the start of daphnid net positive population growth. In both microcosms a Cu concentration from $CuSO_4$ of 0.053 to 0.102 mg/L had a daphnid recovery time of 18 to 21 days (Fig. 2). At higher concentrations of 0.238 from the $CuSO_4$ and 0.163 from the brass mg/L, the daphnid population took 42 and 46 days, respectively, to exhibit net growth (Fig. 3).

The interaction between the algae, daphnia, and toxicant concentration was closely examined. In the controls (Fig. 4) the *D. magna* reached a peak population density within 20 to 25 days followed by a regular decline. During this period total algae stayed low, with minor peaks due to blooms by less available species such as blue-greens. The density of algae at a concentration of 0.01 mg/L brass (Fig. 5a) was not significantly different from controls. As the toxicant dose

TABLE 4—*Recovery of net daphnid population growth in SAM bioassays, concentration in mg/L.*

	Initial Toxicant Concentration	Measured Cu, Day 14	Recovery Time, Days
$CuSO_4$	0.5	0.053	18
	1.0	0.102	18
	2.0	0.238	42
Brass	0.01	BDL[a]	NE[b]
	0.5	0.089	21
	1.0	0.163	46

[a] BDL = Below detectable limits.
[b] NE = No effect.

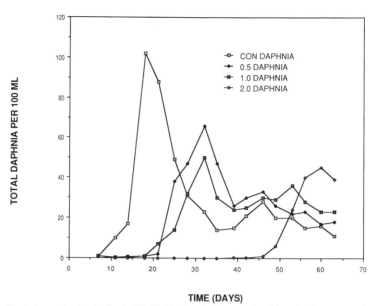

FIG. 2—*Daphnia production in the $CuSO_4$ SAMs. An obvious relationship exists between toxicant dose and the onset of daphnid reproduction, although the 0.5 and 1.0-mg/L concentrations are statistically indistinguishable. At a concentration of 2.0 mg/L, daphnid reproduction was not apparent until 50 days into the experiment.*

increased, the peak algal density also increased, albeit occurring later in the course of the experiment (Fig. 4b, Fig. 5b). As the population of daphnia grew, the algal density sharply declined.

Two differences are readily noted in comparison of the daphnid and algal populations in the copper sulfate and brass microcosms. First, the algal populations were consistently higher in the copper sulfate experiments. Second, the brass was a more potent inhibitor of net growth of the daphnid population than the copper sulfate.

Chemical Fate

The dissolution of copper sulfate into soluble copper occurred fairly quickly. By the end of microcosm Day 14, (seven days post addition of $CuSO_4$), 27% of the initial copper was dissolved. Subsequent weekly analysis of dissolved copper from microcosm Days 21 to 56 showed an average percent dissolution ranging from 12% (high dose), 18% (medium dose), to 24% (low dose). Percent disassociation was determined by comparing the amount of dissolved copper to the total concentration initially introduced from the test compound.

In general, soluble copper decreased from Day 14 to 21 followed by a steady state from Days 21 to 56 (Fig. 6a). The high dose level shows a greater variation in copper levels than either the low or medium doses.

The concentration of dissolved copper (mg/L) disassociated from the brass was monitored from microscosm Day 7 to 63 (Fig. 6b). Analysis directly after the addition of brass showed that it did not disassociate immediately; dissolved copper remained at background levels. By at least Day 14

DAPHNID PRODUCTION BRASS SAM

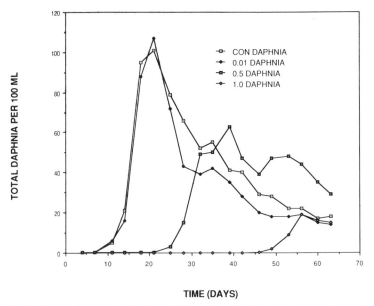

FIG. 3—*Daphnid reproduction in the brass SAMs. Reproduction in the control and the 0.01-mg/L concentration is indistinguishable. A dose-response relationship is apparent in the 0.5 and the 1.0-mg/L replicates.*

(seven days post addition of brass), 23% of the copper had disassociated from the brass at both the medium and high dose levels. The dissolved copper concentration undergoes a slight decrease from Day 21 to 28, then attains equilibrium from Days 35 to 63. During equilibrium, both the medium and high dose levels remained at the same copper concentration. This represented a disassociation of 26% to soluble copper at the medium dose level and 15% at the high dose. Both low dose and control copper levels were below detectable limits, 0.02 mg/L, throughout the study period.

Discussion

An important result from the daphnid acute assays is the fact that the organisms filter the brass particles from the water column, thus providing a route of exposure to a substance not in solution. Daphnia from the SAM was not observed for copper ingestion. However, the disassociation of brass is too slow to account for the toxicity to be explained by soluble copper alone during the early stages of the SAM. It has generally been assumed that once altered to an insoluable precipitate or bound to an organic particle that the metal is removed as a toxicant. Upon resuspension of this material, in nature during the fall or spring turnovers, the material may well be available for ingestion with subsequent toxic effects.

The short-term assays provided a means by which it was possible to successfully select a no-impact, moderate-impact, and severe-impact dosages for the SAM assay. Data crucial to this determination was the fate data estimating release of the Cu and Zn into the water column. Several important interactions would have been difficult to predict using only the short-term data.

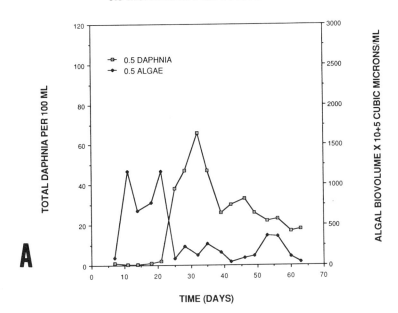

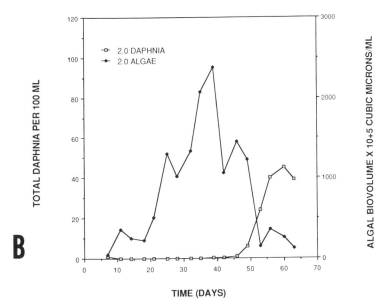

FIG. 4—*Algal and daphnid interactions: CuSO$_4$ SAM. As the concentration of CuSO$_4$ increased so did the maximum biovolume of the algae. As the concentration of toxicant decreased to a level at which the daphnia could reproduce, the algal biovolume quickly decreased.*

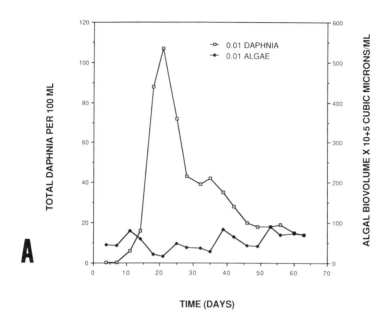

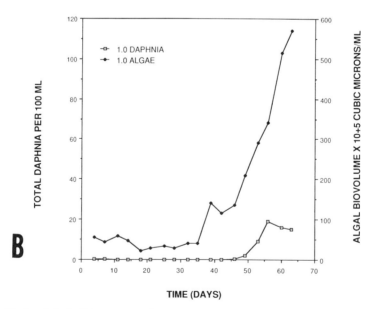

FIG. 5—*Algal and daphnid interactions: brass SAM. As in the $CuSO_4$ SAM, the increase in toxicant concentrations lends to a greater maximum biovolume of algae. However, the maximal algal biovolume never approaches that of the $CuSO_4$ SAM.*

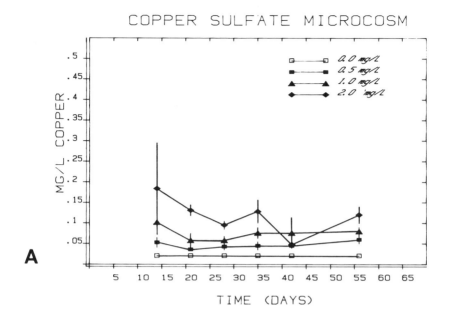

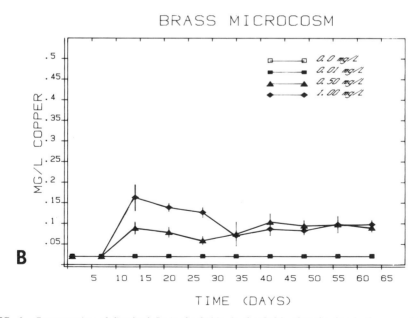

FIG. 6—*Concentration of dissolved Cu in the SAMs. In the CuSO₄ (5A) the dissolved copper decreased until a rough equilibrium was established with treated microcosms exhibiting approximately the same levels of dissolved Cu. In the brass microcosm (5B) the toxicant was released into the water column with an increase until Day 14. After that the highest toxicant level experienced a slow reduction of dissolved Cu until an equilibrium was reached. In both microcosms the concentrations of dissolved Cu reached the same equilibrium.*

Even in the preliminary analysis of the daphnid and algae data, it is obvious that the sensitivity of the daphnia to Cu allowed an increased algal biomass directly proportional to the dose. This effect was obvious in both the copper sulfate and brass SAMs. Algal growth was consistently lower in the brass SAMs. This may be due to the toxic effects of the Zn released in conjunction with the Cu. The algal population rapidly diminished as the daphnid population grew.

During the initial phases of the experiment, the concentration of soluble copper in the microcosm changed due to the chemnical and physical properties of the compound, as well as the physical, chemical, and biological properties of the ecosystem [12]. For instance, copper readily reacts with other chemical species to form insoluble precipitates or soluble oxides, hydroxides, carbonates, and Cu (organic) [13]. The concentration of these species varies with the chemical conditions of the microcosm such as pH, hardness, and nutrient levels. The percentage of the chemical that is not soluboilzed can become associated with sediment and particulate matter (biotic and abiotic), hence they are less available to the biota. Copper levels can also be affected by the algae, which can uptake and excrete copper depending upon their internal cell concentration. Algae also have the ability to release complexing ligands to regulate the concentrations of heavy metals in their environment [14,15]. These and other factors probably were responsible for the levels of dissolved copper present.

Comparison between the copper sulfate and the brass microcosms showed that their percentage of chemical disassociation and dissolved copper equilibrium were very similar. The copper sulfate disassociation, at equilibrium, ranged from 16% (high dose—0.8 mg/L Cu) to 25% (low dose—0.2 mg/L Cu) while the brass disassociation ranged from 15% (high dose—0.68 mg/L Cu) to 29% (medium dose—0.34 mg/L Cu). Both microcosms reached and maintained an equilibrium for dissolved copper. This began approximately midway through each study and continued on to the end. At this point the chemical and biological transformation processes had reached a steady state such that the formation of dissolved copper species balanced their rate of conversion to particulate forms. Bioconcentration mechanisms could also account for the equilibrum. Since the initial amount of available copper was high, it could gradually accumulate in the biota. Consequently, as the concentration of copper in the biota became larger, the net rate of removal from the media declined and eventually reached zero.

An important aspect of this study including both short-term assays and the SAM is the utility of both assays coupled with fate data used in concert. In the short-term assays it soon was apparent that toxicity of brass to the daphnids had to be due to a mechanism other than ionization of the material followed by absorption since fate studies demonstrated very little release of Cu over a 48-h period. Subsequent examination demonstrated the ingest of the particles. On the other hand the interaction between the daphnia and algae leading to an increase in algal biomass with an increase in toxicant is a result difficult to predict from short-term acute data. Additionally, the slow reduction of Cu in the column followed by an equilibrium in both the copper sulfate and brass SAMs indicates that fate processes not available in single species short-term assays are available in multispecies assays.

The importance of understanding interactions among populations is crucial in estimating the long-term impact of the introduction of an xenobiotic. In the algal population of the microcosm a reversal of the expected dose-response relationship was observed. This reversal was due to the reduction of the daphnid populations at higher concentrations of toxicant. A toxicant can affect many levels of an ecosystem, from the molecular biology of the DNA molecule to the spatial and temporal variability of resources [16]. The severest shortcoming of single species assays is that other than toxicity directed against the test organism, all other levels of community interaction are bypassed. At our present level of understanding the dynamics of ecosystems, we are unable to accurately predict multidimensional interactions from acute studies alone.

Finally, metal toxicity is a classical subject of aquatic toxicology. Extensive studies have been conducted, yet major questions as to availability, mode of exposure, synergistic and antagonistic interactions, and cellular mechanisms of toxicity remain. Although trace metals have been consistently demonstrated to impact aquatic systems these questions still must be answered both to aid in the recovery of damaged ecosystems and as a model for studies of other toxic pollutants.

References

[1] Leland, H. V. and Kuwabara, J. S., in *Fundamentals of Aquatic Toxicology: Methods and Applications*, G. M. Rand and S. R. Petrocelli, Eds., Hemisphere Publishing Corp., Washington, DC, 1985, pp. 374–415.
[2] Prevot, P. and Soyer-Gobillard, M. O., *Journal of Protozoology*, Vol. 3367, pp. 42–47.
[3] Lalande, M. and Pinel-Alloul, B., *Environmental Toxicology and Chemistry*, Vol. 5, 1986, pp. 95–102.
[4] Kuwabara, J. S. and Leland, H. V., *Environmental Toxicology and Chemistry*, Vol. 5, 1986, pp. 197–203.
[5] Taub, F. B. and Read, P. L., "Standardized Aquatic Microcosm Protocol," Vol. II, Final Report Contract No. 223-80-2352, Food and Drug Administration, Washington, DC, 1983.
[6] Harrass, M. C. and Taub, F. B.," Comparison of Laboratory Microcosms and Field Responses to Copper," in *Validation and Predictability of Laboratory Methods for Assessing the Fate and Effects of Contaminants in Aquatic Ecosystems, ASTM STP 865*, T. P. Boyle, Ed., American Society for Testing and Materials, Philadelphia, 1985, pp. 57–64.
[7] Stay, F. S., Larsen, D. P., Katko, A., and Rohm, C. M., " Effects of Atrazine on Community Level Responses in Taub Microcosms," *Validation and Predictability of Laboratory Methods for Assessing the Fate and Effects of Contaminants in Aquatic Ecosystems, ASTM STP 865*, T. P. Boyle, Ed., American Society for Testing and Materials, Philadelphia, 1985, pp. 75–90.
[8] Sprague, J. B. in *Fundamentals of Aquatic Toxicology: Methods and Applications*, G. M. Rand and S. R. Petrocelli, Eds., Hemisphere Publishing Corp., Washington, DC, 1985, pp. 124–163.
[9] Haley, M. V., Johnson, D. W., Hart, G. S., Muse, W. T., and Landis, W. G., *Journal of Applied Toxicology*, Vol. 6, No. 4, 1986, pp. 281–285.
[10] Johnson, D. W., Haley, M. V., Hart, G. S., Muse, W. T., and Landis, W. G., *Journal of Applied Toxicology*, Vol. 6, No. 3, 1986, pp. 225–228.
[11] Thomson, S. M., Burnett, D. C., Bergmann, J. D., and Hickson, C. J., *Journal of Applied Toxicology*, Vol. 62, No. 3, 1986, pp. 197–209.
[12] Rand, G. M. and Petrocelli, S. R., *Fundamentals of Aquatic Toxicology—Methods and Applications*, Hemisphere Publishing Corp., Washington, DC, 1985, pp. 1–28.
[13] Chakoumakos, C., Russo, R. C., and Thurston, R. V., *Environmental Science and Technology*, Vol. 13, 1979, pp. 213–219.
[14] Martin, D. F., "Coordination Chemistry of the Oceans," in *Equilibrium Concepts in Natural Water Systems*, R. F. Gould, Ed., ACS *Advances in Chemistry* Series, American Chemical Society, Washington, DC.
[15] Van Den Berg, C., Wong, P., and Chau, Y., *Journal of the Fisheries Research Board of Canada*, Vol. 36, 1979, pp. 901–905.
[16] Landis, W. G., "Resource Competition Modeling of the Impacts of Xenobiotics on Biological Communities," *Aquatic Toxicology and Hazard Assessment: Ninth Symposium, ASTM STP 921*, T. M. Poston and R. Purdy, Eds., American Society for Testing and Materials, Philadelphia, 1986, pp. 55–72.

Waste Site Hazard Assessment and Biodegradation

Charles A. Staples[1]

How Clean Is Clean? A Use of Hazard Assessment in Groundwater for Evaluation of an Appropriate Formaldehyde Spill Remedial Action Endpoint

REFERENCE: Staples, C. A., "**How Clean Is Clean? A Use of Hazard Assessment in Groundwater for Evaluation of an Appropriate Formaldehyde Spill Remedial Action Endpoint,**" *Aquatic Toxicology and Hazard Assessment: 10th Volume, ASTM STP 971,* American Society for Testing and Materials, Philadelphia, 1988, pp. 483–490.

ABSTRACT: A spill of formaldehyde from a pipe was discovered in October 1983 at a manufacturing site. Groundwater investigation revealed that much of the formaldehyde was trapped within a clay till layer 8 ft below the surface at a concentration of up to 9% by weight. The formaldehyde was leaching out of the unsaturated zone and moving into the saturated zone. A remedial action plan was begun that involved the oxidation of formaldehyde with injection of hydrogen peroxide coupled with pumping withdrawal and reuse of surrounding contaminated groundwater.

In the development of any hazardous waste site/groundwater remedial program, one issue that must be addressed is "When is the remediation finished?" or "How clean is clean?" In the case of the above formaldehyde spill, the specific question was how much formaldehyde can remain in the clay till such that leaching of the formaldehyde to groundwater with subsequent degradation, dispersion, and transport to a down gradient surface water with dilution would result in toxicologically safe concentrations for both humans and aquatic life? A remedial action endpoint of 100 ppm formaldehyde was predicted to achieve these human and environmental health goals using classical hazard assessment methodologies. These hazard assessment procedures are discussed in terms of their relative contribution and usefulness in the overall remedial action process.

Two-dimensional solute transport groundwater modeling was used to predict concentrations that may discharge into a down gradient river. Hydrogeologic features, including water level gradients, dispersivities, hydraulic conductivities, porosity, and aquifer material density, were measured with a calculated leaching rate and formaldehyde biodegradation rates for estimating down gradient solute transport times and concentrations. Formaldehyde entering surface waters would be diluted by appropriate stream flows to yield instream formaldehyde exposure concentrations. Resultant formaldehyde brook concentrations were combined with the conservative no impact concentration of 40 ppb to yield a margin of safety (MS). The instream formaldehyde concentration yielding an MS < 1.0 was used to back calculate the formaldehyde level that could be left in groundwater for dispersion and biodegradation by natural processes (here, 100 ppm). These modeling calculations were used to guide the appropriate choice of the target remedial action endpoint by using conservative assumptions and parameter values. Model results were considered sufficiently valid by local regulatory personnel to be of use as guidance for setting the remedial action target endpoint while the model predictions are still being fully verified. Final verification by measuring formaldehyde instream concentrations is part of the longer term remediation plan (<3 years).

KEYWORDS: clean, assessment, groundwater, remedial action.

Hazardous waste handling, disposal, spills, and leaks are covered by this nation's various environmental laws. These laws include various Resource, Conservation, and Recovery Act

---

[1] Environmental research specialist, Monsanto Co., Environmental Sciences Center, St. Louis, MO 63167.

(RCRA) laws and amendments and the Comprehensive Environmental Compensation and Liability Act (CERCLA or Superfund) [1–3]. Additionally, some states have similar legislation to enhance federal rules designed to meet local needs and concerns [4]. Rules allowing for the deciding of a spill's imminence of hazard to human health or environment have become part of this overall set of laws and guidelines. When a leak or spill is judged to be an imminent hazard, certain actions are specified. When a leak or spill is judged not to be imminently hazardous, remedial actions can be developed with consideration of cost and different technical approaches.

The best approach to sound decision making concerning the extent of required remediation of hazardous spills or leaks is to use a hazard assessment approach [5,6]. Hazard assessment is a sequential process of collecting and/or predicting exposure concentrations of chemicals and determining whether or not human populations and the environment will be adversely impacted if exposed to those chemicals at known concentrations. Risk assessment includes the additional concept of probability of the exposure occurring at some given concentration. Determination of "safety" is done by mathematically comparing a toxicologically safe concentration with a measured or predicted exposure concentration. The exposure term is multiplied by a calculated uncertainty factor while the "safe" concentration typically has uncertainty built into its determination. This mathematical comparison is called a margin of safety (MS). The overall equation is as follows

$$\text{Margin of Safety (MS)} = [\text{Safe}]/([\text{Exposure}] \times \text{Uncertainty}) \qquad (1)$$

where

Safe = final "safe" concentration,
Exposure = measured or predicted exposure concentration, and
Uncertainty = calculated exposure prediction uncertainty.

The result of these margin of safety determinations is several. Firstly, a calculated MS can be used to evaluate any release of chemical in terms of real concern. Secondly, possible remediation scenarios can be evaluated in terms of effectiveness of hazard mitigation. Thirdly, completion of the remediation can be evaluated in terms of "How clean is clean?" Fourth, the best use of limited resources is realized. Finally, the approach outlined here and elsewhere allows use of the best available data and best available technology to evaluate MSs, remedial activities, and final resolution of environmental problems.

Formaldehyde Release

Monsanto Co. discovered and reported a release of formaldehyde at its manufacturing facility in Massachusetts in October 1983. Seventeen thousand gallons of 50% formaldehyde were released from a failed underground pipeline. Working with the Massachusetts Department of Environmental Quality and Engineering (DEQE), Monsanto began investigating the extent of the formaldehyde distribution in the subsurface.

The leakage occurred near the center of the 162 ha (400-acre) plant facility. The site was about 1500 ft hydraulically up gradient from the region of groundwater discharge into the nearby Chicopee River. Initial soil borings (October 1983) into the affected area near the pipeline showed formaldehyde levels of up to 9% in soil 2 to 10 ft deep. Additional soil borings (December 1983) and monitoring wells confirmed the extent of contamination horizontally. Vertical movement of the formaldehyde had occurred in an upper sand layer, through a clay till, into underlying groundwater (Fig. 1).

A means of removing the formaldehyde to "acceptable" concentrations was needed. The term "acceptable" refers to formaldehyde concentrations that impact neither human nor environmental health at the appropriate points of exposure. A plan was developed by a technical team including

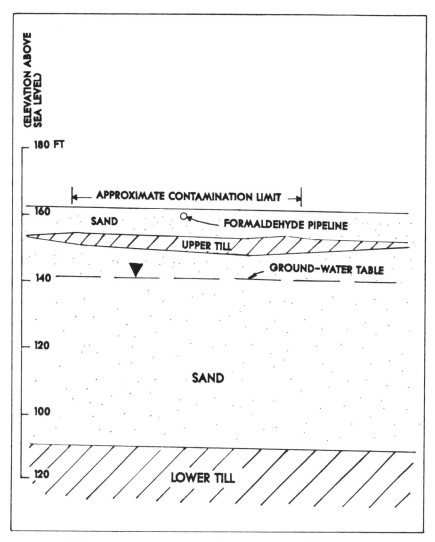

FIG. 1—*Cross section of subsurface in the area of formaldehyde release.*

Monsanto and DEQE personnel and outside consultants. The selected plan involved in-situ treatment of the formaldehyde trapped in the clay till. The in-situ process utilized hydrogen peroxide to oxidize the clay till formaldehyde trapped in pore spaces. Pumping of groundwater by recovery wells beneath the area was carried out to retain excess hydrogen peroxide and any formaldehyde in the groundwater. Due to the low permeability of the clay till (10^{-6} cm/s) thorough flushing was not possible. The clay till could continue to serve as a source of formaldehyde moving to groundwater for a long period of time. A point of remediation completion was required to guide the remediation plan. In other words, what extent of treatment is sufficient for the protection of both human and environmental health?

The approach taken to determine this "how clean is clean" endpoint was to evaluate a safe concentration needed at the point of human contact, the river. The safe concentration was determined to be 40 ppb [7] for human health. The task was to calculate the concentration that could remain in the clay till, move to groundwater, be transported to the river discharge point, be mixed by river flow, and result in the 40 ppb or less.

Modeling Formaldehyde Fate and Transport

Model Description

The model used for this study was a numerical solution to the two-dimensional convective-dispersive mass transport equation (Eq 2) [8].

$$dC/dt + Vx\, dC/dx + Vy\, dC/dY = Dx\, d^2C/dX^2 + Dy\, d^2C/dY^2 - KC \qquad (2)$$

where

C = concentration,
t = time,
V = velocity,
D = dispersion coefficient,
K = sink term, and
x,y = lateral and longitudinal distances.

Boundary conditions are as follows:

Co = initial concentrations,
$C = Coe^{-kt}$, $x = 0$ ($Y^1 \leq Y \leq Y^2$),
$C = 0$, $x = 0$ (all other Y),
$dC/dY \leq 0$, Y approaches $\pm \infty$,
$dC/dx \leq 0$, x approaches $+ \infty$, and
$C = 0$, $t = 0$.

The model assumes a patch source (vertical and horizontal). Aquifer width is considered infinite laterally and semiinfinite longitudinally. Numerical integration was accomplished by a Gauss-Legendre quadrature scheme [8]. Specific data input needs include lateral and longitudinal dispersions and velocities, source decay constant, solute decay constant, and grid layout.

Model Input Parameters

Specific model parameter values are shown in Table 1. Formaldehyde aqueous solubility is in the percent level, so partitioning to soils was considered to be nonexistent. Soil partition coefficients are used to calculate retardation factors. Here, since no partitioning was expected, the retardation factor was calculated to be 1.0.

A critical loss pathway for formaldehyde is biodegradation. Formaldehyde may biodegrade at low concentrations. Although one use of formaldehyde is as a preservative, formaldehyde has been shown to degrade in soil column water [9]. Formaldehyde degradation rate is controlled by several factors. Most studies of biodegradation of formaldehyde used an acclimated "seed" or culture that was conditioned to biodegrade the chemical. Additionally, after "seeding" with acclimated microorganisms, the cultures efficiency decreased in a matter of three weeks so as to require "reseeding." Other evidence of biodegradation [10,11] suggests that formaldehyde may degrade fairly rapidly within sewage treatment systems. These studies also utilized acclimated

TABLE 1—*Specific model input values and pertinent chemical physical property data.*

Parameter	Value
Formaldehyde	
Aqueous solubility	<1%
Soil retardation factor[a]	1.0
Biological decay	0.0036 to 0.36/year
Aquifer	
Permeability	12 to 75 ft/day
Velocity	75 to 450 ft/year
Distance to river	1500 ft
Lateral dispersion	10 ft^2/day
Porosity	0.30
Water gradient	0.005 ft/ft
River	
Flow	1000 ft^3/s

[a] Soil retardation factor (Rf) = water velocity/solute velocity, or $Rf = 1 +$ (soil bulk density/porosity)(partition coefficient). Partition coefficient was essentially zero, so $Rf = 1$.

microbial populations. It can be concluded that formaldehyde will biodegrade in soils but slowly. Degradation can be increased with careful acclimation of the microbial populations [10]. Further considerations are that formaldehyde is toxic to degrading [11] organisms at 130 to 175 ppm. Anaerobic degradation of sewage material can be inhibited at concentrations as low as 100 ppm but frequently between 300 and 1000 ppm [12]. As a conservative value, we will consider the inhibition concentration to be 100 ppm, thus insuring anaerobic and aerobic biodegradation.

A microbial degradation rate of formaldehyde was calculated. A first order kinetic biodegradation rate was used to describe formaldehyde degradation in groundwater. A study of restoration of a hazardous spill site contained formaldehyde degradation rate data in soils of 0.36/year [13]. The starting formaldehyde concentration was less than a few hundred ppm. Another important controlling factor was that microbial populations can control the extent of biodegradation in groundwaters as the populations are sometimes unevenly distributed [14]. To account for population distribution and to be conservative, the degradation rate will be considered here to be between 0.0036 and 0.36/year. The 0.0036/year rate was obtained by arbitrarily dividing the 0.36/year by a factor of 100 for conservatism.

Modeling Assumptions

The two-dimensinal solute transport model used in this study carried several key assumptions. First, use of the model assumed the transport of solute to be two not three dimensional. This was applicable for the extent of aquifer affected. Second, the aquifer was infinite laterally, that is, no recharge boundaries were present. This assumption was valid based on hydrogeology data at this site. Third, homogeneity and isotropy of hydraulic conductivity are assumed. Well log data and pump tests suggested that aquifer materials and their effects on permeability were reasonably uniform across the affected portion of the site. Finally it was assumed that clay till interstitial concentrations were equal to the groundwater concentration. This was a very conservative assumption but was carried through the modeling simulations.

Modeling Results

Formaldehyde concentrations in the river at the point of aquifer discharge were calculated using the ranges of model input values in Table 1. Table 2 shows the calculated concentrations at the source, at river edge, and in the river after dilution for each set of model input data. Maximum biodegradable formaldehyde concentrations were 100 mg/L. The 100 mg/L source concentration resulted in less than the maximum allowable 40 ppb formaldehyde in the river (Table 2). It was considered desirable to have all residual formaldehyde in the groundwater amenable to biodegradation by the aquifer microbial populations to further "cleanse" the aquifer system. Lack of biodegradation of the residual formaldehyde would not change the conclusions.

The river concentrations calculated (assuming decay to occur) were then compared with the "safe" toxicological endpoint of 40 μg/L using Eq 1. The calculated margins of safety (MS) were greater than 1.0 for all runs (333, 57, 9.5 for Runs 1, 2, and 3). To account for modeling uncertainty the upper and lower limits of all available data were used in such a way as to yield best and worst case river formaldehyde concentrations. This minimizes the uncertainty. The resulting MSs were considered acceptable for all runs covering the entire range of aquifer data since the MS needed only to be ≤1.0. The formaldehyde level at the source needs to be less than 100 mg/L for residual formaldehyde to degrade. An acceptable "how clean is clean" endpoint was concluded to be ≤100 mg/L formaldehyde at the source or at least 99.89% removal of the initial clay till formaldehyde concentration.

Examination of measured groundwater formaldehyde concentrations (during remediation) from a series of monitoring wells at the clay till indicates current concentrations to be between 1 and 18 ppm or between 99.98 and 99.999% removal of initial formaldehyde levels which were 9% formaldehyde or 90 000 mg/L. Predicted river concentrations were made using 18 mg/L as the concentration in the groundwater, assuming 100% of groundwater recharge is due to percolation from the clay till. River concentrations would be 0.02, 0.13, 0.76 μg/L with MSs of 1850, 317, 53 for Runs 1, 2, and 3, respectively. Predicted river concentrations were made using 1 mg/L as the actual concentration in the groundwater at the source. River concentrations would be 0.0012,

TABLE 2—*Modeling results for range of input values.*

	Run 1	Run 2	Run 3
PARAMETER			
Lateral dispersion	10 m²/day	10 m²/day	10 m²/day
Velocity	75 ft/year	100 ft/year	450 ft/year
Retardation factor	1.0	1.0	1.0
Total decay	0.36/year	0.036/year	0.0036/year
River flow	1000 ft³/s	1000 ft³/s	1000 ft³/s
FORMALDEHYDE CONCENTRATIONS			
In river[a]	0.12 μg/L	0.70 μg/L	4.20 μg/L
At river edge[b]	17 mg/L	70 mg/L	99 mg/L
At source[b]	100 mg/L	100 mg/L	100 mg/L

[a] Calculated from a real discharge to river of 100-ft-wide plume, 100-ft. half channel width.
[b] Source concentration that results in ≤40 ppb river concentration with biodegradation occurring.

0.007, 0.042 µg/L with MSs of 33300, 5700, 950 for Runs 1, 2, and 3, respectively. Clearly the remediation has already succeeded in reducing formaldehyde concentrations to acceptable levels. Additionally, since the clay till formaldehyde concentrations will be steadily reducing through time given no further input, river concentrations will steadily decrease and corresponding margins of safety increase.

Discussion and Conclusions

A potential problem resulted from the leak of 17 000 gal of formaldehyde from a broken underground pipeline. An investigation revealed the extent of subsurface exposure. Up to 9% formaldehyde was found in a clay till beneath the site and was migrating to the underlying groundwater. Recovery wells near the leakage center recovered several thousand gallons of formaldehyde. The remaining formaldehyde was treated in the clay till by injecting hydrogen peroxide through a series of wells.

A "how clean is clean" endpoint was needed to establish the extent of needed remediation. Solute transport modeling and river dilution calculations were used to back calculate an acceptable source concentration. A "safe" concentration was established to be 40 µg/L. Use of conservative assumptions for modeling yielded acceptable and desirable clay till formaldehyde levels of <100 mg/L or 99.89% removal. This concentration allowed biodegradation of trace quantities of formaldehyde to occur. The current measured clay till concentrations are 1 to 18 mg/L for a 99.98 to 99.999% removal. This would result in a river concentration of between about 0.001 and 1.0 µg/L. Margins of safety would range from 53 to 33 300. These margins of safety are considered sufficient to consider the spill remediated.

Use of the "how clean is clean" endpoint helps to focus available resources (personnel, equipment, time, dollars) on the necessary remediation to a problem. Required remediation of groundwater beyond the plan discussed for the problem presented here would have been more costly and would not have yielded better protection of human or environmental health. This study reports on a technically sound workable solution to a leak of a manufactured material. The approach was innovative, engineering-wise practical, cost effective, and highly successful. The in-situ treatment coupled with natural subsurface attenuation mechanisms yielded a rapidly attainable endpoint that was more than adequately protective of human and environmental health. Personnel involved investigated a potential human health problem, developed and successfully executed a remedial action, and used a technically defensible "how clean is clean" endpoint to that remediation. Clearly this technology is here to stay. These techniques can and should be applied to virtually all types of environmental problems. In all such cases "how clean is clean" endpoints are required to successfully remediate the situation.

References

[1] U.S. Congress, Resource Conservation and Recovery Act, 1976.
[2] U.S. Congress, Amendments to Resource Conservation and Recovery Act, 1984.
[3] U.S. Congress, Comprehensive Environmental Response, Compensation and Liability Act, 1980.
[4] State of Massachusetts, Massachusetts Oil and Hazardous Materials Release Prevention and Response Act, Chapter 21E, 1983.
[5] Staples, C. A. and Kimerle, R. A., 1984. "Integrated Multicompartmental Site Hazard Assessment," Annual Meeting of the Society of Environmental Toxicology and Chemistry, Washington, DC.
[6] Staples, C. A., and Kimerle, R. A., "The Cleanup of Chemical Waste Sites—A Rational Approach," Proceedings of the Society for Risk Analysis Annual Meeting, Washington, DC, in press.
[7] Personal communication, 1984, on acceptable surface water concentration for formaldehyde, Massachusetts Department of Environmental Quality and Engineering, Springfield, MA.
[8] Cleary, R. W. and Ungs, M. J., "Groundwater Pollution and Hydrology—Mathematical Models and Computer Programs," Water Resources Program, Princeton University, Princeton, NJ, 1978.

[9] Wentsel, R. S., Jones, III, W. E., Foutch, R., Wilkenson, M., and Kitchens, J. F., "Accelerated Restoration of Spill Damaged Lands," in *Proceedings of Control of Hazardous Material Spills*, U.S. Environmental Protection Agency, Washington, DC, 1980.

[10] Dickerson, B. W., unknown pages from book but uses this study as the source for cited information: Gellman, I. and Henkelekiwn, H., "Biological Oxidation of Formaldehyde," *Sewage and Industrial Wastes*, Vol. 22, 1950, p. 1321.

[11] Sikes, D. J., McCulloch, M. N., and Blackburn, J. W., "Containment and Mitigation of a Formaldehyde Rail Car Spill Using Novel Chemical and Biological In-Situ Treatment Technique," presented at 1984 Hazardous Material Spills Conference, Nashville, TN, 1984.

[12] Pearson, F., Chang, S. C., and Gautier, M., "Toxic Inhibition of Anaerobic Biodegradation," *Journal of Water Pollution Control Federation*, Vol. 52, No. 3, pp. 472–482.

[13] Wentsel, R. S., Foutch, R. H., Harward, III, W. E., Jones, III, W. E., and Kitchens, J. F., "Restoring Hazardous Spill-Damaged Areas," EPA-600 12-81-208, U.S. Environmental Protection Agency, Cincinnati, OH, 1981.

[14] Wilson, J. T., McNabb, J. F., Balkwill, D. L., and Ghiorse, W. C., "Enumeration and Characterization of Bacteria Indigenous to a Shallow Water-Table Aquifer," *Groundwater*, Vol. 21, No. 2, 1983, pp. 134–142.

Robert M. Desjardins,[1] *Wayne C. Bradbury,*[1] *and Patricia L. Seyfried*[1]

Effects of Metals from Mine Tailings on the Microflora of a Marsh Treatment System

REFERENCE: Desjardins, R. M., Bradbury, W. C., and Seyfried, P. L., "**Effects of Metals from Mine Tailings on the Microflora of a Marsh Treatment System,**" *Aquatic Toxicology and Hazard Assessment: 10th Volume, ASTM STP 971,* W. J. Adams, G. A. Chapman, and W. G. Landis, Eds., American Society for Testing and Materials, Philadelphia, 1988, pp. 491–502.

ABSTRACT: Bacteria isolated from a freshwater marsh treatment system on the site of a mine tailings basin were assessed for heavy metal and antibiotic resistance. All isolates were found to be multiply antibiotic and heavy metal resistant. *Klebsiella* and *Pseudomonas* spp. demonstrated the highest levels of resistance to the antibiotics and metals tested. The fecal coliform group displayed similar resistance patterns, suggesting the presence of a common plasmid. Examination of the *K. oxytoca* marsh system isolate for the presence of plasmid deoxyribonucleic acid (DNA) revealed six plasmids, ranging in size from 3×10^6 to 50×10^6 daltons (Da). The results of this study suggest that the presence of mine tailings in an aquatic environment may promote the development of antibiotic and heavy metal resistance among the microbial flora.

KEYWORDS: bacteria, marsh treatment system, mine tailings, antibiotic and heavy metal resistance, plasmids

The industrial impact of mining has been responsible for exposing metal reserves that would not normally come in contact with the surface environment. The mine tailings, or crushed waste rock, are usually deposited with or in water [1] in order to minimize dust problems due to the fine granular nature of the rock.

Deposition of tailings in the environment has been shown to be hazardous to normal aquatic life [2,3] because tailings from mineral benefaction operations represent a major source of the release of heavy metals [4,5]. These metals dissociate readily in the aqueous environment of biological membranes, facilitating their transport into the cell as metal ions. Once inside the cell, the metal may exert its effect on a multiplicity of target sites, often resulting in cell death [6].

A number of research groups have reported that microorganisms may survive exposure to high levels of heavy metals [3,7–9]; moreover, these heavy metal-resistant bacteria are frequently resistant to one or more antibiotics [10–12]. Several studies have shown that drug and metal resistance is often mediated by the same plasmid [25,30].

It was the objective of this study to assess the comparative heavy metal and antibiotic resistance patterns, as well as the plasmid profiles, of microorganisms exposed to heavy metals from mine tailings. The study site selected was adjacent to the mining town of Cobalt, Ontario, where wastewater from the town is treated in a mine tailings–based marsh treatment system. Bacteria were isolated from the marsh system and identified. The effect of 11 heavy metal ions and 17 antibiotics on the isolates was determined.

---

[1] Graduate research assistant, assistant professor, and professor, respectively, Department of Microbiology, Faculty of Medicine, University of Toronto, Toronto, Canada M5S1A8.

Materials and Methods

Bacterial Strains

Surface water samples were collected from natural and artificial marsh treatment systems situated in the area of a mine tailings basin during November 1983 and June 1984. Appropriate dilutions of each water sample were made and plated (0.1 mL) on casein-peptone-starch (CPS) agar [13] at pH 7.2. Isolates were identified to the species level according to standard methods.

Pseudomonas aeruginosa, Escherichia coli, Enterobacter aerogenes, Klebsiella oxytoca, and *Aeromonas hydrophila* were retained for toxicity testing. These bacteria are frequently isolated from marsh treatment systems and are also common inhabitants of aquatic environments [14].

Heavy Metal Solutions

The following metal salts were used: $Pb(NO_3)_2$, $NiCl_2.6H_2O$, $Na_2HAsO_4.7H_2O$, $AlCl_3.6H_2O$, $FeCl_3$, $CuCl_2.2H_2O$, $CrCl_3.6H_2O$, $CoCl_2.6H_2O$, $ZnSO_4.7H_2O$, $CdCl_2.21/2H_2O$, and $HgCl_2$. Stock solutions of the metals were prepared as follows. Metal salts were added to deionized distilled water and the solutions filtered through a 0.45-μm filter. Appropriate test solutions were prepared on the day of assay by diluting the stock solution in sterile deionized distilled water. Concentrations of metal ions ranged from 2 to 65 mg/L for all metals tested except mercury, which ranged from 0.15 to 0.5 mg/L.

A stock solution mixture was prepared for use in toxicity testing. The mixture contained in mg/L: As 40, Al 20, Fe 35, Zn 40, Ni 45, Cr 40, Co 45, Cd 60, Cu 40, and Hg 7. Dilutions of this mixture were made in sterile distilled deionized water as required.

ATP Bioassy

The firefly assay of intracellular bacterial adenosine triphosphate (ATP), described by Seyfried and Horgan [6], was used to measure the toxic effects of metal ions on the aquatic isolates. The procedure was further modified by spectrophotometrically adjusting the cell suspension to 0.5 optical density (OD) ($\lambda = 625$ nm) for all isolates.

Log phase cells were diluted 1:10 in fresh minimal medium broth (MMB) or MMB supplemented with metal salts and incubated at 25°C in a gyratory water bath. At specified intervals, 1.0 mL of treated or untreated culture was added to an equal volume of 0.2% apyrase (Sigma) in nutrient broth (Difco). A 0.5-mL aliquot was removed after 10 min incubation at 37°C and added to boiling TRIS-EDTA buffer. The ATP was extracted for exactly 2 min after which time the samples were removed and stored on ice.

Sterile HEPES buffer and luciferin-luciferase (Mono Research, Turner Design Reagents) were used to quantitatively measure the amount of heat-extracted ATP. A 50-μL sample was placed in a polystyrene cuvette and 50-μL of the luciferase enzyme added. The cuvette was shaken gently and placed in a Model 3000 integrating photometer (SAI Technology Co.) set at the 10-s integration mode. Light, produced as counts, was converted to ATP μg/L using standard curves. Results presented are the average of duplicate or triplicate assays.

Statistical Analyses

Data obtained from the ATP toxicity bioassays were analyzed to determine the minimal inhibitory concentration (MIC) and minimal bactericidal concentration (MBC) of metals and metal mixtures. The lowest concentration of metal that was able to cause a significant reduction in the amount of

ATP μg/L compared to the control was deemed the MIC. MIC values were determined 21 h post exposure to the metal. Significance was calculated as follows:

$$\text{statistical } t \text{ value} = \frac{(\log \text{ATP}_0 - \log \text{ATP}_{\text{MIC}})}{\sqrt{2} \times \text{SD}}$$

The MBC represented a significant reduction in ATP (μg/L) at $T = 21$ h, compared with the level of ATP (μg/L) at $T = \text{O}$ h. Significance was calculated using

$$\text{statistical } t \text{ value} = \frac{(\log \text{ATP}_{\text{MBC } T0} - \log \text{ATP}_{\text{MBC } T21})}{\sqrt{2} \times \text{SD}}$$

where SD = standard deviation.

Concentration of metals that produced t values <0.05 were selected as the MIC or MBC, respectively.

Antibiotic Sensitivity Testing

All isolates were tested for antibiotic resistance by means of the Kirby-Bauer disk method [15]. The following antibiotic concentrations were employed in the analyses: novobiocin 30 μg, colistin sulphate 10 μg, naladixic acid 30 μg, clindamycin 2 μg, penicillin G 10 units (u), neomycin 30 μg, vancomycin 30 μg, tobramycin 10 μg, bacitracin 10 intenational units (iu), carbenicillin 100 μg, polymyxin B 300 iu, ampicillin 10 μg, lincomycin 2 μg, sulphamethoxazole trimethoprim 25 μg, and kanamycin 30 μg.

Plasmid DNA Analysis

Plasmid DNA was prepared and analyzed by the method of Bradbury et al. [16]. A metal-resistant strain of *Klebsiella oxytoca*, isolated from the Cobalt marsh treatment system, was selected for plasmid extraction and analysis. Cells for the plasmid preparation were grown in trypticase soy broth (Difco) and 1.5-mL volumes used for plasmid analysis. Two *Escherichia coli* strains and one *Klebsiella pneumoniae* strain carrying standard plasmids with the following five molecular weights were used to measure the size (in Da) of the *K. oxytoca* plasmids: 70×10^6, 56×10^6, 9×10^6, 5.3×10^6, and 2×10^6.

Preparation of ^{32}P-labelled Probe DNA

A large molecular weight plasmid, pWB (plasmid Wayne Bradbury) 56 with a molecular weight of 56×10^6 Da, carrying antibiotic resistances to the following antibiotics: gentamicin (4 μg), tobramycin (4 μg), kanamycin (8 μg), tetracycline (4 μg), chloramphenicol (15 μg), and ampicillin (16 μg), was used as a source of probe DNA. The pWB 56 was purified from an *E. coli* transconjugant according to the method of Bradbury et al. [16], and the plasmid was labeled by nick translation according to methods previously described [17].

Southern Transfer and Hybridization to ^{32}P-labelled Probe DNA

Following agarose gel electrophoresis of the plasmid DNAs, the gels were denatured and neutralized and the DNAs plotted onto nitrocellulose filters, baked, and stored until required for the hybridization experiments according to the method of Bradbury et al. [17]. The ^{32}P-labeled pWB 56 probe was hybridized to the filters at approximately 10^8 cpm per μg of DNA. Autoradiography was performed as previously described [17].

Results

Intracellular ATP Measurements Assessing Metal Toxicity

Figure 1 demonstrates the toxic effect of mercury on *K. oxytoca*. Although 0.21 mg/L of Hg appeared to be inhibitory for the first 9 h post exposure, some recovery of the organism was apparent thereafter. A Hg concentration of 0.35 mg/L or greater was necessary to completely inhibit the organism.

The toxic effects of varying concentrations of Hg, Zn, Cu, and Cd at 21 h post metal exposure on *K. oxytoca* and *P. aeruginosa* are shown in Figs. 2 and 3. The *Pseudomonas* isolate was more sensitive to all levels of Hg than the *Klebsiella* organism. Both displayed a similar sensitivity pattern to Cu, Cd, and Zn.

Table 1 lists the MIC and Table 2 the MBC of metals and metal mixtures for each of the Cobalt isolates studied. The standard deviation for all experiments ranged from 0.001 to 0.555. Concentrations of metals that produced t values with p values of <0.05 were selected as the MIC and MBC, respectively.

Mercury was the most toxic and arsenic the least toxic metal based on the intracellular ATP measurements. Lead, aluminum, and arsenic did not show any significant toxic effect at the levels tested ($p < 0.05$).

All the members of the family Enterobacteriaceae, that is, *Klebsiella*, *Enterobacter*, and *Escherichia* spp., displayed similar MIC and MBC patterns. *Klebsiella* was deemed the most metal resistant and *Aeromonas* the least metal resistant.

Antibiotic Resistance

Table 3 lists the antibiotic resistance patterns of several genera isolated from the Cobalt marsh treatment facility. All bacteria tested were multiply drug resistant; the *Pseudomonas* isolate was shown to be resistant to 12 of the 17 antibiotics tested.

The *Klebsiella*, *Enterobacter*, and *Escherichia* spp. exhibited almost identical antibiotic resistance profiles. All were resistant to novobiocin, colistin sulphate, penicillin G, bacitracin, vancomycin, ampicillin, lincomycin, and clindamycin. In addition, the *Escherichia* sp. was resistant to erythromycin. These enteric bacteria were found to be sensitive or intermediate in resistance to the remaining nine antibiotics.

As indicated in the table, all five genera were resistant to lincomycin and 10 units of penicillin G. Sensitivity to kanamycin, neomycin, and tobramycin was evident in all isolates tested.

Plasmid Analysis

A total of six plasmids were identified in the *Klebsiella oxytoca* strain having the following molecular weights: 50×10^6, 40×10^6, 10.5×10^6, 6.6×10^6, 3.5×10^6 and 3×10^6 Da (Fig. 4, Lane 4). Following transfer to nitrocellulose filters, these plasmids were hybridized to the ^{32}P-labeled pWB 56 probe.

Hybridization Analysis

The ^{32}P-labeled pWB 56 probe hybridized only with the largest 50×10^6 *K. oxytoca* plasmid (Fig. 4, Lane 3). The intensity of the hybrid band as determined by densitometry was approximately

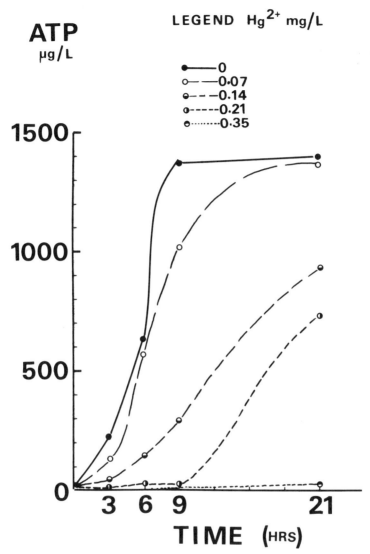

FIG. 1—*Toxic effect of mercury on a marsh treatment system isolate of* K. oxytoca *as measured by the firefly luciferase assay (intracellular ATP micrograms per litre).*

25% the intensity of hybridization observed when the pWB 56 was annealed to itself (compare Fig. 4, Lane 2 with Lane 3). This suggested at least 25% homology between the *K. pneumoniae* pWB 56 sequences and those comprising the 50×10^6 Da *K. oxytoca* plasmid.

Discussion

Multiply heavy metal–resistant bacteria have been isolated from a wide variety of environments [11,24]. Futhermore, Mills and Colwell [20] and Timoney et al. [7] have shown that heavy metal

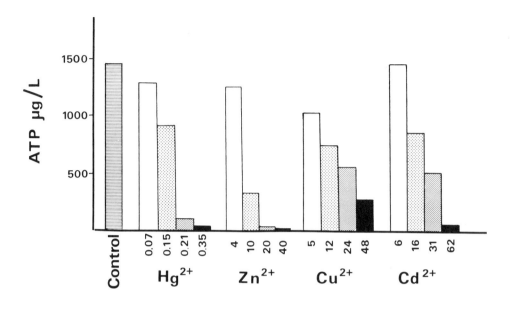

FIG. 2—*Toxic effects of varying concentrations of mercury, zinc, copper, and cadmium on a K. oxytoca marsh treatment system isolate as indicated by intracellular ATP (micrograms per litre) levels 21 h after exposure to the metals.*

contamination of Chesapeake Bay and the New York Bight, respectively, have been exerting a selective pressure for the development of metal-resistant bacterial populations. Correspondingly, it is reasonable to assume that the high levels of metals detected in the Cobalt marsh water may be selecting for metal tolerant microorganisms.

Results of the intracellular ATP assay showed that, among the oganisms tested, *Klebsiella* and *Pseudomonas* were the most metal resistant. Previous studies have indicated that their resistant properties may be governed by a variety of factors. For example, some *Pseudomonas* strains are able to produce an extracellular polymer that prevents the accumulation of metal ions in the cytoplasm of the cell [18]. *Klebsiella* strains are also known to produce a thick polysaccharide capsule which can diffuse into the surrounding liquid medium [14]. Bitton and Freihofer [19] showed that this capsular material was able to reduce the toxic effect of Cu and Cd ions on *Klebsiella* isolates.

Other factors responsible for metal resistance are genetic determinants encoding metal resistance [25,26]. Kelch and Lee [23] state that different genera from a similar environment may possess similar genetic information which may code for metal resistance. The fact that all the Cobalt isolates had similar metal resistance profiles (Tables 1 and 2) suggests that the survival or selection of these strains in the marsh environment was not random. A common pool of resistance factor plasmids coupled with *in situ* plasmid transfer among the Cobalt isolates may account for the similar MIC and MBC patterns. Plasmid transfer between members of the same species or genera and members of different families has been reported [22,23].

Plasmids may also be responsible for the antibiotic resistance patterns observed among the isolates [11,27]. The fact that the fecal coliform organisms, *Escherichia, Enterobacter*, and

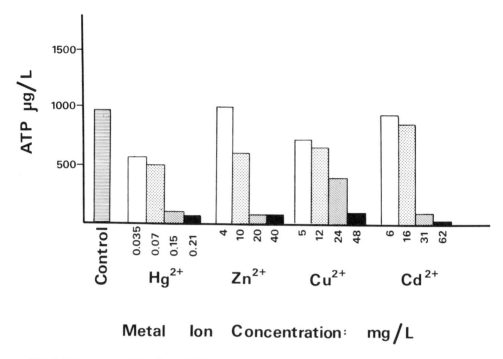

FIG. 3—*Comparison of the effects of different concentrations of mercury, zinc, copper, and cadmium on a marsh isolate of* P. aeruginosa *as determined by the intracellular ATP assay (in micrograms per litre) at 21 h post metal exposure.*

Klebsiella, displayed the same antibiotic-resistant profile suggests that these isolates may have gained their resistance mechanism (plasmids) from a common source. Similar results have been reported by Bell and coworkers [28].

Isolation of multiply-resistant *Pseudomonas* sp. is not uncommon [21]. Marques et al. [10] stated that 6% of their *P. aeruginosa* isolates were resistant to 12 of the 14 antibiotics tested. In this study the *P. aeruginosa* isolate was found to be resistant to 12 of the 17 antibiotics and 9 of the 11 metals tested.

In support of these findings, several research groups have shown that metal and antibiotic resistance is frequently found in sewage-borne organisms [23,27–29]. Other studies have shown that drug and metal resistance is often mediated by the same plasmid or R-factor [25,30]. With genetic information for metal and antibiotic resistance grouped on the same plasmid, it is reasonable to assume that either heavy metals or antibiotics could serve as selective pressures for populations of bacteria hosting these R-factors [7]. In the case of the Cobalt isolates, metal selection due to metal elution from the tailings seems more likely.

Analysis of the *K. oxytoca* marsh isolate for plasmid DNA showed that the organism contained six plasmids, ranging in size from 3×10^6 to 50×10^6 Da. The pWB 56 probe used in the study was a plasmid that had been isolated from a hospital strain of *K. pneumoniae* which carried multiple antibiotic resistance. Homology between the 50×10^6 Da *K. oxytoca* plasmid and the pWB 56 probe suggests sequence similarities that may reflect the presence of multiple antibiotic resistance genes in the environmental *K. oxytoca* strain. This supports the view, mentioned previously, that mine tailings could provide an environmental stress that may lead to the development of heavy metal and antibiotic resistance among aquatic microorganisms.

TABLE 1—*Toxicity of metals for Cobalt marsh isolates as indicated by the MIC and determined by intracellular ATP (in micrograms per litre) measurements.*

Organism	MIC Metal Ion Concentration, mg/L											Mixture[a]
	Co	Cu	Cd	Hg	As	Ni	Zn	Fe	Pb	Al	Cr	
Pseudomonas	<2.5	12 to 24	16 to 31	0.07 to 0.15	>40	<5	10 to 20	26 to 34	>63	>20	10	1 to 10
Escherichia	<2.5	<5	<6	0.07 to 0.15	>40	<5	4 to 10	>34	>63	>20	39	1 to 10
Enterobacter	<2.5	5 to 12	16 to 31	0.07 to 0.15	>40	<5	10 to 20	>34	>63	>20	39	0.01 to 0.1
Aeromonas	2.5 to 5	5 to 12	<6	0.035 to 0.07	>40	34 to 45	4 to 10	26 to 34	>63	>20	>39	1 to 10
Klebsiella	<2.5	<5	16 to 31	0.3 to 0.4	>40	>45	4 to 10	>34	>63	>20	>39	0.1 to 1

[a] Expressed as percent of stock metal mixture. Stock solution contains in mg/L: As 40, Fe 35, Al 20, Zn 40, Cd 60, Cu 50, Cr 40, Co 45, Ni 45, and Hg 7.

TABLE 2—*Toxicity of metals for Cobalt marsh isolates as indicated by the MBC and determined by intracellular ATP (in micrograms per litre) measurements.*

Organism	MBC Metal Ion Concentration, mg/L										
	Co	Cu	Cd	Hg	As	Ni	Zn	Fe	Pb	Al	Cr	Mixture[a]
Pseudomonas	<2.5	>48	46 to 62	0.3 to 0.4	>40	12 to 24	10 to 20	>34	>63	>20	>39	1 to 10
Escherichia	5 to 11	5 to 12	16 to 31	0.07 to 0.15	>40	5 to 12	10 to 20	>34	>63	>20	>39	1 to 10
Enterobacter	5 to 11	36 to 48	46 to 62	0.07 to 0.15	>40	5 to 12	30 to 40	>34	>63	>20	>39	1 to 10
Aeromonas	1 to 2.5	12 to 24	6 to 16	0.035 to 0.07	>40	24 to 45	4 to 10	>34	>63	>20	>39	1 to 10
Klebsiella	5 to 11	>48	46 to 62	>0.4	>40	24 to 45	10 to 20	>34	>63	>20	>39	10 to 100

[a] Expressed as percent of stock metal mixture. Stock solution contains in mg/L: As 40, Fe 35, Al 20, Zn 40, Cd 60, Cu 50, Cr 40, Co 45, Ni 45, and Hg 7.

TABLE 3—*Antibiotic resistance patterns of Cobalt marsh isolates.*

Antibiotic	Pseudomonas	Aeromonas	Escherichia	Klebsiella	Enterobacter
Novobiocin	R	R	R	R	R
Colistin sulphate	S	S	R	R	R
Penicillin G	R	R	R	R	R
Tobramycin	S	S	S	S	S
Bacitracin	R	R	R	R	R
Vancomycin	R	R	R	R	R
Lincomycin	R	R	R	R	R
Naladixic acid	R	S	I	S	S
Clindamycin	R	R	R	R	R
Neomycin	S	S	S	S	S
Carbenicillin	R	S	S	I	I
Polymyxin B	S	S	S	S	S
Erythromycin	R	S	R	I	I
Ampicillin	R	I	R	R	R
Amikacin	R	I	S	S	S
Kanamycin	S	S	S	S	S
Sulpamethoxazole trimethoprim	R	S	S	S	S

NOTE: R = resistance; S = sensitive; I = intermediate.

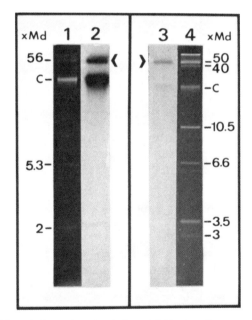

FIG. 4—*Agarose (0.7%) gel electrophoresis of plasmid DNA from* K. pneumoniae *and* K. oxytoca *and hybridization analysis with the* K. pneumoniae–*derived pWB 56 probe. DNA was prepared from each isolate and subjected to gel electrophoresis. The plasmids were transferred by blotting to nitrocellulose filters and hybridized with a high specific activity pWB 56 probe (10^8 cpm ^{32}P/ug DNA). Lanes 1 and 2 illustrate the plasmid profiles of* K. pneumoniae, *and* K. oxytoca, *respectively. Hybridization of the pWB 56 probe annealed to itself (arrowhead, Lane 2) and the 50-Mda (megadalton) plasmid of* K. oxytoca *(arrowhead, Lane 3). The degree of homology was approximately 25% as determined by densitometry. Hybridization of the probe to the chromosomal region of the gel (C) is due to the presence of sheared, homologous plasmid DNA that comigrates with the denatured chromosomal DNA. Other bands observed in the gel are relaxed plasmid DNA ($\times$ Md = times megadaltons).*

Acknowledgments

This work was supported by the Ontario Ministry of the Environment under Project No. 109 RR. The excellent technical assistance of D. Rego was greatly appreciated.

References

[1] Nielson, R. F. and Peterson, H. B. in *Ecology and Reclamation of Devastated Land*, Vol. 2, G. Davis and R. J. Hutnik, Eds., Gordon and Breach Science Publishers, New York, 1973, p. 103.
[2] Gemmell, R. P. in *Symposium Proceedings: International Conference of Heavy Metals in the Environment*, T. C. Hutchinson, Ed., University of Toronto, Toronto, Ontario, 1975, p. 579.
[3] Johnson, R. D., Miller, R. J., Williams, R. E., Wai, C. M., Weise, A. C., and Mitchell, J. E. in *Symposium Proceedings: International Conference on Heavy Metals in the Environment*, T. C. Hutchinson, Ed., University of Toronto, Toronto, Canada, 1975, p. 465–485.
[4] Galbraith, J. H. and Williams, R. E., *Groundwater*, Vol. 10, 1972, p. 33.
[5] Andrews, R. D. in *Symposium Proceedings: International Conference on Heavy Metals in the Environment*, Vol. 2, T. C. Hutchinson, Ed., University of Toronto, Toronto, Ontario, 1975, pp. 645–76.
[6] Seyfried, P. L. and Horgan, C. B. L. in *Aquatic Toxicology and Hazard Assessment: Sixth Symposium, ASTM STP 802*, W. E. Bishop, R. D. Cardwell and B. B. Heidolph, Eds., American Society for Testing and Materials, Philadelphia, 1983, pp. 425–441.
[7] Timoney, J. F., Port, J., Giles, J. and Spanier, J., *Applied and Environmental Microbiology*, Vol. 36, 1978, pp. 465–472.
[8] Sadik, M., Zaidi, T. H., Moda, A. U., and Mian, A. A., *Bulletin of Environmental Contamination and Toxicology*, Vol. 29, 1982, pp. 313–319.
[9] Trevors, J. T., Oddie, K. M., and Belliveau, B. H., *FEMS Microbiological Reviews*, Vol. 32, 1985, pp. 39–54.
[10] Marques, A., Congreado, F., and Simon-Pujol, D., *Journal of Applied Bacteriology*, Vol. 47, 1979, pp. 347–350.
[11] Harnett, N. M. and Gyles, C. L., *Applied and Environmental Microbiology*, Vol. 48, 1984, pp. 930–935.
[12] Belliveau, B. H., Wong, P., and Trevors, J. in *Toxicity Assessment*, in press.
[13] Staples, D. G., and Fry, J. C., *Journal of Applied Bacteriology*, Vol. 36, 1973, pp. 179–191.
[14] *Bergey's Manual of Systematic Bacteriology*, Vol. 1, N. R. Krieg and J. G. Holt, Eds., Williams and Wilkins, Baltimore/London, 1984.
[15] Bauer, A. W., Kirby, W. M., Sherris, J. C., and Turck, M., *American Journal of Clinical Pathology*, Vol. 45, pp. 493–496.
[16] Bradbury, W. C., Marko, M. A., Hennessy, J. N., and Penner, J. L., *Infectious Immunity*, Vol. 40, 1983, pp. 460–463.
[17] Bradbury, W. C., Murray, R. G. E., Mancini, C., and Morris, V. L., *Journal of Clinical Microbiology*, Vol. 21, pp. 24–28.
[18] Corpe, W. A., *Development of Industrial Microbiology*, Vol. 16, 1975, pp. 249–255.
[19] Bitton, G. and Freihofer, V., *Microbial Ecology*, Vol. 4, 1978, pp. 119–125.
[20] Mills, A. L. and Colwell, R. R., *Bulletin of Environmental Contamination and Toxicology*, Vol. 18, 1977, pp. 99–103.
[21] Sonnenwirth, A. C., in *Microbiology*, 3rd ed., B. Davis, R. Dulbecco, H. Eisen and S. Ginsberg, Eds., Harper and Row, New York, 1980.
[22] Olsen, R. H. and Shipley, P., *Journal of Bacteriology*, Vol. 113, 1973, pp. 772–780.
[23] Kelch, W. J. and Lee, J. S., *Applied and Environmental Microbiology*, Vol. 36, 1978, pp. 450–456.
[24] Seyfried, P. L. in *Aquatic Toxicology, ASTM STP 707*, J. G. Eaton, P. R. Parrish, and A. C. Hendricks, Eds., American Society for Testing and Materials, Philadelphia, 1980, pp. 224–232.
[25] Nakahara, H., Ishikawa, T., Sarai, Y., Kondo, I., Kozukue, M., and Silver, S., *Applied and Environmental Microbiology*, Vol. 33, 1977, pp. 975–976.
[26] Houba, C. and Remacle, J., *Microbial Ecology*, Vol. 6, 1980, pp. 55–69.
[27] Varma, M., Thomas, W., and Prasad, C., *Journal of Applied Bacteriology*, Vol. 41, 1976, pp. 347–349.
[28.] Bell, J. B., Macrae, W. R., and Elliott, G. E., *Applied and Environmental Microbiology*, Vol. 40, 1980, pp. 486–491.

[29] Calmoris, J. J., Armstrong, J. L., and Seidler, R. J., *Applied and Environmental Microbiology*, Vol. 47, 1984, pp. 1238–1242.
[30] Nakahara, H., Ishikawa, T., Sarai, Y., Kondo, I., Kozukue, H., and Mitsuhaski, S., *Antimicrobial Agents and Chemotherapy*, Vol. 11, 1977, pp. 999–1103.

Martina A. Bianchini,[1] Ralph J. Portier,[1] Kuniko Fujisaki,[1] Charles B. Henry,[1] Paul H. Templet,[1] and John E. Matthews[2]

Determination of Optimal Toxicant Loading for Biological Closure of a Hazardous Waste Site

REFERENCE: Bianchini, M. A., Portier, R. J., Fujisaki, K., Henry, C. B., Templet, P. H., and Matthews, J. E., **"Determination of Optimal Toxicant Loading for Biological Closure of a Hazardous Waste Site,"** *Aquatic Toxicology and Hazard Assessment: 10th Volume, ASTM STP 971,* W. J. Adams, G. A. Chapman, and W. G. Landis, Eds., American Society for Testing and Materials, Philadelphia, 1988, pp. 503–516.

ABSTRACT: Information will be presented on Phase I and Phase II of a multitask effort to achieve biological closure of an abandoned hazardous waste site on the Mississippi River. Waste materials, consisting primarily of aliphatic and polycyclic aromatic hydrocarbons in the form of buried sludges and lagoon wastes, were examined. Optimal loading levels were evaluated on the basis of biodegradative potential tests and acute toxicity of leachate. Microbial adenosine 5^a triphosphate (ATP) and microbial diversity were used in conjunction with Microtox™ tests to establish an acceptable land treatment experimental design. Methodology is presented on ascertaining optimal waste loading rates, based on percent oil and grease, for both laboratory mesocosm and field plots. Microbial ATP data indicated that both waste types stressed the indigenous microbial populations through Day 10. However, the microbial population showed recovery by Day 17. Lagoon wastes, applied at 4 and 8%, showed minimal stress and high microbial ATP levels by Day 17. Buried wastes showed comparable results for 2 to 4% waste loadings at Day 17. Microbial diversity indices confirmed ATP estimates, which suggested a maximum loading rate of 8% for lagoon wastes and 2 to 4% for buried wastes. EC_{50} levels of leachates predicted 3.0 to 9.0% waste weight (wet basis) for lagoon waste and 3.0 to 6% waste weight (wet basis) for the buried waste. A loading rate of 4% for lagoon waste and 2.5% for buried waste was selected as the acceptable mean loading rate. Gas chromatography/mass spectroscopy (GC/MS) data in subsequent Phase II studies (mesocosms) documented significant biotransformation and biodegradation of the wastes at these optimized loading rates.

KEYWORDS: biodegradation, biotransformation, polycyclic aromatic hydrocarbons (PAHs), *in situ* biological treatment (ISBT)

Industrialized deltas often have the combined problems of improper land management and unwise water resource allocation. Significant threats to both surface and groundwater supplies are posed by the location of numerous hazardous waste sites in close proximity to commercial activities. A case in point is the industrialized lower delta of the Mississippi River in Louisiana. Noteworthy strides have been made along this "chemical corridor" in reducing waste discharges into the Mississippi River. However, many abandoned refineries, hazardous waste pits, and pipelines remain. These facilities represent a continuing threat to land and water resources and human health in the area and also pose a great economic liability for their removal.

---

[1] Institute for Environmental Studies, Louisiana State University, Baton Rouge, LA 70803.
[2] Robert S. Kerr Environmental Research Laboratory, Ada, OK 74820.

In this paper, we present Phase I (screening tests) and Phase II (mesocosm tests) of a multitask effort for achieving biological closure of an abandoned hazardous waste site situated on the Mississippi River. Biological closure is achieved through *in situ* biological treatment (ISBT), a method which favors the growth of the adapted indigenous microflora in the soil and so promotes the degradation of the toxicants. Waste materials from the site were evaluated for biodegradative potential and acute toxicity of leachate. The validity of the Phase I tests for establishing biotransformation/biodegradation kinetics and leachate toxicity were documented in subsequent Phase II laboratory mesocosm studies.

The selected approach, *in situ* biological treatment, is consistent with the National Contingency Plan to provide the lowest cost alternative which provides a permanent remedy for existing or potential threat to public health and environment. On-site biodegradation through land treatment was selected as the least expensive alternative ($3 million) and most applicable to the site. This technique consists of spreading and mixing the oily sludges over land. The decomposition of hydrocarbon components is encouraged by the soil microorganisms [1].

Materials and Methods

Site Characterization

The hazardous waste site investigated was the "Old Inger" CERCLA (Comprehensive Environmental Response, Compensation and Liability Act of 1980) site located on the east bank of the Mississippi River in Louisiana, about 4.5 miles north of the city of Darrow. It is an abandoned oil reclamation facility that was operated between 1967 and 1978. As part of the operation of the plant, sludges were stored in a large open lagoon and buried in a shallow pit; some wastes were spilled into the adjacent swamp. In March 1978, a large spill contaminated a total of 16 acres of the surrounding area. Failure by the owner to clean up the site resulted in the formal declaration by the Louisiana Environmental Control Commission in June 1981 that the site was abandoned.

The wastes identified at the site were consistent with the nature of the oil reclamation plant. They were mixtures of oil refinery oils, motor and automobile oils, and lube oils. As is typical of waste oils, hazardous priority pollutants such as benzene, toluene, and polycyclic aromatic hydrocarbons (PAHs) were present. No polychlorinated biphenyls (PCBs) were found; very low levels of chlorinated hydrocarbons and low levels of heavy metals were found. Tables 1 and 2 present analyses of the "pure wastes," classified as buried waste and lagoon waste, and river silt (control). All were directly collected from the site. By means of gas chromatography (GC) elution time and gas chromatography/mass spectroscopy (GC/MS), 24 PAHs were identified (F-2 fraction, Table 1) and 22 aliphatic hydrocarbons (F-1 fraction, Table 2). Quantitation was by external standard GC in all cases. Detection limit for both fractions was 100 ppm due to the large dilution factors and the lower response factors of the higher molecular weight components. The detection limit for the control soil, however, was 1 ppm.

Microbial Density

The Standard Plate Count Method (SPC) is a direct quantitative measurement of the viable aerobic and facultative anaerobic microflora. Four general groups of microorganisms, that is, bacteria, actinomycetes, yeasts, and filamentous fungi, were enumerated using colony forming units (CFU) and SPC. Replicate 1.0-mL aliquots were inoculated on Jensen's agar medium [2] supplemented with 40 µg/mL cycloheximide (Sigma) to inhibit growth of filamentous fungi. Filamentous fungi and yeasts were enumerated on Martin's agar medium [3] with 30 µg/mL

TABLE 1—*Primary F-2 constituents in CERCLA waste site (Inger Oil Refinery, Darrow, LA)*.

Compound	Buried Wastes	Lagoon Wastes	River Silt (Control)
1. Naphthalene	920	1500	...
2. 1-methylnaphthalene	1200[a]	1700[a]	...
3. 2,3-dimethylnaphthalene	600[a]	1000[a]	...
4. 2,3,5-trimethylnaphthalene	600[a]	700[a]	...
5. Biphenyl	1700[a]	1600[a]	...
6. Fluorene	850	1300	...
7. Dibenzothiophine	800[a]	1200[a]	...
8. Phenanthrene	4528	5757	trace
9. Anthracene	2257	3227	trace
10. 1-methylphenanthrene	800[a]	600[a]	...
11. Fluoranthrene	930	1686	...
12. Pyrene	830	1059	trace
13. Benzo(a) anthracene	470	570	trace
14. Chrysene	340	550	trace
15. Benzo(b) fluoranthrene	ND	ND	trace
16. Benzo(e) pyrene	ND	60	...
17. Benzo(a) pyrene	<100[b]	215	...
18. Perylene	...	...	...
19. Acenapthene	2300	1466	...
20. Substituted benzene	1500[a]	2100[a]	...
21. Indenopyrene	1100	321	...
22. Benzo(g,h,i) perlene	1200	1432	...
23. Dibenzo(a,h) anthracene	100+	357	...
24. Benzo(k) pyrene	ND	457	...
25. Unknown (E.T. = 24.51)	ND	...	<1 ppm

NOTE: Concentration in ppm, [a] semiquantitative, [b] confirmed by GC/MS. Trace = <1 ppm, detection limit 100 ppm for lagoon and buried waste, detection limit 1 ppm for river silt. ND = not detected; E.T. = elution time.

streptomycin (Sigma) and 30 μl/mL chlortetracycline (Sigma) to retard bacterial growth. All plates were incubated for four days at 30°C. Bacteria and actinomycetes were counted on a Biotran III Automatic Colony Counter (New Brunswick Scientific). Yeasts and fungi were enumerated manually.

Adenosine 5' Triphosphate (ATP)

A modification of adenosine 5' triphosphate (ATP) assay as advanced by Holm-Hansen and Booth [4], and further presented by Van de Werf, Verstraete [5], and Karl [6], was used for determination of microbial biomass. 1.0-g aliquots from all mesocosms (wet weight) were transferred into dilution bottles containing 99-mL sterile distilled and deionized water, homogenized on a homogenizer (Janke & Kunkel Ultra Turrax SD 45) for about 45 s. A 100-μL aliquot of that suspension was then transferred to a 3-mL plastic vial. The vial was inserted into a Lumac 3M Biocounter (Biocounter M2010, Lumac Systems, USA); 100μL of buffer and 100-μL NRB reagent (Nucleotide releasing agent for bacterial cells) were added. ATP was released from microbial cells by adding NRB to a 100-μL sample. After application of the reagent, ATP was measured using the following reactions:

$$\text{luciferin} + \text{luciferase} + \text{ATP} - \text{Mg}^{++} \rightarrow (\text{luciferin} + \text{luciferase} + \text{AMP}) + pp_i$$

$$(\text{luciferin} + \text{luciferase} + \text{AMP}) - O_2 \rightarrow \text{decarboxyluciferin} + \text{luciferase} + CO_2 + \text{AMP} + \text{light}$$

TABLE 2—*Primary F-1 constituents in CERCLA waste site (Inger Oil Refinery, Darrow, LA).*

Compound	Buried Wastes, ppm	Lagoon Wastes, ppm
C-12	800	2300
C-13	1100	3600
C-14	1100	4000
C-15	1400	4300
C-16	1700	5000
C-17	1400	4400
Pristane	4300	3100
C-18	1200	4900
Phytane	2700	1500
C-19	750	3300
C-20	370	2900
C-21	240	2000
C-22	a	1400
C-23	a	840
C-24	a	600
C-25	a	890
C-26	a	300
C-27	a	320
C-28	a	b
C-29	a	1200
C-30	a	560
C-31	a	330

a Below detection limit of 100 ppm.
b Lab contaminant.

Following the injection of 100 µL of a luciferin luciferase solution (Lumit, Lumac systems) into the vial, light outputs expressed as relative light units (RLU) were determined over a 10-s integration period. Relative light units were expressed as µg ATP/g dry net weight using standardized 10-µL aliquots of a known ATP standard. Quench corrections for each environmental sample were established with 100-µL aliquots of ATP standard/buffer added in place of the 100-µL buffer addition.

Microtox™

Acute toxicity assessment of waste/soil water soluble fractions (WSF) was determined, following procedures outlined by Matthews et al. and further described in the EPA "Permit Guidance Manual on Hazardous Waste Land Treatment Demonstrations" [7]. A maximum acceptable initial loading rate was determined by calculating the EC_{50} and 95% confidence intervals for the raw waste WSF. The lower EC_{50} limit for the confidence interval provides the upper loading rate for waste/soil toxicity testing. In this case, the EC_{50} was the volume % of the distilled deionized water (DW) extract effecting a 50% decrease in bioluminescence of the marine luminescence bacterium *Photobacterium phosphoreum* during a 5-min test period. The DW extract was obtained by shaking 400-mL DW + 100-g sample for 24 h. All samples were analyzed on a Beckman Microtox Model 2055 toxicity analyzer.

Soil/waste Chromatographic Analyses

The principal analysis performed involved determination of aliphatic (F-1) and aromatic hydrocarbons (F-2). Equal amounts of the waste/soil mixture were used to form a composite. Two grams of that composite were extracted with 10 mL of Dichloromethane (DCM), and a portion of the extract was placed on a silica gel column. The column was eluted, first with hexane to remove and trap the aliphatics (F-1) and then with 30% DCM in hexane to elute the polynuclear aromatics (PAH). The two liquid fractions were analyzed in a GC flame ionization detector (FID) (Hewlett Packard 5890, equipped with a 50-m by 0.32-mm DB-5 phenyl methyl silicone capillary column). Each compound peak was quantitated and identified by GC/MS (Hewlett Packard 5970B), and the identities of several unknowns were elucidated in this manner. A complete protocol for GC determination of the F1 and F2 constituents is presented in Ref 8.

Experimental Design

Screening Tests

Optimal loading rates for the CERCLA site soil/wastes were determined on the basis of both indigenous microbial population performances and acute toxicity leachate analysis.

The described investigations for ISBT involved three major correlative aspects as depicted in Fig. 1. The screening tests consisted of a series of experimental mesocosms with mixture of Mississippi River silt and top soil (sandy clay). Soil mixtures Type I and Type II consisted of one part river silt and two parts top soil, and two parts river silt and one part top soil, respectively. The soils were collected from uncontaminated areas near the contaminated site and air dried prior to analyzation. The soils were passed through a No. 10 sieve (Soiltest, Inc.) to remove larger particles and plant residues, and also to render them more homogenous. Soil Type II was found to be unsuitable because of its lower EC_{50}, detected by the EPA research laboratory in ADA, Oklahoma. Soil Type I, therefore, was selected for the course of the experiment. The loading rates of the wastes mixed with soil Type I were 2, 4, 8, and 12% by weight (based on percent oil and grease) for both lagoon and buried waste. No nutrient amendment was added. The highest level of microbial activity, evaluated 2 h after mixing and reevaluated over a period of 17 days after mixing, determined the optimal loading rate for each waste type.

Mesocosm Tests

After optimum toxicant loading determination, Phase II was set up, a battery of 21 mesocosms to evaluate biotransformation processes under the selected loading rates. Together with the loading of wastes, each mesocosm received a nutrient amendment consisting of 4.0 g KH_2PO_4, 2.0 g K_2HPO_4, 1.0 g $MgSO_4$, and 0.5 g KNO_3 × 4.5 kg^{-1} soil mixture, respectively. Nutrient amendments were incorporated into the mesocosm by hand mixing with the soil mixture and the waste at Day 0. A second nutrient amendment consisting of 2.0 g KH_2PO_4, 2.0 g K_2HPO_4, 1.0 g $MgSO_4$, and 0.5 g KNO_3 × 4.5 kg^{-1} soil mixture, respectively, followed on Day 27. The percent moisture was also adjusted on Day 27 to >10%. Since microbial oxidation of PAH requires molecular oxygen, the whole contents of the mesocosms were remixed weekly to promote oxygenated conditions.

Field Verification Studies

The pilot verification study involved the setup of five 12 × 50 × 2-ft. plots at the Old Inger site. The plots were filled with soil Type I and sequentially loaded with the two waste types. Plot 1 (control) did not receive any toxicant loading. Plots 2 and 3 (Group I) were loaded twice, first with 4% and then with 2.5% buried waste. Plots 4 and 5 (Group II) received four sequential waste loadings of the following type: 4% lagoon, 2.5% lagoon, 2.5% buried, and 2.5% buried waste. There were about six weeks between each application for Group II. At the first application, the

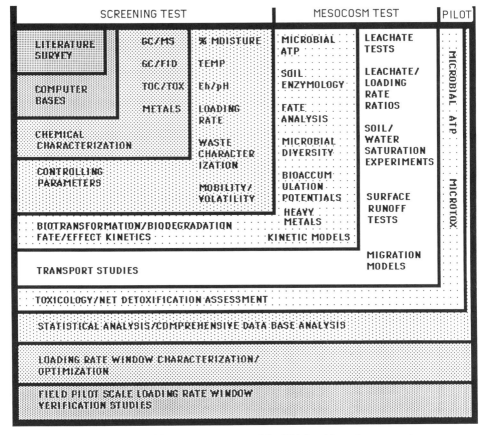

FIG. 1—*Land treatability assessment protocol for EPA-listed hazardous waste sites.*

plots were seeded with adapted microorganisms isolated from the site and cultivated in the laboratory. Between each treatment, the plots were tilled and sampled to a depth of 6 in.

Results and Discussion

Screening Tests

Optimal loading rates for the CERCLA site soil/wastes were determined by careful evaluation of both indigenous microbial population performance and acute toxicity leachate analysis. Microbial ATP showed initial stress on the indigenous microbial populations through Day 10 for both waste types (Fig. 2). However, the microbial population showed recovery by Day 17. Lagoon wastes, applied at 2% and 4%, showed minimal stress and increased microbial ATP levels by Day 17. Buried wastes showed comparable results for 2 to 4% waste loadings at Day 17. Microbial diversity indices agreed with ATP estimates, which suggested a maximum loading rate of 8% for lagoon wastes and 2 to 4% for buried wastes (Fig. 3). EC_{50} levels of leachates (Fig. 4) predicted 3.0 to 9.0% waste weight (wet basis) for lagoon waste and 3 to 6.0% waste weight (wet basis) for the

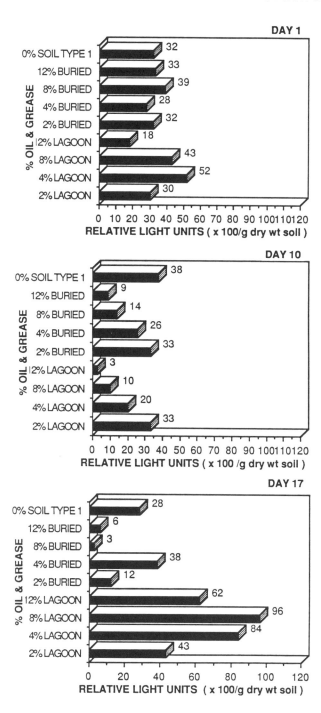

FIG. 2—*Microbial ATP profile of screening test microcosms.*

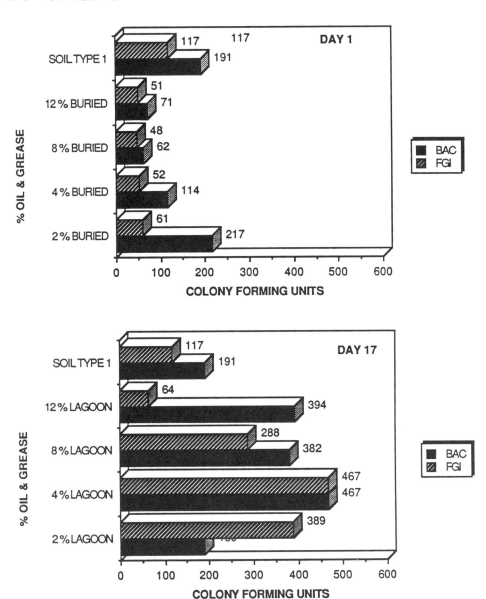

FIG. 3—*Microbial density of screening test mesocosms.*

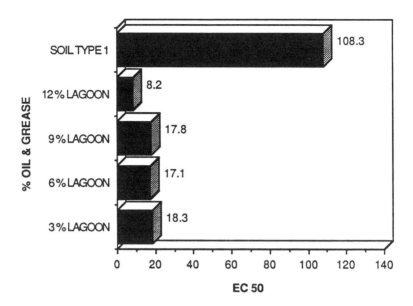

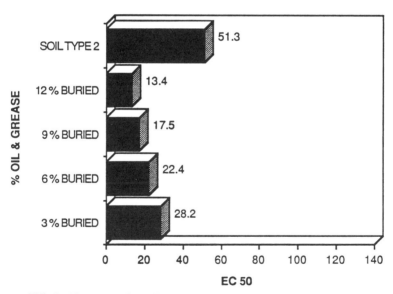

FIG. 4—*Microtox analysis of WSF of screening mesocosm soil/waste loading.*

buried waste. A loading rate of 4% for lagoon waste and 2.5% for the buried waste was found to be the mean acceptable loading rate for the Phase II mesocosm studies.

Mesocosm Tests

Manuscript limitations prevent the showing of all of the data; therefore, for the following results, only lagoon waste was selected for presentation. Figure 5 shows GC/FID chromatograms of the

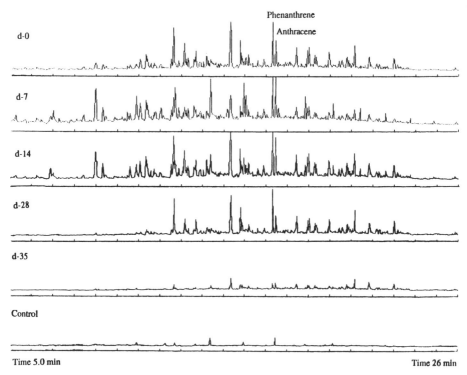

FIG. 5—*GC/FID analysis for 4% lagoon waste, F-2 fraction over a 35-day time frame.*

F-2 fractions for lagoon waste mesocosms with a 4% loading rate and the control soil. All compounds show a decrease over the 35-day incubation period. The total gross removal over this time period was almost 70%. Figure 6 represents GC/FID chromatograms for the aliphatic fraction (F-1) for 4% lagoon mesocosms over the same time frame. Table 3 gives the component concentrations of the F-1 fraction in ppm. The increases in concentrations for C-17, pristane, and C-18 at Day 28 are probably due to sample variability. The total removal was 90.85% The same degradation pattern, but at a faster rate, was observed for the 2.5% lagoon waste loading (Fig. 7). The total removal for the 2.5% loading F-2 fraction was 96.95% (Table 4).

The kinetics of toxicant disappearence for lower molecular weight PAHs, especially naphthalene, cannot be completely attributed to a microbial biotransformation scenario, but may also involve the high volatility of these compounds. Moderate abiotic losses of these PAHs due to volatility has been reported extensively [1], but is believed to not be substantial. Photooxidation of the PAHs was minimized in the mesocosm tests by incubating the experimental units, covered, in the dark. However, some unidentified abiotic mechanisms of PAH loss probably occurred. Additional investigations into identifying intermediate abiotic transformation processes are underway.

Conclusions

Optimal toxicant loading rates windows, determined in screening tests, were shown to be acceptable for inducing microbial biotransformation/biodegradation in mesocosm units and minimizing acute leachate toxicity. The above studies for all of the compounds analyzed exhibited a decrease in concentration over time for both laboratory and field studies. The decreases were

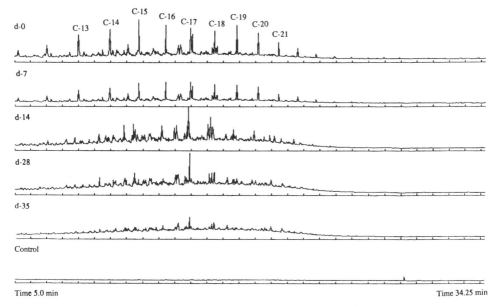

FIG. 6—*GC/FID analysis for 4% lagoon waste, F-1 fraction over a 35-day time frame.*

TABLE 3—*F-1 fraction; 4% lagoon waste.*

	Incubation Period, Days				
Compound	0	7	14	28	35
C-11	83	44	<3.5	<3.5	<3.5
C-12	119	58	40	<3.5	<3.5
C-13	236	113	42	14	<3.5
C-14	202	101	43	19	<3.5
C-15	214	108	10	47	18
C-16	249	133	18	8	6
C-17	221	115	35	217	91
Prystane	141	94	17	29	15
C-18	227	127	140	143	49
Phytane	71	54	14	12	8
C-19	167	91	42	33	21
C-20	135	78	28	<3.5	<3.5
C-21	93	53	8	<3.5	<3.5
C-22	57	34	4	<3.5	<3.5
C-23	29	17	<3.5	<3.5	<3.5
C-24	17	10	<3.5	<3.5	<3.5
C-25	11	7	<3.5	<3.5	<3.5
[a] Total (Gross)	2272	1237	441	522	208

NOTE: Min detection limit 3.5 ppm.
[a] Total of each column, concentration in ppm.

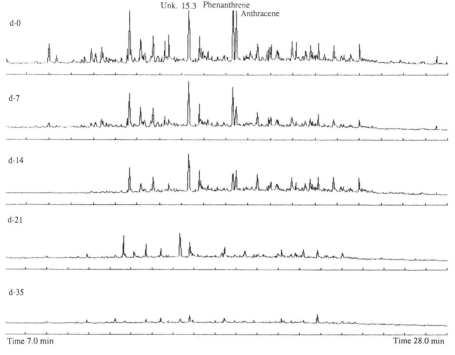

FIG. 7—*GC/FID analysis for 2.5% lagoon waste, F-2 fraction over a 35-day time frame.*

mostly attributed to microbial activity of the indigenous soil microflora. However, undefined abiotic losses were noted and need to be further studied. Both waste types, the lagoon and buried wastes, at both loading rates, 2.5% and 4%, were degraded by the indigenous microflora. As compounds were degraded by microorganisms, the metabolites identified were generally methyl-substituted homologs, which are generally less toxic than the parent compound and usually not found on priority pollutants or toxic substances lists of EPA. Mutagenicity and carcinogenicity of these intermediate degradation products need to be evaluated.

Partial data sets from the pilot verification study (Phase III) are presented to support screening test and mesocosm test data.

Preliminary data from pilot verification studies is shown in Fig. 8. These GC/FID chromatograms represent the first waste application for both Group I and Group II plots over a 25-day time frame. The first chromatogram of each series shows the soil plots before waste application. Complete differentiation into the two fractions, F-1 and F-2, was not possible. A multiple waste inoculum recovered from a different location at the Inger Site was used. Additionally, the waste contained a lesser amount of PAHs than the one tested in the laboratory. However, the general trend of hydrocarbon reduction as observed in Phase II studies was evident. It is important to know that the residual components shown at Day 25 have been determined by GC/MS to be primarily F-1 straight chain hydrocarbons, respectively C-17 and C-19 aliphatics.

In land-farm soil experiments conducted by Bossert et al. [1], analysis of oily sludge at the conclusion of a 1280-day laboratory simulation showed F-2 fractions to have been more rapidly eliminated than F-1 fractions. Although parallels can be drawn between our field experiment, the F-2 fraction constituted only a small portion of the total hydrocarbons present; 47% of the total hydrocarbons applied (seven applications) were eliminated in their bioflask study [1]. Preliminary

TABLE 4—*F-2 fraction; 2.5% lagoon waste.*

Compound	Incubation, Days				
	0	7	14	21	35
Naphthalene	18.0	>0.1	>0.1	>0.1	>0.1
Methylnapht.	19.0	>0.1	>0.1	>0.1	>0.1
Dimethylnap.	13.0	>0.1	>0.1	>0.1	>0.1
UNK (Rt = 11.1)	28.0	16.0	>0.1	>0.1	>0.1
UNK (Rt = 11.4)	25.0	14.0	1.0	>0.1	>0.1
Trimethylnap.	7.0	0.6	>0.1	>0.1	>0.1
Acenaphthalene	20.0	11.0	3.0	1.6	0.2
UNK (Rt = 12.7)	82.0	64.0	42.0	39.0	1.8
UNK (Rt = 13.2)	48.0	39.0	20.0	15.0	0.2
Biphenyl	19.0	12.0	5.0	1.2	>0.1
Fluorene	30.0	13.0	3.8	>0.1	0.5
UNK (Rt = 15.3)	170.0	120.0	119.0	75.0	13.0
UNK (Rt = 15.8)	50.0	32.0	36.0	25.0	>0.1
UNK (Rt = 15.9)	21.0	>0.1	>0.1	>0.1	>0.1
UNK (Rt = 16.2)	24.0	15.0	12.0	>0.1	>0.1
Dibenzothio.	21.0	8.0	7.0	>0.1	>0.1
Phenanthrene	138.0	68.0	42.0	7.8	1.3
Anthracene	97.0	42.0	38.0	22.0	4.2
UNK (Rt = 18.4)	33.0	36.0	44.0	22.0	3.3
UNK (Rt = 18.9)	27.0	16.0	13.0	2.9	0.3
UNK (Rt = 19.0)	33.0	18.0	18.0	5.2	>0.1
1-Methylphen.	17.0	11.0	12.5	4.4	0.7
UNK (Rt = 19.9)	51.0	33.0	39.0	9.5	>0.1
UNK (Rt = 20.8)	28.0	17.0	22.0	9.0	>0.1
Fluoranthene	37.0	22.0	32.0	14.0	4.0
Pyrene	21.0	11.0	20.0	11.0	13.0
Total (Gross)	1900	860	710	340	58

NOTE: The above numbers are accurate to only 2 significant fig. concentr. in ppm. UNK = unknown; Rt = retention time.

data from our field experiments indicate a net total hydrocarbon removal of 80.34% for Group I and 83.40% for Group II by Day 25. The differences in overall microbial performance in the degradation process of the Phase II and Phase III investigations reported above and those of Bossert et al. [1] are due to differences in environmental parameters such as temperature, moisture, soil, microbial diversity/density, and other biotic or abiotic factors.

It is important to emphasize that waste sludges with different F-1 and F-2 content may not necessarily behave similarly under comparable conditions. However, in the above studies, all of the compounds analyzed exhibited decreases in concentration over time for both laboratory and field tests. The final closure activities will probably require one-and-a-half-years time treatment, provided the waste applications and reapplications can be carried out in three-months intervals as presently indicated. The costs of the total procedure are estimated to rise from originally $3 million to now $6 million. Postclosure monitoring of soil and leachate collected from the site is recommended for a time period of 30 years after setup of land-farming facilities.

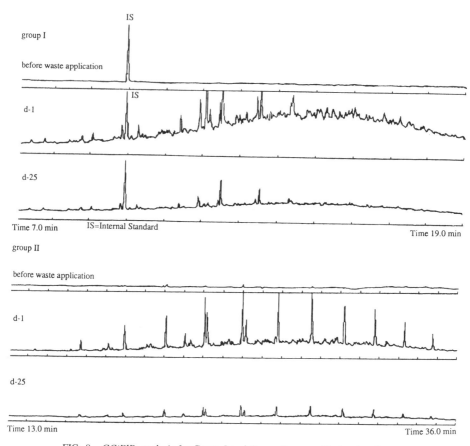

FIG. 8—*GC/FID analysis for Group I and Group II over a 25-day time frame.*

References

[1] Bossert, I., Kachel, W. M., and Bartha, R., *Applied and Environmental Microbiology,* April 1984, pp. 763–767.
[2] Jensen, H. Z., *Soil Science,* Vol. 30, 1930, pp. 59–77.
[3] Martin, J. P., *Soil Science,* Vol. 69, 1950, pp. 215–233.
[4] Holm-Hansen, O. and Booth, C. R., *Limnology and Oceanography,* Vol. 11, 1966, pp. 510–519.
[5] Van de Werf, R. and Verstraete, W. in *Proceedings, International Symposium Analytical Applications of Bioluminescence and Chemiluminescence,* Schram et al., Eds., 1979, pp. 333–338.
[6] Karl, D. M., *Microbiology Reviews,* Vol. 44, 1980, pp. 739–796.
[7] "Permit Guidance Manual on Hazardous Waste Land Treatment Demonstrations," EPA/530-SW-84-015, Dec. 1984.
[8] *Federal Register,* Vol. 49, No. 209, Oct. 1984, Rules and Regulations, Method 610, pp. 533–541.

Ralph J. Portier[1] *and Kuniko Fujisaki*[1]

Enhanced Biotransformation and Biodegradation of Polychlorinated Biphenyls in the Presence of Aminopolysaccharides

REFERENCE: Portier, R. and Fujisaki, K., "**Enhanced Biotransformation and Biodegradation of Polychlorinated Biphenyls in the Presence of Aminopolysaccharides,**" *Aquatic Toxicology and Hazard Assessment: 10th Volume, ASTM STP 971,* W. J. Adams, G. A. Chapman, and W. G. Landis, Eds., American Society for Testing and Materials, Philadelphia, 1988, pp. 517–527.

ABSTRACT: Fate analysis predictions of agricultural or industrial source toxicants in aquatic or marine environments can be affected by the presence or absence of adapted (preexposed) microbial populations and amenable substrates for cometabolic or cooxidative biotransformation. Polychlorinated biphenyls (PCBs) were evaluated in controlled laboratory microcosm systems in the presence of the aminopolysaccharide polymer chitin. Toxicants evaluated included the commercial Aroclors 1232, 1248, and 1254. Crosscouplings of each PCB mixture and polymer were noted, resulting in an optimization of epiphytic microbial metabolic activity. Significant primary degradation was noted for the lesser chlorinated biphenyls, that is, Aroclor 1232. Gas chromatography analysis of Aroclor 1248 and 1254 showed several tetra and penta isomers remaining. However, when polymer epiphytic populations were predominantly adapted microbial strains, that is, a *Pseudomonas* species and an *Acinetobacter* species, additional refractile isomers were metabolized. A comparison of kinetic rate constants based on biotransformation of each commercial mixture showed significant changes in half-life as correlated to isomer mixture, available polymer, and microbial inoculation.

KEYWORDS: polychlorinated biphenyls, chitin, biotransformation, biodegradation, preexposure

Microbial breakdown of any potential toxicant in a natural environment depends primarily on its relative concentration and its availability to the indigenous microbial community [1]. This is particularly true in coastal wetland environments where physicochemical processes and seasonal fluctuations of biomass material play an important role in toxicant concentration and availability. pH, salinity, temperature, and the relative concentration of preexposed microbial strains have also been demonstrated to be important variables to be considered in establishing a quantitative estimate for environmental fate [2]. In laboratory and subsequent field investigations, the aminopolysaccharide chitin has been identified as providing not only an amenable substrate for epiphytic microbial populations to colonize and survive but also can serve as an effective scavenger of dissolved organic components [3,4].

In this research presentation, the role of this sorbing component in affecting the environmental fate of commercial polychlorinated biphenyl (PCB) mixtures was evaluated. Although numerous studies dealing with the microbial metabolism of PCBs have been published over the last decade, most studies have concentrated on the interaction of microorganisms in aquatic and terrestrial systems under oligotrophic conditions. Furthermore, adapted microbial populations were not

---

[1] Associate professor and research associate, respectively, Aquatic/Industrial Toxicology Lab, Institute for Environmental Studies, Louisiana State University, Baton Rouge, LA 70803.

necessarily present. Most reports, that is, axenic and/or mixed culture research, present data on the primary degradation of lower chlorinated biphenyls or specific isomers. An excellent review of these studies was done by Furukawa [5]. The present study was undertaken to determine if an enhanced biodegradation/biotransformation phenomia for PCBs could be seen in a nutrient-rich coastal aquatic environment in the presence of organic amendments such as chitin. Contributions of adapted microbial strains colonizing these adsorptive detrital surfaces during the reductive dechlorination process were investigated.

Materials and Methods

Strain Isolation

A *Pseudomonas* strain IS140 and an *Acinetobacter* strain LS241 were isolated from PCB-contaminated sediments of a freshwater lake in Baton Rouge, Louisiana and colonized in a minimal salt solution broth containing biphenyl [6]. The minimal salts contained, in g/L of distilled water: $MgSO_4$, 10.0 g; $FeSO_4 \cdot 7\ H_2O$ 0.5 g; K_2HPO_4, 25.0 g; KH_2PO_4, 15.0 g; NH_4Cl, 27.7 g. Minimal salt solutions were sterilized by autoclaving. Biphenyl and Aroclors™ used in these investigations as primary carbon sources were filter sterilized (0.21-μm filter, Millipore Corp.) and aseptically added to the sterile broth. Both microbial strains were grown on a rotary platform shaker in the aforementioned mineral salt media containing 0.1% biphenyl. Residual and insoluble biphenyl or Aroclor components were removed by filtration of the culture through glass wool. As recommended by Sayler et al. [7], cells were harvested from the late exponential growth phase by centrifugation at 10 000 × g for 10 min. Cells were resuspended from centrifugation in 0.025 M PO_4 buffer, (pH 7.0) and served as inoculum for all aquatic microcosms.

Polychlorinated Biphenyls

Specific isomers evaluated in these investigations included 4,4' dichlorobiphenyl (4,4' DCB), 2,5,4' trichlorobiphenyl (2,5,4' TCB), and 2,4,5,2'5' pentachlorobiphenyl (2,4,5,2',5' PCB). Aroclor commercial mixtures included Aroclor 1232, Aroclor 1248, and Aroclor 1254. These isomer/congener mixtures represent an adequate sampling of the possible 209 PCB combinations. All toxicants were obtained from Foxboro Analabs.

Polysaccharide Preparation

The aminopolysaccharide, chitin, used in these aquatic microcosm experiments was derived from the freshwater crayfish *Procambarus clarkii*. The chitinous exoskeleton was removed and prepared following procedures previously outlined [4]. This material was dried at 60°C for 24 h. No attempt was made to crush or break up the remaining pieces of chitin so as to provide a natural rather than an artificial substrate composition for exposure to soluble PCB fractions and/or indigenous microbial populations.

Microcosm Systems

To maintain a precise controlled environment, continuous flow microcosm systems [8] were used in toxicant/substrate experiments. The unit consists of a 2400-mL reaction vessel in which toxicants or substrates can be pumped via peristaltic pumps. Accurate flow rates as low as 0.0042 mL/min can be obtained. Temperature is maintained by a heat lamp system regulated by a proportional indicating temperature controller. pH/Eh (redox potential) of the reaction vessel is maintained by a series of controllers connected to the peristaltic pumps or gas regulators. Samples were withdrawn aseptically from the vessel by means of micropipet or syringe. A complete discussion of the microcosm systems used in these investigations is reported elsewhere [8,9].

Polymer Toxicant Cross-Coupling Experiments

Initial microcosm experiments focused on the interaction of chitin in the form of the detrital-like particles with PCBs in sterile microcosm systems. Specific congeners, congener mixtures, and Aroclor mixtures were added in controlled dosages of 10 mg/L to the aquatic microcosm units by means of micropipettes. Each microcosm contained filter-sterilized freshwater collected from a freshwater swamp in the upper Barataria Bay estuary of Louisiana. Chitin at a total dry weight dosage of 0.5 g/L was introduced by means of peristaltic pump. Microcosm operating parameters were analogous to field conditions. Sedimentary materials were not used in these experiments so as to minimize the interference of clay materials occluding the more highly chlorinated congeners. The focus of these experiments was to ascertain relative biphenyl/polymer bioavailability. After exposure to the various PCBs mixtures, the chitin substrate was collected by means of filtration on glass wool filters and extracted using a solvent extraction procedure as outlined previously [8]. Recovery of PCB congeners and Aroclor mixtures on chitin is presented in Table 1. Overall recovery was 94.6% for control microcosms and 89.4% for chitin-amended microcosms.

Polymer/Toxicant/Adapted Microbial Population Experiments

A replicate series of microcosm experiments similar to the sterile experiments mentioned above were then conducted. In these microcosms, however, the preadapted isolates were introduced at a concentration of 10^5 cells/mL. Microbial populations were allowed to reach 10^6 cells/g dry weight of chitin before PCB mixtures were added.

Polymer/Toxicant/Heterotrophic Microbial Population Experiments

In these experiments, microcosm systems were operated under more real world conditions in that water and sedimentary materials collected from the field were used. Aluminum cylinder cores, 60 cm tall by 7.6 cm diameter, were inserted into the sediment to the water level, capped and

TABLE 1—*Gas chromatographic recovery of PCB congeners and Aroclor mixtures on chitinous surfaces.*

Toxicant	Initial Concentration, ppm	Final Concentration	Recovery, %
4,4' DCB	5	4.61 ± 0.2	92.2
2,5,4' TCB	5	4.78 ± 1.7	95.6
2,4,5,2',5' PCB	5	4.66 ± 0.31	93.2
Aroclor 1232[a]	5	4.18 ± 0.26	83.6
Aroclor 1248[b]	5	4.23 ± 0.41	84.6
Aroclor 1254[c]	5	4.38 ± 0.16	87.6

[a] Peak No. 4—(see Fig. 2).
[b] Peak No. 7—(see Fig. 3).
[c] Peak No. 9—(see Fig. 4).

sealed, and stored on ice for transportation to the laboratory for microbial analysis. All samples were processed within 12 h. Surface sediments from the particular sites, that is, the top 15 cm., were used to inoculate individual microcosms and to categorize the major microbial groups present.

Water/sediment phase tests were conducted in which isomers, 4,4' DCB, were introduced into continuous flow 2-L microcosms with sediment/water interfaces analogous to that of the natural environment. Replicate microcosm units were established for each isomer with ($N = 6$) and without ($N = 6$) chitin-amended sediments. An equal battery of microcosms served as controls for both chitin-amended ($N = 6$) and nonamended ($N = 6$) tests, respectively. The total number of all microcosms used for each isomer test was $N = 24$. Freshwater sedimentary materials were amended with 1.0 g of chitin from the freshwater crayfish *Procambarus clarkii*. Saline sedimentary materials were amended with a commercially available chitin (Sigma Chemical) from blue crab waste. Temperature, pH, and flow rate parameters and controls reflected *in situ* conditions. Chitin additions represented *in situ* measurements of chitin in freshwater environments [3,4] and marine environments [3,4].

Radiotracer Studies on Polymer/isomer Interactions

A series of replicate static microcosms [9] were used to evaluate possible changes or interferences in biopolymer recycling in the aquatic environment as a consequence of biphenyl exposure. 14 C-ul-N-acetyl D glucosamine (Pathfinders Laboratories) was introduced into sterile control and inoculated microcosms containing sediment and water from the Barataria Bay estuary. Total radioactive substrate added to each microcosm was 1.2 µCi/L. The isomers, 4,4' dichlorobiphenyl and 2,4,5,2',5' pentachlorobiphenyl, were introduced by means of micropipet prior to biopolymer addition. Chlorinated biphenyl concentrations were 5 mg/L (solvent introduced). Mineralization rates for labelled biopolymer in the presence or absence of these chemicals were determined by means of 14 CO_2 expiration using methods as outlined previously [3,4,8,9].

Chromatographic Methods

Analysis of chitin surfaces and residual water fractions for polychlorinated biphenyls followed methods as outlined for capillary column gas chromatography [10]. Controls, consisting of analysis of biphenyl residuals on glass surfaces and sterile chitinous surfaces, were conducted. Volatilization losses from microcosm units were monitored using resin traps as previously outlined [4,8].

Results and Discussion

Polymer/Toxicant Crosscoupling

Chitin, in the form of detrital material, was seen to be an effective adsorbing surface for water soluble PCB isomers, mixed congeners, and Aroclor commercial mixtures. As shown in Fig. 1, a combined commercial mixture containing both Aroclor 1232 and 1254, representing the possible mono, di, tri, tetra, and penta congeners, was effectively removed by chitin. Residual levels in the aqueous phase comprised less than 1.34 + 0.21% of the total PCB inoculum added. PCB congener distribution patterns have been reported in upper Hudson River sediments [10] as well as other locations worldwide. It is generally recognized that sedimentary materials, particularly fine silts and clays, have played an important role in not only transport of PCBs through the aquatic environment but also in epiphytic microbial colonization with deposition. Chitinous materials, having a pronounced affinity for adsorbing these chlorinated biphenyls in the aqueous phase, may also play an important role in overall transport and extent of contamination in bottom

PCB / POLYMER CROSS COUPLING

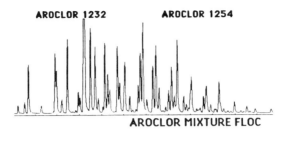

FIG. 1—*Adsorption of a combined Aroclor mixture on a chitinous surface.*

sediments. Such detrital materials, generally exhibiting a greater degree of bouyancy, would be a major toxicant transport mechanism and, as such, must be considered in possible modeling efforts [*11*].

Biotransformation of PCBs on Biopolymer Surfaces

In the above study, sterile detrital particles of chitin were shown to actively adsorb PCBs. In subsequent aqueous phase microcosm studies with adapted microbial populations, enhanced biotransformation of PCBs on these active surfaces was demonstrated. Figure 2 shows relative biotransformation of Aroclor 1232 for control and experimental (inoculated) microcosms afer 48 h. Significant loss of mono, di, and tri chlorobiphenyls was realized. Aroclor 1248 biotransformation in the presence of biphenyl utilizers on chitin is shown in Fig. 3. Again, most of the lesser chlorinated congeners have been biotransformed and/or biodegraded.

Aroclor 1254 was also evaluated for biotransformation on chitinous surfaces (Fig. 4). Trichlorobiphenyl congeners were significantly removed as was noted in the commercial mixtures of lower percent chlorine. Tetra isomers reflected partial biotransformation similar to Aroclor 1248. However, the more recalcitrant congeners, (Peaks 8 and 9), containing five or more chlorines, appear to also have been affected by epiphytic reductive dechlorination.

Chlorinated Biphenyl Influences on Biopolymer Cycling in an Estuary

From the experimental data presented, chitinous surfaces would appear to be a selective interface on which highly adapted microbial populations can actively cometabolize or cooxidize rather recalcitrant components of selected Aroclors and other PCB congeners. However, *in situ* microbial populations on detrital materials are not only heterogeneous but also not necessarily PCB acclimated or sensitive. Thus, epiphytic microbial populations were evaluated in microcosm systems having a sediment/water interface analogous to field conditions. Figure 5 presents toxicant/substrate

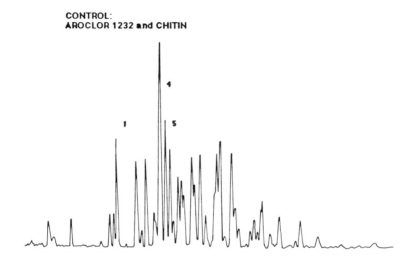

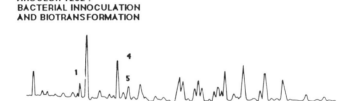

FIG. 2—*Biotransformation of Aroclor 1234 on chitinous surfaces by adapted microbial strains.*

profiles for three isomers at varying salinities in the presence and absence of detrital chitin. In freshwater control microcosms, the di and trichlorinated isomers were significantly degraded after 48 h. In chitin-amended microcosms, 4,4' DCB and 2,5,4' TCB were degraded in excess of 70%. The average % degradation rate for 2,4,5,2'5' PCB ranged from 29.2 + 0.4% degradation in control microcosms to 48.7 + 1.3% degradation in chitin-amended microcosms. With increasing salinity, enhanced biotransformation was still seen, particularly in marine microcosms where minimal preadaptation or preexposure to chlorinated biphenyls by the indigenous microflora occurs. For example, 2,4,5,2'5' PCB was only subjected to appreciable biotransformation in chitin-amended microcosms. Thus, indigenous microflora on chitinous surfaces may undergo a rapid selection resulting in the survival of either preadapted or nonsensitive microbial strains. These populations are subsequently more active in toxicant biotransformation, particularly since the availability of additional carbon, that is, chitin for cometabolic processes is also readily at hand.

Figure 6 presents data on substrate turnover of the chitin monomer N acetyl D glucosamine in the presence of these aforementioned isomers. The selective pressure on epiphytic chitinoclastic populations was noted for both 4,4' DCB and 2,4,5,2',5' PCB at 10 h. However, these populations recovered rapidly and were above initial levels at 30 h. This was also seen in 14 CO_2 expiration

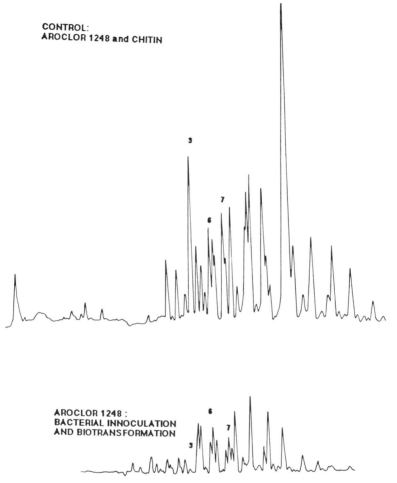

FIG. 3—*Biotransformation of Aroclor 1248 on chitinous surfaces by adapted microbial strains.*

profiles of the monomer substrate. Control microcosms significantly mineralized the N acetyl D glucosamine within a 15-h period. Microbial populations in the presence of 4,4' DCB reached a comparable degree of mineralization of 30 h. Populations exposed to the more highly chlorinated isomer, 2,4,5,2',5' PCB, reached a comparable mineralization of substrate only after 72 h. Thus, each isomer initially inhibited substrate turnover. However, after an initial incubation period, these populations proceeded to actively mineralize the biopolymer.

Conclusions

Fate estimates for important toxic chemical classes, such as polychlorinated biphenyls, may be biased when the availability and influence of naturally occurring substrates such as chitin within the aquatic/marine environment are not considered. Table 2 summarizes half-life estimates of specific isomers and congener mixtures analyzed in these investigations [*12*]. Generally, a more rapid biotransformation was seen in substrate amended experiments. In those experiments in which

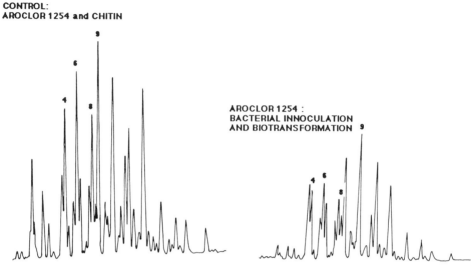

FIG. 4—*Biotransformation of Aroclor 1254 on chitinous surfaces by adapted microbial strains.*

preadapted biphenyl utilizers were in appreciable numbers, rate kinetics were improved. Comparable to laboratory and field studies reported in the literature [12,13], the degree of chlorination also influences rate kinetics. However, in the presence of chitin, the contribution of functional groups, that is, chlorine, attached to a parent molecule structure, that is, biphenyl ring, may be somewhat modified due to adsorption, allowing for enhanced biotransformation. Isomeric configuration may be a significant alteration attributable to chitin. Additional investigations on this aspect of polymer/toxicant crosscoupling is in progress.

Bacterial growth did not appear to be directly affecting toxicant solubilization as a consequence of surfactant production. Levels of chlorinated biphenyls in the aqueous phase were below detection limits (<0.01 mg/L). Residual levels on microcosm glass surfaces accounted for less than 0.5% of total biphenyl added.

Finally, the removal of competing heterotrophic microorganisms by a toxicant such as a chlorinated biphenyl may facilitate two important processes: (1) the rapid colonization of a toxicant specific adapted population microbial; and (2) induction of a subsequent cooxidative or cometabolic process resulting in a more complete biotransformation of the target toxicant. Additionally, this substrate/toxicant interaction may be more pronounced in saline microenvironments than in freshwater environments, particularly where readily available energy substrates such as chitin are in abundance. Thus, examination of changes to the epiphytic microbial community can serve as a basis for modeling the fate of a toxicant in a dynamic estuarine system as well as for forming the basis of interpreting the impact of the compound on primary trophic levels in estuarine food webs. The results of the present study show that polychlorinated biphenyls can be degraded aerobically by cometabolic/cooxidative microbial processes.

Acknowledgments

Portions of the work reported in this paper were funded from grant NA 81-AA-D-0046, National Oceanic and Atmospheric Administration (NOAA) Office of Sea Grant. Additional support was

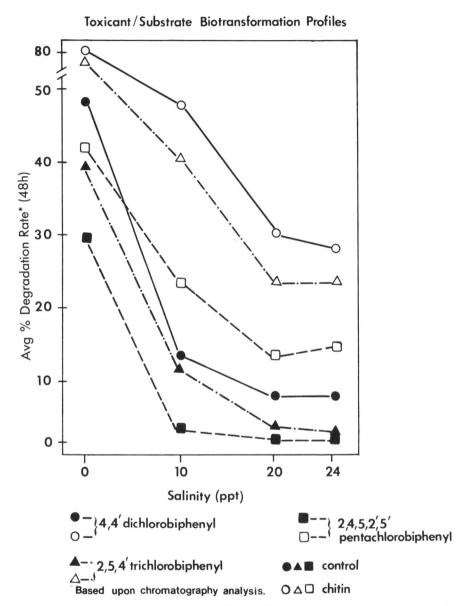

FIG. 5—*Toxicant/substrate biotransformation profiles of PCB isomers as affected by salinity: water/sediment phase microcosms.*

obtained from a research and development program funded by the Louisiana Board of Regents, Contract No. B20. The development of the microcosm systems presented in the paper was supported by Grant No. R-804976, U.S. Environmental Protection Agency.

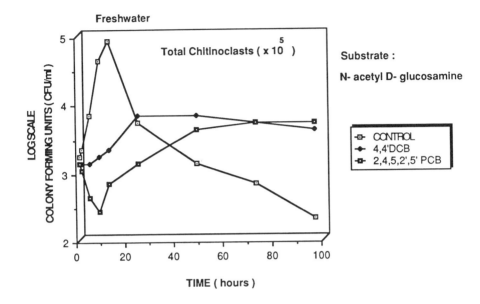
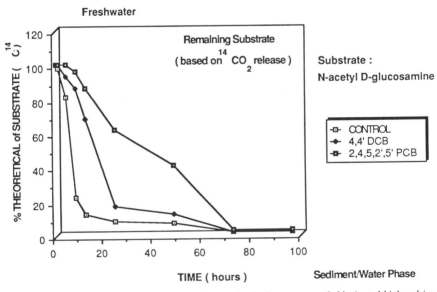

FIG. 6—*Biodegradation of ^{14}C-N-acetyl D-glucosamine in the presence of chlorinated biphenyl isomers.*

TABLE 2—*Half-life estimates[a] for chitin addition and chitin/adapted microbe addition of specific congener peaks.*

Toxicant	Polymer Addition	Microbe Addition	Half-Life, Days	Peak Nos.
4,4′ DCB	−	−	1.42 ± 0.41	1
	+	−	0.98 ± 0.21	1
	+	+	0.46 ± 0.33	1
2,5,4′ TCB	−	−	2.61 ± 0.14	1
	+	−	1.18 ± 0.34	1
	+	−	0.70 ± 0.18	1
2,4,5,2′,5′ PCB	+	−	1.32 ± 0.4	1
	+	+	0.80 ± 0.7	1
Aroclor 1232	−	−	61.4 ± 3.6	1,4,5
	+	−	33.4 ± 0.9	1,4,5
	+	+	26.8 ± 0.7	1,4,5
Aroclor 1248	−	−	77.6 ± 8.2	3,6,7,8
	+	−	38.6 ± 2.4	3,6,7,8
	+	+	31.9 ± 3.6	3,6,7,8
Aroclor 1254	−	−	81.9 ± 7.2	4,6,8,9
	+	−	36.4 ± 3.8	4,6,8,9
	+	+	35.5 ± 2.2	4,6,8,9

NOTE: Plus sign means yes, minus sign means no.
[a] Determined by gas chromatographic analysis of congener peaks with comparable retention times adjusted for % recovery of cumulative peaks.

References

[1] Alexander, M. in *Microbial Degradation of Pollutants in Marine Environments*, A. W. Bourquin and P. H. Pritchard, Eds., EPA-600/9-79-012 Environmental Protection Agency, Washington, DC.
[2] Gillett, J. W. in *Dynamics, Exposure and Hard Assessment of Toxic Chemicals*, R. Haque, Ed., Ann Arbor Science Publications, Ann Arbor, MI, 1980, pp. 231–249.
[3] Portier, R. J. and Meyers, S. P., *Developments in Industrial Microbiology*, Vol. 22, 1982, pp. 459–478.
[4] Portier, R. J., Chen, H. M., and Meyers, S. P., *Developments in Industrial Microbiology*, Vol. 24, 1983, pp. 409–424.
[5] Furukawa, K. in *Biodegradation and Detoxification of Environmental Pollutants*, A. Chakrabarty, Ed., CRC Press, Inc., Boca Raton, FL, 1982, pp. 33–51.
[6] Furukawa, K., Tamizuka, N., and Kamibayashi, A., *Applied Environmental Microbiology*, Vol. 38, pp. 301–310.
[7] Sayler, G. S., Shan, M., and Colwel, R. R., *Microbiology Ecology*, Vol. 3, pp. 241–255.
[8] Portier, R. J. and Meyers, S. P. in *Toxicity Screening Procedures Using Bacterial Systems*, D. Liu and B. Dutka, Eds., Marcel Dekker, Inc., New York, 1984, pp. 345–379.
[9] Portier, R. J. in *Validation and Predictabiity of Laboratory Methods for Assessing the Fate and Effects of Contaminants in Aquatic Ecosystems*, ASTM STP 865, T. P. Boyle, Ed., American Society for Testing and Materials, Philadelphia, 1985, pp. 14–30.
[10] Brown, J. F., Wagner, R. E., Bedard, D. L., Brennan, M. J., Carnahan, J. C., May, R. J. and Tofflemire, T. J., *Northeastern Environmental Science*, Vol. 3, 1984, pp. 231–244.
[11] Portier, R. J. in *Proceedings*, Second International Colloquin of Marine Bacteriology, IFREMER, Actes de Colloques, Vol. 3, 1986, pp. 579–587.
[12] Focht, D. D., and Brunner, W., *Applied Environmental Microbiology*, Vol. 50, 1985, pp. 1058–1063.
[13] Brunner, W., Sutherland, F. H., and Focht, D. D., *Journal of Environmental Quality*, Vol. 14, 1985, pp. 324–328.

Research Beneficial to the Standards Setting Process

Jane S. Hughes,[1] *Meryl M. Alexander,*[1] *and K. Balu*[2]

An Evaluation of Appropriate Expressions of Toxicity in Aquatic Plant Bioassays as Demonstrated by the Effects of Atrazine on Algae and Duckweed

REFERENCE: Hughes, J. S., Alexander, M. M., and Balu, K., "**An Evaluation of Appropriate Expressions of Toxicity in Aquatic Plant Bioassays as Demonstrated by the Effects of Atrazine on Algae and Duckweed,**" *Aquatic Toxicology and Hazard Assessment: 10th Volume, ASTM STP 971*, W. J. Adams, G. A. Chapman, and W. G. Landis, Eds., American Society for Testing and Materials, Philadelphia, 1988, pp. 531–547.

ABSTRACT: This investigation was undertaken to determine the toxicity of the herbicide atrazine to a variety of aquatic plant species as expressed by different endpoints, including the five-day EC_{50}, the five-day no-observed effect concentration (NOEC), and the phytostatic and phytocidal concentrations. The validity of each of these endpoints and the related issues of test duration and biological response are discussed.

Tests were conducted with two freshwater algae *(Anabaena flos-aquae* and *Navicula pelliculosa),* a marine alga *(Dunaliella tertiolecta),* and a freshwater vascular plant, *Lemna gibba* (duckweed). The test procedure consists of a five-day exposure phase followed by a nine-day recovery phase, with standing crop as the measure of effect on growth.

Of the four endpoints, the NOEC is the most conservative and is also the most sensitive to the experimental design and statistical procedures used. A 50% reduction in population growth (EC_{50}) is also conservative and is an acceptable parameter for a screening test. However, determination of the EC_{50} alone does not indicate the lethality of the test material or the recovery potential of the test species. Due to environmental significance and statistical considerations, the phytostatic and phytocidal concentrations are recommended as the primary responses.

KEYWORDS: aquatic plants, algae, toxicity tests, atrazine, phytostatic concentration, phytocidal concentration, *Anabaena flos-aquae, Navicula pelliculosa, Dunaliella tertiolecta, Lemna gibba,* duckweed

In a typical toxicity test with algae or aquatic vascular plants, a known biomass of the test species is inoculated into vessels containing growth medium treated with the test material at several different concentrations. The vessels are incubated under appropriate conditions for a specified amount of time, and the increase in biomass is measured either at intervals during the test and/or at the end of the test.

Disagreement exists, however, over the selection of an appropriate endpoint for these assays. Results have been expressed by determining if test concentrations caused significant differences in growth relative to the control [1–6]. The highest test concentration in which growth is not

---

[1] Project environmental scientist and environmental scientist, respectively, Malcolm Pirnie, Inc., White Plains, NY 10602.

[2] Senior environmental specialist, CIBA-Geigy Corp., Greensboro, NC 27419.

significantly different from that in the control may be termed the no-observed effect concentration (NOEC). Results have also been expressed as the concentration causing 50% inhibition of growth relative to the control (EC_{50}) [3,7–12]. Payne and Hall [13] questioned the validity of the EC_{50} for algal bioassays and developed a test procedure to determine the algistatic and algicidal concentrations. This test procedure employs a five-day exposure to the test material followed by a nine-day recovery period. The algistatic or, more generally, phytostatic concentration is defined as the concentration that allows no net growth of the population of test organisms during exposure to the test material but permits growth when the organisms are placed in test material-free medium. The algicidal or phytocidal concentration is defined as the lowest concentration tested which allows no net increase in population density during either the exposure or recovery period, that is, the organisms do not recover when transferred to test material-free medium.

The selection of an appropriate endpoint is also linked with the selection of an appropriate duration for the test. Test periods ranging from several hours [4,10,14] to several weeks [8,15,16] have been utilized. Another factor to be considered is the biological response to be measured. Standing crop or biomass (typically measured by dry weight, cell counts, or frond counts) is most often used [17–20], but some test methods recommend that results be expressed in terms of growth rate [3,17,21,22].

In this investigation, the toxicity of the herbicide atrazine to four aquatic plant species was expressed by different endpoints, including the EC_{50}, the NOEC, and the phytostatic and phytocidal concentrations. The significance of each of these responses is discussed and the related issues of test duration and biological response are considered.

Methods

Test Organisms

Separate tests were conducted with four aquatic plant species, including freshwater algae (the blue-green *Anabaena flos-aquae* and the diatom *Navicula pelliculosa*), a marine alga (the green flagellate *Dunaliella tertiolecta*), and a freshwater vascular plant (the duckweed *Lemna gibba*). These organisms are widely distributed in the environment, readily available, easily cultured, and have been recommended by the U.S. Environmental Protection Agency for use in toxicity testing [2,19,20,23].

Organisms used for testing came from laboratory stock cultures. The original cultures of the three algal species were obtained from the University of Texas Culture Collection, Austin, Texas. *Lemna gibba* G3 was originally obtained from C. F. Cleland, Smithsonian Institution Radiation Biology Laboratory, Rockville, Maryland. Stock cultures were maintained under the conditions described in Table 1 and transfers were made weekly into fresh medium.

Glassware

Test vessels were Erlenmeyer flasks fitted with foam stoppers to permit gas exchange. All glassware used in testing was cleaned in accordance with ASTM Practice for Algae Growth Potential Testing with *Selenastrum capricornutum* (D 3978–80), with the addition of an acetone rinse to remove organic contamination.

Media

A. flos-aquae was cultured and tested in algal assay procedure (AAP) medium [19] at pH 7.5 ± 0.1. The same medium, with the addition of 20 mg/L silicon [13], was used for culturing and

TABLE 1—*Culturing and testing conditions.*

	Anabaena flos-aquae	*Navicula pelliculosa*	*Dunaliella tertiolecta*	*Lemna gibba*
Medium	AAP	AAP with Si	Modified Burkholders ASW with NAAM	M-Hoaglands without EDTA or sucrose
Temperature	24 ± 2°C	20 ± 2°C	20 ± 2°C	25 ± 2°C
Light intensity, photoperiod, and type	2152 ± 323 lm/m^2 (200 foot-candles), continuous cool-white	4304 ± 650 lm/m^2 (400 foot-candles), continuous cool-white	4304 ± 650 lm/m^2 (400 foot-candles), continuous cool-white	5272 ± 785 lm/m^2 (350 foot-candles), continuous warm-white
Shaking	Manual, twice daily	Continuous, 100 rpm	Continuous, 100 rpm	None
Test vessels and solution volumes	500-mL Erlenmeyer flasks with 150-mL test solution	250-mL Erlenmeyer flasks with 50-mL test solution	250-mL Erlenmeyer flasks with 50-mL test solution	500-mL Erlenmeyer flasks with 200-mL test solution

testing *N. pelliculosa*. *D. tertiolecta* was cultured and tested in modified Burkholder's artificial seawater (ASW) with New Algal Assay Medium (NAAM) nutrients [23] at pH 8.0 ± 0.1. M-type Hoagland's medium without Ethylenediamine tetraacetate (EDTA) or sucrose [24] at pH 5.0 ± 0.1 was used for culturing and testing *L. gibba*. Media for the three algal species were filter-sterilized; medium for *L. gibba* was autoclaved.

Test Material

The test material was technical grade atrazine, (2-chloro-4-ethylamino-6-isopropylamino-S-triazine, CAS No. 1912-24-9). Atrazine, an inhibitor of photosynthesis, is widely used as a preemergence herbicide on broad-leaved plants. For each assay, test concentrations were prepared by adding the required volumes of a stock solution of atrazine (0.02 mg/mL in distilled deionized water without a solvent) to the appropriate test medium in volumetric flasks to yield nominal concentrations of 0.1, 0.2, 0.4, 0.8, 1.6, and 3.2 mg/L. Analytical confirmation of these nominal concentrations was not conducted. After thorough mixing, aliquots of each test treatment were added to each of three replicate test vessels. The control contained test medium with no additions. Test vessel sizes and solution volumes were as described in Table 1.

Inoculum

To begin each algal assay, an inoculum was prepared from a sample of a seven-day-old stock culture of that species. Population density in these samples was determined using an electronic particle counter as described below. For each algal species, a calculated volume of inoculum between 0.1 and 1.0 mL was aseptically added to each test vessel to yield a nominal initial concentration of 20 000 cells/mL.

For the assay with *L. gibba,* plant material used to start the test was obtained from two-day-old stock cultures. Four plants, consisting of four fronds each for a total of sixteen fronds, were aseptically added to each test vessel.

Incubation

Incubation conditions are given in Table 1. On each working day during each assay, temperature was recorded and the flasks were randomly repositioned to minimize spatial differences in the incubator.

Observations

The response of the test organisms was indicated by the increase in population density over time. Cell counts (or frond counts, for the duckweed assay) were performed on Days 3 and 5 of the exposure phase and Days 2, 6, and 9 of the recovery phase.

For the algal assays, cell counts were made using an electronic particle counter (Model ZBI Coulter counter with C-1000 channelyzer and MHR computer). On each counting day, a sample was collected aseptically from each test flask using an automatic micropipette with a sterile tip. The samples were placed in individual particle-free disposable containers and diluted with the electrolyte Isoton II (Coulter Electronics, Inc., Hialeah, Florida). Three counts per replicate were made and were multiplied by the appropriate conversion factors (for sample dilution and volume counted) to yield cells/mL. The Coulter counter was calibrated using an organic calibration material.

A special procedure had to be used to count *A. flos-aquae*. This species grows in filaments, which, during the logarithmic phase of growth, may be hundreds of cells long. These filaments must be broken up prior to counting. An effective method for reducing the length of the filaments without rupturing the cells is sonication. On each counting day, a 5.0-mL sample was collected aseptically from each flask, placed in a particle-free disposable container, and sonicated for 5 min in a water bath ultrasonic cleaning machine. Dilutions were then made of the sonicated samples and cell counts performed as described above.

Observations for the duckweed assay consisted of recording frond production and appearance. In order to eliminate subjective decisions on frond maturity, every frond visibly projecting beyond the edge of the parent frond was counted. As fronds age and die, they lose their pigmentation, becoming chlorotic. The number of chlorotic fronds was recorded, but not included in the total frond count. Frond counting was performed using a lighted magnifying lens.

Recovery Phase

After five days' exposure of the algal and duckweed cultures to atrazine, the recovery phase was initiated. For the algal assays, algal cells from cultures with Day 5 cell concentrations similar to or less than the initial inoculum level were centrifuged and resuspended in test material-free medium, following the procedure of Payne and Hall [*13*] for a nine-day recovery phase. (Due to incubator malfunction, the recovery phase had to be terminated on Day 5 for *D. tertiolecta*.) Duckweed from cultures with Day 5 frond counts similar to or less than the initial inoculum level were washed in successive rinses of filter-sterile distilled deionized water and transferred to test material-free medium, also for a nine-day recovery phase.

From each test concentration, triplicate flasks were inoculated with a calculated cell concentration of 3000 cells/mL for the algal assays and with 16 fronds for the duckweed assay. For *D. tertiolecta*, however, sufficient cells were available only to inoculate two replicates for the recovery phase, one at 2000 and one at 3000 cells/mL. The test vessels, the solution volumes, and the incubation conditions for the recovery phase were identical to those for the exposure phase.

Data Analysis

Mean standing crop values at five days (the end of the exposure period) were computed from the cell count or frond count data. For each assay, the mean values for each test concentration were expressed relative to that in the control.

For the algal assays, percent inhibition, I, was calculated according to the following formula

$$\%I = \frac{C - T}{C} \times 100 \tag{1}$$

where

C = mean cell number in the control, and
T = mean cell number in treated culture.

For the duckweed assay, percent inhibition was calculated according to the following formula

$$\%I = \frac{(C - O) - (T - O)}{C - O} \times 100 \tag{2}$$

where

C = mean frond number in the control,
O = original frond number (inoculum level), and
T = mean frond number in treated culture.

The $\log_{10}$ of concentration (x-axis) was plotted against percent inhibition expressed as probit (y-axis) to determine the five-day EC_{50} values. The line of best fit, the concentration corresponding to 50% inhibition, and the associated 95% confidence interval for this predicted value were determined by "inverse estimation" linear regression [25].

The five-day NOEC is defined as the highest concentration tested that had no significant effect, relative to the control, on the standing crop at five days. Significant differences were identified by one-way analysis of variance (ANOVA) and Duncan's [26] new multiple range test. Data were first tested for homoscedasticity (equality of variances) using the variance ratio test [27], comparing each variance to the control variance. ANOVA was performed on homoscedastic data while individual t-tests were performed on heteroscedastic data [27,28]. All tests of significance were at $\alpha = 0.05$.

For each assay, the phytostatic concentration with its associated 95% confidence interval was determined by "inverse estimation" linear regression analysis of $\log_{10}$ of the cell numbers (or frond numbers) at the end of the exposure period against the $\log_{10}$ of the concentration of test material. The phytostatic concentration is then selected by "inverse estimation" as that concentration of test material at which the $\log_{10}$ of the Day 5 cell or frond number is equivalent to the $\log_{10}$ of the initial inoculum level. At the phytostatic concentration, the ratio of the Day 5 cell (or frond) number to the Day 0 number is one. The phytocidal concentration is identified as the lowest concentration tested in which population growth does not occur during either the exposure or the recovery phase.

Results

The responses of the test species were quite similar (Fig. 1 and Table 2). All four species were significantly affected by the lowest test concentration, 0.1 mg/L, and consequently the NOEC's

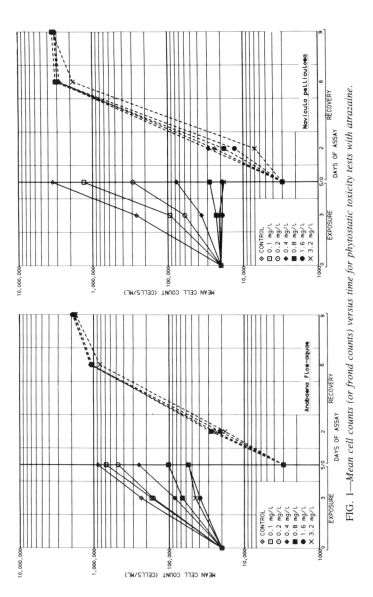

FIG. 1—*Mean cell counts (or frond counts) versus time for phytostatic toxicity tests with atrazine.*

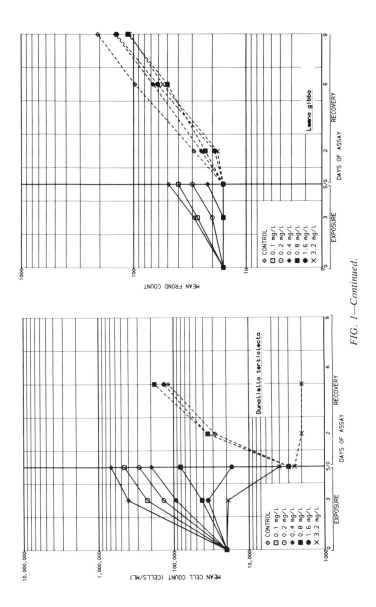

FIG. 1—Continued.

TABLE 2—*Nominal concentrations of atrazine (mg/L) producing effects in aquatic plants (95% confidence limits in parentheses).*

Species	NOEC	EC_{50}[1]	Phytostatic Concentration	Phytocidal Concentration
Anabaena flos-aquae	<0.1	0.23 (0.12 to 0.38)	4.97 (2.39 to 14.19)	>3.2
Navicula pelliculosa	<0.1	0.06 (0.002 to 0.21)	1.71 (0.40 to 13.22)	>3.2
Dunaliella tertiolecta	<0.1	0.17 (0.11 to 0.26)	1.45 (0.44 to 6.72)	3.2
Lemna gibba	<0.1	0.17 (0.13 to 0.23)	1.72 (0.31 to 24.1)	>3.2

for all species are <0.1 mg/L. The five-day EC_{50}s (Fig. 2) ranged from 0.06 to 0.23 mg/L, with *N. pelliculosa* the most sensitive and *A. flos-aquae* the least sensitive species for this parameter. The phytostatic concentrations (Fig. 3) for three of the species (*N. pelliculosa*, *D. tertiolecta*, and *L. gibba*) are alike: 1.71, 1.45, and 1.72 mg/L, respectively. The phytostatic concentration for *A. flos-aquae* was 4.97 mg/L, an extrapolated value higher than any of the test concentrations. Thus, *A. flos-aquae* was again the least sensitive species.

With the exception of *D. tertiolecta*, all species were able to recover from the growth inhibition that occurred during the exposure phase. For *A. flos-aquae*, *N. pelliculosa*, and *L. gibba*, the phytocidal concentrations exceeded the highest nominal test concentration, 3.2 mg/L. *D. tertiolecta* was able to recover from exposure to 1.6 mg/L atrazine but not from exposure to 3.2 mg/L.

Discussion

Comparison to Literature

Although the test methods, species, and endpoints used in this study are not identical to those used by other investigators, the results of this study are in general agreement with the literature. Other investigators have reported the toxicity of atrazine to various species of algae and aquatic macrophytes to be in the range of about 0.04 to 1.1 mg/L [10–12,22,29].

The responses of the algal species tested in this study are within the ranges previously reported for various algae. From the work of Hutber et al. [22] and Larsen et al. [10], it would appear that the blue-green algae are slightly less sensitive than other algal species, and the results of the *A. flos-aquae* assay reported here support this conclusion. The response of *L. gibba* was very similar to that of the algal species tested and also similar to literature values for other species of aquatic macrophytes [10]. The narrow range of atrazine concentrations within which toxicity occurs to aquatic plants is understandable since atrazine acts as an inhibitor of the photosynthetic process.

Test Design, Duration, and Growth Responses

In contrast to acute toxicity tests with aquatic animals, where the death of individuals is measured, the effect measured in aquatic plant toxicity tests is the growth of a population. Algal population growth follows an exponential curve, with a lag phase, a log phase, and a stationary phase. Duckweed population growth, however, is typically linear rather than exponential. Considerations of test design for duckweed assays are different from those for algae in several respects and are discussed separately.

In algal assays, a test material may affect any or all of three parameters: lag phase of growth, growth rate, and standing crop. These effects are illustrated in Fig. 4. Curve A represents a control culture. In Curve B, the test material has caused an increase in the lag phase of growth, but the maximum specific growth rate and the maximum standing crop eventually attained are no different from that in the control. Even in the presence of the test material, this effect is typically of short duration and the population "recovers." The environmental significance of an increase in the duration of the lag phase is unclear, but it is possible that subtle changes in phytoplankton population dynamics could result.

Curve C represents a depression of the maximum specific growth rate relative to the control, but the lag phase and maximum standing crop are unaffected. Miller et al. [19] stated that growth rate should not be used as a parameter in batch cultures since growth rate is indirectly related to external nutrient concentrations. However, Nyholm [30] argues that growth rate determines the competitive success of an algal species in a dynamic natural ecosystem.

In Curve D, neither the lag phase nor the maximum specific growth rate is affected, but the standing crop is reduced. Of the three parameters, the environmental significance of a decrease in standing crop is most easily understood. Curve E represents the type of response most often observed, in which both the lag phase of growth is affected and the maximum standing crop reduced by the test material. The growth rate at various time intervals is reduced, but the maximum specific growth rate is unaffected. In Curve F, all three parameters are affected.

Ideally, it would be desirable to measure effects on all three parameters. However, determination of lag phase and growth rate effects requires frequent time-consuming biomass measurements early in the assay, when they are most difficult to do because of low population densities. For algal toxicity screening tests, standing crop is recommended as the parameter of choice due to practical considerations and environmental significance.

Since standing crop varies over time in an exponentially growing population, the duration of the assay and the selection of the appropriate time interval at which to measure the response of the test species is crucial. If the assay is too short, standing crop effects may not yet be evident. In addition, accurate biomass measurements are most difficult to make during the first 24 h (or even 48 h, for some slower-growing species). If the assay is too long, the fate of the test material (especially in a static system) becomes important. Walsh [3] and Walsh and Merrill [31] observed a loss of toxicity after six or seven days in algal assays with effluents. Algal cultures are incubated under conditions of relatively high temperature and light intensity and often the test vessels are continuously shaken. These conditions would tend to promote any physical, chemical, or biological processes affecting the fate of the test material, particularly photodegradation and volatilization. Therefore, exposure of the algae to the test material for four or five days is recommended. If an additional biomass measurement is made on Day 2 or 3, impacts on the lag phase of growth can also be detected.

The growth of a duckweed population in a bioassay system is typically linear. Thus, a test material which affects the standing crop would also affect the growth rate and vice versa. Standing crop appears to be an appropriate response parameter for duckweeds. Because duckweed populations grow more slowly than algal populations, test duration should be at least four or five days and possibly as long as seven days. For both algal and duckweed toxicity tests, a test design which allows for both an exposure phase and a recovery phase will provide additional significant information, as discussed below.

Comparison of Endpoints

The four endpoints measured in this study can be categorized by their relative sensitivity. The NOEC was the most sensitive, followed by the EC_{50}, the phytostatic concentration, and the phytocidal concentration, in decreasing order. The advantages and disadvantages of each of these endpoints are discussed below.

The NOEC is the most conservative of the parameters. It is determined by an analysis of

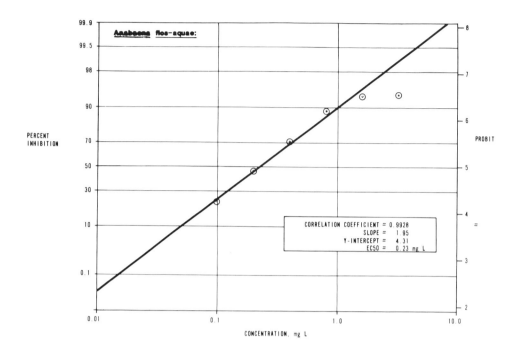
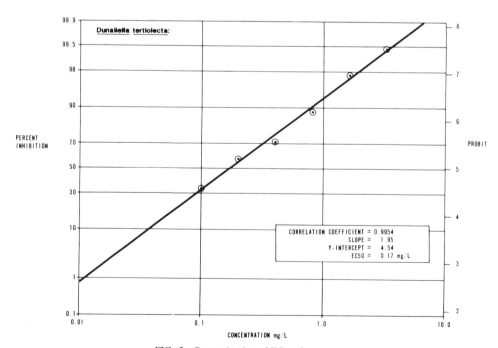

FIG. 2—*Determination of EC_{50} of atrazine.*

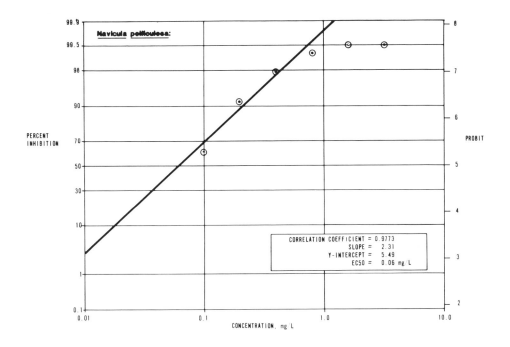

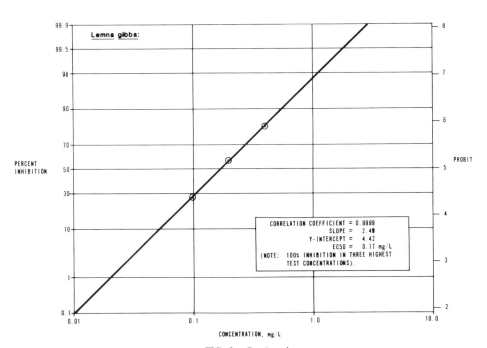

FIG. 2—*Continued.*

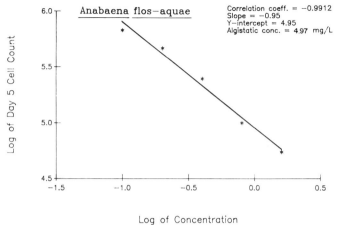
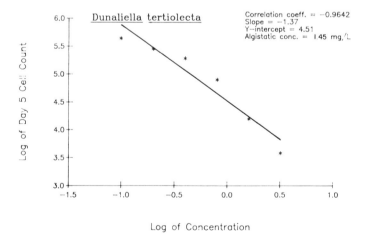

FIG. 3—*Determination of phytostatic concentration of atrazine.*

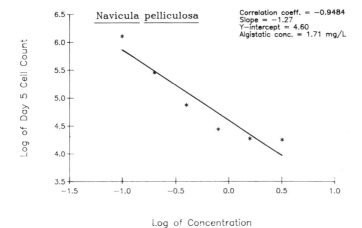

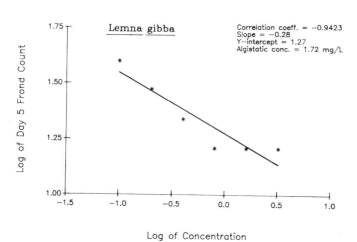

FIG. 3—*Continued.*

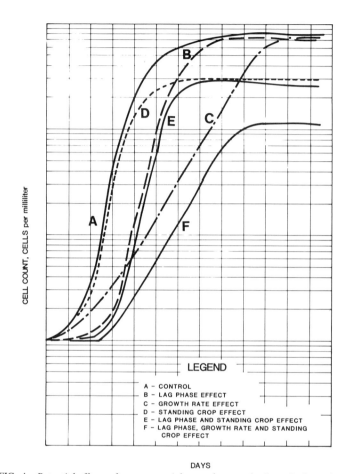

FIG. 4—*Potential effects of a test material upon the growth of an algal population.*

variance and a multiple comparison test as the highest test concentration resulting in no significant differences from the control. The NOEC, more than the other parameters, is sensitive to the selection of the level of significance, the statistical procedures used, sample size, and the selected test concentrations [32]. Whereas the point estimates in the atrazine studies were derived using most if not all of the data, the NOECs were determined in several cases by a t-test, which found that the means from the control and the lowest test concentration were significantly different. However, knowledge of the NOEC may be useful. The concentration of test material that does not affect an algal population during the first four or five days of exposure probably respresents a "safe" concentration.

The EC_{50} concept was designed to express the results of quantal or dichotomous data [1]. Probit or logit models are typically used to describe trends in this type of data, and the results are expressed as a point estimate. The data from an aquatic plant toxicity test is quantitative, since the response (growth) is continuous and not "all or none." However, point estimates can also be made from continuous data [32]. There are some limitations to recognize in the application of the EC_{50} concept to toxicity tests with aquatic plants. A 50% reduction in standing crop after four or five days exposure to the test material may be an overly conservative estimate of potential environmental effects; in a logarithmically growing population, a 50% deficit may be compensated for [13,19]. In addition, EC_{50}s and confidence limits cannot be calculated by the usual methods for aquatic plant toxicity tests. This is because the typical methods (probit, moving average, Spearman-Karber, binomial) require, in the calculations, the number of organisms affected relative to the number of organisms tested. In an aquatic plant toxicity test, the number of test organisms increases over time. To circumvent this problem, EC_{50}s and confidence limits for aquatic plant assays can be calculated using "inverse estimation" linear regression techniques. In the standard application of linear regression, a new value for the dependent variable (the response in a toxicity test) and its confidence limits are predicted. Inverse estimation allows for the prediction of a new value of the independent variable (the dose) and its confidence limits. Despite these limitations, the EC_{50} is an acceptable endpoint to use when comparing the relative toxicities of different test materials or the relative sensitivities of different species of aquatic plants. Caution must be used when comparing aquatic plant EC_{50} values to aquatic animal EC_{50} values, remembering that, at the EC_{50}, half of the aquatic plant population was inhibited but not necessarily killed.

As demonstrated by the results of the assays with atrazine, populations in which growth is completely inhibited by the test material may be able to recover completely in the absence of the test material. This is because the algal cells may be prevented from photosynthesizing (as is the case with atrazine) or dividing or some other physiologic process but they may not be killed. Some test materials may indeed kill the algal cells, but unless 100% of the cells are killed, the population potentially has the ability to recover. The phytostatic test procedure provides for the assessment of recovery potential.

The phytostatic concentration, which, like the EC_{50}, is a point estimate determined by a linear regression technique, has definite environmental significance and its use has been supported by others [17,33,34]. Since some exposure scenarios in the natural environment would consist of intermittent exposures of short duration (for example, seasonal application of a pesticide or herbicide), the ability of an algal population to recover is important. A continuous exposure to the phytostatic concentration would cause complete inhibition of growth. However, in the absence of a continuous input, recovery of the population would be expected as the test material degrades. Determination of the EC_{50} alone does not allow for assessment of recovery potential. The phytostatic procedure also allows for the determination of the phytocidal concentration, the concentration at which the cells are killed and recovery does not occur. The phytocidal concentration is not determined by any regression analysis or other statistical technique but is simply observed to be the lowest test concentration which inhibits growth completely during both the exposure and recovery phase. No confidence limits can be determined for this parameter.

Recommendations

A four- or five-day aquatic plant toxicity test to determine the EC_{50}, based on standing crop, is acceptable as a screening test. The EC_{50} is a conservative measure of toxicity for aquatic plant population growth, and a four- or five-day period (at least for algae, if not duckweed) represents a chronic exposure [34]. Depending upon the magnitude of the EC_{50} in comparison to an expected environmental concentration, further testing may not be necessary. The EC_{50} may also be used to compare the relative toxicity of different test materials or the relative sensitivity of different species. The limitations of the applications of the EC_{50} concept to aquatic plant toxicity testing must be understood and comparisons to EC_{50}s with animals made with caution.

The phytostatic test procedure provides a better assessment of toxic effects on an aquatic plant population. If substantial inhibition is observed from a four-or five day exposure to the test material, a recovery phase should be conducted and the phytostatic and phytocidal concentrations determined. It should be possible, in most cases, to determine the EC_{50} and the phytostatic and phytocidal concentrations from one well-designed assay, although additional test concentrations may be necessary. Such a test would provide a fairly complete assessment of toxicity to the aquatic plant species tested.

Acknowledgments

We would like to express our thanks to CIBA-GEIGY Corp. for supporting this study and Robert Black of Malcolm Pirnie for assistance with the statistical procedures.

References

[1] *Standard Methods for the Examination of Water and Wastewater,* 16th ed., American Public Health Association, American Water Works Association, Water Pollution Control Federation, Washington, DC, 1985.
[2] "Algal Assay Procedure: Bottle Test," U.S. Environmental Protection Agency, National Eutrophication Research Program, 1971.
[3] Walsh, G. E., Duke, K. M., and Foster, R. B., *Water Research,* Vol. 16, 1982, pp. 879–883.
[4] Stratton, G. W., Burrell, R. E., Kurp, M. L., and Corke, C. T., *Bulletin of Environmental Contamination and Toxicology,* Vol. 24, 1980, pp. 562–569.
[5] Gray, R. H., Hanf, R. W., Dauble, D. D., and Skalski J. R., *Environmental Science and Technology,* Vol. 16, No. 4, 1982, pp. 225–229.
[6] Subba-Rao, R. V. and Alexander, M., *Bulletin of Environmental Contamination and Toxicology,* Vol. 25, 1980, pp. 215–220.
[7] Chiaudani, G. and Vighi, M., *Mitteilungen des Internationalen Vereins der Limnolnologie,* Vol. 21, 1978, pp. 316–329.
[8] Vocke, R. W., Sears, K. L., O'Toole, J. J., and Wildman, R. B., *Water Research,* Vol. 14, 1980, pp. 141–150.
[9] Maki, A. W., Geissel, L. D., and Johnson, H. E., in *Investigations in Fish Control,* United States Department of the Interior, Fish and Wildlife Service, Washington, DC, 1975.
[10] Larsen, D. P., deNoyelles, Jr., F., Stay, F., and Shiroyama, T., *Environmental Toxicology and Chemistry,* Vol. 5, No. 2, 1986, pp. 179–190.
[11] Hollister, T. A. and Walsh, G. E., *Bulletin of Environmental Contamination and Toxicology,* Vol. 9, No. 5, 1973, pp. 291–295.
[12] Walsh, G. E., *Aquatic Toxicology,* Vol. 3, 1983, pp. 209–214.
[13] Payne, A. G. and Hall, R. H., in *Aquatic Toxicology (Second Conference), ASTM STP 667,* L. L. Marking and R. A. Kimerle, Eds., American Society for Testing and Materials, Philadelphia, 1979, pp. 171–180.
[14] Hannan, P. J. and Patouillet, C., *Journal of the Water Pollution Control Federation,* Vol. 51, No. 4, 1975, pp. 834–840.
[15] Miller, W. E., Greene, J. C., and Shiroyama, T. in *Proceedings of the Symposium on Terrestrial and Aquatic Ecological Studies of the Northwest,* 26–27 Mar. 1976, EWSC Press, Eastern Washington State College, Cheney, WA, pp. 317–325.
[16] Sturm, R. N. and Payne, A. G. in *Bioassay Techniques and Environmental Chemistry,* G. E. Glass, Ed., Ann Arbor Science Publishers, Inc., 1973, pp. 403–424.
[17] *Federal Register,* Vol. 50, No. 188, Friday, 27 Sept. 1985, Part 797, "Environmental Effects Testing Guidelines."
[18] Joubert, G., *Water Research,* Vol. 14, 1980, pp. 1759–1763.
[19] Miller, W. E., Greene, J. C., and Shiroyama, T., "The *Selenastrum capricornutum* Printz Algal Assay Bottle Test," EPA-600/9-78-018, Environmental Protection Agency, Corvallis, OR, 1978.
[20] Holst, R. W. and Ellwanger, T. C., "Pesticide Assessment Guidelines, Subdivision J, Hazard Evaluation: Nontarget Plants," EPA-540/9-82-020, Environmental Protection Agency, Washington, DC, 1982.
[21] *Guideline for Testing of Chemicals* No. 201, Alga, Growth Inhibition Test, Organization for Economic Cooperation and Development (OECD), 1981.
[22] Hutber, G. N., Rogers, L. J., and Smith, A. J., *Zeitschrift fur Allgemeine Mikrobiologie,* Vol. 19, No. 6, 1979, pp. 397–402.
[23] "Marine Algal Assay Procedure: Bottle Test," U.S. Environmental Protection Agency, National Environmental Research Center, Corvallis, OR, 1974.
[24] Hillman, W. S., *American Journal of Botany,* Vol. 48, No. 5, 1961, pp. 413–419.
[25] Snedecor, G. W. and Cochran, W. G., *Statistical Methods,* 6th ed., Iowa State University Press, Ames, IA, 1967.

[26] Duncan, D. B., *Biometrics,* Vol. II, 1955, pp. 1–42.
[27] Steel, R. G. D. and Torrie, J. H., *Principles and Procedures of Statistics,* McGraw-Hill, NY, 1980.
[28] Remington, R. D. and Schork, M. A., *Statistics with Applications to the Biological and Health Sciences,* Prentice-Hall, Englewood Cliffs, N.J., 1970.
[29] "Effects of Atrazine on Two Freshwater and Five Marine Algae," unpublished, CIBA-GEIGY Corp., Greensboro, NC, 1978.
[30] Nyholm, N., *Water Research,* Vol. 19, No. 3, 1985, pp. 273–279.
[31] Walsh, G. E. and Merrill, R. G. in *Algae as Ecological Indicators,* Academic Press, NY, 1984, pp. 329–360.
[32] Stephan, C. E. and Rogers, J. W. in *Aquatic Toxicology and Hazard Assessment: Eighth Symposium, ASTM STP 891,* R. C. Bahner and D. J. Hansen, Eds., American Society for Testing and Materials, Philadelphia, 1985, pp. 328–338.
[33] Bartlett, L., Rabe, F. W., and Funk, W. H., *Water Research,* Vol. 8, 1974, pp. 179–185.
[34] Elnabaraway, M. T. and Welter, A. N. in *Algae as Ecological Indicators,* Academic Press, NY, 1984, pp. 317–328.

Philip A. Lewis[1] *and William B. Horning*[1]

A Short-Term Chronic Toxicity Test Using *Daphnia Magna*

REFERENCE: Lewis, P. A. and Horning, W. B., **"A Short-Term Chronic Toxicity Test Using *Daphnia Magna*,"** *Aquatic Toxicology and Hazard Assessment: 10th Volume, ASTM STP 971,* W. J. Adams, G. A. Chapman, and W. G. Landis, Eds., American Society for Testing and Materials, Philadelphia, 1988, pp. 548–555.

ABSTRACT: A seven-day short-term chronic toxicity test using ten-day-old *Daphnia magna* is described as a possible alternative to the *Ceriodaphnia* short-term test. The test was begun on Friday so that weekend work was limited to feeding. The test was conducted at 25°C, and the animals were fed daily.

Five tests using sodium pentachlorophenate conducted during the development of the method gave test results similar to those previously reported for *Daphnia magna* chronic tests and comparable to *Ceriodaphnia* short-term tests conducted at the U.S. Environmental Monitoring and Support Laboratory (EMSL), Cincinnati. This test can be completed with about half the labor required for the *Ceriodaphnia* test, and the neonates can be easily counted without the aid of a microscope.

KEYWORDS: toxicity, chronic test, *Daphnia magna*, *Ceriodaphnia*, sodium pentachlorophenate

Measurement of the toxicity of effluents discharged into surface waters is required to meet the objectives of the Clean Water Act of 1977 and subsequent revisions. Since the biological integrity of indigenous communities of aquatic organisms in these receiving waters may be affected by toxic discharges, the toxicity of these effluents must be determined.

Cladocerans are extensively used for aquatic toxicity testing because they are readily available, adapt well to laboratory conditions, require little space, are easy and inexpensive to culture, are sensitive to a wide range of toxic substances, and give reproducible test results [1,2]. *Daphnia magna* have been used extensively for toxicity testing because: (1) they reproduce parthenogenetically, which allows for the establishment of clones with little genetic variability; (2) they can be maintained in a strictly defined media; (3) they reproduce rapidly, so that a good supply can be maintained for continuous testing; and (4) they have a relatively short life cycle, so that tests can be completed in a minimum amount of time [3–5].

Ceriodaphnia dubia has been recommended as a test species in effluent toxicity testing where time and space are critical [1] because its use requires a much smaller volume of water and the test can be completed in one third the time required for a *Daphnia* life cycle (21-day) test. Most of the difficulties caused by *Ceriodaphnia*'s strict nutritional requirements have been solved by using a recently developed diet (Don Mount, U.S. EPA, Duluth, MN, personal communication) consisting of a combination of digested trout chow, yeast, and cerophyl to which selenium and humic acid have been added. Nevertheless, the *Ceriodaphnia* test is labor intensive and counting of the young requires the use of a stereomicroscope.

---

[1] Aquatic biologist and chief, respectively, Aquatic Biology Section, Biological Methods Branch, Environmental Monitoring and Support Laboratory—Cincinnati, Office of Research and Development, U.S. Environmental Protection Agency, Cincinnati, OH 45268.

The method described in this paper is proposed as a possible alternative to the *Ceriodaphnia* short-term test because it is less labor intensive and the young are easily counted without the use of a microscope. The method has been successfully used in five preliminary tests using the reference toxicant sodium pentachlorophenate (NaPCP), and it is being evaluated for possible use in effluent toxicity testing.

The objectives of this paper are to describe the proposed short-term chronic method using *Daphnia magna* and to report on the results of five tests conducted at the U.S. Environmental Monitoring and Support Laboratory, Cincinnati, Ohio (EMSL-Cincinnati) using the reference toxicant NaPCP.

Test Procedures—General

This test method was developed in response to the need for a short-term chronic cladoceran test designed to fit into the normal work week and reduce the amount of weekend work. Accordingly, it is recommended that the test be started on Friday, the adults transferred to fresh test solution and the young counted on Monday and Wednesday, and the test terminated on the following Friday (in effluent testing changing test solutions daily is preferred). The first brood (and often the second brood) occurs sometime before the test solution is changed on Monday and succeeding broods occur at two- to three-day intervals thereafter so that three or four broods are produced during a seven-day test when conducted at 25°C.

Daphnia magna is the recommended test species because of its large size, ease of identification, availability, ease of handling, and past use. However, other species of daphnids, especially *Daphnia pulex,* may also be suitable for use with this method. The identity of any organisms used must be verified regardless of the source.

A culture should be started and maintained in good condition at least one month before the test is to begin in order to insure having healthy animals for the test. Only a few animals will be needed to start the cultures because of their prolific reproductive potential. It is often best to start the culture from one animal which is killed after producing young and slide mounted for positive identification. Culture methods are similar to those used in acute tests [6]. Cultures of *Daphnia magna* and *D. pulex* may be obtained from the U.S. Environmental Monitoring and Support Laboratory, Cincinnati, Ohio 45268 or from commercial sources.

Other procedures similar to the one recommended in this paper could be developed with *D. magna* and other species of daphnids (especially *D. pulex*). The primary objective is to obtain three or four broods and a reasonable number of young within a week or ten days and retain the inherent sensitivity of cladoceran life-cycle tests. The method we propose does achieve this objective in tests with NaPCP but further testing of the procedure, or modifications of it (for example, starting the test with younger animals), may be needed with other chemicals and with effluents to confirm its utility.

Proposed Test Procedures

Facilities and Equipment

Facilities must include equipment for culturing, holding, and acclimating the test organisms as described in ASTM Method for Conducting Static Acute Toxicity Tests on Waste-Waters with *Daphnia* (D 4229-84) and U.S. EPA Acute Toxicity Test Manual [6]. Temperature control may be accomplished by use of water baths, environmental chambers, aquarium water heaters, thermostatically controlled room heaters, and air conditioners. Test facilities must be free from dust and other sources of contamination. A good source of water for culturing and dilution should be readily available.

Dilution Water

One of the advantages of this method is that reconstituted water [hardness 160 to 180 mg/L as calcium carbonate ($CaCO_3$)] can be used for culturing and as the dilution water as well as surface water, well water, or (as a last resort) dechlorinated tap water. Reconstituted hard water is recommended because it is easily prepared, is of known quality, produces predictable results, and allows adequate growth and reproduction. The preparation of reconstituted water from distilled or deionized water is described in most aquatic toxicology test manuals [6,7].

Food and Feeding

The daphnids in the cultures and test vessels are fed a diet of trout chow-cerophyl-yeast suspension prepared by adding 6.3 g of trout chow pellets, 2.6 g of dried yeast, and 0.5 g of cerophyl to 500 mL of distilled or deionized water in a blender. The mixture is stirred at high speed for 5 min and allowed to settle in the refrigerator for 2 h. The supernatant, containing approximately 10 mg (dry weight) per mL, is poured into small (50 to 100 mL) polyethylene bottles with screw caps and frozen until needed [6].

The bottles must be removed from the freezer in time to thaw before feeding time. After thawing, the food mixture may be kept in the refrigerator for a maximum of one week. After one week in the refrigerator, the unused portion is discarded. Feed at the rate of 1.5 mL per litre of media on alternate days in the cultures and 0.05 mL per adult per day in the test vessels.

There are other foods that may be used for *Daphnia* [8,9] but these other foods may result in changes in the development time of the daphnids and require that the test start with younger (or older) animals. The goal is to have the first brood produced soon after the test begins and continue the test for two or three additional broods; however, tests started with animals that have already produced one brood and produce at least three additional broods during the test should be valid.

Obtaining Animals for the Test

In order to produce similar aged animals which will reach reproductive maturity at the proper time to begin the test, it is necessary to transfer ten adult *Daphnia* containing mature eggs from the laboratory culture to each of six 250-mL glass beakers (brood beakers) containing 200 mL of dilution water (for example, reconstituted hard water) on the Monday of the week before the test is to begin (eleven days prior to the day the test is to begin). Food (0.3 mL per beaker) is added and the beakers are incubated at 25°C for 24 h. After 24 h newly hatched neonates are removed from the brood beakers and placed ten to a beaker in other 250-mL beakers (holding beakers) containing 200 mL of dilution water. Adults in the brood beakers are discarded. Neonates in the holding beakers are fed 0.3 mL of food per beaker and incubated at 25°C. These maturing animals are fed every other day and on the day the test begins before the animals are transferred to the test solutions. On the sixth day of the holding period the media is replaced in the holding beakers.

If the test is to be conducted in the field, the beakers containing the maturing daphnids should be emptied into clean 500-mL polyethylene bottles for transport to the test site within 24 h. If more than a day will be spent in travel to the test site, the test animals could be shipped by express mail to arrive after the on-site laboratory has been prepared.

Test Conditions

The test is begun on Friday with each 100-mL glass beaker containing 50 mL of test solution and one ten-day-old animal. Ten animals are used for each treatment and the control. Each beaker is provided with 0.05 mL of trout chow food mixture at the beginning of the test. (Another procedure is to use two 400-mL glass beakers each containing 200 mL of test solution and five animals per treatment, but this procedure gives less statistically powerful results and is not recommended in definitive tests.) Daily feeding of 0.05 mL of food mixture per animal is recommended, however feeding 0.1 mL on Friday and skipping one day over the weekend may be acceptable. The animals are transferred to fresh test solution and the neonates counted and removed from the beakers on Monday and Wednesday. The test ends on Friday. Changing the test solutions daily is often preferred, especially if the toxicant is easily degradable or when testing effluents.

Test temperature is very important because *Daphnia magna* will not produce three broods in seven days at temperatures below 24°C when using trout chow as food. Temperatures over 26°C cause temperature stress; thus a test temperature of 25 ± 1°C is recommended.

At least five concentrations of wastewater or toxicant and a control should be used in conducting the test. These test concentrations should be based on a preliminary range-finding test. At least one concentration should show no effect and one should adversely affect the test organisms reproductive capacity. At least 80% of the animals should survive in the controls [7] and they should produce a minimum of ten young per adult. Lower numbers often occur if the cultures are being stressed, and use of these animals may cause invalid test results.

Temperature and dissolved oxygen (DO) should be monitored daily in at least one beaker from each treatment and control. If DO falls below 4 mg/L, gentle aeration should be provided. Hardness (as $CaCO_3$) and pH should be measured in at least one beaker in the control and high test concentration each time the test solutions are changed.

Test Method Development Procedures

Facilities and Equipment

The tests were conducted at the Newtown Facility, EMSL-Cincinnati using methods for culturing, holding, and acclimating of the test organisms as described in ASTM D 4229-84 and the U.S. EPA Acute Toxicity Test Manual [6]. Temperature control was accomplished by use of a water bath. All glassware was washed thoroughly in soap and water and rinsed with tap water, acetone, hydrochloric acid, and deionized water.

Dilution Water

Reconstituted hard water [160 to 180 mg/L (as $CaCO_3$)] [6], made up from distilled or deionized water that had been run through a Milli-Q Water Purification System, was used for culturing and for test dilution water. The highest concentration of test solution was made up from ampules of NaPCP obtained from the Quality Assurance Branch, EMSL-Cincinnati. The remaining treatment concentrations were prepared by 1:1 dilutions resulting in 0.8, 0.4, 0.2, 0.1, and 0.05-mg/L treatments used in all of the tests reported in this paper.

Food and Feeding

The *Daphnia* were fed a diet of trout chow-cerophyl-yeast suspended in distilled water at the rate of 1.5 mL per litre of media on alternate days in the cultures and 0.05 mL per adult per day in the test vessels. In the first few seven-day tests conducted during the development of this test method the media was changed and the animals were fed 1.0 mL of food every other day, as was done during the 21-day test. However, in all the seven-day tests reported here the *Daphnia* were fed every day and the media was changed every other day.

The *Ceriodaphnia* were fed a diet of digested tropical fish flakes (Mardel Laboratories, Villa Park, IL 60181) mixed with cerophyl and yeast at the rate of 3 mL per litre in the cultures and 0.1 mL per animal per day in the tests. The food was prepared by the following procedure:

1. Digested flake food.
   a. Place 5 g of crushed tropical fish flakes in 500 mL of deionized water and blend at high speed for 5 min.
   b. Add 500 mL of deionized water and aerate for seven days.
   c. Strain through 52-μm netting.
2. Cerophyl.
   a. Place 5 g of cerophyl or alfalfa in 500 mL of deionized water and blend at high speed for 5 min.
   b. Add 500 mL of deionized water and strain through 52-μm netting.
3. Yeast.
   a. Place 5 g of dried yeast in 500 mL of deionized water and blend for 5 min at high speed.
   b. Add 500 mL of deionized water and mix well.
4. Mixing of individually prepared components.
   a. Mix 800 mL of the digested flake food, the cerophyl, and the yeast that were made in Steps 1 to 3 and pour into small capped containers and freeze.

The food was thawed as needed and then kept in the refrigerator, but was not used after having been thawed for more than one week.

Obtaining Animals for the Test

A *Daphnia magna* culture has been maintained at the Newtown Facility for several years in 4-L glass beakers. The animals in each beaker are fed 4.5 mL of trout chow suspension every other day and thinned twice a week. About three fourths of the culture media is replaced with reconstituted hard water each Monday. In order to have ten-day-old animals ready for the test it was necessary to establish holding beakers. Eleven days prior to the day the test was to begin ten adult *Daphnia* containing mature eggs were transferred from the laboratory cultures to each of six 250-mL glass beakers (brood beakers) containing 200 mL of reconstituted hard water. Food (0.3 mL of trout chow) was added and the beakers were incubated at 25°C for 24 h. After 24 h, newly hatched neonates were removed from the brood beakers and placed ten to a beaker in other 250-mL beakers (holding beakers) containing 200 mL of dilution water. Neonates in the holding beakers were fed 0.3 mL of food per beaker and incubated at 25°C. These maturing animals were fed every other day. Food was also added on Friday before the animals were transferred to the test solutions. On Monday of test week the media was replaced in the holding beakers. During the holding period of the early tests, the media was not changed in the holding beakers, but this resulted in retardation of maturation. The media was changed on Day 6 (Monday) in the holding beakers of all of the tests reported in this paper.

The 21-day *Daphnia magna* test was started with less than 24-h-old animals obtained by the same procedure used for the acute tests during a previous study [*10*].

The animals used to start the *Ceriodaphnia* test were obtained from established cultures at the Newtown Facility using the procedure described in EPA Method 1002.0 [7].

Test Conditions

All of the *Daphnia* and *Ceriodaphnia* tests described here were begun on Friday using one animal per test vessel and ten animals for each treatment and the control. The tests were conducted at 25 ± 1°C. Five concentrations of the toxicant and a control treatment were used in conducting all of the tests. All tests had at least 80% control survival and a minimum of ten young per adult.

Temperature and dissolved oxygen (DO) were monitored in at least one beaker from each treatment and the control daily. Hardness (as $CaCO_3$) and pH were measured in at least one beaker in the control and high test concentration each time the test solutions were changed.

The 21-day test was conducted according to the procedure described by Stephan [11] except that the trout chow used was mixed with cerophyl and yeast as described in this paper and fed at a rate of 0.05 mL (0.5 mg) per animal per day.

The same procedure proposed in this paper was used for all the 7-day *Daphnia magna* tests and was very similar to the procedure used for the 21-day test, except that the *Daphnia* were ten days old when the test began. Most of these animals had already produced one brood before the test began. During the early stages of the development of this test method we conducted a test using two 400-mL beakers containing five *Daphnia* each per treatment. This resulted in poor survival and reproduction and was abandoned in favor of the method proposed in this paper.

The *Ceriodaphnia* tests were conducted according to EPA Method 1002.0 [7] except that the tropical fish flakes was used in place of trout chow in the food mixture.

Data Analysis

The LC_{50} and NOEC (No observed effect concentration) were determined for the test results. The seven-day LC_{50} was obtained using the Moving Average Angle Method [6]. NOEC was obtained by use of Dunnett's procedure [7]. The average number of young per adult was found by the method described by Hamilton [12] for the seven-day *Ceriodaphnia* test. The number of young per live adult was determined each day and, at the end of the test, these were added to give the number of young per adult based on the average number of live adults. If an adult died between one day and the next and produced young during that time, it is assumed to be half a live adult.

Results of Preliminary Tests

Results of five short-term *Daphnia magna* tests, using the method described in this paper with the reference toxicant NaPCP, appear in Table 1. Although not specifically designed as precision tests (the test concentrations were not verified analytically), the test data indicate that variability of results using this test method is in the normal range for Cladoceran toxicity tests [10].

These results compare well with NaPCP data for chronic toxicity tests with *Daphnia magna* reported in the literature, a 21-day chronic test with *D. magna* conducted at EMSL-Cincinnati, and eleven short-term *Ceriodaphnia* tests conducted at EMSL-Cincinnati as shown in Table 2.

Discussion

Geiger et al. [13] described a seven-day minichronic test that was started with 24-h neonate *D. pulex* to be used for predicting the effects of water-soluble fractions of hydrocarbons and a simulated refinery effluent. Growth (length of the preadult) was predictive of the chronic effect with the same degree of accuracy as reproduction in the 21-day test.

Adams and Heidolph [14] reported the results of using application factors with short-term chronic toxicity tests as a means of estimating the 21-day MATC for 20 chemicals. *D. magna* were used in 24- and 48-h acute and 7-, 14-, and 21-day chronic tests starting with less than 24-h old neonates. Both the 48-h acute test and the 7-day test could be used with an application factor to estimate the 21-day results using survival and reproduction, respectively. They also

TABLE 1—*Results of five seven-day short-term chronic tests with* Daphnia magna *using NaPCP.*

Test Number	Control Survival, %	X̄ Young Per Adult In Control	LC_{50}, mg/L	LOEC, mg/L	$NOEC^a$, mg/L
1	100	31.4	0.25	0.40	0.20
2	90	17.5	0.46	0.80	0.40
3	100	19.1	0.41	0.40	0.20
4	100	22.7	0.50	0.80	0.40
5	100	37.9	0.40	0.40	0.20
Mean	98	25.7	0.40	0.56	0.28
SD	4.5	7.8	0.09	0.20	0.10
CV, %	4.6	30.2	23.5	35.0	35.0

NOTE: SD = standard deviation and CV = coefficient of variation.

[a] NOEC is survival NOEC. Reproductive NOEC was the same as for survival except for Test 1 where reproductive NOEC was 0.40 mg/L.

determined that growth was a more precise measurement than reproduction but a larger application factor was necessary.

Nebeker et al. [15] described a ten-day *D. magna* chronic test started with five-day-old *Daphnia* for use in screening freshwater sediments for toxicity.

The test described in this paper has advantages over all these daphnid tests. It lasts only seven days, the tedious work of measuring and weighing the organisms is eliminated, and the results are more accurate than when an application factor is used.

Advantages of using *D. magna* as an alternate to *Ceriodaphnia* include: (1) they are much better known both taxonomically and ecologically because of their long use as test organisms; (2) they are much easier to handle and counting of the neonates is much faster; (3) culturing and testing methods are better defined; (4) food requirements are better understood; (5) test media need not be changed daily; and (6) the test can be conducted with about half the man hours needed for the *Ceriodaphnia* test (the *Ceriodaphnia* test takes about 15 to 20 man hours, while the *Daphnia magna* test requires about 9 or 10 man hours).

The seven-day short-term *Daphnia magna* toxicity test using ten-day-old preadults and survival or reproduction as end points as described in this paper has proven to be useful in assessing chronic effects of the reference toxicant NaPCP and may prove to be useful in predicting chronic

TABLE 2—*Comparison of NaPCP chronic test data (mg/L) on Cladocerans.*

Species	Duration	No.	LC_{50} Mean	CV, %	NOEC Mean	CV, %	Reference
Daphnia magna	7-day	5	0.40 (0.25 to 0.50)	23.5	0.28 (0.20 to 0.40)	35.0	This paper
	21-day	2	0.43 (0.40 to 0.47)	11.4	0.32	. . .	[5]
	21-day	$?^a$	0.40	. . .	0.34	. . .	[16]
	21-day	1	0.15	. . .	0.20	. . .	This paper
Ceriodaphnia	7-day	11	0.32 (0.18 to 0.50)	41.1	0.23 (0.10 to 0.40)	34.5	This paper

NOTE: The numbers in parentheses give the range.

[a] The authors in Ref. *16* did not give any information on number of tests or the test precision.

effects of effluents and other toxic materials. These results are also comparable to data generated with the seven-day *Ceriodaphnia* test using the reference toxicant NaPCP.

References

[1] Mount, D. I. and Norberg, T., *Environmental Toxicology and Chemistry*, Vol. 3, 1984, pp. 425–434.
[2] Elnabarawy, M. T., Welter, A. N., and Robideau, R. R., *Environmental Toxicology and Chemistry*, Vol. 5, 1986, pp. 393–398.
[3] Lal, H., Misra, V., Viswanathan, P. N., and Murti, C. R. R., *Ecotoxicology and Environmental Safety*, Vol. 8, 1984, pp. 447–450.
[4] *Standard Methods for the Examination of Water and Wastewater*, 14th ed., American Public Health Association, Washington, DC, 1975, pp. 762–766.
[5] Adema, D. M. M., *Hydrobiologia*, Vol. 59, 1978, pp. 125–134.
[6] Peltier, W. and Weber, C. I., "Methods for Measuring the Acute Toxicity of Effluents to Aquatic Organisms (Third Edition)," EPA/600/4-85/013, Environmental Monitoring and Support Laboratory—Cincinnati, U.S. Environmental Protection Agency, Cincinnati, OH, 1985.
[7] Horning, W. B., II and Weber, C. I., "Short-Term Methods for Estimating the Chronic Toxicity of Effluents and Receiving Waters to Freshwater Organisms," EPA/600/4-85/014, Environmental Monitoring and Support Laboratory—Cincinnati, U.S. Environmental Protection Agency, Cincinnati, OH, 1985.
[8] Cowgill, U. M., Emmel, H. W., Hopkins, D. L., Takahashi, I. T., and Parker, W. M., *Internationale Revue Gesamten Hydrobiologie*, Vol. 71, 1986, pp. 79–99.
[9] Taylor, M. J. in *Aquatic Toxicology and Hazard Assessment: Seventh Symposium, ASTM STP 854*, R. D. Cardwell, R. Purdy and R. C. Bahner, Eds., American Society for Testing and Materials, Philadelphia, 1985, pp. 53–72.
[10] Lewis, P. A. and Weber, C. I. in *Aquatic Toxicology and Hazard Assessment: Seventh Symposium*, ASTM STP 854, R. D. Cardwell, R. Purdy and R. C. Bahner, Eds., American Society for Testing and Materials, Philadelphia, 1985, pp. 73–86.
[11] Stephan, C. E., "Proposed Standard Practice for Conducting Renewal Life Cycle Toxicity Tests with the Daphnid, *Daphnia magna*," Draft No. 2, 12 Oct. 1977, American Society for Testing and Materials, Philadelphia.
[12] Hamilton, M. A., *Environmental Toxicology and Chemistry*, Vol. 5, 1986, pp. 205–212.
[13] Geiger, J. G., Buikema, A. L., Jr., and Cairns, J., Jr. in *Aquatic Toxicology: Third Symposium, ASTM STP 707*, J. C. Eaton, P. R. Parrish, and A. C. Hendricks, Eds., American Society for Testing and Materials, Philadelphia, 1980, pp. 13–26.
[14] Adams, W. J. and Heidolph, B. B. in *Aquatic Toxicology and Hazard Assessment: Seventh Symposium, ASTM STP 854*, R. D. Cardwell, R. Purdy and R. C. Bahner, Eds., American Society for Testing and Materials, Philadelphia, 1985, pp. 87–103.
[15] Nebeker, A. V., Cairns, M. A., Gakstatter, J. H., Malueg, K. W., Schuytema, G. S., and Krawczyk, D. F., *Environmental Toxicology and Chemistry*, Vol. 3, 1984, pp. 617–630.
[16] Adema, D. M. M. and Vink, G. J., *Chemosphere*, Vol. 10, 1981, pp. 533–554.

A Look to the Future

John Cairns, Jr.[1]

Integrated Resource Management: The Challenge of the Next Ten Years

REFERENCE: Cairns, J., Jr., **"Integrated Resource Management: The Challenge of the Next Ten Years,"** *Aquatic Toxicology and Hazard Assessment: Tenth Volume, ASTM STP 971,* W. J. Adams, G. A. Chapman, and W. G. Landis, Eds., American Society for Testing and Materials, Philadelphia, 1988, pp. 559–566.

ABSTRACT: The past ten years have been notable because of the development of the process of hazard evaluation in which chemical fate was coupled in a systematic and orderly way with biological effects. This development was amplified in the five-book Pellston series that began with *Estimating the Hazard of Chemical Substances to Aquatic Life, ASTM STP 657* and in the ASTM Aquatic Toxicology and Hazard Assessment series, of which this is the 10th volume. The future, however, is intriguing, so my comments will focus on that.

The next ten years will bring a concerted effort to use the results of hazard evaluation and aquatic toxicology, as well as other types of environmental and ecological information, in the process of integrated resource management. The purpose of the hazard assessment activities of Committee E-47 is to help prevent ecosystem damage. The purpose of integrated resource management, on the other hand, is to optimize performance of the ecosystem, which requires different information and a different management strategy than hazard assessment. Although much is written about the management of large natural systems such as drainage basins, little effective action has been taken. Institutional arrangements fragment the responsibility so that federal and state regulatory agencies and other organizations are often at odds and sometimes in direct conflict in their attempts to optimize that portion of resource management assigned to them.

Integrated resource management will require that some of the enormous resources now being spent on litigation be redirected into problem solving. Some of the prescriptive regulations, which now pit the regulators against the regulated in such ways that obvious solutions to simple problems cannot be implemented, must be changed. Since there is very little scientific and institutional information about integrated resource management, both of which will be essential for systems level quality control, and since funding to generate this information is not likely to come from traditional sources, some acknowledgement of the value of generating this information must be incorporated in the planning.

It is essential that committees of standards-setting organizations engaged in environmental assessment and related activities become much more active in developing integrated resource management approaches. Unless this is done soon and skillfully, even the finest standard methods will not be used effectively. Without integrated resource management, information will be fragmented, often inappropriate for the decision being made, or in apparent conflict with other equally valid information. In short, those engaged in environmental surveillance, monitoring, and hazard prediction must pay more attention to the structure of the management systems into which their methods will flow.

KEYWORDS: environmental management, interdisciplinary studies, multimedia projects, resource utilization, multiple use

---

[1] University distinguished professor, Department of Biology, and director, University Center for Environmental and Hazardous Materials Studies, Virginia Polytechnic Institute and State University, Blacksburg, VA 24061.

The following is a list of assumptions on which my views are based.

1. We are in a major transitional period comparable to the agricultural revolution. The agricultural revolution developed because the unmanaged environment was incapable of delivering food in either the quality or quantity desired by society. The *environmental management revolution* is developing because the mismanaged environment (managed not integrated) is clearly incapable of meeting the demands placed on it. Natural systems may not even retain a semblance of their present form in the absence of effective system level management.

2. Fragmented management simply does not work. Loss of critical wetlands and other surface changes has reduced groundwater recharge rate and quality in some areas at the very time when demands on the groundwater supply are already exceeding natural recharge rates. Nutrient, suspended solids, toxic waste discharges, and runoff into the Chesapeake Bay system with the resulting threat to the seafood and recreational industries of that area could be cited as a result of fragmented management. Some of these changes in the Chesapeake Bay are affecting the social structure of the Bay. An example is the loss of commercial fishing and oyster harvesting that have been carried out continuously by generations of certain families. The economic loss, though locally severe, may be less important than the cultural changes that will occur.

3. Whatever priorities are set for the use of natural systems, they will prove ineffective if the aggregate assimilative capacity of these systems for societal wastes is not central to the development of the management plan. Since each ecosystem is unique and varies seasonally and cyclically over the years, each region will have to develop its own management plan. This plan needs to consider not only the needs and interests of the citizens of the area but also the capacity of the effected ecosystems.

4. A major opportunity for reducing pressure on ecosystems exists in the rehabilitation of damaged ecosystems. For example, wetlands can be created on surface mine ecosystems where no wetlands existed before [1]. There is persuasive evidence that these ecosystems can provide wildlife habitats of considerable recreational and aesthetic quality and simultaneously markedly improve the recharge water quality entering the groundwaters and/or surface waters. This is not to denigrate the important effort to preserve the remaining wetlands but to show the opportunity for partially replacing lost ecosystem amenities that also provide socioeconomic benefits.

5. As always, individual attitudes are a key to the resolution of these problems, whatever management strategies are contemplated. It is my contention that the reductionist philosophy that permeates educational institutions is the primary obstacle to achieving the desired goals. It is worth emphasizing that the reductionist approach (that is, determining how things work by isolating the components from each other) has been enormously successful in science and engineering. This philosophy has, however, isolated the disciplines from each other physically by housing them in different buildings on university campuses, by creating language barriers due to the highly technical jargon that has developed in disciplines and subdisciplines, and, finally, by the fact that reductionists dominate tenure and promotion committees in the specific disciplines and as "gatekeepers" to the profession exclude those engaging in interdisciplinary activities because they do not fit the classical role model. There is no major problem facing society in matters of the environment, population, energy, etc. that can be resolved by a single discipline. Yet the educational system isolates students from each other in a direct relationship to the length of their educational career. Thus, the more advanced the degree, the less likelihood that their investigations will be understood by any but a few people in their peer group. Anyone doubting this can read either the dissertation titles at almost any commencement in the United States or a list of Senator Proxmire's golden fleece awards. Despite the insecurity felt by most faculty in an era of declining enrollments and diminishing resources, it is important that the educational system examine the educational process so that a holistic view can coexist with the reductionist approach. Since most of the faculty members now tenured are reductionists, it is not likely that they will initiate this examination.

The purpose of the hazard assessment activities of ASTM and many other organizations is to prevent ecosystem damage. The purpose of integrated resource management is to obtain optimal performance from the system, which requires both different information and a different strategy. Those engaged in environmental assessment and related activities must become much more active in developing integrated resource management approaches. Unless this is done soon and skillfully, even the finest standard methods will not be used effectively for this new purpose. Without integrated resource management, information will be fragmented, often inappropriate for the decision being made, or in apparent conflict with other equally valid information. In short, organizations producing standard methods for environmental surveillance, monitoring, and hazard prediction must pay more attention to the structure of the management systems into which their methods will flow. I believe that if these organizations do not meet this challenge the probability that the methods produced by the hardworking task forces, subcommittees, and committees will frequently be misused. It is for this reason that I have chosen to look at the next ten years in the context of what is needed to ensure that standard methods will be properly and effectively utilized.

The benefits of integrated resource management are so evident after even the most superficial examination that one wonders why we have had such difficulty in implementation. The more obvious benefits are:

1. Cost-effectiveness.
2. A long-term protection of the resource.
3. Enchanced possibility for effective multiple use.
4. More rapid and effective restoration of damaged resources to usable condition.
5. Reduced expenditure of energy and money on conflicts over use and the possibility of redirection of these energies and funds to resource management.

Scientists and engineers involved with water resource problems have long recognized that the institutional problems associated with managing water resources are invariably more aggravating and intractable than the scientific and technical problems. Typically, it is not the lack of methodology that impedes more effective use of water and other natural resources, but it is the failure of the many institutions (each charged with a fragment of resource use) to integrate resource management responsibilities. This situation was created as part of national policy during an era of specialization. The field of medicine has been recently redirected toward a holistic approach, but with specialists still playing a critical role. Resource management must do the same.

The great cost of inadequate resource management is not fully appreciated. Resource management strategies endorsed by professionals often cannot be implemented because of unanticipated or unfortunate institutional blocks.

Coping With Conflicting and Competing Demands

How does one cope with conflicting and competing demands on natural resources? One of the major obstacles in this regard is the large segment of the public and their legislative representatives to whom management is a dirty word.

Preservationists and many environmental activists still believe that a "hands off" policy will ensure that the status quo of the remaining natural or seminatural ecosystems will be maintained. While this policy may be true for limited time periods where the ecological conditions are moderately stable and where colonization by outside invading species does not occur, for many areas there is persuasive evidence that management is needed even for preservation. For example, early settlers in the northern midwestern United States plowed substantial portions of the prairies. Naturally, plowing destroyed the resident prairie ecosystem, but, astonishingly, it also destroyed the prairies that were not plowed. The reason is that the prairies were fire-dependent ecosystems. Plowing interrupted the formerly unbroken expanse of prairie, significantly reducing the probability that any particular area would be burned with the same frequency that it had before. As a consequence of the reduced number of fires, small oaks began to grow in many areas that had escaped the plow, and oak forests replaced the tall grass prairies [2].

Ironically, two of the best examples of prairies now remaining are managed systems at the University of Wisconsin Arboretum and at the Argonne National Laboratory. There is even evidence that the prairies themselves were not "natural" before the arrival of the European colonists. Their persistence may have been the result of deliberate burning by aboriginal cultures who did this to prevent invasion of trees that would have destroyed their way of life [3]. Coevolution of plants and animals has intrigued many ecologists, so it should not be surprising that there has been a coevolution of a human cultural system and a "natural" ecosystem. Therefore, the first step in reducing conflicting and competing demands is to develop a management program to preserve ecosystem integrity and communicate the essence of this plan to the general public.

Determining the Conditions Necessary for Maintaining Ecosystem Integrity

From a management standpoint, it is important to recognize that some ecosystems are perturbation dependent while others are perturbation independent [3]. Certain ecosystems, such as the New Jersey pine barrens, prairie grasslands, and the bottomland hardwood forests in the flood plains of the Mississippi depend on periodic "catastrophic events" such as fires and floods to maintain the species composition and ecosystem and community dynamics that characterize them. The devastation caused by these natural perturbations may be unsightly to members of society, and the unknowledgeable may appeal to governmental agencies to prevent these events from occurring. In fact, enlightened management might require just the opposite. Fires should be set if this is justified to protect ecosystem integrity or for other reasons so that the perturbations would occur under more controlled and supervised conditions. Figure 1 (from Vogl [3]) illustrates these points. Systems such as the tropical forests in the Amazon Basin appear to be perturbation independent. These ecosystems should be managed in an entirely different way; namely, they should be protected from system-wide perturbations as much as possible.

The bottom line to all of this is that conflicting and competing demands on an ecosystem will be reduced if the nature of the ecosystem and the management practices are clearly understood and effectively communicated at the earliest possible time. For example, it is not a good idea to issue housing construction permits in fire-dependent ecosystems. Nor is it a good idea to construct hazardous waste sites or oil refineries on flood-dependent ecosystems. Yet, if these activities are permitted, the choice will then be either the protection of the societal artifacts or the protection of the integrity of the ecosystem. Halfway measures will result in neither.

Reducing Conflicts By Understanding Ecosystem Integrity and Assimilative Capacity

When ecosystems are under stress, important processes become erratic and difficult to predict [4]. Up to a point, ecosystems are capable of assimilating anthropogenic wastes or activities without severe disequilibrium. In short, ecosystems can appear to be maintaining normal structure and function with all of the customary measures of ecosystem health, including a natural degree of variability. While there is some scientific disagreement about whether there are clearcut or fairly discernible thresholds of response (threshold equals a concentration of a chemical or some other potential stressor below which there are no observable deleterious effects), there is enough evidence that response thresholds exist to justify their use [5]. Of course, there is always the possibility that responses exist for which present assessment methodology is not adequate, but, for management purposes, we must assume (as does a physician giving an annual physical) that if the generally accepted standard methods for examination reveal no observable deleterious effects then no further action is required except continued monitoring (for humans, in the form of a repeat

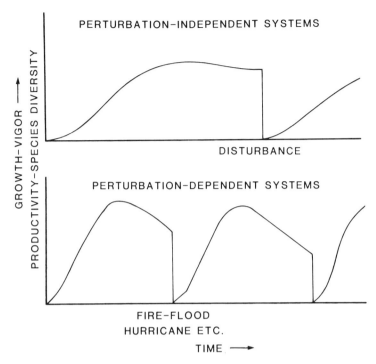

FIG. 1—*Disturbances in general ecosystems create vegetational setbacks and complete recovery is slow, whereas disturbances in perturbation-dependent ecosystems usually stimulate pulses of growth which rapidly decline unless disturbed again.*

physical). A system remaining in the nominative (normal state including natural variation) state needs little or no attention. If the system can be kept free of observable deleterious effects as a consequence of anthropogenic waste discharges or activities, management involvement will be relatively minor. This is particularly true if one stays well below the perceived deleterious response threshold. A constant problem is the more diverse the array of ecosystem uses, the greater the possibility that the aggregate of anthropogenic stresses that may be individually harmless may be collectively deleterious. Each of the waste dischargers (or other stressors) may well not be causing environmental stress, but collectively they may be damaging the environment. Such cases are not well documented with scientific evidence because there are few system-wide studies of the type likely to detect such broad effects unless they are quite severe. There are a number of instances where circumstantial evidence demonstrates this clearly [6].

Management costs increase dramatically as one approaches or crosses the response threshold. Even the cost of effective monitoring increases as well. As a consequence, as one approaches or crosses the threshold, the probability of conflicts and disagreements is high.

Problems of Scale

Carrying out the activities just described is achievable scientifically with presently available methodologies. Although, of course, deteriorating ecological conditions now much in the news, such as those in the Chesapeake Bay, are markedly influenced not only by the anthropogenic effects directly on the Bay, but by inputs from the tributary streams, which frequently have their origins on the entire land mass comprising the Chesapeake Bay drainage basin [6]. The ecosystem boundaries can be fairly well described since the drainage basin is well known. Within this geographic boundary, one must decide the types of evidence to be gathered. These might include present, past, and future land and water use patterns within the system; past, present, and future population density; past and present environmental quality as measured by a number of key characteristics; and environmental quality conditions that are desirable and achievable in the future. The level of detail, the frequency of the sampling, the statistical reliability desired, and so on should be determined by the type of decision being made rather than the ability for precision that present sampling permits and the staff available to do the sampling. If one excludes stresses originating outside the drainage basins, such as acid rain and other airborne contaminants, the geographic boundaries of a drainage basin study are fairly easily defined. One must then determine sampling frequency, location, and precision within subunits of this larger system. There are four pertinent considerations that provide additional helpful insights.

Management Interactions Using the Integrated Resource Management Approach

During initial stages, this may appear to increase conflict, since identifying important characteristics and uses for an area must be carried out at the earliest stages of planning to develop an integrated management plan. Use of system "loading capacity" limits should reduce conflicts in the long term. Such limits for an area should be understood better by the public than are strictly "political" decisions by an agency.

Integrated resource management for an area should enhance multiple uses and reduce conflicts by distributing use patterns and making compromises more likely for each particular use. A central computer-based data bank is essential for maintaining effective quality control. By avoiding inevitable, costly resource degradation/restoration costs associated with fragmented decisions, integrated resource management should produce more public benefits on a sustained basis for less money and maintain overall resource quality, thereby yielding public satisfaction.

Different Levels of Risk in Integrated Resource Management

Yes, because existing data bases are incomplete and capabilities of predicting results from management efforts involve some uncertainties. It seems best to discuss the likelihood for achieving results from management in three categories: (a) outcome predictable and fairly certain; usually involves a low level of risk and generates only limited new information; (b) outcome intermediate in certainty and predictability; an intermediate level of risk also is involved; however, there is good potential to generate some new information needed to improve management; and (c) outcome highly uncertain; a high level of risk is involved; there is considerable potential for obtaining much new and presumably useful information.

It is important to recognize that integrated resource management will improve as the information base for it expands. At this time, some rewards are needed for those who plan and conduct research to generate information required to improve predictable outcomes from integrated resource management. New approaches are needed to stimulate more such research during this period of financial stress.

Maintaining Resource Structure (Form) and Function (Services)

Resource management plans must have responsibilities clearly identified for maintaining (a) resource structure or form and (b) functions or services. It seems to be more costly to restore and maintain a historic landscape condition (structure and services) than to restore and/or maintain the services alone from it. There is some indication that the public is interested primarily in services, rather than both structure and services. If this is true, the number of options available for integrated resource management is increased markedly. Included would be some options that involve the best predictions of outcomes and are least costly.

In certain cases (for example, rare and unusual ecosystems, aquifer recharge areas, national parks, and so on), both structure and services should be maintained. For other units of the landscape (for example, some mined areas), restoring and maintaining ecosystem services may be adequate and only those should be required.

Recordkeeping in Resource Management

Records of periodic events, such as floods, forest fires, landslides, prescribed fires, timber harvests on rotations up to 80 or more years, and so on are essential because many such natural events and management actions occur at intervals greater than the length of duty tour by agency personnel, particularly administrators (decision makers). Records help to avoid inappropriate responses and provide "institutional or agency memory" required to maintain natural resource integrity and perpetuate the flow of services from units of the landscape.

Getting from Here to There

Most institutions, including academic, will not make the needed changes enthusiastically, and the resistance to change will be fierce in many institutions. The worst case scenario will be the production of a facade to give the illusion of change while there is business as usual behind it. Those who control funding can institute real change and make it more than a gesture. It will not be easy, but it is possible. A number of farsighted people have already identified the goals and objectives. Industrialists tired of prescriptive regulations are at least willing to try a different approach. Regulators tired of debilitating court battles are as well. It is the right time for a holistic approach.

Summary

My key points are:

1. To move from fragmented or piecemeal management to integrated resource planning and management.
2. To approach integrated resource management on a multidisciplinary basis.
3. To focus on attributes and health of an aquatic or terrestrial ecosystem—such as hydraulic regime, energy flow, and nutrient cycling—and not merely on features of one or more individual species.
4. To identify the assimilative capacities of aquatic and terrestrial ecosystems and to use multiple thresholds in management, such as for volumes and types of outdoor recreation, industrial wastes, and so on.
5. To avoid fragmentary, incremental decision making, thereby helping to assure that units of the landscape yield multiple benefits on a continuing basis.
6. To recognize that units of the landscape altered inappropriately provide numerous opportunities and challenges for restoration. For example, in the 1950s, England's Thames River lacked species of wild living resources. After the Queen demanded and had installed cost-effective integrated resource management for some years, conditions were restored, with the river now supporting 100 species.
7. To recognize that in restoration efforts it may be difficult to identify and determine historic

conditions, because records of those conditions usually are sketchy or lacking, and choosing which of several conditions were prevalent at an area decades ago is difficult. In some cases, such as mined areas, it may not be possible to restore the landscape to historic (premined) conditions. A more practical alternative is to consider restoring an area to a desirble or prescribed condition.

8. To encourage universities to focus on interdisciplinary resource training and management, rather than continuing the subdivision of topics into even smaller units. Demands of employers and stipulations accompanying funds made available to higher institutions of learning need to emphasize the need for interdisciplinary approaches.

9. To conduct surveys to gain understanding of citizen attitudes, demands, and support for integrated natural resource management. Results from such customer or marketing evaluations are critical in helping program administrators and managers understand the public pulse.

Acknowledgments

This discussion was originally the keynote address, Second Natural Resources Leaders Seminar, Charleston, South Carolina, 12–14 December 1985. My perception of the problem was significantly modified by the discussions in the working sessions that followed. The support for the basic concept by the administrators in attendance was particularly heartening. I am indebted to Laurence R. Jahn of the Wildlife Management Institute for major assistance in crystallizing some of the major issues and to Gerald Purvis, Seminar Chairman, Walter A. Denero, Director, J. W. Fanning Community Leadership Center, University of Georgia, and his capable assistant, Michele Carter, for establishing conditions that enhanced discussion.

This manuscript was typed several times by Rebecca Allen, typed in final form by Betty Higginbotham, and edited by Darla Donald, editorial assistant, University Center for Environmental Studies.

References

[1] *Wetlands and Water Management on Mined Lands*, Brooks, R. P., Samuel, D. E., and Hill, J. B., Eds., Pennsylvania State University, University Park, PA, 1985.
[2] Jordan, W. R., III, "Ecological Restoration: Healing the Reentry of Nature," *Orion*, in press.
[3] Vogl, R. in *The Recovery Process in Damaged Ecosystems*, J. Cairns, Jr., Ed., Ann Arbor Science Publishers, Inc., Ann Arbor, MI, 1980, pp. 63–94.
[4] Odum, E. P., Fin, J. T., and Franz, E. H., *BioScience*, Vol. 29, 1979, pp. 349–352.
[5] May, R. M., *Nature*, Vol. 269, 1977, pp. 471–477.
[6] *Land Use and the Chesapeake Bay*, Coale, C. W., Marshall, J. P., and Kerns, W. R., Eds., Publ. 305-003, Virginia Cooperative Extension Service, Virginia Polytechnic Institute and State University, Blacksburg, VA, 1985.

SUBJECT INDEX

A

Acanthomyosis sculpta, 447–53
Acid rain
  New York State lakes, 103–109
  northeastern U.S.A. and Canada, 296–99
Adsorption (*See* Sorption)
Agriculture
  slash and burn, 58–59
Algae
  atrazine toxicity, 533–45
  in microcosm toxicity tests, 387–400, 402–03, 468–79
  oil toxicity, 437–39, 442–44
Aminopolysaccharides
  effect on PCBs, 518–27
Anabaena flos-aquae, 531–45
Antibiotics
  cyclophosphamide, 318–25
  effect on bacteria, 493, 494–501, 500(table)
  use in FETAX, 322–23
Apalachicola Bay, 370–71
Aquatic plants (*See* Plants, aquatic)
Aroclor 1254
  biotransformation on chitin, 521
  comparative cytotoxicity, 455–57
  liver microsome activation, 323–24
Aroclors
  biodegradation of, 517–27
  cytotoxicity tests, 455, 457–58, 458(table), 459–60
Arsenic
  fish vertebral anomalies, 411–14, 412(table)
  sea water contamination, 410
Asterionella formosa, 65–70
ASTM Standard D 4429–84: 549
ASTM STP 865: 362
ASTM Symposia on Aquatic Toxicology, 8–15, 23–24
  analysis of papers, 9(table)
ATP
  toxicity testing, 505
Atrazine
  phytostatic/phytocidal effects, 531–45

B

Bacteria
  biodegradation of PCBs, 517–27
  deep sea
    genetic transfer in, 39–40
  effluent toxicity tests, 190–202
  marsh, 492–501
  Photobacterium phosphoreum, 192, 439, 444–46
  plasmids, 37–40, 399–40, 493, 496–97
  pollution of water, 45
  and toxic effects of,
    antibiotics, 493, 494–501, 500(table)
    heavy metals, 494–500, 498, 499(tables)
    oil, 439, 444–46
    PCBs, 517–27
Behavioral responses
  fish, 327–39
Benthic invertebrates
  bryozoa, 72–75, 82–85
  chaoborus, 72, 73, 75–77, 82–85
  chironomids, 72, 73, 77–82, 85
  in ecosystems,
    estuarine, 373–75, 382
    freshwater, 72–85, 117–26
    saltwater, 117–26, 149–57, 161–75
  in lake sediments, 72–85
  multiparameter studies, 82–85
  and organic contaminants, 115–26, 119(table), 121(tables), 166–72, 167(table)
  PCB uptake, 151–57, 152(table)
  recruitment of, 378–79
Benzo(a)pyrene
  partitioning of, 236–41
BF–2 cells, 454–61
Bioaccumulation
  heavy metals, 299–301
  organic contaminants, 150–57
  tin, 220–22, 225–26
Bioassay end points, 345–59, 531–32, 539–45
Bioassays (*See also* Microcosms, Standardized Aquatic)
  aquatic plant, 531–45

field validation of, 361–69
firefly, 492, 505
oil-contaminated water, 447–53, 503–16
short term vs standardized aquatic microcosm, 471–49
Bioavailability
    organic contaminants, 149–57, 241–46
    zinc, 433
Biochemistry
    and acid stress monitoring, 297–99
Biodegradation
    cationic surfactant, 167
    2,4-dicholorophenol(DCP), 179, 182(table)
    PCBs, 518, 520–25
    polycyclic aromatic hydrocarbons, 512–16
    tributylin, 222, 226–27, 229(table)
Biomass production
    mathematical modeling, 264–66
    naphthalene toxicity, 266–71
Biomonitoring
    biochemical endpoints, 298–99
    ecosystems, 562–63
    gene probes, 33–35
    laboratory vs field results, 361–68
    NPDES, 161–75
    overview of, 2
    use of oysters, 169–72, 74–75
    ventillatory tests on fish, 307–14
Biotechnology [See also Gene probes; Genetically engineered microorganisms(GEM)]
    and aquatic toxicology, 1
    and waste cleanup, 1
Biotransformation
    polycyclic aromatic hydrocarbons, 507, 512–16
Bivalves (See also Oysters)
    and PCB uptake, 151–57, 152(table)
Bluegill sunfish
    BF–2 cells, 454–61
    1,3,5-trinitrobenzene toxicity, 307–14
        cough test, 311–12
        movement test, 311–12
        ventilation test, 307–14
Brass, 468–79

C

Cadmium, 57(table), 94
Calcium, 297–98
Cationic surfactant
    partitioning of, 139–48

Ceriodaphnia dubia, 16, 18, 548, 551–54
Chelatins, 301
Chemical mixtures, 2, 24
Chesapeake Bay
    environmental damage, 560, 563–64
    tributylin biodegradation, 222, 226–28, 229(table)
Chitin
    and PCB biodegradation, 518, 520–25
Chitinase, 39–40
Chlorine
    toxicity persistence, 181
Chromatography
    organometals, 220
Chromium, 212–13
Cladocerans (See *Ceriodaphnia dubia*; *Daphnia*; *Daphnia magna*; *Daphnia pulex*)
Computers
    programs
        DRASTIC, 47
        Sandia model, 47
        TOX-SCREEN model, 249–50, 252–54
    role in aquatic toxicology, 16–17
Consensus reports
    on standard test methods, 18
Copper, 385–404, 473–74
Cyclophosphamide, 318–25
Cyprinodon variegatus, 192
Cytotoxicity
    BF–2 cells,
    neutral red assay, 455–56

D

Daphnia
    in microcosm laboratory tests, 385–400, 402–03
Daphnia magna
    ASTM Standard D 4229:549
    short-term chronic toxicity test, 548–55
    and toxic effects of,
        brass dust, 468–79, 472(table)
        copper sulfate, 468–79, 472(table)
        red phosphorous, 252
Daphnia pulex
    short-term chronic toxicity test, 548–55
    and toxic effects of,
        chromium, 212
        naphthenic acid, 212
        wastewater effluent, 204–14
        zinc, 425–34

toxic fraction identification,
    204–14, 208(table), 212(illus)
Data aggregation
    centroid method, 347–51
    least squares method, 350–51
    maximum likelihood method, 350–51
Deforestation
    and lake eutrophication, 66–68
DDT (*See* Dichlorophenyl trichloro ethane)
Diatoms
    and lake pH, 89–90, 94–97, 99–101, 103
    use in paleolimnology, 63–70
3,4-Dichloroaniline, 463–67, 466(table)
2,4-Dichlorophenol(DCP), 178–88
Dichlorophenyl trichloroethane(DDT), 119–26
Dissolved organic macromolecules (DOM)
    effect on contaminant bioavailability/transport, 233–46
DNA
    aquatic organism toxicity, 300
    bacterial plasmid analysis, 493–95
    sequence detection, 30–35
Dodecyl sodium sulfate
    avoidance behavior test, fish, 331–32
Dodecyl trimethyl ammonium chloride (TMAC)
    biodegradation of, 147
    partitioning of, 139–48
    sediment sorption/desorption, 145–47
Dominance relationships
    in marine macroinvertebrate systems, 374–75, 377, 382
Dose-response relationships
    brass dust on daphnids, 468–79
    naphthalene in freshwater aquatic system, 264–71, 266(table)
    state-space analysis, 282
    theoretical aspects, 343–59
    toxic substances in rainbow trout, 327–39
Drilling fluid
    toxicity tests, 447–53
Duckweed
    effect of atrazine, 533–45
Dunaliella tertiolecta, 531–45
Dye waste effluent, 166–72

E

EC_{50}
    interlaboratory test precision, 191–93
Ecosystems
    regional management of, 560–65

Ecotoxicology
    interstitial water sampling, 138–48
    state-space analysis, 275–84
    validation of laboratory data, 361–68
Effluent contamination
    laboratory simulation, 177–88, 408–17
    toxic fraction identification(TFIP), 204–14
    toxicity test variability, 190–202, 192(table)
Effluents, complex
    from dye production, 161–75
    fractionation of, 204–14, 206(illus), 212(illus)
    from munitions testing, 247–59
    National Pollutant Discharge Elimination System(NPDES) monitoring, 161–88
    from ore smelter, 411–17
    from pulp mill, 411–17, 412(table)
Environment preservation, 561–62
Environmental fate
    formaldehyde spill, 483–89
    naphthalene,
        mathematical modeling, 261–73
    phosphorous burn products, 247–59
    smoke screen products, 247–59
Environmental management, 559–65
Environmental Protection Agency(EPA), 115–16, 177–88, 204–05, 423–34, 507–08
    TOX-SCREEN model, 249–50, 252–54
Environmental simulation, 177–88, 212–13, 361–68
Environmental toxicity persistence unit(ETPU), 179–81, 183–88
Environmental transport
    organic contaminants, 242–46
Enzymes
    acid effect on fish, 297
    metal effects on fish, 299–300
EPA (*See* Environmental Protection Agency)
Erosion
    control measures, 135–37
    measurement of, 129
Estuarine ecosystem, 161–75, 370–71
ETPU (*See* Environmental toxicity)
    persistence unit
Eutrophication
    lakes, 63–70, 66–68, 71–86

F

Facilities, research, 15–16
Fate modeling, 14, 17

polycylic aromatic hydrocarbons, 263–64, 263(illus), 269–70, 272–73
Fathead minnows
  breeding of, 16
FETAX (*See* Frog Embryo Teratogenesis Assay—*Xenopus*)
Field study
  laboratory validation, 363–64, 366, 369–83, 406–17
  problems of 290–91, 361–67
  site-specific validation, 366, 369–83
Fish (*See also* Bluegill sunfish; Fathead minnows; Fourhorn sculpin; Sheepshead minnows; Rainbow trout)
  enzymes, 297, 299–300
  tests of,
    behavior, 327–39
    cough, 311–12
    growth, 463–67
    movement, 311–14
    ventilation, 307–14
  and toxic effects of,
    acid, 296–99
    arsenic, 411–14, 412(table)
    heavy metals, 299–301
    1,3,5-trinitrobenzene, 307–14
    vertebral anomalies, 406–17, 412, 415(tables)
Flocculation, 128, 136–37
Fluorescence detection
  in organometals, 220–222, 225–26, 231
Formaldehyde, 483–89
Fossils, 63–70, 72–85, 91–102 (*See also* Paleolimnology)
Fourhorn sculpin, 407–17
Fragilaria crotonensis, 65
Freshwater ecosystems (*See also* Lakes; Ponds; Rivers)
  acidification, 296–99
  PCB contamination, 117–26
  sediment organic carbon concentration, 115–26
Frog Embryo Teratogenesis Assay—*Xenopus*(FETAX), 316–25

G

Gas exchange
  acid effect on fish, 296–98
GEM (*See* Genetically engineered microorganisms)
Gene probes, 29–35, 31, 32(tables)
  for chitinase, 39–40
  for 16S ribosomal DNA, 40
  detection of, 29–35
Genetic exchange
  bacteria,
    in freshwater environment, 39
    in ocean environment, 40
Genetically engineered microorganisms, (GEM), 1, 41–42, 43–49
Glycogen
  monitor for metal toxicity, 301
Groundwater contamination
  formaldehyde, 483–89
Growth rate
  fish, 463–64
Gulf of Bothnia
  ecotoxicological study,
    fish, 406–17

H

Hazard assessment
  EPA protocol, 508(illus)
  field vs laboratory evidence, 361–68
  formaldehyde in groundwater, 483–89
  future of, 19, 20–22, 23–24, 559–65
  genetically engineered microorganisms in water, 43–49, 44(illus)
  and management systems, 561, 564–66
  naphthalene
    mathematical modeling, 267–69
  oil sand extraction waste, 436–46
  papers on, 12–13
  via sediment analysis, 128–37
Heavy metals (*See* Metals, heavy)
Herbicides
  atrazine, 531–45
Histology
  acid effect on fish, 297
Hormesis, 327–39
  definition, 327–28
  tests on fish, 328–39
Humic acid, 234–35
Hybridization, DNA, 30, 35, 493–95
Hydrometer, 129

I

In situ biological treatment (ISBT), 503–16
Indicator species procedure, 423–34
Integration of information, 14–15
Interdisciplinary studies, 560

SUBJECT INDEX 571

Interstitial water
  sampling methods, 138–48
Invertebrate subfossils
  in lake sediments, 72–85
Invertebrates, benthic (*See* Benthic invertebrates)
Iron
  in lake muds, 54
ISBT (*See In situ* biological treatment)

K

Kinetics
  polycyclic aromatic hydrocarbons, 512
  zinc, 433

L

Laboratory vs field data, 14–15
Laguna de Petenxil, Guatamala, 59
Lakes (*See also* New York State lakes; Laguna de Petenxil; Linsley Pond; Meridian lake; Washington State lakes)
  acidification of, 103–09, 296–99
  eutrophication, 63–70, 71–86
  mud
    chemical/mineralogical/biological analyses, 54–60, 63–70
  sediment (*See* Sediments)
  water quality, 86, 91
LC_{50}
  comparison with metal-binding protein concentration, 310
  interlaboratory test precision, 190–93
  limitations of, 343–45
  mysid shrimp, 449
  value in hazard assessment, 21
Lead
  in lake sediment, 65–68, 67(table), 93, 102
Lemna gibba, 533–45
Lepomis macrochirus, 311–14
Linsley pond, Connecticut, 54
  eutrophication of, 83, 85(illus)
  invertebrate subfossils, 74
  mud composition, 55(table)

M

Macroinvertebrates (*See also* Benthic invertebrates)
  toxicity tests,
    estuarine ecosystem, 369–83
    waste effluent, dye plant, 166–72

Management of resources
  integration of, 560–61, 564–65
  lakes, 86
  role of standards committees, 560
Marsh treatment system, 491–501
Mathematical modeling
  formaldehyde in groundwater, 486–89, 487, 488(tables)
  genetically engineered microorganisms, 45–49
  integrated fates and effects model (IFEM), 262–66, 272
  naphthalene in aquatic system, 261–73
  phosphorous compounds in sand/gravel, 249–50
Maya culture
  decline of, 58–60
Maximum acceptable toxicant concentration(MATC), 467
Mercury
  in lake mud, 55–57, 57(table)
Meridian Lake, Washington, 64–70
Mesocosms
  oil waste toxicity test, 507, 511–12
Metabolic activation system, 316–25
Metal-binding proteins
  as toxicity indicators, 301–02
Metal-specific imaging, 226
  oil contamination of freshwater, 275–84
Metallic compounds
  toxicity persistence, 183, 184
Metallothioneins, 301
Metals (*See* Brass; Cadmium; Copper; Iron; Lead; Mercury; Metals, heavy; Metals, trace; Organotins; Silver; Tins; Zinc)
Metals, heavy
  effect on marsh microflora, 491–501, 498–99(tables)
  fish biochemistry, 229–302
  fish vertebral anomalies, 411–14, 412(table)
  Standardized Aquatic Microcosm, 468–79
  sea water, 408, 411–14
Metals, trace
  brass dust, 468–79
Methods (*See also* Bioassays; Toxicity tests)
  papers on, 9–11
Microcosms, multispecies, 369–83
  in laboratory, 382–404, 518–19
  Standardized Aquatic Microcosm (SAM), 384–404, 470–49

Microorganisms (*See also* Bacteria)
   gene probes for detection, 29–35
   genetically engineered, 41–42, 43–49
   oil waste site, 504–05
   river, 439–42
Microsomes, rat liver
   use in chemical screening, 316–25
Mictrotox, 178–88, 192, 439, 444–46, 506
Midges, 72, 73, 77–85
Mine tailings, 491–501
Minnows
   fathead, 425–34
   sheepshead, 192
Mississippi river delta, 503–04
Molecular toxicity predictors, 223–24, 223(table)
Monitoring (*See* Biomonitoring)
Monomers
   and fish avoidance behavior, 332–34
Multimedia projects, 564–65
Myoxocephalus quadricornis, 407–17
Mysidopsis bahia, 192

N

Naphthalene
   environmental fate, mathematical model, 261–73
   fish production, 266–69
   freshwater toxicity, dose-response, 264–66
Naphthenic acids, 212–13
Narragansett bay, 151–57
National Pollutant Discharge Elimination System(NPDES), 161–75, 177–78
Navicula pelliculosa, 531–45
New York State Lakes, 89–108
Nonpolar organic contaminants, screening method, 115–26
NPDES (*See* National Pollutant Discharge Elimination System)

O

Obscurant smokes
   environmental fate, 247–59
Ocean
   plasmid mobility in, 39–40
Oil
   extraction process waste, 436–46
   offshore drilling, 447–53
   reclamation wastes, 503–16
   synthetic,

effect on freshwater systems, 275–84
Optical imaging
   tin, 220–22, 225–26, 219–31
Ore smelter effluent, 408, 411–14
Organic contaminants (*See also* Benzo(a)pyrene; 3,4-Dichloroanilene; 2,4-Dichlorophenol; Dichlorophenyl trichloroethanol; Formaldehyde; Naphthalene; Naphthenic acid; Polychlorinated biphenyls; Polycyclic aromatic hydrocarbons; Toluenes; Tributylin; 1,3-Trinitrobenzene)
   bioavailability of, 241–46
   in esturarine stream, 164
   fish vertebral anomalies, 414–16
   partitioning of, 233–46
   toxicity persistence, 183, 185
Organometallic compounds, 219–31
Organotins, 219–31
   structure-activity relationships, 223–24
   toxicity, 224
   uptake mechanisms, 223–24, 229–30
Osmoregulation
   freshwater organisms, 296–97, 299–301
Overview of aquatic toxicology, 1–3
Oysters
   as biomonitoring tool, 169–72, 174–75
   chemical analysis, 164
   larvae toxicity tests, 170, 172

P

Pacific Coast ecosystem, 447–53
PAHs (*See* Polycyclic aromatic hydrocarbons)
Paleoecology, 53–60
Paleolimnology, 1
   benthic invertebrates, 72–85
   lake muds, 53–60, 63–70
   lake sediments, 71–86, 89–108
   Linsley Pond, Connecticut, 54
   New York State lakes, 89–108
   and study of Mayan culture, 58–60
   Washington State lakes, 63–70
Particle size distribution(PSD)
   erosion materials, 128–37
   sediment analysis, 128–37
Particles
   and binding of benzo(a)pyrene, 236–41
   and binding of organic contaminants, 236–46
Particulate organic carbon, 242–45
Partitioning

benzo(a)pyrene, 236–41
  organic contaminants, 233–46
PCBs (*See* Polychorinated biphenyls)
pH
  and diatom analysis, 89–108
Phosphates, 252
Phosphorus
  burn chemistry, 248(illus)
  environmental fate, 247–59
  in lake sediment, 65, 68–70, 82–83
Photobacterium phosphoreum, 439, 444–46
Photolysis
  2,4-dichlorophenol, 180, 182–83
Phthalates
  in lake mud, 57–58
Pimephales promelas, 16, 425–34
Plankton, 437–39, 442–44
Plants, aquatic
  atrazine toxicity, 531–45
Plasmids, bacterial,
  heavy metal toxicity test, 493, 496–97
  mobility in ocean, 399–40
  selection/exchange, 37–39
Pollen
  fossils in lake sediment, 66–67, 91–93, 97–99, 101–02
Polychaetes
  PCB uptake, 151–57, 152(table)
Polychlorinated biphenyls(PCBs)
  bacterial effects, 517–27
  bioavailability tests, 149–57
  biodegradation, 517–27
  biotransformation, 517–27
    freshwater/saltwater ecosystems, 119–26
  environmental fate, 58
  in sediments, 119–26
  toxic effects on BF–2 cells, 455, 457–58, 458(table), 459–60
Polycyclic aromatic hydrocarbons(PAHs)
  bioavailability/transport, 236–46
  biodegradation, 503–16
  environmental fate,
    mathematical modeling, 261–73
  in situ biological treatment, 503–16
  oil waste site, 505(table)
  partitioning of, 234–46
Polymers
  fish avoidance behavior, 332–34
Ponds
  effect of synthetic oil, 279–84

Prairies
  preservation of, 560–61
Prediction (*See* Hazard assesment; Validation)
Preference factors
  bioaccumulation in benthic organisms, 149, 156–57
Priority pollutants
  sediments/shellstock, 165
Pulp mill effluent
  chemical characterization, 406, 409(table), 411

R

Rainbow trout, 192, 327–39, 463–67
Recruitment
  in marine macroinvertebrate systems, 374–75, 378–79
Remedial action
  endpoint determination, 483–89
  formaldehyde spill, 483–89
Research facilities, 15–16
Resource utilization, 561, 564
Results of studies
  papers on, 11
Review
  aquatic toxicology, 1–3, 7–24
  inter/intralaboratory toxicity tests, 190–202, 192(table)
Risk assessment (*See* Hazard assessment)
Rivers, 424–34, 503–04
RNA
  aquatic organism toxicity, 300

S

Salmo gairdneri 192, 327–39, 463–67
Saltwater ecosystems
  dye waste effluent, 166–72
  PCB contamination, 117–26
  sediment organic carbon concentration, 115–26
SAM (*See* Microcosms, Standardized Aquatic)
Sand, sorption of, 250–52, 254–57
Screening assay
  teratogenesis, 316–25
Screening level concentration(SLC)
  organic carbon concentration, freshwater/saltwater, 115–26
Sediments
  benthic invertebrates, 71–86
  chemical analysis, 164

diatom analysis, 89–108
dichlorophenyl trichloroethane, 119–26
dodecyl trimethyl ammonium chloride sorption, 138–47
estuarine ecosystem, 370–73
and hazard assessment, 128–37
interstitial water sampling, 139–48
lead in, 65–68, 67(table), 93, 102
organic contaminants, 115–26, 151–57, 242–46
paleolimnology studies, 53–60, 63–70, 71–86, 89–108
particle size distribution, 128–37
PCBs, 119–26
phosophorus, 65, 68–70, 82–83
phthalates, 57–58
pollen fossils, 66–67, 91–93, 97–102
subfossil analysis, 73
Selenastrum capricornatum, 437–39, 442–44
Sheepshead minnows, 192
Shrimp
  grass, 166–68
  mysid, 447–53
  opossum, 192
Sludge
  oil sand extraction waste, 436–46
Silver, 57
Sorption
  benzo(a)pyrene, 236–41
  2,4-dichlorophenol, 180
  dissolved organic macromolecules (DOM), 242–46
  PCBs in sediments, 145–47
  phosphorus burn products, 250–52, 254–57
  sand, 250–52, 254–57
Spills
  remedial action, 483–89
State space analysis
  coal-derived synthetic oil,
    effect on freshwater system, 275–84
Storm-induced effect
  on marine macroinvertebrate ecosystem, 375
Stress, xenobiotic
  in aquatic organisms, 289–903
Subfossils
  in lake sediment, 72–85
Sulfur, 54
Swamps, seasonal, 59–60

T

Tailings
  mine, 491–501
  oil sand extraction, 436–46
Teratogenesis assay
  chemical screening, 316–25
  cyclophosphamide, 318–25
  Frog Embryo Teratogenesis Assay—*Xenopus*(FETAX), 316–25
  screening assay, 316–25
Tin, 219–31
Toluenes, 458–59, 460
Topology
  organometals, 220
  organotins, 223–24
Toxicity (*See also* Cytotoxicity)
  acid precipitation, 296–99
  antibiotics, 318–25, 493–501, 500(table)
  Aroclor compounds, 455, 458(table), 457–60
  arsenic, 410–14, 412(table)
  atrazine, 531–45
  brass dust, 468–79
  cadmium, 57(table), 94
  chlorine, 181, 185, 187–88
  chromium, 212
  copper sulfate, 385–404, 468–79
  3,4-dichloroaniline, 463–67, 466(tables)
  2,4-dichlorophenol, 178–88
  dichlorophenyl trichloroethane, 119–26
  dodecyl sodium sulfate, 331–32
  dodecyl trimethyl ammonium chloride, 139–48
  drilling fluids, 448–53
  formaldehyde, 483–89
  heavy metals, 299–302, 468–79, 491–501
  lead, 65–68, 93, 102
  mercury, 55–57, 57(table)
  mine tailings, 491–501, 498–99(tables)
  naphthalene, 266–71
  naphthenic acid, 212
  oil, 275–84, 436–46, 447–53, 503–16
  organotins, 224–25
  PCBs, 119–26, 455, 457–58, 458(table), 459–60, 517–27
  polycyclic aromatic hydrocarbons, 236–46, 261–73, 503–16
  phosphorus, 247–59
  pulp mill effluent, 406, 409(table), 411
  smoke screens, 247–59

sulfur, 54
tin, 219–31
toluenes, 458–460
tributylin, 226–28, 229(table)
1,3,5-trinitrobenzene, 307–14
zinc, 424–34, 430(table)
Toxicity persistence
chlorine, 181, 185, 187–88
2,4-dichlorophenol, 178–88, 182(table)
metallic compounds, 183, 184
organic contaminants, 183, 185
Toxicity tests (*See also* Bioassays; Biomonitoring; Teratogenesis assays)
acute, 190–202, 192(table)
algal assay, 531–45
aquatic plants, 532, 533–39
ATP assay, 505
clinical measurements, 294
Daphnia microcosm tests, 385–400, 402–03
Daphnia magna
ASTM Standard D 4229–84:549
interlaboratory precision, 193
short-term chronic tests, 548–55
Daphnia pulex
short-term chronic tests, 548–55
differentiation from stress effect, 291–94, 292(illus)
effluents, 190–202
field validation, 185–88,186, 187(tables), 507–08
fish,
avoidance behavior, 327–39
cough/movement/ventilation, 307–14
growth rate, 463–67
vertebral anomalies, 406–17, 412(table), 415(table)
fraction identification, 204–14, 206(illus), 212(illus)
historical perspective, 19–22, 23
in vitro alternatives, 454–61
integration of data, 343–59
interlaboratory/intralaboratory, 190–202; 384–404
laboratory simulation, 212–13, 423–34
microbial population, 508–11
microcosms, 369–404
multiple species, 370–83
Ohio River Valley Water Sanitation Commission (ORSANCO), 365
precision of, 190–202,

interlaboratory, 193, 194–95(table), 384–404
intralaboratory, 193, 196–97(table)
reproducibility of, 384–404
review of, 190–202, 192(table)
short-term chronic, 548–55
single species, 361–62, 364, 425–26, 426–34, 428, 431(tables)
single vs multiple species, 369–70, 384–404, 468–79
standards for, 364–66
ASTM Standard D 4229–84:549
statement of purpose, 365–66
for toxicity reduction evaluation, 205
validation,
ASTM STP 865, 362
variability, 198–200, 199,202(tables)
ventilatory, 310–14
Toxicologists, aquatic
recruitment of, 2
Toxicology
comparative, 343–59
scalers, 345–50, 355–59
TOX-SCREEN, 249–50, 252–54
Tributylin, 226–28, 229(table)
Trinitrobenzene, 307–14
Trinity river, Texas, 414–34
Trophic interactions (*See* Microcosms, Standardized Aquatic)

U

Urban development
and lake eutrophication, 68–70

V

Validation
toxicity tests, 361–67, 507–08
Ventillatory patterns
bluegills exposed to 1,3,5-trinitrobenzene, 307–14, 314(table)
Vertebral anomalies
fourhorned sculpin, 410–17, 412, 415(tables)
Viviforms, 39
Volatilization,
2,4-dicholorphenol, 180, 182

W

Washington State lakes, 63–70
Waste
lagoon/sludge, 503–16

management, 503–04
oil, 436–46, 503–16
optimal loading of, 508
Wastewater
   effluents (*See* Effluents, complex)
   oil sand extraction waste, 436–46
   treatment of, 162, 172
Water (*See also* Groundwater)
   partitioning of surfactants, 138–48
   resources, 561
Water fleas (See *Daphnia; Daphnia magna; Daphnia pulex*)
Water quality
   lakes, 71, 86
   site-specific testing, 423
   use of laboratory microcosms, 369–83
Wetlands, 560

X

Xenobiotic stress
   aquatic organisms, 289–303
Xenopus, 316–25

Z

Zinc, 423–34, 474, 478–79

Author Index

A

Adams, W. J., 3
Alexander, M. M., 531
Anderson, D. S., 89
Arthur, M. F., 177

B

Babich, H., 454
Balu, K., 531
Bantle, J. A., 316
Barkay, T., 29
Bartell, S. M., 261
Bean, D. J., 115
Bengtsson, B.-E., 406
Bengtsson, A., 406
Bianchini, M. A., 503
Black, M. C., 233
Blair, W. R., 219
Borenfreund, E., 454
Bradbury, W. C., 491
Brinckman, F. E., 219
Brugam, R. B., 63

C

Cairns, J., Jr., 361, 559
Chapman, G. A., 3
Charles, D. F., 89
Clement, W. H., 177
Colwell, R. R., 37
Conquest, L. L., 384
Cooney, J. D., 247
Cornaby, B. W., 115
Cowgill, U. M., 53
Crisman, T. L., 71
Crossland, N. O., 463

D

Davidson, L. F., 138
Davis, R. B., 89
Dawson, D. A., 316
Degraeve, G. M., 177
Desjardins, R. M., 491
Dickson, K. L., 7, 204, 423
Durell, G. S., 247
Dysart, B. C., 128

F

Fava, J. A., 190
Frauenthal, M., 247
Fujisaki, K., 503, 517

G

Galloway, J. N., 89
Gardner, R. H., 261
Gasith, A., 204
Giesy, J. P., 289
Gillespie, R. B., 177
Graney, R. L., 289
Grimes, D. J., 37
Grothe, D. R., 190
Gulbransen, T. C., 115

H

Hadjinicolaou, J., 327
Haley, M. V., 468
Hamelink, J. L., 7
Hamilton, S. J., 406
Hand, V. C., 138
Henry, C. B., 503

Horning, W. B., 548
Hughes, J. S., 531

J

Johnson, A. R., 275
Johnson, D. W., 468
Jop, K. M., 204

K

Kaczmarek, S. A., 204
Kimerle, R. A., 7
Kindig, A. C., 384
Knapp, K. A., 128

L

Landis, W. G., 3, 468
LaRoche, G., 327
Lewis, P. A., 548
Livingston, R. J., 369
Lowrie, L. N., 343
Ludwig, R. D., 436

M

MacDonell, M. T., 37
Macek, K. J., 7
Machuzak, M. J., 447
Marcus, J. M., 161
Masters, J. A., 138
Matthews, J. E., 503
Mayer, F. L., Jr., 7, 406
McCarthy, J. F., 233
McElroy, A. E., 149
Means, J. C., 149
Mikel, T. K., Jr., 447
Molak, V., 43
Mount, D. I., 7
Muse, W. T., Jr., 468

N

Neff, J. M., 115

O

O'Brien, G. K., 177
Olson, G. J., 219
O'Neill, R. V., 261
Ortiz-Conde, B. A., 37

P

Parkerton, T. F., 204, 423
Parks, E. J., 219
Parrish, P. R., 7
Pittinger, C. A., 138
Pollack, A. J., 247
Portier, R. J., 503, 517

R

Rodgers, J. H., Jr., 423
Roszak, D. B., 37
Rue, W. J., 190

S

Saleh, F. Y., 423
Sayler, G. S., 29
Scanlon, J. S., 115
Scott, G. I., 161
Seyfried, P. L., 491
Shedd, T. R., 307
Shirazi, M. A., 343
Shook, T., 247
Somerville, C. C., 37
Staples, C. A., 483
Stara, J. F., 43
Stewart, S. S., 423
Straube, W., 37
Swearingen, G. R., 161

T

Taub, F. B., 384
Templet, P. H., 503

V

Vaga, R. M., 115
van der Schalie, W. H., 307
Versteeg, D. J., 289
Vigon, B. W., 247

W

Wickramanayake, G. B., 247

Y

Yong, R. N., 436

Z

Zeeman, M. G., 307